T0212908

Modern Birkhäuser Classics

Many of the original research and survey monographs in pure and applied mathematics published by Birkhäuser in recent decades have been groundbreaking and have come to be regarded as foundational to the subject. Through the MBC Series, a select number of these modern classics, entirely uncorrected, are being re-released in paperback (and as eBooks) to ensure that these treasures remain accessible to new generations of students, scholars, and researchers.

Six Lectures on Commutative Algebra

J. Elias
J.M. Giral
R.M. Miró-Roig
S. Zarzuela
Editors

Reprint of the 1998 Edition

Birkhäuser Verlag
Basel · Boston · Berlin

Editors:

J. Elias
J.M. Giral
R.M. Miró-Roig
S. Zarzuela
Departament d'Àlgebra i Geometria
Universitat de Barcelona
Gran Via de les Corts Catalanes, 585
08007 Barcelona
Spain
e-mails: elias@ub.edu
           giral@ub.edu
           miro@ub.edu
           szarzuela@ub.edu

Originally published under the same title as volume 166 in the Progress in
Mathematics series by Birkhäuser Verlag, Switzerland, ISBN 978-3-7643-5951-5
© 1998 Birkhäuser Verlag, P.O. Box 133, CH-4010 Basel, Switzerland

---

1991 Mathematics Subject Classification 13-02, 14-02, 18-02, 13D02, 13A15,
13F20, 13A35, 13D40, 13D45, 13H15, 13D99

Library of Congress Control Number: 2009937806

Bibliographic information published by Die Deutsche Bibliothek
Die Deutsche Bibliothek lists this publication in the Deutsche Nationalbibliografie;
detailed bibliographic data is available in the Internet at <http://dnb.ddb.de>.

ISBN 978-3-0346-0328-7 Birkhäuser Verlag AG, Basel · Boston · Berlin

© 2010 Birkhäuser Verlag AG
Basel · Boston · Berlin
P.O. Box 133, CH-4010 Basel, Switzerland
Part of Springer Science+Business Media
Printed on acid-free paper produced of chlorine-free pulp. TCF ∞

ISBN 978-3-0346-0328-7   e-ISBN 978-3-0346-0329-4

9 8 7 6 5 4 3 2 1   www.birkhauser.ch

Contents

Luchezar L. Avramov
Infinite Free Resolutions

Mark L. Green
Generic Initial Ideals

Craig Huneke
Tight Closure, Parameter Ideals, and Geometry

Peter Schenzel
On the Use of Local Cohomology in Algebra and Geometry

Giuseppe Valla
Problems and Results on Hilbert Functions of Graded Algebras

Wolmer V. Vasconcelos
Cohomological Degrees of Graded Modules

Preface

A "Summer School on Commutative Algebra" was held from July 16 to 26, 1996, at the Centre de Recerca Matemàtica (CRM) in Bellaterra, a mathematical research institute sponsored by the Institut d'Estudis Catalans. It was organized by CRM and supported by CRM, Ministerio de Educación y Ciencia, Comissió Interdepartamental de Recerca i Innovació Tecnològica and Universitat de Barcelona. The School organizers were J. Elias, J. M. Giral, R. M. Miró-Roig and S. Zarzuela.

The School mainly consisted in 6 series of 5 lectures surveying and highlighting recent developements in Commutative Algebra, as well as 9 addressed talks and 28 short communications. Over 150 participants attended the School and contributed to its success.

This volume contains texts grown out of the six series of lectures given by the authors in the Summer School on Commutative Algebra. We are deeply indebted to them for their effort and excellent job.

The editors want to express their gratitude to the Director of the CRM, M. Castellet, and its staff for their support to our project without which it would not have taken place. We also want to thank our graduate students who helped in several tasks, specially J. Martínez for his unvaluable assistance in the preparation of this volume. Finally, thanks are due to Birkhäuser Verlag for their interest in it.

J. Elias
J.M. Giral
R.M. Miró-Roig
S. Zarzuela

Barcelona, September 1997.

Infinite Free Resolutions

Luchezar L. Avramov

This text is based on the notes for a series of five lectures to the Barcelona *Summer School in Commutative Algebra* at the Centre de Recerca Matemàtica, Institut d'Estudis Catalans, July 15–26, 1996.

Joan Elias, José Giral, Rosa Maria Miró-Roig, and Santiago Zarzuela were successively fantastic organizers, graceful hosts, and tactful editors. I am extremely grateful to them for everything.

It was a pleasure to address the receptive audience, and to be part of it for the lectures of Mark Green, Craig Huneke, Peter Schenzel, Tito Valla, Wolmer Vasconcelos, and the talks of many others. I learned a lot, both from and at the blackboard. Dave Jorgensen supplied a list of typos and corrections, and Alberto Corso introduced me to Xy-pic; their help is much appreciated.

Srikanth Iyengar has read more versions than he might care to recall. For all those corrected errors and pertinent suggestions, he has my special thanks.

Finally, I wish to acknowledge my most obvious debt: these notes would not have been written without Zoya's understanding and support.

Introduction

A free resolution of a module extends a presentation in terms of generators and relations, by choosing generators for the relations, then generators for the relations between relations, *ad infinitum*. Two questions arise:

How to write down a resolution?

How to read the data it contains?

These notes describe results and techniques for *finite* modules over a *commutative noetherian* ring R. An extra hypothesis that R is *local* (meaning that it has a unique maximal ideal) is often imposed. It allows for sharper formulations, due to the existence of unique up to isomorphism *minimal* free resolutions; in a global situation, the information can be retrieved from the local case by standard means (at least when all finite projective modules are free).

*   *   *

A finite free resolution is studied 'from the end', using linear algebra over the ring(!) R. The information carried by matrices of differentials is interpreted in

---

The author was partially supported by a grant from the National Science Foundation.

terms of the arithmetic of determinantal ideals. When R is a polynomial ring over a field each module has a finite free resolution; in this case, progress in computer algebra and programming has largely moved the construction of a resolution for any concrete module from algebra to hardware.

The monograph of Northcott [125] is devoted to finite free resolutions. Excellent accounts of the subject can be found in the books of Hochster [89], Roberts [136], Evans and Griffith [61], Bruns and Herzog [46], Eisenbud [58].

∗   ∗   ∗

Here we concentrate on modules that admit no finite free resolution.

There is no single approach to the construction of an infinite resolution and no single key to the information in it, so the exposition is built around several recurring themes. We describe them next, in the local case.

∗   ∗   ∗

To resolve a module of infinite projective dimension, one needs to solve an infinite string of iteratively defined systems of linear equations. The result of each step has consequences for eternity, so a measure of control on the propagation of solutions is highly desirable. This may be achieved by endowing the resolution with a *multiplicative structure* (a natural move for algebraists, accustomed to work with algebras and modules, rather than vector spaces). Such a structure can always be put in place, but its internal requirements may prevent the resolution from being minimal. Craft and design are needed to balance the diverging constraints of multiplicative structure and minimality; determination of cases when they are compatible has led to important developments.

Handling of resolutions with multiplicative structures is codified by *Differential Graded* homological algebra. Appearances notwithstanding, this theory precedes the familiar one: *Homological Algebra* [51] came after Cartan, Eilenberg, and MacLane [56], [50] developed fundamental ideas and constructions in DG categories to compute the homology of Eilenberg-MacLane spaces. An algebraist might choose to view the extra structure as an extension of the domain of rings and modules in a new, homological, dimension.

DG homological algebra is useful in *change of rings* problems. They arise in connection with a homomorphism of rings $Q \to R$ and an R–module M, when homological data on M, available over one of the rings, are needed over the other. A typical situation arises when R and M have finite free resolutions over Q, for instance when Q is a *regular ring*. It is then possible to find multiplicative resolutions of M and R over Q, that are 'not too big', and build from them resolutions of M over R. Although not minimal in general, such constructions are often useful, in part due to their functoriality and computability.

Multiplicative structures on a resolution are inherited by derived functors. It is a basic observation that the induced *higher structures in homology* do not depend of the choice of the resolution, and so are invariants of the R–module (or R–algebra) M. They suggest how to construct multiplicative structures of the resolution, or yield *obstructions* to its existence. Sometimes, they provide *criteria* for a ring- or module-theoretic property, e.g. for R to be a *complete intersection*. All known proofs that this property localizes depend on such characterizations – as all proofs that regularity localizes use the homological description of regular rings.

The behavior of resolutions at infinity gives rise to intriguing results and questions. New notions are needed to state some of them; for instance, *complexity* and *curvature* of a module are introduced to differentiate between the two known types of asymptotic growth. A striking aspect of infinite resolutions is the *asymptotic stability* that they display, both numerically (uniform patterns of Betti numbers) and structurally (high syzygies have similar module-theoretic properties). In many cases this phenomenon can be traced to a simple principle: the beginning of a resolution is strongly influenced by the defining relations of the module, that can be arbitrarily complicated; far from the source, the relations of the ring (thought of as a residue of some regular local ring) take over. In other words, the singularity of the ring dominates asymptotic behavior.

Part of the evidence comes from information gathered over specific classes of rings. At one end of the spectrum sit the complete intersections, characterized by asymptotically polynomial growth of all resolutions. The other end is occupied by the *Golod rings*, defined by an extremal growth condition on the resolution of the residue field; all resolutions over a Golod ring have asymptotically exponential growth; higher order homology operations, the *Massey products*, play a role in constructions. Results on complete intersections and Golod rings are presented in detail; generalizations are described or sketched.

A basic problem is whether some form of the polynomial/exponential dichotomy extends to all modules over local rings. No intermediate growth occurs for the residue field, a case of central importance. This result and its proof offer a glimpse at a mutually beneficial interaction between local algebra and rational homotopy theory, that has been going on for over a decade.

A major link in that connection is the *homotopy Lie algebra* of a local ring; it corresponds to the eponymous object attached to a CW complex, and has a representation in the cohomology of every R–module. Its structure affects the asymptotic patterns of resolutions, and is particularly simple when R is a complete intersection: each cohomology module is then a finite graded module over a polynomial ring, that can be investigated with all the usual tools. This brings up a strong connection with modular representations of finite groups.

*   *   *

Proving a result over local rings, we imply that a corresponding statement holds for graded modules over graded rings. Results specific to the graded case are mostly excluded; that category has a life of its own: after all, Hilbert [88] introduced resolutions to study graded modules over polynomial rings.

*   *   *

The notes assume a basic preparation in commutative ring theory, including a few homological routines. Modulo that, complete proofs are given for all but a couple of results used in the text. Most proofs are second or third generation, many of them are new. Constructions from DG algebra are developed from scratch. A bonus of using DG homological algebra is that spectral sequences may be eliminated from many arguments; we have kept a modest amount, for reasons of convenience and as a matter of principle.

$$*\qquad*\qquad*$$

The only earlier monographic exposition specifically devoted to infinite resolution is the influential book of Gulliksen and Levin [83], which concentrates on the residue field of a local ring. The overlap is restricted to Sections 6.1 and 6.3, with some differences in the approach. Sections 6.2, 7.2, and 8.2 contain material that has not been presented systematically before.

1. Complexes

This chapter lays the ground for the subsequent exposition. It fixes terminology and notation, and establishes some basic results.

All rings are assumed[1] commutative. No specific references are made to standard material in commutative algebra, for which the books of Matsumura [117], Bruns and Herzog [46], or Eisenbud [58], provide ample background.

For technical reasons, we choose to work throughout with algebras over a ubiquitous commutative ring $\Bbbk$, that will usually be unspecified and even unmentioned[2] (think of $\Bbbk = \mathbb{Z}$, or $\Bbbk = k$, a field).

1.1. Basic constructions.   Let R be a ring.

A (*bounded below*) *complex* of R–modules is a sequence

$$\cdots \longrightarrow F_n \xrightarrow{\ \partial_n\ } F_{n-1} \xrightarrow{\ \partial_{n-1}\ } F_{n-2} \longrightarrow \cdots$$

of R–linear maps with $\partial_{n-1}\partial_n = 0$ for $n \in \mathbb{Z}$ (and $F_n = 0$ for $n \ll 0$). The *underlying R–module* $\{F_n\}_{n\in\mathbb{Z}}$ is denoted $F^\natural$. Modules over R are identified with complexes concentrated in degree zero (that is, having $F_n = 0$ for $n \neq 0$); $|x|$ denotes the degree of an element x; thus, $|x| = n$ means $x \in F_n$.

Operations on complexes.   Let E, F, G be complexes of R–modules.

A *degree d homomorphism* $\beta\colon F \to G$ is simply a collection of R-linear maps $\{\beta_n\colon F_n \to G_{n+d}\}_{n\in\mathbb{Z}}$. All degree d homomorphisms from F to G form an R-module, $\mathrm{Hom}_R\,(F,G)_d$. This is the degree d component of a complex of R–modules

---

[1] Following a grand tradition of making blanket statements, to be violated down the road.
[2] Thus, the unqualified word 'module' stands for '$\Bbbk$–module'; homomorphisms are $\Bbbk$-linear; writing 'ring' or 'homomorphism of rings', we really mean '$\Bbbk$–algebra' or 'homomorphism of $\Bbbk$–algebras'. The convention is only limited by your imagination, and my forgetfulness.

$\mathrm{Hom}_R\,(F,G)$, in which the boundary of β is classically defined by

$$\partial(\beta) = \partial^G \circ \beta - (-1)^{|\beta|}\beta \circ \partial^F \, .$$

The power of -1 is a manifestation of the sign that rules over homological algebra: Each transposition of homogeneous entities of degrees i, j is 'twisted' by a factor $(-1)^{ij}$, and permutations are signed accordingly.

The cycles in $\mathrm{Hom}_R\,(F,G)$ are those homomorphisms β that satisfy $\partial \circ \beta = (-1)^{|\beta|}\beta \circ \partial$; they are called *chain maps*. Chain maps $\beta, \beta' \colon F \to G$ are *homotopic* if there exists a homomorphism $\sigma \colon F \to G$ of degree $|\beta| + 1$, called a *homotopy* from β to β', such that

$$\beta - \beta' = \partial \circ \sigma + (-1)^{|\beta|}\sigma \circ \partial \, ;$$

equivalently, $\beta - \beta'$ is the boundary of σ in the complex $\mathrm{Hom}_R\,(F,G)$.

A chain map β induces a natural homomorphism $\mathrm{H}(\beta)\colon \mathrm{H}(F) \to \mathrm{H}(G)$ of degree $|\beta|$; homotopic chain maps induce the same homomorphism. Chain maps of degree zero are the *morphisms* of the category of complexes. A *quasi-isomorphism* is a morphism that induces an isomorphism in homology; the symbol $\simeq$ next to an arrow identifies a quasi[3]-isomorphism.

The *tensor product* $E \otimes_R F$ has $(E \otimes_R F)_n = \sum_{i+j=n} E_i \otimes_R F_j$ and

$$\partial(e \otimes f) = \partial^E(e) \otimes f + (-1)^{|e|}e \otimes \partial^F(f) \, .$$

The *transposition map* $\tau(e \otimes f) = (-1)^{|e||f|}f \otimes e$ is an isomorphism of complexes $\tau \colon E \otimes_R F \to F \otimes_R E$.

A degree d homomorphism $\beta \colon F \to G$ induces degree d homomorphisms

$$\mathrm{Hom}_R\,(E,\beta) \colon \mathrm{Hom}_R\,(E,F) \to \mathrm{Hom}_R\,(E,G) \, ,$$

$$\mathrm{Hom}_R\,(E,\beta)\,(\alpha) = \beta \circ \alpha \, ;$$

$$\mathrm{Hom}_R\,(\beta,E) \colon \mathrm{Hom}_R\,(G,E) \to \mathrm{Hom}_R\,(F,E) \, ,$$

$$\mathrm{Hom}_R\,(\beta,E)\,(\gamma) = (-1)^{|\beta||\gamma|}\gamma \circ \beta \, ;$$

$$\beta \otimes_R E \colon F \otimes_R E \to G \otimes_R E \, , \quad (\beta \otimes_R E)(f \otimes e) = \beta(f) \otimes e \, ;$$

$$E \otimes_R \beta \colon E \otimes_R F \to E \otimes_R G \, , \quad (E \otimes_R \beta)(e \otimes f) = (-1)^{|\beta||e|}e \otimes \beta(f) \, ,$$

with signs determined by the Second Commandment[4]. All maps are natural in both arguments. If β is a chain map, then so are the induced maps.

The *shift* ΣF of a complex F has $(\Sigma F)_n = F_{n-1}$ for each n. In order for the degree 1 bijection $\Sigma^F \colon F \to \Sigma F$, sending $f \in F_n$ to $f \in (\Sigma F)_{n+1}$, to be a chain map the differential on ΣF is defined by $\partial^{\Sigma F}(\Sigma^F(f)) = -\Sigma^F(\partial^F(f))$.

The *mapping cone* of a morphism $\beta \colon F \to G$ is the complex $\mathrm{C}(\beta)$ with underlying module $G^\natural \oplus (\Sigma F)^\natural$, and differential $\begin{pmatrix} \partial^G & (\Sigma^G)^{-1}\Sigma(\beta) \\ 0 & \partial^{\Sigma F} \end{pmatrix}$. The connecting

---

[3]The symbol $\cong$ is reserved for the real thing.

[4]Obey the Sign! While orthodox compliance is a nuisance, transgression may have consequences ranging anywhere from mild embarrassment (confusion of an element and its opposite) to major disaster (creation of complexes with $\partial^2 \neq 0$).

homomorphism defined by the exact *mapping cone sequence*

$$0 \to G \to \mathrm{C}(\beta) \to \Sigma F \to 0$$

is equal to $\mathrm{H}(\beta)$. Thus, β is a quasi-isomorphism if and only if $\mathrm{H}(\mathrm{C}(\beta)) = 0$.

A *projective* (respectively, *free*) *resolution* of an R–module M is a quasi-isomorphism $\epsilon^F \colon F \to M$ from a complex F of projective (respectively, free) modules with $F_n = 0$ for $n < 0$. The *projective dimension* of M is defined by $\mathrm{pd}_R M = \inf\{p \,|\mathrm{M}$ a projective resolution with $F_n = 0$ for $n > p\}$.

Example 1.1.1. In the *Koszul complex* $K = K(\boldsymbol{g}; R)$ on a sequence $\boldsymbol{g} = g_1, \dots, g_r$ of elements of R the module K_1 is free with basis $x_1, \dots, x_r$, the module K_n is equal to $\bigwedge_R^n K_1$ for all n, and the differential is defined by

$$\partial(x_{i_1} \wedge \cdots \wedge x_{i_n}) = \sum_{j=1}^n (-1)^{j-1} g_{i_j}\, x_{i_1} \wedge \cdots \wedge x_{i_{j-1}} \wedge x_{i_{j+1}} \wedge \cdots \wedge x_{i_n}\,.$$

For each R–module M, set $K(\boldsymbol{g}; M) = K(\boldsymbol{g}; R) \otimes_R M$, and note that $\mathrm{H}_0(K(\boldsymbol{g}; M)) = M/(\boldsymbol{g})M$. A crucial property of Koszul complexes is their *depth[5]-sensitivity*: If M is a finite module over a noetherian ring R, then

$$\sup\{i \mid \mathrm{H}_i(K(\boldsymbol{g}; M)) \neq 0\} = r - \mathrm{depth}_R((\boldsymbol{g}), M)\,.$$

In particular, if $g_1, \dots, g_r$ is an R–regular sequence, then $K(\boldsymbol{g}; R)$ is a free resolution of $R' = R/(\boldsymbol{g})$ over R, and $\mathrm{H}(K(\boldsymbol{g}; M)) \cong \mathrm{Tor}^R(R', M)$.

Minimal complexes. A *local ring* $(R, \mathfrak{m}, k)$ is a noetherian ring R with unique maximal ideal $\mathfrak{m}$ and residue field $k = R/\mathfrak{m}$. A complex of R–modules F, such that $\partial_n(F_n) \subseteq \mathfrak{m}F_{n-1}$ for each n, is said to be *minimal*. Here is the reason:

Proposition 1.1.2. *If F is a bounded below complex of finite free modules over a local ring $(R, \mathfrak{m}, k)$, then the following conditions are equivalent.*

(i) *F is minimal.*
(ii) *Each quasi-isomorphism $\alpha \colon F \to F$ is an isomorphism.*
(iii) *Each quasi-isomorphism $\beta \colon F \to F'$ to a bounded below minimal complex of free modules is an isomorphism.*
(iv) *Each quasi-isomorphism $\beta \colon F \to G$ to a bounded below complex of free modules is injective, and $G = \mathrm{Im}\,\beta \oplus E$ for a split-exact subcomplex E.*

Proof. (i) $\implies$ (iii). The mapping cone $\mathrm{C}(\beta)$ is a bounded below complex of free modules with $\mathrm{H}(\mathrm{C}(\beta)) = 0$. Such a complex is split-exact, hence so is $\mathrm{C}(\beta) \otimes_R k \cong \mathrm{C}(\beta \otimes_R k)$. Thus, $\mathrm{H}(\beta \otimes_R k)$ is an isomorphism. Since both $F \otimes_R k$ and $F' \otimes_R k$ have trivial differentials, each $\beta_n \otimes_R k$ is an isomorphism. As F_n and F_n' are free modules, β_n is itself an isomorphism by Nakayama.

---

[5]Recall that the *depth* of an ideal $I \subseteq R$ on M, denoted $\mathrm{depth}_R(I, M)$, is the maximal length of an M-regular sequence contained in I.

(iii) $\implies$ (iv). Choose a subset $Y \subseteq G$ such that $\partial(Y) \otimes_R 1$ is a basis of the vector space $\partial(G \otimes_R k)$. As $Y \cup \partial(Y)$ is linearly independent modulo $\mathfrak{m}G$, it extends to a basis of the R–module $G^{\natural}$. Thus, the complex

$$E: \quad \bigoplus_{y \in Y} (0 \to Ry \to R\partial(y) \to 0) \qquad (*)$$

is split-exact and a direct summand of G, hence $G \to G/E = G'$ is a quasi-isomorphism onto a bounded below free complex, that is minimal by construction. As the composition $\beta' : F \to G \to G'$ is a quasi-isomorphism, it is an isomorphism by our hypothesis. The assertions of (iv) follow.

(iv) $\implies$ (ii): the split monomorphisms $\alpha_n : F_n \to F_n$ are bijective.

(ii) $\implies$ (i). Assume that (i) fails, and form a surjective quasi-isomorphism $\beta : F \to F/F' = G$ onto a bounded below free complex as in the argument above. There is then a morphism $\gamma : G \to F$ with $\beta\gamma = \mathrm{id}^G$, cf. Proposition 1.3.1. Such a γ is necessarily a quasi-isomorphism, hence $\gamma\beta = \alpha$ is a quasi-isomorphism $F \to F$ with $\mathrm{Ker}\,\alpha \supseteq F' \neq 0$, a contradiction. $\qquad\square$

1.2. Syzygies. In this section $(R, \mathfrak{m}, k)$ is a local ring. Over such rings, projectives are free; for finite modules, this follows easily from Nakayama.

A *minimal resolution* F of an R–module M is a free resolution, that is also a minimal complex. If $m_1, \dots, m_r$ minimally generate M, then the Third Command-ment[6] prescribes to start the construction of a resolution by a map $F_0 = R^r \to M$ with $(a_1, \dots, a_r) \mapsto a_1 m_1 + \cdots + a_r m_r$. This is a surjection with kernel in $\mathfrak{m}R^r$; iterating the procedure, one sees that M has a minimal resolution. As any two resolutions of M are linked by a morphism that induces the identity of M, Proposition 1.1.2 completes the proof of the following result of Eilenberg from [55], where minimal resolutions are introduced.

Proposition 1.2.1. *Each finite R–module M has a minimal resolution, that is unique up to isomorphism of complexes. A minimal resolution F is isomorphic to a direct summand of any resolution of M, with complementary summand a split-exact free complex. In particular, $\mathrm{pd}_R M = \sup\{n \mid F_n \neq 0\}$.* $\qquad\square$

The 'uniqueness' of a minimal resolution F implies that each R–module $\mathrm{Syz}_n^R(M) = \mathrm{Coker}(\partial_{n+1} : F_{n+1} \to F_n) \cong \partial_n(F_n)$ is defined uniquely up to a (non-canonical) isomorphism; it is called the n'th *syzygy* of M; note that $\mathrm{Syz}_0^R(M) = M$, and $\mathrm{Syz}_n^R(M) = 0$ for $n < 0$.

The number $\beta_n^R(M) = \mathrm{rank}_R F_n$ is called the n'th *Betti number* of M (over R). The complexes $F \otimes_R k$ and $\mathrm{Hom}_R(F, k)$ have zero differentials, so $\mathrm{Tor}_n^R(M, k) \cong F_n \otimes_R k$ and $\mathrm{Ext}_R^n(M, k) \cong \mathrm{Hom}_R(F_n, k)$; in other words:

Proposition 1.2.2. *If M is a finite R–module, then*

$$\beta_n^R(M) = \nu_R\big(\mathrm{Syz}_n^R(M)\big) = \dim_k \mathrm{Tor}_n^R(M, k) = \dim_k \mathrm{Ext}_R^n(M, k)$$

and $\mathrm{pd}_R M = \sup\{n \mid \beta_n^R(M) \neq 0\}$. $\qquad\square$

---

[6]Resolve minimally!

Syzygies behave well under certain base changes by *local homomorphisms*, that is, homomorphisms of local rings $\varphi(R, \mathfrak{m}) \to (R', \mathfrak{m}')$ with $\varphi(\mathfrak{m}) \subseteq \mathfrak{m}'$.

Proposition 1.2.3. *If M is a finite R–module and $\varphi\colon R \to R'$ is a local homomorphism such that $\mathrm{Tor}_n^R(R', M) = 0$ for $n > p$, then*

$$\mathrm{Syz}_{n-p}^{R'}\left(R' \otimes_R \mathrm{Syz}_p^R(M)\right) \cong R' \otimes_R \mathrm{Syz}_n^R(M) \qquad \textit{for} \quad n \geq p.$$

Proof. Let F be a minimal free resolution of M over R.

Since $\mathrm{H}_n(R' \otimes_R F) \cong \mathrm{Tor}_n^R(R', M) = 0$ for $n > p$, the complex of R'–modules $(R' \otimes_R F)_{\geqslant p}$ is a free (and obviously minimal) resolution of the module $\mathrm{Coker}(R' \otimes_R \partial_{p+1}) \cong R' \otimes_R \mathrm{Coker}\,\partial_{p+1} \cong R' \otimes_R \mathrm{Syz}_p^R(M)$. $\square$

Corollary 1.2.4. *Let M be a finite R–module.*

(1) *If $R \to R'$ is a faithfully flat homomorphism of local rings, then*

$$\mathrm{Syz}_n^{R'}\left(R' \otimes_R M\right) \cong R' \otimes_R \mathrm{Syz}_n^R(M) \qquad \textit{for} \quad n \geq 0.$$

(2) *If a sequence $g_1, \ldots, g_r \in R$ is both R–regular and M–regular, then*

$$\mathrm{Syz}_n^{R'}(M') \cong R' \otimes_R \mathrm{Syz}_n^R(M) \qquad \textit{for} \quad n \geq 0$$

with $R' = R/(g_1, \ldots, g_r)R$, and $M' = M/(g_1, \ldots, g_r)M$. $\square$

Proof. The proposition applies with $p = 0$, twice: by definition in case (1), and by Example 1.1.1 in case (2). $\square$

When R is local, $\mathrm{depth}_R M = \mathrm{depth}_R(\mathfrak{m}, M)$ and $\mathrm{depth}\,R = \mathrm{depth}_R R$.

Corollary 1.2.5. *If R is a direct summand of $\mathrm{Syz}_n^R(M)$, then $n \leq m$, where $m = \max\{0, \mathrm{depth}\,R - \mathrm{depth}\,M\}$.*

Proof. Let $n > m$, and assume that $\mathrm{Syz}_n^R(M)$ has R as a direct summand. Choose a maximal $(R \oplus M)$–regular sequence, and complete it (if necessary) to a maximal R–regular sequence $\boldsymbol{g}$. By Example 1.1.1 and Proposition 1.2.3, $R' = R/(\boldsymbol{g})$ is a direct summand of a syzygy of the R'–module $R' \otimes_R \mathrm{Syz}_m^R(M)$, hence sits in $\mathfrak{m}F'$, where F' is a free R'–module. As $\mathrm{depth}\,R' = 0$, the ideal $(0 \colon_{R'} \mathfrak{m}) \neq 0$ annihilates R'; this is absurd. $\square$

Depth can be computed cohomologically, by the formula

$$\mathrm{depth}_R M = \inf\{n \mid \mathrm{Ext}_R^n(k, M) \neq 0\}.$$

The following well known fact is recorded for ease of reference.

Lemma 1.2.6. *If M is a finite R–module, then*

$$\mathrm{depth}_R \mathrm{Syz}_1^R(M) = \begin{cases} \mathrm{depth}_R M + 1 & \textit{when} \quad \mathrm{depth}_R M < \mathrm{depth}\,R; \\ g \geq \mathrm{depth}\,R & \textit{when} \quad \mathrm{depth}_R M = \mathrm{depth}\,R; \\ \mathrm{depth}\,R & \textit{when} \quad \mathrm{depth}_R M > \mathrm{depth}\,R. \end{cases}$$

Proof. Track the vanishing of $\mathrm{Ext}_R^n(k, -)$ through the long exact cohomology sequence induced by the exact sequence $0 \to \mathrm{Syz}_1^R(M) \to F_0 \to M \to 0$. $\square$

Proposition 1.2.7. *If M is a finite R–module with $\mathrm{pd}_R M < \infty$, then*

(1) $\mathrm{pd}_R M + \mathrm{depth}_R M = \mathrm{depth}\, R$.

(2) $(0 :_R M) = 0$, or $(0 :_R M)$ *contains a non-zero-divisor on R.*

(3) $\sum_{n \geq 0} (-1)^n \beta_n^R(M) \geq 0$, *with equality if and only if* $(0 :_R M) \neq 0$.

Proof. Set $m = \mathrm{pd}_R M$, $g = \mathrm{depth}_R M$, and $d = \mathrm{depth}\, R$.

(1) As $\mathrm{Syz}_m^R(M) \neq 0$ is free, Corollary 1.2.5 yields $m + g \leq d$. Thus, $g \leq d$, and if $g = d$, then M is free and (1) holds. If $g < d$, then assume by descending induction that (1) holds for modules of depth $> g$, and use the lemma:

$$m + g = (\mathrm{pd}_R \mathrm{Syz}_1^R(M) + 1) + (\mathrm{depth}_R \mathrm{Syz}_1^R(M) - 1) = d\,.$$

(2) and (3). If F is a minimal resolution of M and $\mathfrak{p} \in \mathrm{Spec}\, R$, then

$$0 \to (F_m)_\mathfrak{p} \to (F_{m-1})_\mathfrak{p} \to \cdots \to (F_1)_\mathfrak{p} \to (F_0)_\mathfrak{p} \to M_\mathfrak{p} \to 0$$

is exact. If $\mathfrak{p} \in \mathrm{Ass}\, R$, then $\mathrm{depth}\, R_\mathfrak{p} = 0$, hence $M_\mathfrak{p}$ is free by (1).

Counting ranks, we get $\sum_n (-1)^n \beta_n^R(M) = \mathrm{rank}_{R_\mathfrak{p}} M_\mathfrak{p} \geq 0$.

If $\sum_n (-1)^n \beta_n^R(M) = 0$, then $M_\mathfrak{p} = 0$ for all $\mathfrak{p} \in \mathrm{Ass}\, R$, so $(0 :_R M) \not\subseteq \bigcup_{\mathfrak{p} \in \mathrm{Ass}\, R} \mathfrak{p}$, that is, $(0 :_R M)$ contains a non-zero-divisor.

If $(0 :_R M) \neq 0$, then $(0 :_{R_\mathfrak{p}} M_\mathfrak{p}) = (0 :_R M)_\mathfrak{p} \neq 0$ for $\mathfrak{p} \in \mathrm{Ass}(0 :_R M) \subseteq \mathrm{Ass}\, R$; as $M_\mathfrak{p}$ is free, this implies $M_\mathfrak{p} = 0$, and so $\sum_n (-1)^n \beta_n^R(M) = 0$. $\square$

The arguments for (2) and (3) are from Auslander and Buchsbaum's paper [17]; there is a new twist in the proof of their famous equality (1). It computes depths of syzygies when $\mathrm{pd}_R M < \infty$; otherwise, Okiyama [126] proves:

Proposition 1.2.8. *If M is a finite R–module with $\mathrm{pd}_R M = \infty$, then*

$$\mathrm{depth}\, \mathrm{Syz}_n^R(M) \geq \mathrm{depth}\, R \qquad for \quad n \geq \max\{0, \mathrm{depth}\, R - \mathrm{depth}_R M\}$$

with at most one strict inequality, at $n = 0$ or at $n = \mathrm{depth}\, R - \mathrm{depth}_R M + 1$.

Proof. Set $M_n = \mathrm{Syz}_n^R(M)$ and $d = \mathrm{depth}\, R$. Iterated use of Lemma 1.2.6 yields the desired inequality, and reduces the last assertion to proving that inequalities for n and $n + 1$ imply equality for $n + 2$.

Break down a minimal free resolution F of M into short exact sequences $E^i \colon 0 \longrightarrow M_{i+1} \xrightarrow{\iota_{i+1}} F_i \xrightarrow{\pi_i} M_i \longrightarrow 0$. If $\mathrm{depth}_R N_{n+2} > d$, then the cohomology exact sequence of E^{n+1} implies that the homomorphism

$$\mathrm{Ext}_R^d(k, \pi_{n+1}) \colon \mathrm{Ext}_R^d(k, F_{n+1}) \to \mathrm{Ext}_R^d(k, M_{n+1})$$

is injective. As $\mathrm{depth}_R M_n \geq d$, the cohomology sequence of E^n shows that

$$\mathrm{Ext}_R^d(k, \iota_{n+1}) \colon \mathrm{Ext}_R^d(k, M_{n+1} R M) \to \mathrm{Ext}_R^d(k, F_n)$$

is injective. Since $\iota_n \circ \pi_{n+1} = \partial_{n+1} \colon F_{n+1} \to F_n$, we see that the map

$$\mathrm{Ext}_R^d(k, \partial_{n+1}) = \mathrm{Ext}_R^d(k, \iota_n) \circ \mathrm{Ext}_R^d(k, \pi_{n+1})$$

is injective as well. But $\partial_{n+1} \colon F_{n+1} \to F_n$ is a matrix with elements in $\mathfrak{m}$, so $\mathrm{Ext}_R^d(k, \partial_{n+1}) = 0$, hence $\mathrm{Ext}_R^d(k, F_{n+1}) = 0$. Since $\mathrm{depth}_R F_{n+1} = d$ this is impossible, so $\mathrm{depth}\, M_n \leq d$, as desired. $\square$

Remark 1.2.9. By the last two results, there exists an R–regular sequence $\boldsymbol{g}$ of length $d = \operatorname{depth} R$, that is also regular on $N = \operatorname{Syz}_m^R(M)$, where $m = \max\{0, d - \operatorname{depth} M\}$. Proposition 1.2.4.2 yields $\beta_{n+m}^R(M) = \beta_n^R(N) = \beta_n^{R/(\boldsymbol{g})}(N/(\boldsymbol{g})N)$ for $n \geq 0$. When k is infinite, a sequence $\boldsymbol{g}$ may be found that also preserves multiplicity: $\operatorname{mult}(R) = \operatorname{mult}(R/(\boldsymbol{g}))$; if k is finite, then $R' = R[t]_{\mathfrak{m}[t]}$ has $\operatorname{mult}(R') = \operatorname{mult}(R)$, and $\beta_n^R(N) = \beta_n^{R'}(N \otimes_R R')$, for $n \geq 0$.

Remark 1.2.10. The name *graded ring* is reserved[1] for rings equipped with a direct sum decomposition $R = \bigoplus_{i \geqslant 0} R_i$, and having $R_0 = k$, a field. For such a ring we denote $\mathfrak{m}$ the *irrelevant maximal ideal* $\bigoplus_{i > 0} R_i$. An R–module M is *graded* if $M = \bigoplus_{j \in \mathbb{Z}} M_j$ and $R_i M_j \subseteq M_{i+j}$ for all $i, j \in \mathbb{Z}$. To minimize confusion with gradings arising from complexes, we say that $a \in M_i$ has *internal degrees* i, and write $\deg(a) = i$. The d'th *translate* of M is the graded R–module $M(d)$ with $M(d)_j = M_{j+d}$. A degree zero homomorphism $\alpha \colon M \to N$ of graded R–modules is an R–linear map such that $\alpha(M_j) \subseteq N_j$ for all j.

The free objects in the category of graded modules and degree zero homomorphisms are isomorphic to direct sums of modules of the form $R(d)$. Each graded R–module M has a *graded resolution* by free graded modules with differentials that are homomorphisms of degree zero. If $M_j = 0$ for $j \ll 0$ (in particular, if M is finitely generated), then such a resolution F exists with $\partial(F_n) \subseteq \mathfrak{m} F_n$ for all n. This *minimal graded resolution* is unique up to isomorphism of complexes of graded R–modules, so the numbers β_{nj} appearing in isomorphisms $F_n \cong \bigoplus_{j \in \mathbb{Z}} R(-j)^{\beta_{nj}}$ are uniquely defined, and finite if M is a finite R–module; these *graded Betti numbers* of M over R are denoted $\beta_{nj}^R(M)$.

1.3. Differential graded algebra. The term refers to a hybrid of homological algebra and ring theory. When describing the progeny[7], we systematically replace the compound 'differential graded' by the abbreviation DG.

DG algebras. A *DG algebra* A is a complex (A, ∂), with an element $1 \in A_0$ (the *unit*), and a morphism of complexes (the *product*)

$$A \otimes_k A \to A, \qquad a \otimes b \mapsto ab,$$

that is unitary: $1a = a = a1$, and associative: $a(bc) = (ab)c$. In addition, we assume the A is *(graded) commutative*:

$$ab = (-1)^{|a||b|} ba \quad \text{for } a, b \in A \qquad \text{and} \qquad a^2 = 0 \quad \text{when } |a| \text{ is odd},$$

and that $A_i = 0$ for $i < 0$; without them, we speak of *associative* DG algebras.

The fact that the product is a chain map is expressed by the *Leibniz rule*:

$$\partial(ab) = \partial(a)b + (-1)^{|a|} a\partial(b) \qquad \text{for} \quad a, b \in A,$$

Its importance comes from a simple observation: The cycles $Z(A)$ are a graded subalgebra of A, the boundaries $\partial(A)$ are an ideal in $Z(A)$, hence the canonical

---

[7]It is not that exotic: a commutative ring is precisely a DG algebra concentrated in degree zero, and a DG module over it is simply a complex. A prime example of a 'genuine' DG algebra is a Koszul complex, with multiplication given by wedge product.

projection $Z(A) \to H(A)$ makes the homology $H(A)$ into a graded algebra. In particular, each $H_n(A)$ is a module over the ring $H_0(A)$.

A *morphism of DG algebras* is a morphism of complexes $\phi \colon A \to A'$, such that $\phi(1) = 1$ and $\phi(ab) = \phi(a)\phi(b)$; we say that A' is a *DG algebra over A*.

If A and A' are DG algebras, then the tensor products of complexes $A \otimes_{\Bbbk} A'$ is a DG algebra with multiplication $(a \otimes a')(b \otimes b') = (-1)^{|a'||b|}(ab \otimes a'b')$.

A *graded algebra* is a DG algebra with zero differential, that is, a *family*[8] $\{A_n\}$, rather than a direct sum $\bigoplus_n A_n$.

A *DG module* U over the DG algebra A is a complex together with a morphism $A \otimes U \to U$, $a \otimes u \mapsto au$, that satisfies the Leibniz rule

$$\partial(au) = \partial(a)u + (-1)^{|a|}a\partial(u) \qquad \text{for} \quad a \in A \text{ and } u \in U$$

and is unitary and associative in the obvious sense. A *module* is a DG module with zero differential; $U^{\natural}$ is a module over $A^{\natural}$, and $H(U)$ is a module over $H(A)$.

Let U and V be DG modules over A.

A homomorphism $\beta \colon U \to V$ of the underlying complexes is *A–linear* if $\beta(au) = (-1)^{|\beta||a|}a\beta(u)$ for all $a \in A$ and $u \in U$. The A–linear homomorphisms form a subcomplex $\operatorname{Hom}_A(U,V) \subseteq \operatorname{Hom}_{\Bbbk}(U,V)$. The action

$$(a\beta)(u) = a(\beta(u)) = (-1)^{|a||\beta|}\beta(au)$$

turns it into a DG module over A. Two A–linear chain maps $\beta, \beta' \colon U \to V$ that are homotopic by an A–linear homotopy are said to be *homotopic over A*. Thus, $H_d(\operatorname{Hom}_A(U,V))$ is the set of *homotopy classes* of A–linear, degree d chain maps. DG modules over A and their A–linear morphisms are, respectively, the objects and morphisms of the *category of DG modules* over A.

The residue complex $U \otimes_A V$ of $U \otimes_{\Bbbk} V$ by the subcomplex spanned by all elements $au \otimes_{\Bbbk} v - (-1)^{|a||u|}u \otimes_{\Bbbk} av$, has an action

$$a(u \otimes_A v) = au \otimes_A v = (-1)^{|a||u|}u \otimes_A av.$$

It is naturally a DG module over A, and has the usual universal properties.

The shift ΣU becomes a DG module over A by setting $a\Sigma^U(u) = (-1)^{|a|}\Sigma^U(au)$; the map $\Sigma^U \colon U \to \Sigma U$ is then an A–linear homomorphism.

The mapping cone of a morphism $\beta \colon U \to V$ of DG modules over A is a DG module over A, and the maps in the mapping cone sequence are A–linear.

Semi-free modules. A bounded below DG module F over A is *semi-free* if its underlying $A^{\natural}$–module $F^{\natural}$ has a basis[9] $\{e_\lambda\}_{\lambda \in \Lambda}$. Thus, for each $f \in F$ there are unique $a_\lambda \in A$ with $f = \sum_{\lambda \in \Lambda} a_\lambda e_\lambda$; we set $\Lambda_n = \{\lambda \in \Lambda : |e_\lambda| = n\}$.

---

[8] This convention reduces the length of the exposition by 1.713%, as it trims from each argument all sentences starting with 'We may assume that the element x is homogeneous'; note that by Remark 1.2.10 above, a *graded ring* is the usual thing.

[9] Over a ring, a such a DG module is simply a bounded below complex of free modules. For arbitrary DG modules over any graded associative DG algebras, the notion is defined by a different condition: cf. [33], where the next three propositions are established in general.

Note that F is *not* a free object on $\{e_\lambda\}$ in the category of DG modules over A: As $\partial(e_\lambda)$ is a linear combination of basis elements e_μ of lower degree, the choice for the image of e_λ is restricted by the choices already made for the images of the e_μ. What freeness remains is in the important *lifting property*:

Proposition 1.3.1. *If F is a semi-free DG module over a DG algebra A, then each diagram of morphisms of DG modules over A represented by solid arrows*

$$
\begin{array}{ccc}
 & & U \\
 & \nearrow & \downarrow \simeq \beta \\
F & \xrightarrow{\ \alpha\ } & V
\end{array}
$$

with a surjective quasi-isomorphism β can be completed to a commutative diagram by a morphism γ, that is defined uniquely up to A–linear homotopy.

Remark. A degree d chain map $F \to V$ is nothing but a morphism $F \to \Sigma^{-d}V$, so the proposition provides also a 'unique lifting property' for chain maps.

Proof. Note that $F^n = \bigoplus_{|e_\lambda| \leqslant n} Ae_\lambda$ is a DG submodule of F over A, and $F^n = 0$ for $n \ll 0$. By induction on n, we may assume that $\gamma^n \colon F^n \to U$ has been constructed, with $\alpha|_{F^n} = \beta \circ \gamma^n$.

For each $\lambda \in \Lambda_{n+1}$, we have $\partial(\alpha\partial(e_\lambda)) = \alpha(\partial^2(e_\lambda)) = 0$, so $\alpha\partial(e_\lambda)$ is a cycle in V. Since β is a surjective quasi-isomorphism, there exists a *cycle* $z'_\lambda \in U$, such that $\beta(z'_\lambda) = \alpha\partial(e_\lambda)$. Thus, $z_\lambda = \gamma^n\partial(e_\lambda) - z'_\lambda \in U$ satisfies

$$\partial(z_\lambda) = \gamma^n\partial^2(e_\lambda) - \partial(z'_\lambda) = 0 \quad \text{and} \quad \beta(z_\lambda) = \alpha\partial(e_\lambda) - \beta(z'_\lambda) = 0,$$

that is, z_λ is a cycle in $W = \mathrm{Ker}\,\beta$. The homology exact sequence of the short exact sequence of DG modules $0 \to W \to U \to V \to 0$ shows that $\mathrm{H}(W) = 0$, hence $z_\lambda = \partial(y_\lambda)$ for some $y_\lambda \in W$. In view of our choices, the formula

$$\gamma^{n+1}\left(f + \sum_{\lambda \in \Lambda_{n+1}} a_\lambda e_\lambda\right) = \gamma^n(f) + \sum_{\lambda \in \Lambda_{n+1}} a_\lambda y_\lambda \qquad \text{for} \quad f \in F^n$$

defines a morphism of DG modules $\gamma^{n+1} \colon F^{n+1} \to U$, with $\gamma^{n+1}|_{F^n} = \gamma^n$, and completes the inductive construction. As $F = \bigcup_{n \in \mathbb{Z}} F^n$, setting $\gamma(f) = \gamma^n(f)$ whenever $f \in F^n$, we get a morphism $\gamma \colon F \to U$ with $\alpha = \beta\gamma$.

If $\gamma' \colon F \to U$ is a morphism with $\alpha = \beta\gamma'$, then $\beta(\gamma - \gamma') = 0$, hence there exists a morphism $\delta \colon F \to W$ such that $\gamma - \gamma' = \iota\delta$, where $\iota \colon W \subseteq U$ is the inclusion. Again, we assume by induction that a homotopy $\sigma^n \colon F^n \to W$ between $\delta|_{F^n}$ and 0 is available: $\delta|_{F^n} = \partial\sigma^n + \sigma^n\partial$. As

$$\partial(\delta(e_\lambda) - \sigma^n\partial(e_\lambda)) = \delta\partial(e_\lambda) - (\partial\sigma^n)(\partial(e_\lambda))$$
$$= (\delta - \partial\sigma^n)(\partial(e_\lambda)) = (\sigma^n\partial)(\partial(e_\lambda)) = 0$$

and $\mathrm{H}(W) = 0$, there is a $w_\lambda \in W$ such that $\partial(w_\lambda) = \delta(e_\lambda) - \sigma^n\partial(e_\lambda)$. Now

$$\sigma^{n+1}\left(f + \sum_{\lambda \in \Lambda_{n+1}} a_\lambda e_\lambda\right) = \sigma^n(f) + \sum_{\lambda \in \Lambda_{n+1}} (-1)^{|a_\lambda|} a_\lambda w_\lambda$$

is a degree 1 homomorphism $\sigma^{n+1} \colon F^{n+1} \to W$, with $\delta|_{F^{n+1}} = \partial\sigma^{n+1} + \sigma^{n+1}\partial$, and $\sigma^{n+1}|_{F^n} = \sigma^n$. In the limit, we get a homotopy $\sigma \colon F \to W$ from δ to 0, and then $\sigma' = \iota\sigma \colon F \to U$ is a homotopy from γ to γ'. $\square$

Proposition 1.3.2. *If F is a semi-free DG module, then each quasi-isomorphism $\beta \colon U \to V$ of DG modules over A induces quasi-isomorphisms*

$$\mathrm{Hom}_A\,(F,\beta) \colon \mathrm{Hom}_A\,(F,U) \to \mathrm{Hom}_A\,(F,V)\,; \qquad F \otimes_A \beta \colon F \otimes_A U \to F \otimes_A V\,.$$

Proof. To prove that $\mathrm{Hom}_A\,(F,\beta)$ is a quasi-isomorphism, we show the exactness of its mapping cone, which is isomorphic to $\mathrm{Hom}_A\,(F,\mathrm{C}(\beta))$. Thus, we want to show that each chain map $F \to \mathrm{C}(\beta)$ is homotopic to 0. Such a chain map is a lifting of $F \to 0$ over the quasi-isomorphism $\mathrm{C}(\beta) \to 0$. Since $0 \colon F \to \mathrm{C}(\beta)$ is another such lifting, they are homotopic by Proposition 1.3.1.

To prove that $\beta \otimes_A F$ is a quasi-isomorphism, we use the exact sequences $0 \to F^n \to F^{n+1} \to \overline{F}^{n+1} \to 0$ of DG modules over A, involving the submodules F^n from the preceding proof. The sequences split over $A^\natural$, and so induce commutative diagrams with exact rows

$$
\begin{array}{ccccccccc}
0 & \longrightarrow & U \otimes_A F^n & \longrightarrow & U \otimes_A F^{n+1} & \longrightarrow & U \otimes_A \overline{F}^{n+1} & \longrightarrow & 0 \\
& & {\scriptstyle \beta\otimes_A F^n}\big\downarrow & & {\scriptstyle \beta\otimes_A F^{n+1}}\big\downarrow & & {\scriptstyle \beta\otimes_A \overline{F}^{n+1}}\big\downarrow & & \\
0 & \longrightarrow & V \otimes_A F^n & \longrightarrow & V \otimes_A F^{n+1} & \longrightarrow & V \otimes_A \overline{F}^{n+1} & \longrightarrow & 0\,.
\end{array}
$$

By induction on n, we may assume that $\beta \otimes_A F^n$ is a quasi-isomorphism. As

$$\overline{F}^{n+1} \cong \bigoplus_{\lambda \in \Lambda_{n+1}} Ae_\lambda \qquad \text{with} \quad \partial(e_\lambda) = 0 \text{ for all } \lambda \in \Lambda_{n+1}\,,$$

the map $\beta \otimes_A \overline{F}^{n+1}$ is a quasi-isomorphism. By the Five-Lemma, $\beta \otimes_A F^{n+1}$ is one as well, hence so is $\beta \otimes_A F = \beta \otimes (\mathrm{inj\,lim}_n\, F^n) = \mathrm{inj\,lim}_n (\beta \otimes_A F^n)$. $\square$

Proposition 1.3.3. *Let U be a DG module over a DG algebra A. Each quasi-isomorphism $\gamma \colon F \to G$ of semi-free modules induces quasi-isomorphisms*

$$\mathrm{Hom}_A\,(\gamma,U) \colon \mathrm{Hom}_A\,(G,U) \to \mathrm{Hom}_A\,(F,U)\,; \qquad \gamma \otimes_A U \colon F \otimes_A U \to G \otimes_A U.$$

Proof. The mapping cone $C = \mathrm{C}(\gamma)$ is exact. It is semi-free, so by the preceding proposition the quasi-isomorphism $C \to 0$ induces a quasi-isomorphism $\mathrm{Hom}_A\,(C,C) \to 0$. Thus, there is a homotopy σ from id^C to 0^C. It is easily verified that $\mathrm{Hom}_A\,(\sigma,U)$ and $\sigma \otimes_A U$ are null-homotopies on $\mathrm{Hom}_A\,(C,U)$ and $C \otimes_A U$, so these complexes are exact. They are isomorphic, respectively, to $\Sigma^{-1}\mathrm{C}(\mathrm{Hom}_A\,(\gamma,U))$ and $\mathrm{C}(\gamma \otimes_A U)$, which are therefore exact. We conclude that $\mathrm{Hom}_A\,(\gamma,U)$ and $\gamma \otimes_A U$ are quasi-isomorphisms. $\square$

The preceding results have interesting applications even for complexes over a ring. A first illustration occurs in the proof of Proposition 1.1.2. Another one is in the following proof of the classical *Künneth Theorem*.

Proposition 1.3.4. *If G is a bounded below complex of free R–modules, such that $F = \mathrm{H}(G)$ is free, then the Künneth map*

$$\kappa^{GU} \colon \mathrm{H}(G) \otimes_R \mathrm{H}(U) \to \mathrm{H}(G \otimes_R U),$$

$$\kappa^{GU}(\mathrm{cls}(g) \otimes \mathrm{cls}(u)) = \mathrm{cls}(g \otimes u),$$

is an isomorphism for each complex of R–modules U.

Proof. Set $F = \mathrm{H}(G)$. The composition of an R– linear splitting of the surjection $\mathrm{Z}(G) \to F$ with the injection $\mathrm{Z}(G) \to G$ is a quasi-isomorphism $\gamma \colon F \to G$ of semi-free DG modules over R. By the last proposition, so is $\gamma \otimes_R U \colon F \otimes_R U \to G \otimes_R U$. The Künneth map being natural, it suffices to show that κ^{FU} is bijective. As $\partial^F = 0$ and each F_n is free, this is clear. □

2. Multiplicative Structures on Resolutions

Is it possible to 'enrich' resolutions over a commutative ring Q, by endowing them with DG module or DG algebra structures? The rather complete – if at first puzzling – answer comes in three parts:

- For residue rings of Q, algebra structures are carried by essentially all resolutions of length ≤ 3; for finite Q–modules, DG module structures exist on all resolutions of length ≤ 2.
- Beyond these bounds, not all resolutions support such structures.
- There always exist resolutions, that do carry the desired structure.

In this chapter we present in detail the results available on all three counts. Most developments in the rest of the notes are built on resolutions that comply with the Fourth Commandment[10].

2.1. DG algebra resolutions. Let Q be a commutative ring and let R be a Q–algebra. A *DG algebra resolution* of R over Q consists of a (commutative) DG algebra A, such that A_i is a projective Q–module for each i, and a quasi-isomorphism $\epsilon^A \colon A \to R$ of DG algebras over Q.

The next example is the grandfather of all DG algebra resolutions.

Example 2.1.0. If $R = Q/(\boldsymbol{f})$ for a Q–regular sequence $\boldsymbol{f}$, then the Koszul complex on $\boldsymbol{f}$ is a DG algebra resolution of R over Q.

'Short' projective resolutions often carry DG algebra structures.

---

[10]Resolve in kind!

Example 2.1.1. If $R = Q/I$ has a resolution A of length 1 of the form

$$0 \to F_1 \to Q \to 0$$

then the only product that makes it a graded algebra over Q is defined by the condition $F_1 \cdot F_1 = 0$; it clearly makes A into a DG algebra.

If ϕ is a matrix, then for $J, K \subset \mathbb{N}$ we denote ϕ_J^K the submatrix obtained by deleting the rows with indices from J and the columns with indices from K.

Example 2.1.2. If $R = Q/I$ has a free resolution of length 2, then by the Hilbert-Burch Theorem there exist a non-zero-divisor a and an $r \times (r-1)$ matrix ϕ, such that a free resolution of R over Q is given by the complex

$$A: \quad 0 \to \bigoplus_{k=1}^{r-1} Q f_k \xrightarrow{\partial_2} \bigoplus_{j=1}^{r} Q e_j \xrightarrow{\partial_1} Q \to 0$$

$$\partial_2 = \phi \qquad \partial_1 = a\big(\det(\phi_1), \ldots, (-1)^{j-1} \det(\phi_j), \ldots, (-1)^{r-1} \det(\phi_r) \big)$$

Herzog [86] shows that there exists a unique DG algebra structure on A, namely:

$$e_j \cdot e_k = -e_k \cdot e_j = -a \sum_{\ell=1}^{r-1} (-1)^{j+k+\ell} \det(\phi_{jk}^{\ell}) f_\ell \qquad \text{for} \quad j < k; \qquad e_j \cdot e_j = 0.$$

Example 2.1.3. An ideal I in a local ring $(Q, \mathfrak{n})$ is *Gorenstein* if $R = Q/I$ has $\mathrm{pd}_Q R = p < \infty$, $\mathrm{Ext}_Q^n (R, Q) = 0$ for $n \neq p$, and $\mathrm{Ext}_Q^p (R, Q) \cong R$; thus, when Q is regular, I is Gorenstein if and only if R is a Gorenstein ring.

If I is Gorenstein, $\mathrm{pd}_Q R = 3$, and I is minimally generated by r elements, then J. Watanabe [157] proves that the number r is odd, and Buchsbaum-Eisenbud [47] show that there exists an alternating $r \times r$ matrix ϕ with elements in $\mathfrak{n}$, such that a minimal free resolution A of R over Q has the form

$$A: \quad 0 \to Q g \xrightarrow{\partial_3} \bigoplus_{k=1}^{r} Q f_k \xrightarrow{\partial_2} \bigoplus_{j=1}^{r} Q e_j \xrightarrow{\partial_1} Q \to 0$$

$$\partial_2 = \phi \qquad \partial_1 = \big(\mathrm{pf}(\phi_1^1), \ldots, (-1)^{j-1} \mathrm{pf}(\phi_j^j), \ldots, (-1)^{r-1} \mathrm{pf}(\phi_r^r) \big) = \partial_3^*$$

where $\mathrm{pf}(\alpha)$ is the Pfaffian of α. A DG algebra structure on A is given by

$$e_j \cdot e_k = -e_k \cdot e_j = \sum_{\ell=1}^{r} (-1)^{j+k+\ell} \rho_{jk\ell} \, \mathrm{pf}(\phi_{jk\ell}^{jk\ell}) f_\ell \qquad \text{for} \quad j < k;$$

$$e_j \cdot e_j = 0; \qquad e_j \cdot f_k = f_k \cdot e_j = \delta_{jk} g,$$

where $\rho_{jk\ell}$ is equal to -1 if $j < \ell < k$, and to 1 otherwise, cf. [21].

More generally, Buchsbaum and Eisenbud [47] prove that DG algebra structures always exist in projective dimension ≤ 3:

Proposition 2.1.4. *If A is a projective resolution of a Q–module R, such that $A_0 = Q$ and $A_n = 0$ for $n \geq 4$, then A has a structure of DG algebra.*

Proof. For the construction, consider the complex $S^2(A)$, that starts as

$$\cdots \to (A_1 \otimes A_3) \oplus S^2(A_2) \oplus A_4 \xrightarrow{\delta_4} (A_1 \otimes A_2) \oplus A_3$$

$$\xrightarrow{\delta_3} (\wedge^2 A_1) \oplus A_2 \xrightarrow{\delta_2} A_1 \xrightarrow{\delta_1} Q \to 0$$

with differentials defined by the condition $\delta_n|_{A_n} = \partial_n$ and the formulas

$$\delta_2(a \wedge b) = \partial_1(a)b - \partial_1(b)a; \qquad \delta_3(a \otimes b) = -a \wedge \partial_1(b) + \partial_1(a)b;$$
$$\delta_4(a \otimes b) = \partial_1(a)b - a * \partial_3(b); \qquad \delta_4(a * b) = \partial_2(a) \otimes b + \partial_2(b) \otimes a,$$

where $*$ denotes the product in the symmetric algebra. The complex $S^2(A)$ is projective and naturally augmented to R, so by the Lifting Theorem there is a morphism $\mu \colon S^2(A) \to A$ that extends the identity map of R.

Define a product on A (temporarily denoted $\cdot$) by composing the canonical projection $A \otimes A \to S^2(A)$ with μ. With the unit $1 \in Q = A_0$, one has all the properties required from a DG algebra except, possibly, associativity. Because $A_n = 0$ for $n \geq 4$, this may be an issue only for a product of three elements a, b, c, of degree 1. For them we have

$$\partial_3((a \cdot b) \cdot c) = \partial_2(a \cdot b) \cdot c + (a \cdot b)\partial_1(c)$$
$$= (\partial_1(a)b) \cdot c - (\partial_1(b)a) \cdot c + (a \cdot b)\partial_1(c)$$
$$= \partial_1(a)(b \cdot c) - \partial_1(b)(a \cdot c) + \partial_1(c)(a \cdot b).$$

A similar computation of $\partial_3(a \cdot (b \cdot c))$ yields the same result.

As ∂_3 is injective, we conclude that $(a \cdot b) \cdot c = a \cdot (b \cdot c)$.   □

Next we describe two existence results in projective dimension 4.

Example 2.1.5. If Q is local, $\mathrm{pd}_Q Q/I = 4$, and I is Gorenstein, then Kustin and Miller [101] prove that the minimal free resolution of R over Q has a DG algebra structure if $Q \ni \frac{1}{2}$, a restriction removed later by Kustin [97].

Example 2.1.6. If I is a grade 4 perfect ideal generated by 5 elements in a local ring Q containing $\frac{1}{2}$, then Kustin [98], building on work of Palmer [127], constructs a DG algebra structure on the minimal free resolution of Q/I.

The question naturally arises whether it is possible to put a DG algebra structure on each *minimal* resolution of a residue ring of a local ring. As far as the projective dimension is concerned, the list above turns out to be essentially complete: for (perfect) counterexamples in dimension 4, cf. Theorem 2.3.1.

On the positive side, each Q–algebra has *some* DG algebra resolution, obtained by a universal construction: Given a cycle z in a DG algebra A, we embed A into a DG algebra A' by freely adjoining a variable y such that $\partial y = z$. In A' the cycle z has been killed: it has become a boundary.

Construction 2.1.7. Exterior variable. When $|z|$ is even, $\Bbbk[y]$ is the *exterior algebra* over $\Bbbk$ of a free $\Bbbk$–module on a generator y of degree $|z| + 1$; the differential of $A[y]^\natural = A^\natural \otimes_\Bbbk \Bbbk[y]$ is given by

$$\partial(a_0 + a_1 y) = \partial(a_0) + \partial(a_1)y + (-1)^{|a_1|}a_1 z \, ;$$

thus, when A is concentrated in degree zero, $A[y]$ is the Koszul complex $K(z; A)$.

Construction 2.1.8. Polynomial variable. When $|z|$ is odd, $\Bbbk[y]$ is the *polynomial ring* over $\Bbbk$ on a variable y of degree $|z| + 1$, $A[y]^\natural = A^\natural \otimes_\Bbbk \Bbbk[y]$, and

$$\partial\Big(\sum_i a_i y^i \Big) = \sum_i \partial(a_i)y^i + \sum_i (-1)^{|a_i|}i a_i z y^{i-1} \, .$$

In either case, ∂ is the unique differential on $A[y]^\natural$ that extends the differential on A, satisfies $\partial(y) = z$, and the Leibniz formula; we call y a *variable* over A, and often use the more complete notation, $A[y \,|\, \partial(y) = z]$.

Let $u = \mathrm{cls}(z) \in \mathrm{H}(A)$ be the class of z. The quotient complex $A[y]/A$ is trivial in degrees $n \le |z|$ and is equal to $A_0 y$ in degree $n = |z|+1$, so the homology exact sequence shows that the inclusion $A \hookrightarrow A[y]$ induces a morphism of graded algebras $\mathrm{H}(A)/u\,\mathrm{H}(A) \to \mathrm{H}(A[y])$ that is bijective in degrees $\le |z|$.

A *semi-free extension* of A is a DG algebra A' obtained by repeated adjunction of free variables. If Y is the set of all variables adjoined in the process, then we write $A[Y]$ for A'; we also set $Y_n = \{y \in Y \mid |y| = n\}$ and $Y_{\le n} = \bigcup_{i=0}^n Y_i$. Semi-free algebra extensions have a *lifting property*:

Proposition 2.1.9. *If $A[Y]$ is a semi-free extension of a DG algebra A, then each diagram of morphisms of DG algebras over A represented by solid arrows*

with a surjective quasi-isomorphism β can be completed to a commutative diagram by a morphism γ, that is defined uniquely up to A–linear homotopy.

Proof. Set $A^i = A[Y_{\le i}]$. Starting with the structure map $A \to B$, we assume that for some $n \ge -1$ we have a morphism $\gamma^n \colon A^n \to B$ of DG algebras over A, with $\beta\gamma^n = \alpha|_{A^n}$. Over A^n, the set $\{1\} \cup Y_{n+1}$ generates a semi-free submodule F^{n+1} of $A[Y]$. By Theorem 1.3.1, γ^n extends to a morphism $\delta^{n+1} \colon F^{n+1} \to B$ of DG modules over A^n. The graded commutative algebra $(A^{n+1})^\natural$ is freely generated over $(A^n)^\natural$ by Y_{n+1}, so δ^{n+1} extends uniquely to a homomorphisms of graded algebras $\gamma^{n+1} \colon A^{n+1} \to B$. Since δ^{n+1} is a morphism of DG modules, γ^{n+1} is necessarily a morphism of DG algebras. In the limit, one gets a morphism of DG algebras $\gamma \colon A[Y] \to B$, with the desired properties. $\qquad\square$

A *resolvent* of Q–algebra R is a DG algebra resolution of R over Q, that is a semi-free DG algebra extension of Q.

Proposition 2.1.10. *Each (surjective) homomorphism* $\psi\colon Q \to R$ *has a resolvent* $Q[Y]$ *(with* $Y_0 = \varnothing$*). When* Q *is noetherian and* R *is a finitely generated* Q*–algebra there exists a resolvent with* Y_n *finite for each* n.

Proof. Factor ψ as an inclusion $Q \hookrightarrow Q[Y_0]$ into a polynomial ring on a set Y_0 of variables and a surjective morphism $\psi'\colon Q[Y_0] \twoheadrightarrow R$ that maps Y_0 to a set of generators of the Q–algebra R. The Koszul complex $Q \hookrightarrow Q[Y_0][[Y_{\leqslant 1}]$ on a set of generators of $\operatorname{Ker} \psi'$ is a semi-free extension of Q, with $\operatorname{H}_0(Q[Y_{\leqslant 1}]) \cong R$.

By induction on i, assume that consecutive adjunctions to $Q[Y_{\leqslant 1}]$ of sets Y_j of variables of degrees $j = 2, \ldots, n$ have produced an extension $Q \hookrightarrow Q[Y_{\leqslant n}]$ with $\operatorname{H}_i(Q[Y_{\leqslant n}]) = 0$ for $0 < i < n$. Adjoin to $Q[Y_{\leqslant n}]$ a set Y_{n+1} of variables of degree $n+1$ that kill a set of generators of the $Q[Y_0]$–module $\operatorname{H}_n(Q[Y_{\leqslant n}])$. As observed above, we then get $\operatorname{H}_i(Q[Y_{\leqslant n+1}]) = 0$ for $0 < i < n+1$.

Going over the induction procedure with a noetherian hypothesis in hand, it is easy to see that a finite set Y_n suffices at each step. $\qquad\square$

2.2. DG module resolutions.

There is nothing esoteric about DG module structures on resolutions of modules. In fact, they offer a particularly adequate framework for important commutative algebra information.

Remark 2.2.1. Let U be a free resolution of a Q–module M. If $f \in (0 :_Q M)$, then both $f\operatorname{id}^U$ and 0^U induce the zero map on M, hence they are homotopic, say $f\operatorname{id}^U = \partial\sigma + \sigma\partial$. A homotopy σ such that $\sigma^2 = 0$ exists if and only if U can be made a DG module over the Koszul complex $A = Q[y \,|\, \partial(y) = f]$: just set $yu = \sigma(u)$, and note that the homotopy condition for σ translates precisely into the Leibniz rule $fu = \partial(yu) + y\partial(u)$ for the action of y.

In some cases one can prove the existence of a square-zero homotopy.

Proposition 2.2.2. *If* $(Q, \mathfrak{n}, k)$ *is a local ring,* $f \in \mathfrak{n} \setminus \mathfrak{n}^2$, *and* M *is a* Q*–module such that* $fM = 0$, *then the minimal free resolution* U *of* M *over* Q *has a structure of semi-free DG module over the Koszul complex* $A = Q[y \,|\, \partial(y) = f]$.

Proof. Setting $f_j = f\operatorname{id}^{U_j}$, we restate the desired assertion as follows: For each j there is a homomorphism $\sigma_j\colon U_j \to U_{j+1}$, such that:

$$\partial_{j+1}\sigma_j + \sigma_{j-1}\partial_j = f_j \,; \qquad \sigma_{j-1}(U_{j-1}) = \operatorname{Ker} \sigma_j \,;$$

$$\sigma_{j-1}(U_{j-1}) \text{ is a direct summand of } U_j \,.$$

Indeed, by the preceding remark the first two conditions define on U a structure of DG module over A; the third one is then equivalent to an isomorphism of $A^\natural$–modules $U^\natural \cong A^\natural \otimes_Q V$, with $V = U^\natural/yU^\natural$.

The map $\sigma_j = 0$ has the desired properties when $j < 0$, so we assume by induction that σ_j has been constructed for $j \leq i$, with $i \geq -1$. Since

$$\partial_{i+1}(f_{i+1} - \sigma_i\partial_{i+1}) = f\partial_{i+1} - \partial_{i+1}\sigma_i\partial_{i+1} = f\partial_{i+1} - f\partial_{i+1} + \sigma_{i-1}\partial_i\partial_{i+1} = 0$$

and U is acyclic, there exists a map σ_{i+1} such that $\partial_{i+2}\sigma_{i+1} = f_{i+1} - \sigma_i\partial_{i+1}$.

Furthermore, as $\sigma_i \sigma_{i-1} = 0$ by the induction hypothesis, we have

$$(f_{i+1} - \sigma_i \partial_{i+1})\sigma_i = f\sigma_i - \sigma_i \partial_{i+1}\sigma_i = f\sigma_i - f\sigma_i + \sigma_i \sigma_{i-1}\partial_i = 0 \,.$$

Thus, we can arrange for σ_{i+1} to be zero on the direct summand $\operatorname{Im}\sigma_i$ of U_{i+1}.

Let V_{i+1} be a complementary direct summand of $\operatorname{Im}\sigma_i$ in U_{i+1}. For $v \in V_{i+1} \smallsetminus \mathfrak{n}V_{i+1}$, we have $\partial_{i+2}\sigma_{i+1}(v) = fv - \sigma_i\partial_{i+1}(v)$. The two terms on the right lie in distinct direct summands, and $fv \notin \mathfrak{n}^2 V_{i+1}$, so $\partial_{i+2}\sigma_{i+1}(v) \notin \mathfrak{n}^2 U_{i+1}$, and thus $\sigma_{i+1}(v) \notin \mathfrak{n}U_{i+2}$. This shows that $\sigma_{i+1} \otimes_Q k \colon V_{i+1} \otimes_Q k \to U_{i+2} \otimes_Q k$ is injective, so σ_{i+1} is a split injection, completing the induction step. □

The construction of σ above is taken from Shamash [142]; it is implicit in Nagata's [124] description of the syzygies of M over $Q/(f)$ in terms of those over Q, presented next; neither source uses DG module structures.

Theorem 2.2.3. *Let $(Q, \mathfrak{n}, k)$ be a local ring, let $f \in \mathfrak{n} \smallsetminus \mathfrak{n}^2$ be Q–regular, and let M be a finite module over $R = Q/(f)$. If U is a minimal free resolution of M over Q, then there exists a homotopy σ from id^U to 0^U, such that*

$$U' : \qquad \cdots \to \frac{U_n}{fU_n + \sigma(U_{n-1})} \to \cdots \to \frac{U_1}{fU_1 + \sigma(U_0)} \to \frac{U_0}{fU_0} \to 0 \qquad (*)$$

is a minimal R–free resolution of M and $\operatorname{rank}_Q U_n = \operatorname{rank}_R U'_n + \operatorname{rank}_R U'_{n-1}$.

Proof. Set $A = Q[y \,|\, \partial(y) = f]$. By the preceding proposition, U is a semi-free DG module over $A = Q[y \,|\, \partial(y) = f]$. Let σ be the homotopy given by left multiplication with y. By Proposition 1.3.2 the quasi-isomorphism $A \to R$ induces a quasi-isomorphism $U \to U \otimes_A R = U'$. The complex $U/(f, y)U = U/(fU + \sigma(U)) = U'$ is obviously minimal, so we are done. □

DG module structures are more affordable than DG algebra structures.

Remark 2.2.4. Let $S = Q/J$, and let B be a DG algebra resolution of S over Q. If A is the Koszul complex on a sequence $\boldsymbol{f} \subset J$, then by Proposition 2.1.9 the canonical map $A \to Q/(\boldsymbol{f}) \to S$ lifts over the surjective quasi-isomorphism $B \to S$ to a morphism $A \to B$ of DG algebras over Q.

Thus, any DG algebra resolution of S over Q is a DG module over each Koszul complex $K(\boldsymbol{f}; Q)$. However, DG module structures may exist even when DG algebra structures do not: for an explicit example, cf. Srinivasan [146].

Short projective resolutions always carry DG module structures: Iyengar [92] notes that a modification of the argument for Proposition 2.1.4 yields

Proposition 2.2.5. *Let $R = Q/I$, and let M be an R–module. If U is a projective resolution of M, and $U_n = 0$ for $n \geq 3$, then U has a structure of DG module over each DG algebra resolution A of R over Q.* □

On the other hand, not all minimal resolutions of length ≥ 3 support DG module structures over DG algebras $A \neq Q$, cf. Theorem 2.3.1.

Let A be a DG algebra resolution of R over Q, and let M be an R–module. A *DG module resolution* of M over A is a quasi-isomorphism $\epsilon^U \colon U \to M$ of DG modules over A, such that for each n the Q–module U_n is projective. To construct such resolutions in general, we describe a 'linear' adjunction process.

Construction 2.2.6. Adjunction of basis elements. Let V be a DG module over A, and $Z = \{z_\lambda \in V\}_{\lambda \in \Lambda}$ be a set of cycles. For a linearly independent set $Y = \{y_\lambda : |y_\lambda| = |z_\lambda| + 1\}_{\lambda \in \Lambda}$ over the graded algebra $A^\natural$ underlying A, set

$$\partial\left(v + \sum_{\lambda \in \Lambda} a_\lambda y_\lambda\right) = \partial(v) + \sum_{\lambda \in \Lambda} \partial(a_\lambda) y_\lambda + (-1)^{|a_\lambda|} \sum_{\lambda \in \Lambda} a_\lambda z_\lambda \,.$$

This is the unique differential on $V \bigoplus_{\lambda \in \Lambda} A y_\lambda$ which extends that of V, satisfies the Leibniz rule, and has $\partial(y_\lambda) = z_\lambda$ for $\lambda \in \Lambda$.

Proposition 2.2.7. *If A is a DG algebra resolution of R over Q and M is an R–module, then M has a semi-free resolution U over A.*

Proof. Pick a surjective homomorphism $F \to M$ from a free Q–module, and extend it to a chain map of DG modules $\epsilon^0 \colon U^0 = A \otimes_Q F \to M$. Clearly, $H_0(\epsilon^0)$ is surjective. If Z^0 is a set of cycles whose classes generate $\operatorname{Ker} H_0(\epsilon^0)$, then let U^1 be the semi-free extension of U^0, obtained by adjunction of a linearly independent set Y^1 that kills Z^0. Extend ϵ^0 to $\epsilon^1 \colon U^1 \to M$ by $\epsilon^1(Y^1) = 0$, and note that $H_0(\epsilon^1)$ is an isomorphism. Successively adjoining linearly independent sets Y^n, of elements of degree $n = 2, 3, \ldots$ that kill sets Z^{n-1} of cycles generating $H_{n-1}(U^{n-1})$, we get a semi-free DG module $U = \bigcup_n U^n$ over A, with a quasi-isomorphism $\epsilon^U \colon U \to M$. □

In an important case, the constructions are essentially finite.

Proposition 2.2.8. *Let Q be a noetherian ring*

If R is a finite Q–algebra and M is a finite R–module, then there exist a DG algebra resolution A of R over Q and a DG module resolution U of M over A, such that the Q–modules $\operatorname{Coker}(\eta^A \colon Q \to A_0)$, A_n, and U_n, are finite projective for all n and are trivial for $n > \max\{\operatorname{pd}_Q R, \operatorname{pd}_Q M\}$.

Proof. If $r_1, \ldots, r_s$ generate R as a Q–module, then each r_j is a root of a monic polynomial $f_j \in Q[x_j]$, hence R is a residue of $Q' = Q[x_1, \ldots, x_s]/(f_1, \ldots, f_s)$, which is a free Q–module. Use Proposition 2.1.10 to pick a resolvent $A' = Q'[Y]$ such that each A'_n is a finite free module over Q'. Then use Construction 2.2.7 to get a semi-free resolution U' of M over A' with each U'_n a finite free Q'–module; in particular, A'_n and U'_n are finite free Q–modules.

If $\max\{\operatorname{pd}_Q R, \operatorname{pd}_Q M\} = m < \infty$, then define a Q–submodule V of U' by setting $V_{<m} = 0$, $V_m = \partial(U'_{m+1})$, and $V_{>m} = U'_{>m}$. It is easy to check that $V = \{V_n\}$ is a DG A'–submodule with $H(V) = 0$, hence $U = U'/V$ has $H(U) = M$. The assumption on $\operatorname{pd}_Q M$ implies that the Q–module U_m is projective, so U is a DG module resolution of M over A'. Similarly, one sees that $J \subseteq A'$ defined by $J_{<m} = 0$, $J_m = \partial(A'_{m+1})$, and $J_{>m} = A'_{>m}$, is a DG ideal of A, such that $A = A'/J$ is a DG algebra resolution of R. Finally, the Leibniz formula shows that $JU' \subseteq V$, so U is a DG module over A.

The fact that $\operatorname{Coker} \eta$ is projective can be checked locally; Nakayama's Lemma then shows that $\operatorname{Im} \eta$ is a direct summand of the free Q–module A_0. $\qquad\square$

2.3. Products versus minimality. Our goal is the following *non-existence*

Theorem 2.3.1. *Let k be a field, and Q be the polynomial ring $k[s_1, s_2, s_3, s_4]$ with the usual grading, or the power series ring $k[[s_1, s_2, s_3, s_4]]$. There exists no DG algebra structure on the minimal Q–free resolution U of the residue ring*

$$S = Q/I \qquad where \quad I = (s_1^2, \, s_1 s_2, \, s_2 s_3, \, s_3 s_4, \, s_4^2)$$

or on the minimal Q–free resolution U' of the Cohen-Macaulay residue ring

$$S' = Q/I' \qquad where \quad I' = I + (s_1 s_3^6, \, s_2^7, \, s_2^6 s_4, \, s_3^7).$$

If A is a DG algebra over Q, and U or U' is a DG module over A, then $A = Q$.

Remark. To prove the theorem, we check by a direct computation the non-vanishing of certain obstructions introduced by Avramov [22], and described in Theorem 3.2.6 below. Both the examples and the computations simplify those appearing in [22], and were developed in conversations with S. Iyengar.

As in [22], the examples can be used to generate, in any local ring Q with depth $Q = g \geq 4$ (respectively, ≥ 6) perfect ideals of prescribed grades between 4 and g (respectively, Gorenstein ideals of prescribed grades between 6 and g), whose minimal free resolution admits no DG algebra structure.

Gorenstein ideals of grade 5 with this property had been missing, until the paper of Srinivasan [147]. The last open question, whether the minimal resolution of each non-cyclic module of projective dimension 3 (recall Proposition 2.1.4 and Proposition 2.2.5) carries a structure of DG module over some DG algebra $A \neq Q$, was answered by Iyengar [92] with perfect counter-examples.

Construction 2.3.2. Tor algebras. Let $S \leftarrow R \rightarrow k$ be homomorphisms of rings. If D is a DG algebra resolution D of S over R that is a resolution of S by free R–modules, cf. Proposition 2.1.10, then $\operatorname{Tor}^R(S, k) = \operatorname{H}(D \otimes_R k)$ inherits a structure of graded algebra. It can be computed also from a DG algebra resolution E of the second argument, or from resolvents of both arguments, due to the quasi-isomorphisms of DG algebras

$$D \otimes_R k \xleftarrow{\;D \otimes_R \epsilon^E\;} D \otimes_R E \xrightarrow{\;\epsilon^D \otimes_R E\;} S \otimes_R E.$$

Varying in these isomorphisms one varies one of D or E, while keeping the other fixed, one sees that the algebra structure on Tor does not depend on the choice of a DG algebra resolution. It can even be computed[11] from projective resolutions D' and E' with no multiplicative structure: the unique up to homotopy lifting of

---

[11]In fact, this is how they were originally *introduced* by Cartan and Eilenberg [51].

$\mu^S \colon S \otimes_R S \to S$ to a morphism $\mu^{D'} \colon D' \otimes_R D' \to D'$, that conspires with the Künneth map of Proposition 1.3.4 to produce

$$\mathrm{H}(D' \otimes_R k) \otimes_R \mathrm{H}(D' \otimes_R k) \xrightarrow{\kappa} \mathrm{H}\left((D' \otimes_R k) \otimes_R (D' \otimes_R k)\right)$$

$$\cong \mathrm{H}\left((D' \otimes_R D') \otimes_R (k \otimes_R k)\right) \xrightarrow{\mathrm{H}(\mu^{D'} \otimes_R \mu^k)} \mathrm{H}(D' \otimes_R k).$$

As the multiplication $\mu^D \colon D \otimes_R D \to D$ also is a comparison map, the unique isomorphism $\mathrm{H}(D' \otimes_R k) \cong \mathrm{H}(D \otimes_R k)$ transforms products into each other.

There is a related structure in the case of R–modules.

Construction 2.3.3. Tor modules. Let $\psi \colon Q \to R$ and $R \to k$ be homomorphisms of rings, and let M be an R–module.

Choose a DG algebra resolution $\epsilon^A \colon A \to R$ over Q, by Proposition 2.1.10, and a semi-free resolution $\epsilon^U \colon U \to M$ over A, by Proposition 2.2.7. As both A and U are free over Q, we see that $\mathrm{Tor}^Q(M, k) = \mathrm{H}(U \otimes_Q k)$ is a module over the graded algebra $\mathrm{Tor}^Q(R, k) = \mathrm{H}(A \otimes_Q k)$ from the preceding construction. Recycling the discussion there, we verify that this structure is unique, and natural with respect to the module arguments.

The constructions also have a less well known naturality with respect to the *ring* argument; it is the one that we need.

Construction 2.3.4. Naturality. If $\psi \colon Q \to R$ is a ring homomorphism, then k becomes a Q–algebra, so pick a DG algebra resolution C of k over Q. By Proposition 2.1.9, there is a morphism of DG algebras $\gamma \colon C \to E$ over the identity map of k, that is unique up to Q–linear homotopy. The induced map $\mathrm{H}(M \otimes_\psi \gamma) \colon \mathrm{H}(M \otimes_Q C) \to \mathrm{H}(M \otimes_R E)$ is linear over $\mathrm{H}(R \otimes_Q C)$, and does not depend on the choice of γ. Thus, there is a natural homomorphism $\mathrm{Tor}^\psi(M, k) \colon \mathrm{Tor}^Q(M, k) \to \mathrm{Tor}^R(M, k)$ of $\mathrm{Tor}^Q(R, k)$–modules.

If $U' \to M \leftarrow V'$ and $C' \to k \leftarrow E'$ are arbitrary free resolutions, respectively over Q and over R, and $\beta' \colon U' \to V'$ and $\gamma' \colon C' \to E'$ are morphisms inducing id^M and id^k, then $\mathrm{Tor}^\psi(M, k) = \mathrm{H}(\beta' \otimes_\psi k) = \mathrm{H}(M \otimes_\psi \gamma')$.

Remark 2.3.5. Let $(Q, \mathfrak{n}, k)$ be a local (or graded) ring, let $f \in \mathfrak{n}$ be a (homogeneous) regular element, let $\psi \colon Q \to Q/(f) = R$ be the natural projection, and let A be the Koszul complex $Q[y \mid \partial(y) = f]$.

For a DG module resolution U of M over A, set $\overline{U}\langle x \rangle = \bigoplus_{i \geqslant 0} Rx^{(i)} \otimes_Q \overline{U}$, with $|x^{(i)}| = 2i$ and $\partial(x^{(i)} \otimes u) = x^{(i-1)} \otimes yu + x^{(i)} \otimes \partial(u)$. In Example 3.1.2, we show that $\overline{U}\langle x \rangle$ is a resolution of M over R. By Theorem 3.2.6 and Remark 3.2.7, if U is minimal, then $\mathrm{Ker}(\mathrm{Tor}_n^\psi(M, k)) = \mathrm{cls}(y)\,\mathrm{Tor}_{n-1}^Q(M, k)$ for $n \geq 1$.

Proof of Theorem 2.3.1. First we look at the residue ring S. By Remark 2.2.4, it suffices to prove that its minimal free resolution U, over $Q = k[s_1, s_2, s_3, s_4]$ or over $Q = k[[s_1, s_2, s_3, s_4]]$, has no DG module structure over the Koszul complex $A = Q[y \mid \partial(y) = f]$, where $f = s_1^2 + s_4^2$. By the preceding remark, this will follow from

$\mathrm{cls}(y)\,\mathrm{Tor}_3^Q\,(S,k)\not\subseteq\mathrm{Ker}(\mathrm{Tor}_4^\psi\,(S,k))$. The Tor's involved do not change under completion, so we restrict to the graded polynomial ring.

The Koszul resolvent $C = Q[y_1,y_2,y_3,y_4\,|\,\partial(y_i)=s_i]$ of k over Q is a DG algebra over A, via the map $y\mapsto s_1y_1+s_4y_4$; by Remark 2.3.5, the complex $\overline{C}\langle x\rangle$, is a resolution of k over $R = Q/(f)$. For $K = S\otimes_Q C$, we have

$$\mathrm{Tor}^Q\,(S,k) = \mathrm{H}(U\otimes_Q k)\cong\mathrm{H}(K) \qquad (*)$$

as modules over $\mathrm{Tor}^Q\,(R,k) = k[y\,|\,\partial(y)=0]$, cf. Construction 2.3.3. The isomorphisms takes $\mathrm{Ker}\,\mathrm{Tor}^\psi\,(S,k)$ to $\mathrm{Ker}\,\mathrm{H}(\iota)$, where ι is the inclusion $K\subset L = S\otimes_R\overline{C}\langle x\rangle$. So it suffices to exhibit an element

$$w\in\mathrm{Ker}\,\mathrm{H}_4(\iota)\setminus\mathrm{cls}(y)\,\mathrm{H}_3(K). \qquad (\dagger)$$

It is easy to check that the cycles

$$z_1 =(s_1s_4)\,y_1\wedge y_2\wedge y_3\,,\qquad z_2 =(s_1s_4)\,y_1\wedge y_2\wedge y_4\,,$$
$$z_3 =(s_1s_4)\,y_1\wedge y_3\wedge y_4\,,\qquad z_4 =(s_2s_4)\,y_1\wedge y_3\wedge y_4\,,$$

are linearly independent modulo $S\partial(y_1\wedge\cdots\wedge y_4)$. MACAULAY [40] shows that the minimal graded resolution U of the S over Q has the form

$$0\to Q(-6)\xrightarrow{\partial_4}Q(-5)^4\xrightarrow{\partial_3}Q(-4)^3\oplus Q(-3)^4\xrightarrow{\partial_2}Q(-2)^5\xrightarrow{\partial_1}Q\to 0 \qquad (\ddagger)$$

with differentials given by the matrices

$$\partial_1 = \begin{pmatrix} s_1^2 & s_1s_2 & s_2s_3 & s_3s_4 & s_4^2 \end{pmatrix}$$

$$\partial_2 = \begin{pmatrix} 0 & 0 & 0 & -s_2 & -s_4^2 & 0 & -s_3s_4 \\ 0 & 0 & -s_3 & s_1 & 0 & -s_4^2 & 0 \\ 0 & -s_4 & s_1 & 0 & 0 & 0 & 0 \\ -s_4 & s_2 & 0 & 0 & 0 & 0 & s_1^2 \\ s_3 & 0 & 0 & 0 & s_1^2 & s_1s_2 & 0 \end{pmatrix}$$

$$\partial_3 = \begin{pmatrix} 0 & 0 & s_1s_2 & s_1^2 \\ s_1^2 & 0 & s_1s_4 & 0 \\ s_1s_4 & 0 & s_4^2 & 0 \\ s_3s_4 & s_4^2 & 0 & 0 \\ 0 & -s_2 & 0 & -s_3 \\ 0 & s_1 & -s_3 & 0 \\ -s_2 & 0 & 0 & s_4 \end{pmatrix} \qquad \partial_4 = \begin{pmatrix} -s_4 \\ s_3 \\ s_1 \\ -s_2 \end{pmatrix}$$

We conclude from $(*)$ and $(\ddagger)$ that $\mathrm{rank}_k\,\mathrm{H}_3(K)=4$, and thus that $\mathrm{cls}(z_1),...,\mathrm{cls}(z_4)$ is a basis of $\mathrm{H}_3(K)$. Because $yz_i = (s_1y_1+s_4y_4)z_i = 0$ for $i = 1,\ldots,4$, we see that $\mathrm{cls}(y)\,\mathrm{H}_3(K) = 0$. On the other hand,

$$z = (s_1s_4)\,y_1\wedge y_2\wedge y_3\wedge y_4\in K_4$$

is a non-zero cycle, and so not a boundary in K, but becomes one in L:

$$z = \partial\big(s_4\,y_2\wedge y_3\wedge y_4\,y + s_2\,y_3\,x^{(2)}\big)\in L_4\,.$$

We have proved that $w = \mathrm{cls}(z)$ satisfies $(\dagger)$.

Turning to $S' = S/(s_1 s_3^6, s_2^7, s_2^6 s_4, s_3^7)$, consider the commutative square

$$
\begin{array}{ccc}
K & \xrightarrow{\;\iota\;} & L \\
{\scriptstyle \pi}\big\downarrow & & \big\downarrow{\scriptstyle \pi'} \\
K' = S' \otimes_S K & \xrightarrow{\;\iota'\;} & S' \otimes_S L = L'
\end{array}
$$

of morphisms of complexes, with $\iota' = S' \otimes_S \iota$. For $w' = \mathrm{H}(\pi)(w)$ it yields

$$\mathrm{H}(\iota')(w') = \mathrm{H}(\pi')\,\mathrm{H}(\iota)(w) = 0\,.$$

Assume $w' \in \mathrm{cls}(y)\,\mathrm{H}_3(K')$, set $\mathfrak{m} = (t_1, \ldots, t_4)$, and consider the subcomplex

$$J: \quad 0 \to \mathfrak{m}^3 K_4 \to \mathfrak{m}^4 K_3 \to \mathfrak{m}^5 K_2 \to \mathfrak{m}^6 K_1 \to \mathfrak{m}^7 K_0 \to 0$$

of K. Since $(*)$ and $(\ddagger)$ show that the non-zero homology of K is concentrated in internal degrees ≤ 6, cf. Remark 1.2.10, we conclude that $\mathrm{H}(J) = 0$, so the projection $\xi \colon K \to K/J$ is a quasi-isomorphism. On the other hand, it is clear that J is a DG ideal of K, such that $\mathrm{Ker}\,\pi = \mathfrak{m}^7 K \subset J$. Thus, $\xi = \rho\pi$, where $\rho \colon K' \to K/J$ is the canonical map. By the surjectivity of $\mathrm{H}(\xi)$, we have

$$\mathrm{H}(\xi)(w) = \mathrm{H}(\rho)(w') \subseteq \mathrm{cls}(y)\,\mathrm{H}(\xi)\big(\mathrm{H}_3(K)\big) = \mathrm{H}(\xi)(\mathrm{cls}(y)\,\mathrm{H}_3(K))\,.$$

The injectivity of $\mathrm{H}(\xi)$ implies $w \in \mathrm{cls}(y)\,\mathrm{H}_3(K)$, violating $(\dagger)$.

Thus, we have found $w' \in \mathrm{Ker}\,\mathrm{H}_4(\iota') \smallsetminus \mathrm{cls}(y)\,\mathrm{H}_3(K')$. As above, this implies that U' carries no structure of DG module over A. $\square$

3. Change of Rings

Fix a homomorphism of rings $\psi \colon Q \to R$, and an R–module M.

We consider various aspects of the problem: How can homological information on R and M over Q be used to study the module M over R?

3.1. Universal resolutions.
A recent result of Iyengar [92] addresses this problem on the level of resolutions[12].

Theorem 3.1.1. *Let $\epsilon^A \colon A \to R$ be a DG algebra resolution over Q with structure map $\eta^A \colon Q \to A$, and let $\epsilon^U \colon U \to M$ be a DG module resolution of M over A.*

---

[12]In view of Proposition 2.2.8, if Q is noetherian and finite projective Q–modules are free, then a resolution of M over R is *finitistically* determined by matrix data over Q, namely, the multiplication tables of the algebra A and the module U, and the differentials in these finite complexes. With the help of computer algebra systems, such as MACAULAY [40], these data can be *effectively* gathered, at least when Q is a polynomial ring over a (small) field.

With $\overline{A} = (R \otimes_Q \operatorname{Coker} \eta^A)$ and $\overline{U} = (R \otimes_Q U)$, set

$$F_n(A, U) = \bigoplus_{p+i_1+\cdots+i_p+j=n} \overline{A}_{i_1} \otimes_R \cdots \otimes_R \overline{A}_{i_p} \otimes_R \overline{U}_j \, ;$$

$$\partial'(\overline{a}_1 \otimes \cdots \otimes \overline{a}_p \otimes \overline{u}) = \sum_{r=1}^{p} (-1)^{r+i_1+\cdots+i_{r-1}} \, \overline{a}_1 \otimes \cdots \otimes \partial(\overline{a}_r) \otimes \cdots \otimes \overline{a}_p \otimes u$$
$$+ (-1)^{p+i_1+\cdots+i_p} \, \overline{a}_1 \otimes \cdots \otimes \overline{a}_p \otimes \overline{\partial(u)} \, ;$$

$$\partial''(\overline{a}_1 \otimes \cdots \otimes \overline{a}_p \otimes \overline{u}) = \sum_{r=1}^{p-1} (-1)^{r+i_1+\cdots+i_r} \, \overline{a}_1 \otimes \cdots \otimes \overline{a_r a_{r+1}} \otimes \cdots \otimes \overline{a}_p \otimes u$$
$$+ (-1)^{p+i_1+\cdots+i_{p-1}} \, \overline{a}_1 \otimes \cdots \otimes \overline{a}_{p-1} \otimes \overline{a_p u} \, .$$

The homomorphisms $\partial = \partial' + \partial'' \colon F_n(A, U) \to F_{n-1}(A, U)$ make $F(A, U)$ into a complex of R–modules. If the Q–modules $\operatorname{Coker} \eta_i^A$ and U_i are free for all i, then it is a free resolution of M over R.

When Q is a field, $A = A_0$, and U is an A–module, this is the well known *standard resolution*. The First Commandment[13] points the way to generalizations: this is the philosophy of the proof presented at the end of this section.

Example 3.1.2. Let $R = Q/(f)$ for a non-zero-divisor f. If U has a homotopy $\partial\sigma + \sigma\partial = f \operatorname{id}^U$ with $\sigma^2 = 0$, then by Remark 2.2.1 it is a DG module over $A = Q[y \mid \partial(y) = f]$. As $\overline{A}_1 = R\overline{y}$ and $\overline{A}_i = 0$ for $i \neq 1$, all but the last summands in the expressions for ∂' and ∂'' vanish, so $F(A, U)$ takes the form

$$\cdots \longrightarrow \bigoplus_i Rx^{(i)} \otimes_Q U_{n-2i} \xrightarrow{\ \partial\ } \bigoplus_i Rx^{(i)} \otimes_Q U_{n-1-2i} \longrightarrow \cdots$$

where $x^{(i)} = \overline{y} \otimes \cdots \otimes \overline{y}$ (i copies), and $\partial\big(x^{(i)} \otimes u\big) = x^{(i-1)} \otimes \sigma(u) + x^{(i)} \otimes \partial(u)$.

In the setup of the example, a resolution of M over R can be constructed even if no square-zero homotopy is available: this is the contents of the next theorem, due to Shamash [142]; the proof we present is from [25].

Theorem 3.1.3. If $R = Q/(f)$ for a non-zero-divisor f, M is an R–module, and U is a resolution of M by free Q–modules, then there exists a family of Q–linear homomorphisms $\boldsymbol{\sigma} = \big(\sigma^{[j]} \in \operatorname{Hom}_Q(U, U)_{2j-1}\big)_{j \geqslant 0}$, such that

$$\sigma^{[0]} = \partial; \qquad \sigma^{[0]}\sigma^{[1]} + \sigma^{[1]}\sigma^{[0]} = f \operatorname{id}^U; \qquad \sum_{j=0}^{n} \sigma^{[j]}\sigma^{[n-j]} = 0 \quad \text{for } n \geq 2.$$

---

[13]Resolve!

If $\{x^{(i)} : |x^{(i)}| = 2i\}_{i \geqslant 0}$ is a linearly independent set over R, then

$$\cdots \longrightarrow \bigoplus_{i=0}^{n} Rx^{(i)} \otimes_Q U_{n-2i} \xrightarrow{\partial} \bigoplus_{i=0}^{n-1} Rx^{(i)} \otimes_Q U_{n-1-2i} \longrightarrow \cdots$$

$$\partial\big(x^{(i)} \otimes u\big) = \sum_{j=0}^{i} x^{(i-j)} \otimes \sigma^{[j]}(u)$$

is a free resolution $\mathrm{G}(\boldsymbol{\sigma}, U)$ of M over R.

Remark. Clearly, $\sigma = \sigma^{[1]}$ is a homotopy between $f\,\mathrm{id}^U$ and 0^U. If $\sigma^2 = 0$, then one can take $\sigma^{[n]} = 0$ for $n \geq 2$, and both proposition and example yield the same resolution. In general, rewriting the condition for $n = 2$ in the form $\partial\sigma^{[2]} + \sigma^{[2]}\partial = -\sigma^2$, we see that $\sigma^{[2]}$ is a *homotopy* which corrects the failure of σ^2 to be actually 0. A similar interpretation applies to all $\sigma^{[n]}$ with $n \geq 3$, so $\boldsymbol{\sigma}$ is a *family of higher homotopies* between $f\,\mathrm{id}^U$ and 0^U.

Proof. Note that $\sigma^{[0]}$ is determined, let $\sigma^{[1]}$ be any homotopy such that $f\,\mathrm{id}^U = \partial\sigma^{[1]} + \sigma^{[1]}\partial$, and assume by induction that maps $\sigma^{[j]}$ with the desired properties have been defined when $1 \leq j < n$ for some $n \geq 2$. Setting $\tau^{[1]} = f\,\mathrm{id}^U$ and $\tau^{[j]} = -\sum_{h=1}^{j-1} \sigma^{[h]}\sigma^{[j-h]}$ for $j \geq 2$, we have $\partial\sigma^{[j]} = \tau^{[j]} - \sigma^{[j]}\partial$, whence

$$\partial\sigma^{[j]}\sigma^{[n-j]} = \tau^{[j]}\sigma^{[n-j]} - \sigma^{[j]}\tau^{[n-j]} + \sigma^{[j]}\sigma^{[n-j]}\partial \quad \text{for } j = 1, \ldots, n-1\,.$$

Summing up these equalities, we are left with $\partial\tau^{[n]} = \tau^{[n]}\partial$, so $\tau^{[n]}$ is a cycle of degree $2n - 2$ in the complex $\mathrm{Hom}_Q(U, U)$. By Proposition 1.3.2, $U \xrightarrow{\simeq} M$ induces a quasi-isomorphism $\mathrm{Hom}_Q(U, U) \to \mathrm{Hom}_Q(U, M)$; the latter complex is zero in positive degrees, hence $\tau^{[n]}$ is a boundary. Thus, there is a homomorphism $\sigma^{[n]} \colon U \to U$ of degree $2n - 1$, such that $\tau^{[n]} = \partial\sigma^{[n]} + \sigma^{[n]}\partial$. This finishes the inductive construction of the family $\boldsymbol{\sigma}$.

A direct computation shows that $\partial^2 = 0$. Set $G = \mathrm{G}(\boldsymbol{\sigma}, U)$, and note that there is an exact sequence $0 \to R \otimes_Q U \to G \to \Sigma^2 G \to 0$ of complexes of free R-modules. As $\mathrm{H}_i(R \otimes_Q U) \cong \mathrm{Tor}_i^Q(R, M) = 0$ for $i \neq 0, 1$, it yields

$$M \cong \mathrm{H}_0(R \otimes_Q U) \cong \mathrm{H}_0(G), \quad \mathrm{H}_{n+2}(G) = \mathrm{H}_n(\Sigma^2 G) \cong \mathrm{H}_n(G) \quad \text{for } n \geq 1\,,$$

and an exact sequence

$$0 \to \mathrm{H}_2(G) \to \mathrm{H}_2(\Sigma^2 G) \xrightarrow{\eth} \mathrm{H}_1(R \otimes_Q U) \to \mathrm{H}_1(G) \to 0\,.$$

Acyclicity of G will follow by induction on n, once we prove that $\eth$ is bijective.

If $z \in (\Sigma^2 G)_2 = Rx \otimes_Q U_0$ is a cycle, then $\eth(\mathrm{cls}(z))$ is the class of $\partial(x \otimes z) = 1 \otimes \sigma^{[1]}(z) \in R \otimes_Q U_1$. To show that $\mathrm{H}_0(R \otimes_Q \sigma^{[1]}) \colon \mathrm{H}_0(R \otimes_Q U) \to \mathrm{H}_1(R \otimes_Q U)$ is bijective, note that $\sigma^{[1]}$ is a homotopy between $f\,\mathrm{id}^U$ and 0^U, so we may replace U by any Q-free resolution V of M, and show that for some homotopy σ between $f\,\mathrm{id}^V$ and 0^V the map $\mathrm{H}_0(R \otimes \sigma)$ is bijective. Take V to be a semi-free resolution of M over $A = Q[y \mid \partial(y) = f]$, and σ to be left multiplication by y, so that $\mathrm{H}_0(R \otimes \sigma)$ is the action of $1 \otimes y \in \mathrm{H}_1(R \otimes_Q A) = \mathrm{Tor}_1^Q(R, R)$ on $\mathrm{H}_0(R \otimes_Q V) = \mathrm{Tor}_0^Q(R, M)$. By Construction 2.3.3, it can be computed from the resolution A of R over Q, as the multiplication of $\mathrm{H}(A \otimes_Q M) = R[y] \otimes_R M$ with $-(y \otimes 1) \in \mathrm{H}(A \otimes_Q R) = R[y]$; this is obviously bijective. $\qquad\qquad\qquad\qquad\qquad\qquad\qquad\qquad\qquad\square$

We now turn to the proof of Theorem 3.1.1. It uses a nice tool popular with algebraic topologists, cf. [50], [115], but neglected by commutative algebraists.

Construction 3.1.4. Bar construction. Consider a DG module U over a DG algebra A (as always, defined over $\Bbbk$), and set $\widetilde{A} = \mathrm{Coker}(\eta^A \colon \Bbbk \to A)$. Let

$$S_p^{\Bbbk}(A, U) = A \otimes_{\Bbbk} \underbrace{\widetilde{A} \otimes_{\Bbbk} \cdots \otimes_{\Bbbk} \widetilde{A}}_{p \text{ times}} \otimes_{\Bbbk} U, \qquad \text{for} \quad p \geq 0,$$

be the DG module, with action of A on the leftmost factor and tensor product differential ∂^p, and set $S_p^{\Bbbk}(A, U) = 0$ for $p < 0$. The expression on the right in

$$\delta^p(a \otimes \widetilde{a}_1 \otimes \cdots \otimes \widetilde{a}_p \otimes u) = (a a_1) \otimes \widetilde{a}_2 \otimes \cdots \otimes \widetilde{a}_p \otimes u$$

$$+ \sum_{i=1}^{p-1} (-1)^{i-1} a \otimes \widetilde{a}_1 \otimes \cdots \otimes (\widetilde{a_i a_{i+1}}) \otimes \cdots \otimes \widetilde{a}_p \otimes u$$

$$(-1)^p a \otimes \widetilde{a}_1 \otimes \cdots \otimes \widetilde{a}_{p-1} \otimes (a_p u)$$

is easily seen to be well-defined. Another easy verification yields

$$\delta^{p+1} \delta^p = 0 \qquad \text{and} \qquad \partial^{p-1} \delta^p = \delta^p \partial^p \qquad \text{for all} \quad p.$$

Thus, $(S^k(A, U), \delta)$ is a complex of DG modules[14], called the *standard complex*. It comes equipped with $\Bbbk$-linear maps

$$\pi' \colon S_0^{\Bbbk}(A, U) \to U, \quad a \otimes u \mapsto au;$$

$$\iota' \colon U \to S_0^{\Bbbk}(A, U), \quad u \mapsto 1 \otimes u;$$

$$\sigma^p \colon S_p^{\Bbbk}(A, U) \to S_{p+1}^{\Bbbk}(A, U),$$

$$a \otimes \widetilde{a}_1 \otimes \cdots \otimes \widetilde{a}_p \otimes u \mapsto 1 \otimes \widetilde{a} \otimes \widetilde{a}_1 \otimes \cdots \otimes \widetilde{a}_p \otimes u,$$

that (are seen by another direct computation to) satisfy the relations

$$\delta^1 \sigma^0 = \mathrm{id}^{S_0^{\Bbbk}(A,U)} - \iota' \pi';$$

$$\delta^{p+1} \sigma^p + \sigma^{p-1} \delta^p = \mathrm{id}^{S_p^{\Bbbk}(A,U)} \qquad \text{for} \quad p \geq 1.$$

In particular, when A and U have trivial differentials, $\mathrm{H}(S^{\Bbbk}(A, U)) \cong U$. If, furthermore, the $\Bbbk$–modules $\widetilde{A}_i$ and U_i are free for all i, then this *free* resolution is known as the *standard resolution* of U over A.

Returning to the DG context, we reorganize $S^k(A, U)$ into a DG module over A, by the process familiar 'totaling' procedure. The resulting DG module, with

---

[14]Of course, a complex in the category of DG modules is a sequence of morphisms of DG modules $\delta^p \colon C^p \to C^{p-1}$, such that $\delta^{p-1} \delta^p = 0$.

the action of A defined in Section 1.3, is the (normalized) bar construction:

$$B^{\Bbbk}(A, U) = \bigoplus_{p=0}^{\infty} \Sigma^p\left(S_p^{\Bbbk}(A, U)\right) \qquad \text{with} \quad \partial = \partial' + \partial''$$

where ∂' denotes the differential of the DG module $\Sigma^p S_p^{\Bbbk}(A, U)$, and ∂' is the degree -1 map induced by the boundary δ of the complex of DG modules $S^{\Bbbk}(A, U)$; the equality $\partial^2 = $ results from the relations

$$\partial'\partial' = 0, \qquad \partial''\partial'' = 0, \qquad \partial'\partial'' + \partial''\partial' = 0,$$

of which the first two are clear, and the last one is due to the difference by a factor $(-1)^p$ of the differentials of $S_p^{\Bbbk}(A, U)$ and $\Sigma^p S_p^{\Bbbk}(A, U)$. Furthermore, the maps π', ι', σ^p, total to maps

$$\pi \colon B^{\Bbbk}(A, U) \to U, \qquad \iota \colon U \to B^{\Bbbk}(A, U), \qquad \sigma \colon B^{\Bbbk}(A, U) \to B^{\Bbbk}(A, U).$$

Clearly, π is a morphism of DG modules over A, ι is a morphism of complexes over $\Bbbk$, σ is a degree 1 morphism of complexes over $\Bbbk$, and they are related by

$$\pi\iota = 1_U \qquad \text{and} \qquad \partial\sigma + \sigma\partial = \mathrm{id}^{B^{\Bbbk}(A,U)} - \iota\pi.$$

Thus, $H(\pi)$ and $H(\iota)$ are inverse isomorphisms, so π is a quasi-isomorphism.

The canonical isomorphism of DG modules over the DG algebra A,

$$\Sigma^p\left(S_p^{\Bbbk}(A, U)\right) \cong A \otimes_k \underbrace{\Sigma(\widetilde{A}) \otimes_{\Bbbk} \cdots \otimes_{\Bbbk} \Sigma(\widetilde{A})}_{p \text{ times}} \otimes_{\Bbbk} U,$$

expresses the degree n component of the bar construction as

$$B_n(A, U) = \bigoplus_{h+p+i_1+\cdots+i_p+j=n} A_h \otimes_{\Bbbk} \widetilde{A}_{i_1} \otimes_{\Bbbk} \cdots \otimes_{\Bbbk} \widetilde{A}_{i_p} \otimes_{\Bbbk} U_j.$$

The signs arising from the application of the shift, cf. Section 1.3, then yield the following expressions for the two parts of the differential:

$$\partial'(a \otimes \widetilde{a}_1 \otimes \ldots \otimes \widetilde{a}_p \otimes u) = \partial(a) \otimes \widetilde{a}_1 \otimes \cdots \otimes \widetilde{a}_p \otimes u$$

$$+ \sum_{r=1}^{p} (-1)^{r+h+i_1+\cdots+i_{r-1}} a \otimes \widetilde{a}_1 \otimes \cdots \otimes \partial(\widetilde{a}_r) \otimes \cdots \otimes \widetilde{a}_p \otimes u$$

$$+ (-1)^{p+h+i_1+\cdots+i_p} a \otimes \widetilde{a}_1 \otimes \cdots \otimes \widetilde{a}_p \otimes \partial(u)$$

$$\partial''(a \otimes \widetilde{a}_1 \otimes \ldots \otimes \widetilde{a}_p \otimes u) = (-1)^h (aa_1) \otimes \widetilde{a}_2 \otimes \cdots \otimes \widetilde{a}_p \otimes u$$

$$+ \sum_{r=1}^{p-1} (-1)^{r+h+i_1+\cdots+i_r} a \otimes \widetilde{a}_1 \otimes \cdots \otimes \widetilde{a_r a}_{r+1} \otimes \cdots \otimes \widetilde{a}_p \otimes u$$

$$+ (-1)^{p+h+i_1+\cdots+i_{p-1}} a \otimes \widetilde{a}_1 \otimes \cdots \otimes \widetilde{a}_{p-1} \otimes (a_p u).$$

This finishes our description of the bar construction.

Proof of Theorem 3.1.1. We apply Construction 3.1.4 to the DG algebra A and its DG module U, considered as complexes over the base ring $\Bbbk = Q$. Thus, we get quasi-isomorphisms

$$\mathrm{B}^Q(A, U) \xrightarrow{\ \pi\ } U \xrightarrow{\ \epsilon^U\ } M \, .$$

On the other hand, as the DG module $\mathrm{B}^Q(A, U)$ is semi-free over A, so

$$\mathrm{B}^Q(A, U) = \mathrm{B}^Q(A, U) \otimes_A A \xrightarrow{\ \mathrm{B}^Q(A,U) \otimes_A \epsilon^A\ } \mathrm{B}^Q(A, U) \otimes_A R$$

is a quasi-isomorphism by Proposition 1.3.2. Viewed as a complex of R–modules, $\mathrm{B}^Q(A, U) \otimes_A R$ is precisely the complex $\mathrm{F}(A, U)$ described in the statement of the theorem, so $\mathrm{F}(A, U)$ is a free resolution of M over R. $\qquad\square$

3.2. Spectral sequences. Various spectral sequences relate (co)homological invariants of M over R and over Q. Those presented below are *first quadrant*, that is, have ${}^r\mathrm{E}_{p,q} = 0$ when $p < 0$ or $q < 0$, and *of homological type of* , meaning that their differentials follow the pattern ${}^r d_{p,q} \colon {}^r\mathrm{E}_{p,q} \to {}^r\mathrm{E}_{p-r,q+r-1}$.

For starters, here is a classical *Cartan-Eilenberg spectral sequence* [51].

Proposition 3.2.1. *For each Q–module N there exists a spectral sequence*

$$ {}^2\mathrm{E}_{p,q} = \mathrm{Tor}_p^R \left(M, \mathrm{Tor}_q^Q (R, N) \right) \implies \mathrm{Tor}_{p+q}^Q (M, N) \, . $$

Proof. Let $V \to M$ be a free resolution over R, and $W \to N$ be a free resolution over Q. By Proposition 1.3.2, the induced map $V \otimes_Q W \to M \otimes_Q W$ is a quasi-isomorphism, hence $\mathrm{H}(V \otimes_Q W) \cong \mathrm{Tor}^Q (M, N)$. As $V \otimes_Q W \cong V \otimes_R (R \otimes_Q W)$, the filtration $(V_{\leqslant p}) \otimes_R (R \otimes_Q W)$ yields a spectral sequence

$$ {}^2\mathrm{E}_{p,q} = \mathrm{H}_p(V \otimes_R \mathrm{H}_q(R \otimes_Q W)) \implies \mathrm{Tor}_{p+q}^Q (M, N) $$

where $\mathrm{H}_q(R \otimes_Q W) = \mathrm{Tor}_q^Q (R, N)$ and $\mathrm{H}_p(V \otimes_R L) = \mathrm{Tor}_p^R (M, L)$. $\qquad\square$

In simple cases, this spectral sequence degenerates to an exact sequence. This may be used to prove the next result, but we take a direct approach.

Proposition 3.2.2. *If f is a non-zero-divisor on Q, and N is a module over $R = Q/(f)$, then there is a long exact sequence*

$$ \cdots \longrightarrow \mathrm{Tor}_{n-1}^R (M, N) \longrightarrow \mathrm{Tor}_n^Q (M, N) \xrightarrow{\ \mathrm{Tor}_n^\psi (M,N)\ } \mathrm{Tor}_n^R (M, N) $$

$$ \xrightarrow{\ \vartheta_n\ } \mathrm{Tor}_{n-2}^R (M, N) \longrightarrow \mathrm{Tor}_{n-1}^Q (M, N) \longrightarrow \cdots \, . $$

Proof. In the description of $\mathrm{F}(A, U)$ given in example 3.1.2, define a morphism $\iota \colon U \to \mathrm{F}(A, U)$ with $\iota(u) = x^{(0)} \otimes u$, and note that $\mathrm{Coker}\, \iota \cong \Sigma^2 \mathrm{F}(A, U)$. Thus, we have a short exact sequence of complexes of free R–modules

$$ 0 \longrightarrow R \otimes_Q U \xrightarrow{\ R \otimes \iota\ } \mathrm{F}(A, U) \xrightarrow{\ \vartheta\ } \Sigma^2 \mathrm{F}(A, U) \longrightarrow 0 \, . $$

Tensoring it with N over R, and writing down the homology exact sequence of the resulting short exact sequence of complexes, we get what we want. $\qquad\square$

The next spectral sequence is introduced by Lescot [107].

Proposition 3.2.3. *If ψ is surjective and k is a residue field of R, then there is a spectral sequence*

$$^2E_{p,q} = \operatorname{Tor}_p^Q(k, M) \otimes_k \operatorname{Tor}_q^R(k, k) \implies \left(\operatorname{Tor}^Q(k, k) \otimes_k \operatorname{Tor}^R(M, k) \right)_{p+q}.$$

Proof. Choose free resolutions: $U \to k$ over Q; $V \to M$ and $W \to k$ over R. As $U \otimes_Q V$ is a bounded below complex of free R–modules, Proposition 1.3.2 yields the first isomorphisms below; the third ones comes from Proposition 1.3.4:

$$\operatorname{H}(U \otimes_Q V \otimes_R W) \cong \operatorname{H}((U \otimes_Q V) \otimes_R k) \cong \operatorname{H}((U \otimes_Q k) \otimes_k (V \otimes_R k))$$

$$\cong \operatorname{H}(U \otimes_Q k) \otimes_k \operatorname{H}(V \otimes_R k) = \operatorname{Tor}^Q(k, k) \otimes_k \operatorname{Tor}^R(M, k).$$

Thus, the spectral sequence of the filtration $(U \otimes_Q V) \otimes_R (W_{\leqslant p})$ has

$$^2E_{p,q} = \operatorname{H}_p(\operatorname{H}_q(U \otimes_Q V) \otimes_R W) \implies \left(\operatorname{Tor}^Q(k, k) \otimes_k \operatorname{Tor}^R(M, k) \right)_{p+q}.$$

Since U is a bounded below complex of free Q–modules, $U \otimes_Q V \to U \otimes_Q M$ is a quasi-isomorphism, so $^2E_{p,q} \cong \operatorname{H}_p(\operatorname{Tor}_q^Q(k, M) \otimes_R W)$, and that module is equal to $\operatorname{Tor}_p^R\left(\operatorname{Tor}_q^Q(k, M), k\right) \cong \operatorname{Tor}_p^Q(k, M) \otimes_k \operatorname{Tor}_q^R(k, k)$.   □

For us, the preceding sequences have the drawback of going in the 'wrong' direction: they require input of data over the ring of interest, R. Next we describe a sequence where the roles of Q and R are reversed. It belongs to the family of *Eilenberg-Moore spectral sequences*, cf. [122].

By Construction 2.3.3, $\operatorname{Tor}^Q(M, k)$ is a module over the graded algebra $\operatorname{Tor}^Q(R, k)$. A homogeneous free resolution of the former over the latter provides an 'approximation' of a resolution of M over R: this is the contents of the following special case of a result of Avramov [22].

Proposition 3.2.4. *When k is a residue field of R, there is a spectral sequence*

$$^2E_{p,q} = \operatorname{Tor}_p^{\operatorname{Tor}^Q(R,k)}\left(\operatorname{Tor}^Q(M, k), k\right)_q \implies \operatorname{Tor}_{p+q}^R(M, k).$$

Proof. In the notation of Construction 3.1.4, set $B = A \otimes_Q k$ and $V = U \otimes_Q k$. The filtration $\bigoplus_{i \leqslant p} \left(\widetilde{B}^{\otimes i} \otimes V\right)$ of $\operatorname{B}^k(B, V) \otimes_B k$ yields a spectral sequence

$$^0E_{p,q} = \left(S_p^k(B, V) \otimes_B k\right)_q \implies \operatorname{H}_{p+q}(\operatorname{B}^k(B, V) \otimes_B k)$$

with $^0d_{p,q}$ equal to the tensor product differential. By Künneth,

$$^1E_{p,q} = \operatorname{H}_q(S_p^k(B, V) \otimes_B k)_q \cong \left(S_p^k(\operatorname{H}(B), \operatorname{H}(V)) \otimes_{\operatorname{H}(B)} k\right)_q, \quad {}^1d_{p,q} = \delta_q^{[p]} \otimes_R k.$$

As k is a field, $S^k(\operatorname{H}(B), \operatorname{H}(V))$ is a resolution of $\operatorname{H}(V)$ over $\operatorname{H}(B)$, so

$$^2E_{p,q} = \operatorname{Tor}_p^{\operatorname{H}(B)}(\operatorname{H}(V), k)_q.$$

Since $\operatorname{H}(B) = \operatorname{Tor}^Q(R, k)$ and $\operatorname{H}(V) = \operatorname{Tor}^Q(M, k)$, the second page of the spectral sequence has the desired form. The isomorphisms

$$\operatorname{B}^k(B, V) \otimes_B k \cong \left(\operatorname{B}^Q(A, U) \otimes_A R\right) \otimes_R k = \operatorname{F}(A, U) \otimes_R k$$

and Theorem 3.1.1 identify its abutment as $\operatorname{Tor}^R(M, k)$.   □

Corollary 3.2.5. *If ψ is surjective, then for each $n \geq 0$ there is an inclusion*

$$\sum_{i=1}^{n} \operatorname{Tor}_i^Q(R,k) \cdot \operatorname{Tor}_{n-i}^Q(M,k) \subseteq \operatorname{Ker} \operatorname{Tor}_n^\psi(M,k) \ .$$

Proof. The natural map $U \otimes_Q k \to \mathrm{F}(A,U)$, that is, $V \to \mathrm{B}^k(B,V) \otimes_B k$, identifies V with $F^1 \subseteq \mathrm{B}^k(B,V) \otimes_B k$. Thus, the map $\operatorname{Tor}_n^\psi(M,k)$ that it induces in homology factors through

$$\nu_n \colon \operatorname{Tor}_n^Q(M,k) = \mathrm{H}_n(V) \to {}^2\mathrm{E}_{0,n} = \frac{\operatorname{Tor}_n^Q(M,k)}{\sum_{i=1}^n \operatorname{Tor}_i^Q(R,k) \cdot \operatorname{Tor}_{n-i}^Q(M,k)} \ .$$

We get $\operatorname{Ker} \nu_n \subseteq \operatorname{Ker} \operatorname{Tor}_n^\psi(M,k)$, which is the desired inclusion. $\square$

The next theorem shows that the vector spaces

$$o_n^\psi(M) = \frac{\operatorname{Ker} \operatorname{Tor}_n^\psi(M,k)}{\sum_{i=1}^n \operatorname{Tor}_i^Q(R,k) \cdot \operatorname{Tor}_{n-i}^Q(M,k)}$$

are *obstructions* to DG module structures; they were found in [22].

Theorem 3.2.6. *Let $(Q, \mathfrak{n}, k)$ be a local ring, let $\psi \colon Q \to R$ be a surjective homomorphism of rings, and let M be a finite R–module.*

If the minimal free resolution A of R over Q has a structure of DG algebra, and the minimal free resolution U of M over Q admits a structure of DG module over A, then $o_n^\psi(M) = 0$ for all n.

Proof. Under the hypotheses of the theorem, $\partial' \otimes_Q k = 0$ in the bar construction $\mathrm{B}^k(B,V) \otimes_B k$ of Proposition 3.2.4, so the only non-zero differential in the spectral sequence constructed there acts on the first page. Thus, the sequence stops on the second page, yielding $\operatorname{Ker} \operatorname{Tor}_n^\psi(M,k) = \operatorname{Ker} \nu_n$. $\square$

Remark 3.2.7. Let $\boldsymbol{f}$ be a Q–regular sequence. Computing the Tor algebra for $R = Q/(\boldsymbol{f})$ with the help of the Koszul complex $A = K(\boldsymbol{f}; Q)$ we get

$$\operatorname{Tor}^Q(R,k) = \mathrm{H}(A \otimes_R k) = A \otimes_R k = \bigwedge(A_1 \otimes_R k) = \bigwedge \operatorname{Tor}_1^Q(R,k)$$

and hence $\sum_{i=1}^n \operatorname{Tor}_i^Q(R,k) \cdot \operatorname{Tor}_{n-i}^Q(M,k) = \operatorname{Tor}_1^Q(R,k) \cdot \operatorname{Tor}_{n-1}^Q(M,k)$.

3.3. Upper bounds. In this section $\psi \colon Q \to R$ is a finite homomorphism of local rings that induces the identity on their common residue field k, and M is a finite R–module. We relate the Betti numbers of M over R and Q.

Often, such relations are expressed in terms of the formal power series

$$\mathrm{P}_M^R(t) = \sum_{n=0}^{\infty} \beta_n^R(M) t^n \in \mathbb{Z}[[t]] \ ,$$

known as the *Poincaré series* of M over R, and the corresponding series over Q. Results then take the form of coefficientwise inequalities (denoted $\preccurlyeq$ and $\succcurlyeq$) of formal power series; equalities are significant.

Spectral sequence generate inequalities, by an elementary observation:

Remark 3.3.1. In a spectral sequence of vector spaces ${}^r E_{p,q} \implies E$, $r \geq a$, the space ${}^{r+1} E_{p,q}$ is a subquotient of ${}^r E_{p,q}$ for $r \geq a$, and the spaces ${}^\infty E_{p,q}$ are the subfactors of a filtration of E_{p+q}. Thus, there are (in)equalities

$$\dim_k E_n = \sum_{p+q=n} \mathrm{rank}_k \, {}^\infty E_{p,q} \leq \sum_{p+q=n} \mathrm{rank}_k \, {}^r E_{p,q} \leq \sum_{p+q=n} \mathrm{rank}_k \, {}^a E_{p,q} \ .$$

Multiplying the n'th one by t^n, and summing in $\mathbb{Z}[[t]]$, we get inequalities

$$\sum_{n \geq 0} \dim_k E_n t^n \preccurlyeq \sum_n \left(\sum_{p+q=n} \mathrm{rank}_k \, {}^r E_{p,q} \right) t^n \qquad \text{for} \quad r \geq a \ .$$

The next result was initially deduced by Serre from the sequence in Proposition 3.2.1. It is more expedient to get it from that in Proposition 3.2.4.

Proposition 3.3.2. *There is an inequality* $\mathrm{P}^R_M(t) \preccurlyeq \dfrac{\mathrm{P}^Q_M(t)}{1 - t\big(\mathrm{P}^Q_R(t) - 1\big)}$.

Proof. The spectral sequence of Proposition 3.2.4 has

$$\sum_n \left(\sum_{p+q=n} \mathrm{rank}_k \, {}^1 E_{p,q} \right) t^n = \sum_p \left(\sum_q \mathrm{rank}_k \, {}^1 E_{p,q} \, t^q \right) t^p$$

$$= \sum_p \Big(\big(\mathrm{P}^Q_R(t) - 1 \big)^p \, \mathrm{P}^Q_M(t) \Big) t^p$$

$$= \mathrm{P}^Q_M(t) \sum_p \big(\mathrm{P}^Q_R(t) - 1 \big)^p t^p = \frac{\mathrm{P}^Q_M(t)}{1 - t\big(\mathrm{P}^Q_R(t) - 1\big)}$$

so the desired inequality follows from the preceding remark.   □

Remark. If equality holds with $M = k$, then ψ is called a *Golod homomorphism*. These maps, introduced by Levin [109], are studied in detail in [110], [24]; they are used in many computations of Poincaré series. The 'absolute case', when Q is a regular local ring, is the subject of Chapter 5.

Directly from the spectral sequence in Proposition 3.2.3, we read off

Proposition 3.3.3. *There is an inequality* $\mathrm{P}^R_M(t) \, \mathrm{P}^Q_k(t) \preccurlyeq \mathrm{P}^Q_M(t) \, \mathrm{P}^R_k(t)$.   □

Remark. If equality holds, then the module M is said to be *inert* by ψ: these modules are introduced and studied by Lescot [107].

In special cases, universal bounds can be sharpened.

Recall [46] that a finite Q–module N has *rank* if for each prime ideal $\mathfrak{q} \in \mathrm{Ass}\, Q$ the $Q_\mathfrak{q}$–module $N_\mathfrak{q}$ is free, and its rank does not depend on $\mathfrak{q}$. The common rank of these free modules is called the Q–rank of N, and denoted $\mathrm{rank}_Q N$; we write $\mathrm{rank}_Q N \geq 0$ to indicate that the rank of N is defined.

Proposition 3.3.4. *If f is Q–regular and $R = Q/(f)$, then*

$$0 \preccurlyeq \sum_{n=1}^{\infty} \mathrm{rank}_Q \, \mathrm{Syz}_n^Q(M) \, t^{n-1} = \frac{\mathrm{P}_M^Q(t)}{(1+t)} \preccurlyeq \mathrm{P}_M^R(t) \preccurlyeq \frac{\mathrm{P}_M^Q(t)}{(1-t^2)} \, .$$

Proof. Let U be a minimal resolution of M over Q. For each $\mathfrak{q} \in \mathrm{Ass}\, Q$ we have $f \notin \mathfrak{q}$. Thus, $M_\mathfrak{q} = 0$, and so for each n there is an exact sequence

$$0 \to \mathrm{Syz}_{n+1}^Q(M)_\mathfrak{q} \to (U_n)_\mathfrak{q} \to \cdots \to (U_0)_\mathfrak{q} \to 0 \, .$$

It follows that $\mathrm{Syz}_{n+1}^Q(M)_\mathfrak{q}$ is free of rank $\sum_{i \geqslant 0}(-1)^i \beta_{n-i}^Q(M)$: this establishes the equality, and the first inequality.

For the second inequality, apply the exact sequence of Proposition 3.2.2, to get $\beta_n^Q(M) \leq \beta_{n-1}^R(M) + \beta_n^R(M)$ for all n.

The third inequality results from counting the ranks of the free modules in the resolution of M over R given by Theorem 3.1.3. $\qquad\square$

There are useful sufficient conditions for equalities. One is essentially contained in Nagata [124]; the other, from Shamash [142], is given a new proof.

Proposition 3.3.5. *Let f be a Q–regular element, and $R = Q/(f)$.*

(1) *If $f \notin \mathfrak{n}^2$, then $\mathrm{P}_M^R(t) = \mathrm{P}_M^Q(t)/(1+t)$.*

(2) *If $f \in \mathfrak{n}(0 :_Q M)$, then $\mathrm{P}_M^R(t) = \mathrm{P}_M^Q(t)/(1-t^2)$.*

Proof. (1) This is just the last assertion of Theorem 2.2.3.

(2) Let $s_1, \ldots, s_e$ be a minimal set of generators of $\mathfrak{n}$, and write $f = \sum_{j=1}^e a_j s_j$ with $a_j \in (0 :_Q M)$. In a DG algebra resolution $V \to k$ over Q, pick $y_1, \ldots, y_e \in V_1$ such that $\partial(y_j) = s_j$. As $\partial\big(\sum_{j=1}^e a_j y_j\big) = f$, Example 3.1.2 yields a resolution G of k over R, with $G_n = \bigoplus_{i=0}^n Rx^{(i)} \otimes_Q V_{n-2i}$ and

$$\partial\big(x^{(i)} \otimes_Q v\big) = \sum_{j=1}^e a_j x^{(i-1)} \otimes_Q y_j v + x^{(i)} \otimes_Q \partial(v) \, .$$

For $b \in M$, the induced differential of $M \otimes_R G$ then satisfies

$$\partial\big(bx^{(i)} \otimes_Q v\big) = \sum_{j=1}^e a_j bx^{(i-1)} \otimes_Q y_j v + bx^{(i)} \otimes_Q \partial(v) = bx^{(i)} \otimes_Q \partial(v) \, ,$$

so $M \otimes_R G \cong \bigoplus_{i=0}^\infty \Sigma^{2i}(M \otimes_Q V)$ as complexes of R–modules. This yields

$$\mathrm{Tor}^R(M, k) = \mathrm{H}(M \otimes_R G) = \bigoplus_{i=0}^\infty \Sigma^{2i} \mathrm{Tor}^Q(M, k) \, .$$

The desired equality of Poincaré series is now obvious. $\qquad\square$

Remark. The resolution $G(\sigma, U)$ of Theorem 3.1.3 has $\sum \mathrm{rank}_R G_n t^n = \big(\sum \mathrm{rank}_Q U_i t^i\big)/(1-t^2)$. Thus, if U is a minimal resolution of M over Q and $f \in \mathfrak{n}(0 :_Q M)$, then by (2) $G(\sigma, U)$ is a *minimal* R–free resolution of M. Another case of minimality is given by Construction 5.1.2, which shows that if $\mathrm{pd}_Q M = 1$, then $\mathrm{Syz}_n^R(M)$ has a minimal resolution of that form for each $n \geq 1$. Quite the opposite happens 'in general': it is proved in [32] that if R is a complete intersection and the Betti numbers of M are not bounded, then $\mathrm{Syz}_n^R(M)$ has such a minimal resolution for at most one value of n.

4. Growth of Resolutions

The gap between regularity and singularity widens to a chasm in homological local algebra: Minimal resolutions are always finite over a regular local ring, and (very) rarely over a singular one. A bridge[15] is provided by the Cohen Structure Theorem: the completion of each local ring is a residue of a regular ring, so by change of rings techniques homological invariants over the singular ring may be approached from those – essentially finite – over the regular one.

To describe and compare resolutions of modules over a singular local ring, we go beyond the primitive dichotomy of finite versus infinite projective dimension, and analyze infinite sequences of integers, such as ranks of matrices, or Betti numbers. For that purpose there in no better choice than to follow the time-tested approach of calculus, and compare sizes of resolutions to the functions we[16] understand best: polynomials and exponentials.

4.1. Regular presentations. Let I be an ideal in a noetherian ring R. Recall that the *minimal number of generators* $\nu_R(I)$, the height, and the depth of I are always related by inequalities, due to Rees and to Krull:

$$\mathrm{depth}_R(I, R) \leq \mathrm{height}\, I \leq \nu_R(I).$$

For the rest of this section, $(R, \mathfrak{m}, k)$ is a local ring; $\nu_R(\mathfrak{m})$ is then known as its *embedding dimension*, denoted $\mathrm{edim}\, R$, and the inequalities read

$$\mathrm{depth}\, R \leq \dim R \leq \mathrm{edim}\, R.$$

Discrepancies between these numbers provide measures of irregularity:

- $\mathrm{cmd}\, R = \dim R - \mathrm{depth}\, R$ is the *Cohen-Macaulay defect* of R;
- $\mathrm{codim}\, R = \mathrm{edim}\, R - \dim R$ is the *codimension*[17] of R;
- $\mathrm{codepth}\, R = \mathrm{edim}\, R - \mathrm{depth}\, R$ is the *codepth*[17] of R.

---

[15]Warning: crossing may take an infinite time.

[16]Algebraists.

[17]Because 'codimension' has been used to denote depth, and 'codepth' to denote Cohen-Macaulay defect, the notions described here are sometimes qualified by 'embedding'; it would be too cumbersome to stick to that terminology, and to devise new notation.

We use the vanishing of codepth R to define[18] the *regularity* of R. Thus, if R is regular, then each minimal generating set of $\mathfrak{m}$ is a regular sequence.

Two cornerstone results of commutative ring theory determine the role of regular rings in the study of free resolutions.

The *Auslander-Buchsbaum-Serre Theorem* describes them homologically.

Theorem 4.1.1. *The following conditions are equivalent.*

(i) R *is regular.*

(ii) $\mathrm{pd}_R M < \infty$ *for each finite R–module M.*

(iii) $\mathrm{pd}_R k < \infty$.

Proof. (i) $\implies$ (iii). By Example 1.1.1, the Koszul complex on a minimal set of generators for $\mathfrak{m}$ is a free resolution of k.

(iii) $\implies$ (ii). As $\mathrm{Tor}_n^R(M, k) = 0$ for $n \gg 0$, apply Proposition 1.2.2.

(ii) $\implies$ (i). We prove that codepth $R = 0$ by induction on $d = \mathrm{depth}\, R$. If $d = 0$, then k is free by Proposition 1.2.7.1, hence $\mathfrak{m} = 0$. If $d > 0$, then a standard prime avoidance argument yields a regular element $g \in \mathfrak{m} \smallsetminus \mathfrak{m}^2$. Set $R' = R/(g)$, and note that codepth $R = $ codepth R'. As $\mathrm{pd}_{R'} k < \infty$ by Theorem 2.2.3, we have codepth $R' = 0$ by the induction hypothesis. $\qquad\square$

Corollary 4.1.2. *If R is regular, then so is $R_\mathfrak{p}$ for each $\mathfrak{p} \in \mathrm{Spec}\, R$.*

Proof. By the theorem, $R/\mathfrak{p}$ has a finite R–free resolution F; then $F_\mathfrak{p}$ is a finite $R_\mathfrak{p}$–free resolution of $k(\mathfrak{p}) = R_\mathfrak{p}/\mathfrak{p} R_\mathfrak{p}$, so $R_\mathfrak{p}$ is regular by the theorem. $\qquad\square$

The *Cohen Structure Theorem* establishes the dominating position of regular rings. A *regular presentation* of R is an isomorphism $R \cong Q/I$, where Q is a regular local ring. Many local rings (for example, all those arising in classical algebraic or analytic geometry), come equipped with such a presentation. By Cohen's theorem, *every complete local ring has a regular presentation.*

Here is how the two theorems above apply to the study of resolutions.

Since the $\mathfrak{m}$-adic completion $\widehat{R}$ is a faithfully flat R–module, a complex of R–modules F is a (minimal) free resolution of M over R if and only if $\widehat{F} = \widehat{R} \otimes_R F$ is a (minimal) free resolution of $\widehat{M} = \widehat{R} \otimes_R M$ over $\widehat{R}$; in particular, $\mathrm{P}_M^R(t) = \mathrm{P}_{\widehat{M}}^{\widehat{R}}(t)$. The point is: as $\widehat{M}$ and $\widehat{R}$ have *finite* free resolutions over the ring Q, all change of rings results apply with *finite* entry data.

If $(Q, \mathfrak{n}, k)$ is a regular local ring, $R = Q/I$, and $f \in I \smallsetminus \mathfrak{n}^2$, then $Q/(f)$ is also regular, and maps onto R. Iterating, one sees that if R has some regular presentation, then it has a *minimal* one, with $\mathrm{edim}\, R = \mathrm{edim}\, Q$.

---

[18] This clearly implies the usual definition, in terms of the vanishing of codim R. Conversely, if R is regular, then the associated graded ring $S = \bigoplus_n \mathfrak{m}^n/\mathfrak{m}^{n+1}$ is the quotient of a polynomial ring in $e = \mathrm{edim}\, R$ variables. Since S and R have equal Hilbert-Samuel functions, $\dim S = \dim R = e$, so S is the polynomial ring on the classes in S_1 of a minimal set of generators t of $\mathfrak{m}$. In particular, these classes form an S–regular sequence, and then a standard argument shows that t is an R–regular sequence.

Minimal presentations often fade into the background, because some of their invariants can be computed directly over the ring R, from Koszul complexes on minimal sets of generators t of $\mathfrak{m}$. Different choices of t lead to isomorphic complexes, so when we do not need to make an explicit choice of generators, we write K^R instead of $K(t; R)$, and set $K^M = K^R \otimes_R M$.

Lemma 4.1.3. *If K^R is a Koszul complex on a minimal set of generators of $\mathfrak{m}$, and $\widehat{R} = Q'/I'$ (respectively, $R = Q/I$) is a minimal regular presentation, then*

(1) $\mathrm{H}(K^M) \cong \mathrm{H}(K^{\widehat{M}}) \cong \mathrm{Tor}^{Q'}(\widehat{M}, k)$ $\left(\cong \mathrm{Tor}^Q(M, k) \right)$.

(2) $\sup\{i \mid \mathrm{H}_i(K^M) \neq 0\} = \sup\{i \mid \mathrm{H}_i(K^{\widehat{M}}) \neq 0\} = \mathrm{pd}_{Q'} \widehat{M}$ $\left(= \mathrm{pd}_Q M \right)$.

(3) $\mathrm{H}_1(K^R) \cong \mathrm{H}_1(K^{\widehat{R}}) \cong I' \otimes_{Q'} k$ $\left(\cong I \otimes_Q k \right)$.

Proof. The fact that $\widehat{R}$ is faithfully flat over R, and $\mathfrak{m}\widehat{R}$ is its maximal ideal yields the relations on both ends, so we argue for those in the middle. By Example 1.1.1, $K^{Q'}$ is a minimal free resolution of k over Q': this yields (1), and then (2) follows from Proposition 1.2.2. For (3), use the long exact sequence of $\mathrm{Tor}^{Q'}(-, k)$ applied to the exact sequence $0 \to I' \to Q' \to \widehat{R} \to 0$. $\qquad\square$

In view of the lemma, Proposition 3.3.2 translates into:

Proposition 4.1.4. *For each finite R–module M there is an inequality*

$$
\mathrm{P}_M^R(t) \preccurlyeq \frac{\displaystyle\sum_{i=0}^{\mathrm{edim}\,R - \mathrm{depth}\,M} \mathrm{rank}_k\,\mathrm{H}_i(K^M)t^i}{1 - \displaystyle\sum_{j=1}^{\mathrm{codepth}\,R} \mathrm{rank}_k\,\mathrm{H}_j(K^R)t^{j+1}}.
\qquad\square
$$

Corollary 4.1.5. *There is an $\alpha \in \mathbb{R}$, such that $\beta_n^R(M) \leq \alpha^n$ for $n \geq 1$.* $\qquad\square$

A local ring homomorphism $\varphi: R \to R'$, such that $\mathfrak{m}R'$ is the maximal ideal of R', may be lifted – in more than one way – to a morphism of DG algebras $K^\varphi: K^R \to K^{R'}$ (we use a 'functorial' notation, because in the cases treated below the choice of a specific lifting will be of no consequence, while it helps to distinguish K^φ from $K^R \otimes_R \varphi: K^R \to K^R \otimes_R R'$).

The following easily proved statements have unexpectedly strong consequences. The second is the key to Serre's original proof that $\mathrm{pd}_R k$ characterizes regularity in [140]. The third, also due to Serre, is important in the study of multiplicities, cf. [141], [17]; its proof below is from Eagon and Fraser [54].

Lemma 4.1.6. *The complexes K^R and K^M have the following properties.*

(1) *If $g \in \mathfrak{m} \smallsetminus \mathfrak{m}^2$ is R–regular, then the homomorphism $\varphi: R \to R/(g)$ induces a surjective quasi-isomorphism $K^\varphi: K^R \to K^{R/(g)}$.*

(2) *If $a \in K^R$ satisfies $\partial(a) \in \mathfrak{m}^2 K^R$, then $a \in \mathfrak{m}K^R$.*

(3) *For each finite R–module M there is an integer s such that the complex*

$$C^i: \quad 0 \to \mathfrak{m}^{i-e} K_e^M \to \cdots \to \mathfrak{m}^{i-1} K_1^M \to \mathfrak{m}^i K_0^M \to 0 \qquad (*)$$

is exact for $i \geq s$; for each i, C^i is a DG submodule of K^M over K^R.

Proof. (1) Choose $y_1, \ldots, y_e \in K_1^R$, such that $\{\partial(y_i) = t_i \mid j = 1, \ldots, e\}$ is a minimal generating set for $\mathfrak{m}$. We may assume that $g = t_e$, and set $D = R[y \mid \partial(y) = g]$. As K^R is a semi-free DG module over D, by Proposition 1.3.2 the quasi-isomorphism $D \to R/(g)$ induces a quasi-isomorphism $K^R \to R/(g) \otimes_D K^R = K^{R/(g)}$.

(2) The statement may be rephrased as follows: the map $K_n^R / \mathfrak{m} K_n^R \to \mathfrak{m} K_{n-1}^R / \mathfrak{m}^2 K_{n-1}^R$ induced by the differential ∂_n of K^R is injective for all $n \geq 1$. This is a direct consequence of the formula for the Koszul differential, and the minimality of the generating set $t_1, \ldots, t_e$.

(3) As $C^1 = \mathrm{Ker}(K^R \to k) \subset K^R$ is a DG ideal, $C^i = (C^1)^i K^M$ is a DG submodule of K^M. For each $1 \leq n \leq e$ and $i \gg 0$, we have equalities

$$Z_n(C^i) = Z_n(K^M) \cap \mathfrak{m}^{i-n} K_n^M = \mathfrak{m}\big(Z_n(K^M) \cap \mathfrak{m}^{i-n-1} K_n^M \big)$$

(the first by definition, the second by Artin-Rees). Increasing i, we may assume they hold simultaneously for all n. Thus, each $z \in Z_n(C^i)$ can be written as $z = \sum_{j=1}^e t_j v_j$ with $v_j \in \mathfrak{m}^{i-n-1} K_n^M$, so $z = \partial y$ for $y = \sum_{j=1}^e y_j v_j \in C_{n+1}^i$. $\square$

Next we present a result of [21], which shows that the minimal resolution of each R–module M is part of that of 'most' of its residue modules. It is best stated in terms of a property of minimal complexes.

Remark 4.1.7. Let F be a minimal complex of free R–modules, with $F_n = 0$ for $n < 0$. If $\epsilon: F \to N$ is a morphism to a finite R–module N, let $\alpha: F \to G$ be a lifting of ϵ to a minimal resolution G of N. Any two liftings are homotopic, so the homomorphism $k \otimes_R \alpha = \mathrm{H}(k \otimes_R \alpha)$ depends only on ϵ.

We say that ϵ is *essential* if $k \otimes_R \alpha_n$ is injective for each n. In that case, each α_n is a split injection, hence α maps F isomorphically onto a subcomplex of G, that splits off as a graded R–module; the entire resolution of N is obtained then from $\alpha(F)$ by adjunction of basis elements, as in Construction 2.2.6.

When $\mathrm{pd}_R M$ is finite the next result is a straightforward application of the Artin-Rees Lemma. Replacing resolutions over R by resolutions over K^R, we make Artin-Rees work simultaneously in infinitely many dimensions.

Theorem 4.1.8. *Let F be a minimal free resolution of a finite R–module M. There exists an integer $s \geq 1$, such that for each submodule $M' \subseteq \mathfrak{m}^s M$ the augmentation $\epsilon: F \to M'' = M/M'$ is essential. In particular,*

$$\mathrm{P}_{M''}^R(t) = \mathrm{P}_M^R(t) + t\,\mathrm{P}_{M'}^R(t).$$

Proof. Choose s as in Lemma 4.1.6.3, so that $C^s \subseteq K^M$ is an exact DG submodule. The projection $\rho: K^M \to K^M/C^s$ is then a quasi-isomorphism of DG modules over

K^R. Let U be a semi-free resolution of k over K^R, let M' be a submodule in $\mathfrak{m}^s M$, let $\pi \colon M \to M''$ be the canonical map. The composition

$$U \otimes_{K^R} \left(K^R \otimes_R M \right) \xrightarrow{\pi'} U \otimes_{K^R} \left(K^R \otimes_R M'' \right) \longrightarrow U \otimes_{K^R} \left(K^M / C^s \right)$$

of morphisms of DG modules, where $\pi' = U \otimes_{K^R} (K^R \otimes_R \pi)$, is equal to $U \otimes_{K^R} \rho$, and so is a quasi-isomorphism by Proposition 1.3.2. It follows that $H(\pi')$ is injective. As U is a free resolution of k over R and $U \otimes_{K^R} (K^R \otimes_R \pi) = U \otimes_R \pi \colon U \otimes_R M \to U \otimes_R M''$, we see that $H(\pi') = \operatorname{Tor}^R(k, \pi)$. If F'' is a minimal free resolution of M'', then $k \otimes_R \pi' \colon k \otimes_R F \to k \otimes_R F''$ is another avatar of the same map, so it is injective, as desired. For the equality of Poincaré series, apply $\operatorname{Tor}^R(k, -)$ to $0 \to M' \to M \to M'' \to 0$. □

In a precise sense, the residue field has the 'largest' resolution.

Corollary 4.1.9. *For each finite R-module M there exists an integer $\ell \geq 1$, such that* $\operatorname{P}_M^R(t) \preccurlyeq \ell \operatorname{P}_k^R(t)$.

Proof. By the theorem, $\operatorname{P}_M^R(t) \preccurlyeq \operatorname{P}_{M/\mathfrak{m}^s M}^R(t)$ for some s, so we may assume that $\operatorname{length}_R M < \infty$. The obvious induction on length, using the exact sequence of $\operatorname{Tor}^R(-, k)$ then establishes the inequality with $\ell = \operatorname{length}_R M$. □

For another manifestation of the ubiquity of $\operatorname{P}_k^R(t)$, cf. Theorem 6.3.6.

4.2. Complexity and curvature. In this section we introduce and begin to study measures for the asymptotic size of resolutions.

On the polynomial scale, the *complexity* of M over R is defined by

$$\operatorname{cx}_R M = \inf \left\{ d \in \mathbb{N} \,\middle|\, \begin{array}{c} \text{there exists a polynomial } f(t) \text{ of degree } d-1 \\ \text{such that } \beta_n^R(M) \leq f(n) \text{ for } n \geq 1 \end{array} \right\};$$

this is an adaptation from [25], [26], of a concept originally introduced by Alperin and Evens [2] to study modular representations of finite groups; clearly, one gets the same concept by requiring inequalities for $n \gg 0$.

Example 4.2.1. (1) If $R' = k[s_1]/(s_1^2)$, and $t_1 \in R'$ is the image of s_1, then

$$F' \colon \quad \cdots \to R' \xrightarrow{t_1} R' \xrightarrow{t_1} \cdots \xrightarrow{t_1} R' \xrightarrow{t_1} R' \xrightarrow{t_1} R' \to 0. \qquad (*)$$

is a minimal free resolution of $k = R'/(t_1)$. Thus, $\beta_n^{R'}(k) = 1$ for $n \geq 0$, and $\operatorname{cx}_{R'} k = 1$. Each finite module M over the principal ideal ring $k[t_1]$ is a finite direct sum of copies of k and of R', hence $\beta_n^{R'}(M) = \beta_1^{R'}(M)$ for $n \geq 1$.

(2) Set $R = k[s_1, s_2]/(s_1^2, s_2^2) = k[t_1, t_2]$. As R is a free module over its subring R', the complex $R \otimes_{R'} F'$ is a minimal resolution of the cyclic module $M = R/(t_1)$ over R. Thus, $\beta_n^R(M) = 1$ for $n \geq 0$, and $\operatorname{cx}_R M = 1$.

On the other hand, note that $\operatorname{Syz}_1^R (R/(t_1 t_2)) \cong k$, and let F'' be the complex corresponding to F' over $R'' = k[t_2]$. By the Künneth Theorem 1.3.4, $F' \otimes_k F''$ is a resolution of $k \otimes_k k = k$ over $R = R' \otimes_k R''$, and is obviously minimal. Thus, $\beta_n^R(k) = n + 1$ for $n \geq 0$, and $\operatorname{cx}_R \left(R/(t_1 t_2) \right) = \operatorname{cx}_R \left(R/(t_1, t_2) \right) = 2$.

In particular, over the ring R there exist modules of complexity 0, 1, and 2; in fact, these are all the possible values: cf. Proposition 4.2.5.4.

Complexity may itself be infinite.

Example 4.2.2. Let $R = k[s_1, s_2]/(s_1^2, s_1 s_2, s_2^2)$. The isomorphism $\mathfrak{m} \cong k^2$ and the exact sequence $0 \to \mathfrak{m} \to R \to k \to 0$ show that $\beta_{n+1}^R(k) = 2\beta_n^R(k)$ for $n \geq 1$, hence $\beta_n^R(k) = 2^n$ for $n \geq 0$, and $\operatorname{cx}_R k = \infty$. If M is any finite R–module, then $\operatorname{Syz}_1^R(M) \subseteq \mathfrak{m}F_0$ is isomorphic to a finite direct sum of copies of k, so $\operatorname{cx}_R M = \infty$ unless M is free.

For modules of infinite complexity, the exponential scale is used in [29] to introduce a notion of *curvature*[19] by the formula

$$\operatorname{curv}_R M = \limsup_{n \to \infty} \sqrt[n]{\beta_n^R(M)}.$$

Thus, in the last example, either M is free or $\operatorname{curv}_R M = 2$.

Some relations between these *asymptotic invariants* follow directly from the definitions, except for (5), which is a consequence of Corollary 4.1.5:

Remark 4.2.3. For a finite R–module M the following hold:

(1) $\operatorname{pd}_R M < \infty \iff \operatorname{cx}_R M = 0 \iff \operatorname{curv}_R M = 0$.
(2) $\operatorname{pd}_R M = \infty \iff \operatorname{cx}_R M \geq 1 \iff \operatorname{curv}_R M \geq 1$.
(3) $\operatorname{cx}_R M \leq 1 \iff M$ has bounded Betti numbers.
(4) $\operatorname{cx}_R M < \infty \implies \operatorname{curv}_R M \leq 1$.
(5) $\operatorname{curv}_R M < \infty$.

With respect to change of modules, properties of complexity and curvature mirror well known properties of projective dimension.

Proposition 4.2.4. *When M is a finite R–module the following hold.*

(1) $\operatorname{cx}_R M \leq \operatorname{cx}_R k$ *and* $\operatorname{curv}_R M \leq \operatorname{curv}_R k$.
(2) *For each n there are equalities*

$$\operatorname{cx}_R M = \operatorname{cx}_R \operatorname{Syz}_n^R(M) \quad \text{and} \quad \operatorname{curv}_R M = \operatorname{curv}_R \operatorname{Syz}_n^R(M).$$

(3) *If M' and M'' are R–modules, then*

$$\operatorname{cx}_R(M' \oplus M'') = \max\{\operatorname{cx}_R M', \operatorname{cx}_R M''\};$$
$$\operatorname{curv}_R(M' \oplus M'') = \max\{\operatorname{curv}_R M', \operatorname{curv}_R M''\}.$$

(4) *If $0 \to M' \to M \to M'' \to 0$ is an exact sequence of R–modules, then*

$$\operatorname{cx}_R M \leq \operatorname{cx}_R M' + \operatorname{cx}_R M'' \text{ and } \operatorname{curv}_R M \leq \operatorname{curv}_R M' + \operatorname{curv}_R M''.$$

(5) *If a sequence $\boldsymbol{g} \subset R$ is regular on R and on M, then*

$$\operatorname{cx}_R(M/(\boldsymbol{g})M) = \operatorname{cx}_R M \quad \text{and} \quad \operatorname{curv}_R(M/(\boldsymbol{g})M) = \operatorname{curv}_R M.$$

---

[19]So called because it is the *inverse of the radius* of convergence of $\operatorname{P}_M^R(t)$.

(6) *If N is a finite R–module such that $\operatorname{Tor}_n^R(M, N) = 0$ for $n > 0$, then*

$$\max\{\operatorname{cx}_R M, \operatorname{cx}_R N\} \leq \operatorname{cx}_R(M \otimes_R N) \leq \operatorname{cx}_R M + \operatorname{cx}_R N\,;$$
$$\operatorname{curv}_R(M \otimes_R N) = \max\{\operatorname{curv}_R M, \operatorname{curv}_R N\}\,.$$

Proof. (1) comes from Proposition 4.1.9; (2), (3), and (4) are clear.

(5) By Example 1.1.1, $\operatorname{Tor}_n^R(M, R/(\boldsymbol{g})) = \operatorname{H}_n(K(\boldsymbol{g}; M))$, and the latter module vanishes for $n > 0$. Thus, the desired equalities follow from (6).

(6) Let F and G be minimal free resolutions of M and N, respectively; as $\operatorname{H}_n(F \otimes_R G) = \operatorname{Tor}_n^R(M, N) = 0$ for $n \geq 1$, the complex $F \otimes_R G$ is a free resolution of $M \otimes_R N$. It is obviously minimal, so

$$\max\{\beta_n^R(M), \beta_n^R(N)\} \leq \sum_{p+q=n} \beta_p^R(M)\beta_q^R(N) = \beta_n^R(M \otimes_R N)\,.$$

The inequalities for complexities follow. To get the equality for curvatures, rewrite the relations above in terms of Poincaré series:

$$\max\{\operatorname{P}_M^R(t), \operatorname{P}_N^R(t)\} \preccurlyeq \operatorname{P}_M^R(t)\operatorname{P}_N^R(t) = \operatorname{P}_{M \otimes_R N}^R(t)\,.$$

A product converges in the smaller of the circles of convergence of its factors, so the equality of power series yields $\operatorname{curv}_R(M \otimes_R N) \leq \max\{\operatorname{curv}_R M, \operatorname{curv}_R N\}$; the converse inequality comes from the inequality of power series. □

The finiteness of projective dimension of a module is notoriously unstable under change of rings. Complexity and curvature fare better:

Proposition 4.2.5. *Let M be a finite R–module, let $\varphi\colon R \to R'$ be a homomorphism of local rings, and set $M' = R' \otimes_R M$.*

(1) *For each prime ideal $\mathfrak{p}$ in R there are inequalities*

$$\operatorname{cx}_{R_\mathfrak{p}} M_\mathfrak{p} \leq \operatorname{cx}_R M \qquad \text{and} \qquad \operatorname{curv}_{R_\mathfrak{p}} M_\mathfrak{p} \leq \operatorname{curv}_R M\,.$$

(2) *If φ is a flat local homomorphism, then*

$$\operatorname{cx}_{R'} M' = \operatorname{cx}_R M \qquad \text{and} \qquad \operatorname{curv}_{R'} M' = \operatorname{curv}_R M\,.$$

(3) *If $\operatorname{Tor}_n^R(R', M) = 0$ for $n \gg 0$, then*

$$\operatorname{cx}_{R'} M' \leq \operatorname{cx}_R M \qquad \text{and} \qquad \operatorname{curv}_{R'} M' \leq \operatorname{curv}_R M\,.$$

(4) *If $\boldsymbol{g}$ is an R–regular sequence of length r and $R' = R/(\boldsymbol{g})$, then*

$$\operatorname{cx}_R M \leq \operatorname{cx}_{R'} M' \leq \operatorname{cx}_R M + r\,;$$
$$\operatorname{curv}_{R'} M' = \operatorname{curv}_R M \quad \text{when} \quad \operatorname{pd}_R M = \infty\,.$$

Proof. Both (1) and (2) are immediate. (3) results from Proposition 1.2.3. (4) follows from the inequalities of Poincaré series in Proposition 3.3.4. □

Finally, we note a result specific to Cohen-Macaulay rings; it fails in general, as shown by the ring $R = k[[s_1, s_2]]/(s_1^2, s_1 s_2)$ of multiplicity 1.

Proposition 4.2.6. *If R is Cohen-Macaulay, then* $\operatorname{curv}_R M \leq \operatorname{mult} R - 1$; *equality holds when R has minimal multiplicity (defined in Example 15.2.8)*

This follows from a lemma, that sharpens the result of Ramras [135].

Lemma 4.2.7. *If R is Cohen-Macaulay, $\operatorname{mult} R = l$, and $\operatorname{type} R = s$, then*

$$(l-1)\beta_n^R(M) \geq \beta_{n+1}^R(M) \geq \frac{s}{l-s}\beta_n^R(M) \quad \text{for } n > \operatorname{depth} R - \operatorname{depth} M .$$

Proof. By Remark 1.2.9, we may assume that R is artinian, with length $R = l$, and length$(0 :_R \mathfrak{m}) = s$. In a minimal resolution F of M, $\operatorname{Syz}_{n+1}^R(M) \subseteq \mathfrak{m}F_n$, so

$$(l-1)\beta_n^R(M) = \operatorname{length} \mathfrak{m}F_n \geq \operatorname{length} \operatorname{Syz}_{n+1}^R(M) \geq \beta_{n+1}^R(M) \quad \text{for } n \geq 1 .$$

As $\partial((0 :_R \mathfrak{m})F_i) \subseteq (0 :_R \mathfrak{m})\mathfrak{m}F_{i-1} = 0$, we have $(0 :_R \mathfrak{m})F_i \subseteq \operatorname{Syz}_{i+1}^R(M)$, and hence length $\operatorname{Syz}_{i+1}^R(M) \geq s\beta_i^R(M)$. The exact sequence

$$0 \to \operatorname{Syz}_{n+2}^R(M) \to F_{n+1} \to \operatorname{Syz}_{n+1}^R(M) \to 0$$

now yields $l\beta_{n+1}^R(M) = \operatorname{length} F_{n+1} \geq s\beta_{n+1}^R(M) + s\beta_n^R(M)$. $\qquad\square$

4.3. Growth problems. As of today, a distinctive feature of the state of knowledge of infinite free resolutions is that tantalizing questions on the behavior of basic invariants can be stated in very simple terms. We present four groups of interrelated problems, that set benchmarks for many results in the text. Some are (variants) of problems discussed in more detail in [27], others are new. They could be viewed in the broader context of growth of algebraic structures, to which the survey of Ufnarovskij [153] is a good introduction.

For the rest of this section, $(R, \mathfrak{m}, k)$ is a local ring, and M is a finite R–module; to avoid distracting special cases, we assume that $\operatorname{pd}_R M = \infty$.

All problems discussed below have positive answers for modules over Golod rings and over complete intersections. These two cases, for which exhaustive information is available, are treated in detail in later chapters.

Growth. The easily proved inequality over Cohen-Macaulay rings suggests

Problem 4.3.1. Is $\displaystyle\limsup_{n\to\infty} \frac{\beta_{n+1}^R(M)}{\beta_n^R(M)}$ always finite?

There is no problem when the Betti numbers of M are bounded, but little is known on when such modules occur. The only general construction that I am aware of, presented in Example 5.1.3, yields modules with periodic of period 2 minimal resolutions. Such a periodicity implies constant Betti numbers, but not conversely, cf. Remark 5.1.4. On the other hand, there exist rings over which all infinite Betti sequences are unbounded, cf. Theorem 5.3.3.

The simplest pattern of bounded Betti numbers is highlighted in

Problem 4.3.2. If a bounded sequence $\{\beta_n^R(M)\}$ eventually constant?

Problem 4.3.2, proposed in [134], is subsumed in the next one, from [23]:

Problem 4.3.3. Is the sequence $\{\beta_n^R(M)\}$ eventually non-decreasing?

An early class of artinian examples is given by Gover and Ramras [75]. Many more can be found in the papers discussed after Problem 4.3.9.

Complexity. Complexity is a measure for infinite projective dimensions, in the polynomial range of the growth spectrum. It is very interesting to know whether it satisfies an analogue of the Auslander-Buchsbaum result: depth R is an upper bound for all *finite* projective dimensions. More precisely, we propose

Problem 4.3.4. Does $\mathrm{cx}_R M < \infty$ imply $\mathrm{cx}_R M \leq \mathrm{codepth}\, R$?

Over a complete intersection the inequality holds for all modules, and all values between 0 and codepth R occur. More generally, for modules of finite CI-dimension[20] the problem has a positive solution, that comes with a bonus: if R is Cohen-Macaulay (respectively, Gorenstein), but *not* a complete intersection, then $\mathrm{cx}_R M \leq \mathrm{codepth}\, R - 2$ (respectively, $\leq \mathrm{codepth}\, R - 3$).

Finite complexity only imposes an upper bound on the Betti numbers, prescribing no asymptote. As in calculus, when $\{b_n\}$ and $\{c_n\}$ are sequences of positive real numbers, we write $b_n \sim c_n$ to denote $\lim_n b_n/c_n = 1$.

Problem 4.3.5. If $\mathrm{cx}_R M = d < \infty$, is then $\beta_n^R(M) \sim \alpha n^{d-1}$ for some $\alpha \in \mathbb{R}$?

Would Betti sequences turn up that have subpolynomial but not asymptotically polynomial growth, then complexity should be refined by the number $\limsup_n \ln\left(\beta_n^R(M)\right)/\ln(n)$, modeled on[21] *Gelfand-Kirillov dimension*, cf. [71].

Curvature. By d'Alembert's convergence criterion, $\mathrm{curv}_R M$ is contained between $\liminf_n \beta_{n+1}^R(M)/\beta_n^R(M)$ and $\limsup_n \beta_{n+1}^R(M)/\beta_n^R(M)$. To quantify the observation that 'at infinity' Betti numbers display uniform behavior, we risk

Problem 4.3.6. Is $\mathrm{curv}_R M = \lim_{n \to \infty} \dfrac{\beta_{n+1}^R(M)}{\beta_n^R(M)}$?

A positive answer would solve Problem 4.3.2, and also Problem 4.3.3 when $\mathrm{curv}_R M > 1$. If $\mathrm{curv}_R M = 1$, then the Betti sequence of M grows subexponentially; none is known to grow superpolynomially, so we propose

Problem 4.3.7. Does $\mathrm{curv}_R M = 1$ imply $\mathrm{cx}_R M < \infty$?

For k the answer is positive, cf. Theorem 8.2.1. The dichotomy implied by a general positive answer would be all the more remarkable for the fact, that intermediate growth does occur in most (even finitely presented!) algebraic systems: associative algebras, Lie algebras, and groups, cf. [153].

---

[20]That class, introduced by Avramov, Gasharov, and Peeva [32], includes the modules of finite virtual projective dimension of [25], and hence all modules over a complete intersection.

[21]This *is* a GK-dimension: that of $\mathrm{Ext}_R(M,k)$ over $\mathrm{Ext}_R(k,k)$, cf. Chapter 10.

Remark 4.2.3 and Proposition 4.2.4 show that the curvatures of all R–modules (of infinite projective dimension) lie on $[1, \operatorname{curv}_R k]$, but their actual distribution is a mystery. Obviously, the first question to ask is:

Problem 4.3.8. Is the set $\{\operatorname{curv}_R M \mid M \text{ a finite } R\text{–module}\}$ finite?

We also propose an exponential version of Problem 4.3.5:

Problem 4.3.9. If $\operatorname{curv}_R M = \beta > 1$, is then $\beta_n^R(M) \sim \alpha\beta^n$ for some $\alpha \in \mathbb{R}$?

This is proved by Sun [150] over generalized Golod rings, cf. Theorem 10.3.3.2. Intermediate results are known in many other cases. Gasharov and Peeva [70], [130], prove that if R is Cohen-Macaulay of multiplicity ≤ 8, then there is a real number $\gamma > 1$, such that $\beta_{n+1}^R(M) > \gamma\beta_n^R(M)$ for $n \gg 0$. This property is called 'termwise exponential growth' by Fan [63], who extends some methods of [70] to handle artinian rings of 'large' embedding dimension. For a Cohen presentation $\widehat{R} = Q/I$ Choi [52], [53] shows that $c = \operatorname{rank}_K \left(\mathfrak{n}(I : \mathfrak{n})/\mathfrak{n}I\right)$ is an invariant of R, and that $\beta_{n+2}^R(M) > c\beta_n^R(M)$ for $n \geq 1$.

A weaker property, called 'strongly[22] exponential growth' in [26], is established when $\mathfrak{m}^3 = 0$ by Lescot [106]: for each M there is a real number $\gamma > 1$, such that $\beta_n^R(M) \geq \gamma^n$ for $n \geq 0$. This special case is significant, as Anick and Gulliksen [13] prove that each $\mathrm{P}_k^R(t)$ is rationally related to $\mathrm{P}_k^S(t)$ for some artinian k–algebra $(S, \mathfrak{n}, k)$ with $\mathfrak{n}^3 = 0$.

Rationality. The study of infinite resolutions over local rings was triggered by a question, variously and appropriately linked to the names of Kaplansky, Kostrikin, Serre, and Shafarevich: *Does $\mathrm{P}_k^R(t)$ represent a rational function?*

Gulliksen [79] raised the stakes, proving that a positive answer for all $(R, \mathfrak{m}, k)$ would imply that $\mathrm{P}_M^R(t)$ is rational for all R and all M. Anick [10], [11] answered the original question, with a graded artinian local k–algebra[23], such that $\mathrm{P}_k^R(t)$ is transcendental, but the following is widely open:

Problem 4.3.10. Over which R do all modules have rational Poincaré series?

Anick's construction, as reworked by Löfwall and Roos [114], is described in detail by Roos [138], Babenko [39], and Ufnarovskij [153]; these surveys, and that of Anick [12], also describe a finite CW complex, whose loop space homology has a transcendental Poincaré series. Bøgvad [44] uses the artinian examples to produces a Gorenstein ring with irrational $\mathrm{P}_k^R(t)$. Fröberg, Gulliksen and Löfwall [69] show that the rationality of the Poincaré series of the residue fields is not preserved in flat families of local rings.

Jacobsson [93] provides the shortest path to irrational Poincaré series. He also shows that the rationality of $\mathrm{P}_k^R(t)$ does not imply that of all $\mathrm{P}_M^R(t)$, hence:

Problem 4.3.11. Do all rational $\mathrm{P}_M^R(t)$ over R have a common denominator?

---

[22] The terminological discrepancy reflects the growth of expectations, over a decade.
[23] Of rank 13.

A result of Levin [111], cf. Corollary 6.3.7, shows that for the last two problems it suffices to consider modules of finite length. Positive solutions are obtained over generalized Golod rings in [28], cf. Theorem 10.3.3.1.

Besides the aesthetic of the formula in 'closed form' that it embodies, a rational expression for a Poincaré series has practical applications. First, it provides a recurrent relation for Betti numbers that can be useful in constructing a minimal resolution. Second, it allows for efficient estimates of the asymptotic behavior of Betti sequences; for instance, rationality implies a positive solution to Problem 4.3.7, and yields some information on Problem 4.3.2: a bounded Betti sequence is eventually periodic, cf. e.g. [26].

5. Modules over Golod Rings

In this chapter $(R, \mathfrak{m}, k)$ is a local ring.

Golod [74] characterized those rings R over which the resolution of k has the fastest growth allowed by Proposition 4.1.4, that is, which have

$$
\mathrm{P}_k^R(t) = \frac{(1 + t)^{\mathrm{edim}\, R}}{1 - \sum_{j=1}^{\mathrm{codepth}\, R} \mathrm{rank}_k\, \mathrm{H}_j(K^R) t^{j+1}} \ . \tag{5.0.1}
$$

They are now known as *Golod rings*. We present some highlights of the abundant information available on resolutions of modules over them. It neatly splits into two pieces, corresponding to codepth $R \leq 1$ and codepth $R \geq 2$.

5.1. Hypersurfaces. A local ring with codepth $R \leq 1$ is called a *hypersurface*.

To account for the name, consider a minimal presentation $\widehat{R} \cong Q/I$. Lemma 4.1.3.2 yields $\mathrm{pd}_Q Q/I \leq 1$, so I is principal, say $I = (f)$; in particular, regular rings are hypersurfaces. If R is a hypersurface, then $\mathrm{P}_k^R(t) = (1 + t)^{\mathrm{edim}\, R}$ by Example 1.1.1 when it is regular, and $\mathrm{P}_k^R(t) = (1+t)^{\mathrm{edim}\, R}/(1-t^2)$ by Proposition 3.3.5.2 when it is singular. Thus, a hypersurface is a Golod ring.

Resolutions over hypersurface rings have a nice periodicity, discovered by Eisenbud [57]. In particular, $\mathrm{cx}_R M \leq 1$ for each R–module M. Remark 8.1.1.3 contains a strong converse: if $\mathrm{cx}_R k \leq 1$, then R is a hypersurface.

Theorem 5.1.1. *If M is a finite module over a hypersurface ring R, then $\beta_{n+1}^R(M) = \beta_n^R(M)$ for $n > m = \mathrm{depth}\, R - \mathrm{depth}\, M$. The minimal free resolution of M becomes periodic of period 2 after at most $m + 1$ steps. It is periodic if and only if M is maximal Cohen-Macaulay without free direct summand.*

Proof. By the uniqueness of the minimal free resolution F, to say that it is periodic of period 2 after s steps amounts to saying that the syzygy modules $\mathrm{Syz}_s^R(M)$ and $\mathrm{Syz}_{s+2}^R(M)$ are isomorphic. We may test the isomorphism of these finite modules after tensoring by the faithfully flat R–module $\widehat{R}$. Since $\widehat{R} \otimes_R \mathrm{Syz}_n^R(M) \cong \mathrm{Syz}_n^{\widehat{R}}(\widehat{M})$, we may assume that the ring R is complete, and hence of the form $Q/(f)$ for some regular local ring Q.

For $m = \operatorname{depth} R - \operatorname{depth} M + 1$, Proposition 1.2.8 and Corollary 1.2.5 show that $\operatorname{Syz}_m^R(M)$ is maximal Cohen-Macaulay without free direct summand; this establishes the 'only if' part of the last assertion. As Q is regular, and $\operatorname{depth} \operatorname{Syz}_m^R(M) = \operatorname{depth} R = \operatorname{depth} Q - 1$, we have $\operatorname{pd}_Q \operatorname{Syz}_m^R(M) = 1$, so the other assertions follow from the next construction.   □

Construction 5.1.2. Periodic resolutions. Let $(Q, \mathfrak{n}, k)$ be a local ring, let $f \in \mathfrak{n}$ be a regular element, and let M be a finite module over $R = Q/(f)$, with $\operatorname{pd}_Q M = 1$. By Proposition 1.2.7, its minimal resolution U over Q is of the form $0 \to U_1 \to U_0 \to 0$ with $U_1 \cong U_0 \cong Q^b$. Since $fM = 0$, the homothety $f \operatorname{id}_{U_0}$ lifts to a homomorphism $\sigma \colon U_0 \to U_1$.

Considered as a degree 1 map of complexes $\sigma \colon U \to U$, it is a homotopy between $f \operatorname{id}^U$ and 0. If $\delta \colon U_1 \to U_0$ is the differential of U, this means that $\delta\sigma = f \operatorname{id}^{U_0}$ and $\sigma\delta = f \operatorname{id}^{U_1}$. Thinking of σ and δ as $b \times b$ matrices, we get matrix equalities $\delta\sigma = fI_b = \sigma\delta$, called by Eisenbud [57] a *matrix factorization* of f. Clearly, this defines an infinite complex of free R–modules

$$F(\delta, \sigma): \quad \ldots \xrightarrow{R \otimes_Q \delta} R \otimes_Q U_0 \xrightarrow{R \otimes_Q \sigma} R \otimes_Q U_1 \xrightarrow{R \otimes_Q \delta} R \otimes_Q U_0 \to \cdots \quad (*)$$

If $1 \otimes u_1 \in \operatorname{Ker}(R \otimes \delta)$ for $u_1 \in U_1$, then $\delta(u_1) = fu_0 = \delta\sigma(u_0)$ with $u_0 \in U_0$. As δ is injective, $u_1 = \sigma(u_0)$, and thus $\operatorname{Ker}(R \otimes \delta) \subseteq \operatorname{Im}(R \otimes \sigma)$. By a symmetric argument $\operatorname{Ker}(R \otimes \sigma) \subseteq \operatorname{Im}(R \otimes \delta)$, so $F(\delta, \sigma)$ resolves $\operatorname{Coker}(R \otimes_Q \delta) = M$.

When M has a free direct summand, its minimal resolution is not periodic, so $F(\delta, \sigma)$ is not minimal. Conversely, if $F(\delta, \sigma)$ is not minimal, then $N = \operatorname{Im}(R \otimes \sigma) \not\subseteq \mathfrak{m}(R \otimes_Q U_1)$, so R^b has a basis element in N. Thus, N has a free direct summand; since $N \cong \operatorname{Coker}(R \otimes \delta) \cong M$, so does M.

Iyengar [92] notes an alternative approach to the preceding construction:

Remark. The shortness of U forces $\sigma^2 = 0$; reversing Remark 2.2.1, we define a DG module structure of U over the Koszul complex $A = Q[y \mid \partial(y) = f]$, by setting $yu = \sigma(u)$ for $u \in U$. Comparison of formulas then shows that $F(\delta, \sigma)$ coincides with the resolution $F(A, U)$ of Example 3.1.2.

Modules with periodic resolutions exist over all non-linear *hypersurface sections*: the relevant example, due to Buchweitz, Greuel, and Schreyer [48], is adapted for the present purpose by Herzog, Ulrich, and Backelin [87].

Example 5.1.3. In the notation of Construction 5.1.2, assume that $f \in \mathfrak{n}^2$, so that $f = \sum_{i=1}^e a_i s_i$, with $a_i, s_i \in \mathfrak{m}$. Let K be the Koszul complex on $s_1, \ldots, s_e$, let $x_1, \ldots, x_e$ be a basis of K_1, such that $\partial(x_i) = s_i$ for $i = 1, \ldots, e$. Left multiplication with $\sum_{i=1}^e a_i x_i \in K_1$ yields a map τ such that $\partial\tau + \tau\partial = f \operatorname{id}_K$, cf. 2.2.1. Setting $U_0 = \bigoplus_i K_{2i}$ and $U_1 = \bigoplus_i K_{2i+1}$, one then sees that

$$\delta = \sum_i (\partial_{2i+1} + \tau_{2i+1}) \colon U_1 \to U_0 \qquad \text{and} \qquad \sigma = \sum_i (\partial_{2i} + \tau_{2i}) \colon U_0 \to U_1$$

is a matrix factorization of f over Q, so the preceding remark provides a periodic of period 2 minimal resolution of the R–module $\operatorname{Coker}(R \otimes_Q \delta) = M$.

Remark 5.1.4. A ring R *satisfies the Eisenbud conjecture* if each module of complexity 1 has a resolution that is eventually periodic of period 2. This is known for various R: complete intersections (Eisenbud, [57]); with codepth $R \leq 3$, (Avramov, [26]); with codepth $R \leq 4$ that are Gorenstein ([26]) or Cohen-Macaulay almost complete intersections (Kustin and Palmer, [102]); Cohen-Macaulay of multiplicity ≤ 7, or Gorenstein of multiplicity ≤ 11, (Gasharov and Peeva, [70]); some 'determinantal' cases, ([26]; Kustin [99], [100]).

On the other hand, Gasharov and Peeva [70] introduce a series of graded rings of embedding dimension 4 and multiplicity 8, setting

$$R = k[s_1, s_2, s_3, s_4]/(as_1s_3 + s_2s_3, s_1s_4 + s_2s_4, s_3s_4, s_1^2, s_2^2, s_3^2, s_4^2)$$

for some non-zero element a in a field k. It is easy to check that

$$\cdots \to R^2(-n) \xrightarrow{\partial_n} R^2(-n+1) \to \cdots \qquad \text{with} \quad \partial_n = \begin{pmatrix} t_1 & a^n t_3 + t_4 \\ 0 & t_2 \end{pmatrix}$$

is a minimal exact complex of graded R–modules, hence $M = \operatorname{Coker} \partial_1$ has $\beta_n^R(M) = 2$ for all $n \geq 0$. It is proved in [70] that this complex is periodic of period q if and only if q is the order of a in the multiplicative group of k, so Eisenbud's conjecture fails when $q > 2$; similar examples over Gorenstein rings of embedding dimension 5 and multiplicity 12 are also given.

5.2. Golod rings.

To investigate the rings satisfying (5.0.1), Golod [74] introduced in commutative algebra certain higher order homology operations, that have become an important tool for the construction and study of resolutions.

In this section, we use the shorthand notation $\bar{a} = (-1)^{|a|+1}a$.

Remark 5.2.1. Let A be a DG algebra, with $\mathrm{H}_0(A) \cong k$. We say that A admits a *trivial Massey operation*, if for some k–basis $\boldsymbol{b} = \{h_\lambda\}_{\lambda \in \Lambda}$ of $\mathrm{H}_{\geq 1}(A)$ there exists a function $\mu \colon \bigsqcup_{i=1}^{\infty} \boldsymbol{b}^i \to A$, such that

$$\mu(h_\lambda) = z_\lambda \in \mathrm{Z}(A) \qquad \text{with} \quad \mathrm{cls}(z_\lambda) = h_\lambda \,;$$

$$\partial \mu(h_{\lambda_1}, \ldots, h_{\lambda_p}) = \sum_{j=1}^{p-1} \overline{\mu(h_{\lambda_1}, \ldots, h_{\lambda_j})} \mu(h_{\lambda_{j+1}}, \ldots, h_{\lambda_p}) \,.$$

The fact that A admits a trivial Massey operation means that the algebra structure on $\mathrm{H}(A)$ is highly trivial. For example, $z_{\lambda_1} z_{\lambda_2} = \pm \partial \mu(h_{\lambda_1}, h_{\lambda_2})$ implies that $\mathrm{H}_{\geq 1}(A) \cdot \mathrm{H}_{\geq 1}(A) = 0$. Furthermore, as

$$|\mu(h_{\lambda_1}, \ldots, h_{\lambda_p})| = |h_{\lambda_1}| + \cdots + |h_{\lambda_p}| + p - 1 \,,$$

if $\mathrm{H}_i(A) = 0$ for $i \gg 0$, then there are only finitely many obstructions – known as *Massey products* – to the construction of a trivial Massey operation.

Theorem 5.2.2. *If the Koszul complex K^R of a local ring $(R, \mathfrak{m}, k)$ admits a trivial Massey operation μ, defined on a basis $\boldsymbol{b} = \{h_\lambda\}_{\lambda \in \Lambda}$ of $\mathrm{H}_{\geq 1}(K^R)$, then*

$$G_n = \bigoplus_{p+h+i_1+\cdots+i_p=n} K_h^R \otimes_R V_{i_1} \otimes \cdots \otimes_R V_{i_p} \,,$$

where each V_n is an R–module with basis $\{v_\lambda : n = |v_\lambda| = |h_\lambda| + 1\}_{\lambda \in \Lambda}$, and

$$\partial(a \otimes v_{\lambda_1} \otimes \cdots \otimes v_{\lambda_p}) = \partial(a) \otimes v_{\lambda_1} \otimes \cdots \otimes v_{\lambda_p}$$
$$+ (-1)^{|a|} \sum_{j=1}^{p} a\mu(h_{\lambda_1}, \ldots, h_{\lambda_j}) \otimes v_{\lambda_{j+1}} \otimes \cdots \otimes v_{\lambda_p}$$

is a minimal free resolution (G, ∂) of k. In particular, the ring R is Golod.

Remark 5.2.3. The theorem contains one direction of Golod's result: in [74] he also shows that, conversely, the equality of Poincaré series implies that each basis of $H(K^R)$ has a trivial Massey operation, cf. also [83]. The proof given below is taken from Levin [110], and uses an idea of Ghione and Gulliksen [72].

The condition for R to be Golod means that that the differentials ${}^r d_{p,q}$ of the spectral sequence 3.2.4 vanish for $r \geq 1$. Thus, the theorem is 'explained' by Guggenheim and May's description [76] of the differentials of Eilenberg-Moore spectral sequences in terms of certain *matric Massey products*, introduced by May [118]; for proofs using that approach, cf. [18], [24].

Proof. The verification that $\partial^2 = 0$ is direct, using the formulas in Remark 5.2.1. To see that G is exact, note that $K = K^R$ is naturally a subcomplex of G, and $G/K \cong G \otimes_R V$ with $\partial(g \otimes v) = \partial(g) \otimes v$. The exact sequence

$$0 \to K \to G \to G \otimes_R V \to 0$$

of complexes of R–modules has a homology exact sequence

$$\cdots \to \bigoplus_j \left(H_{n-j}(G) \otimes_R V_{j+1} \right) \xrightarrow{\eth_{n+1}} H_n(K) \to H_n(G)$$
$$\to \bigoplus_j \left(H_{n-j}(G) \otimes_R V_j \right) \xrightarrow{\eth_n} H_{n-1}(K) \to \cdots$$

It is clear that $H_0(G) = k$, and that $\eth_{n+1}(1 \otimes v_\lambda) = h_\lambda$. Thus, $\eth_{n+1}$ is surjective, and the homology sequence splits. In the exact sequence

$$0 \to H_1(G) \to H_0(K) \otimes_R V_2 \xrightarrow{\eth_1} H_1(K) \to 0$$

we have $H_0(K) \otimes_R V_2 \cong H_1(K)$, hence $H_1(G) = 0$. Working backwards from here, we see that $H_n(G) = 0$ for $n \geq 1$.

We induce on p to prove that $\mu(h_{\lambda_1}, \ldots, h_{\lambda_p}) \in \mathfrak{m}K^R$ for all sequences $h_{\lambda_1}, \ldots, h_{\lambda_p}$. If $p = 1$, then $\mu(h_\lambda) \subseteq Z(K^R)$, and $Z(K^R) \subseteq \mathfrak{m}K^R$ by Lemma 4.1.6.2. If $p > 1$, then the definition and the induction hypothesis imply $\partial\mu(h_{\lambda_1}, \ldots, h_{\lambda_p}) \in \mathfrak{m}^2K^R$, so $\mu(h_{\lambda_1}, \ldots, h_{\lambda_p}) \in \mathfrak{m}K^R$, again by *loc. cit.*

Thus, G is a minimal resolution of k. A computation identical with the one for the upper bound in the proof of Proposition 3.3.2 shows that $\sum_n \mathrm{rank}_R G_n t^n$ is given by the right hand side of (5.0.1), so R is Golod. $\qquad\square$

The following conditions often suffice to recognize a Golod ring,

Proposition 5.2.4. *A local ring R is Golod if some of the following hold:*

(1) $H_{\geqslant 1}(K^R)$ *is generated by a set of cycles Z, such that $Z^2 = 0$.*

(2) $R/(g)$ *is Golod for a regular element $g \in \mathfrak{m} \smallsetminus \mathfrak{m}^2$.*

(3) $R \cong Q/(f)$ *for a Golod ring $(Q, \mathfrak{n}, k)$ and a regular element $f \in \mathfrak{n} \smallsetminus \mathfrak{n}^2$.*

(4) $\widehat{R} \cong Q/I$, *where $(Q, \mathfrak{n}, k)$ is regular, $\mathrm{edim}\, Q = \mathrm{edim}\, R$, and the minimal resolution A of $\widehat{R}$ over Q is a DG algebra with $A_{\geqslant 1} \cdot A_{\geqslant 1} \subseteq \mathfrak{n} A$.*

Proof. (1) Select a subset $\{z_\lambda \in Z\}_{\lambda \in \Lambda}$ such that $\boldsymbol{b} = \{\mathrm{cls}(z_\lambda)\}_{\lambda \in \Lambda}$ is a basis of $H_{\geqslant 1}(K^R)$. A trivial Massey operation can then be defined by setting $\mu(h_{\lambda_1}, \ldots, h_{\lambda_j}) = 0$ for $j \geq 2$, so R is Golod by Theorem 5.2.2.

(2) and (3). We note that: $\mathrm{edim}\, R = \mathrm{edim}(R/(g)) - 1$ (obvious); $\mathrm{P}_k^R(t) = (1+t)\,\mathrm{P}_k^{R/(g)}(t)$ (Proposition 3.3.5.1); $\mathrm{rank}_k\, \mathrm{H}_n(K^R) = \mathrm{rank}_k\, \mathrm{H}_n(K^{R/(g)})$ for each n (Lemma 4.1.6.1). Putting these equalities together, we see that R satisfies the defining equality (5.0.1) if and only if $R/(g)$ does.

(4) The Golod conditions for R and $\widehat{R}$ are equivalent, so we may assume that Q/I is a regular presentation of R. By Theorem 5.2.2, it suffices to show that K^R admits a trivial Massey operation.

Choose a set of cycles $\{z_\lambda \in A \otimes_Q K^Q\}_{\lambda \in \Lambda}$, such that $\boldsymbol{b} = \{h_\lambda = \mathrm{cls}(z_\lambda)\}_{\lambda \in \Lambda}$ is a basis of $H_{\geqslant 1}(A \otimes_Q K^Q)$. Set $\mu(h_\lambda) = z_\lambda$, and assume by induction that a function $\mu \colon \bigsqcup_{i=1}^{p-1} \boldsymbol{b}^i \to A \otimes_Q K^Q$ has been constructed for some $p \geq 2$, and satisfies the conditions of Remark 5.2.1. The element

$$z_{\lambda_1, \ldots, \lambda_p} = \sum_{j=1}^{p-1} \overline{\mu(h_{\lambda_1}, \ldots, h_{\lambda_j})} \mu(h_{\lambda_{j+1}}, \ldots, h_{\lambda_p})$$

is then a cycle in $A \otimes_Q K^Q$. If $\epsilon \colon K^Q \xrightarrow{\simeq} k$ is the augmentation, then

$$(A \otimes_Q \epsilon)(z_{\lambda_1, \ldots, \lambda_p}) \in (A \otimes_Q k)_{\geqslant 1} \cdot (A \otimes_Q k)_{\geqslant 1} = 0 \,.$$

Since $(A \otimes_Q \epsilon)$ is a quasi-isomorphism, $z_{\lambda_1, \ldots, \lambda_p}$ is a boundary, so extend μ to $\bigsqcup_{i=1}^{p} \boldsymbol{b}^i$ by choosing $\mu(h_{\lambda_1}, \ldots, h_{\lambda_p})$ such that $\partial(\mu(h_{\lambda_1}, \ldots, h_{\lambda_p})) = z_{\lambda_1, \ldots, \lambda_p}$. This completes the inductive construction of a trivial Massey operation on $A \otimes_Q K^Q$. The augmentation $A \to R$ induces a surjective quasi-isomorphism $\phi \colon A \otimes_Q K^Q \to K^R$, so to get a trivial Massey operation on the basis $H(\phi)(\boldsymbol{b})$ of $H_{\geqslant 1}(K^R)$, set $\mu'(\phi(h_{\lambda_1}), \ldots, \phi(h_{\lambda_p})) = \phi(\mu(h_{\lambda_1}, \ldots, h_{\lambda_p}))$. $\qquad\square$

As a first application, we present a result of Shamash [142].

Proposition 5.2.5. *If $\mathrm{codim}\, R \leq 1$, then R is Golod.*

Proof. We may assume that R is a complete, and so has a minimal regular presentation $R \cong Q/I$. As Q is a catenary domain, height $I \leq 1$; since Q is factorial, there is an $f \in \mathfrak{n}$ such that $I = fJ$. In $K^R = R \otimes_Q K^Q$ each cycle of degree ≥ 1 has the form $1 \otimes z$, where $z \in K^Q$ satisfies $\partial(z) = fv$ for some $v \in K^Q$. Also, $f\partial(v) = \partial(fv) = \partial^2(z) = 0$, and so $\partial(v) = 0$. Choosing $u \in K^Q$ such that

$\partial(u) = f$, we get $\partial(z - uv) = 0$. As K_1^Q is acyclic, $z - uv = \partial(w)$ for some $w \in K^Q$. Thus, $\mathrm{cls}(1 \otimes z) = \mathrm{cls}(1 \otimes u)\,\mathrm{cls}(1 \otimes v)$, so $\mathrm{H}_{\geqslant 1}(K^R)$ is generated by $(1 \otimes u)\mathrm{Z}(K^R)$. As $u^2 = 0$, case (1) of Proposition 5.2.4 applies.   □

Remark. It is not clear how the Golod property of a local ring relates to other characteristics of its singularity. For one thing, it does not fit in the hierarchy

$$\text{regular} \implies \text{complete intersection} \implies \text{Gorenstein} \implies \text{Cohen-Macaulay}.$$

The only Golod rings that are Gorenstein are the hypersurfaces: compare Remark 5.2.3 with the fact that if R is Gorenstein, then $\mathrm{H}(K^R)$ has Poincaré duality. On the other hand, a Golod ring may or may not be Cohen-Macaulay: compare the preceding proposition with Example 5.2.8. Furthermore, the Golod condition is not stable under localization, as demonstrated by the next example.

Example 5.2.6. Let Q be the regular ring $k[s_1, s_2, s_3]_{(s_1, s_2, s_3)}$. The ring $R = Q/(s_1^2 s_3, s_2^2 s_3)$ is Golod by Proposition 5.2.5. On the other hand, its localization at $\mathfrak{p} = (s_1, s_2)$ is the ring $\ell[s_1, s_2]/(s_1^2, s_2^2)$, where $\ell = k(s_3)$. By example 4.2.1.2, $\mathrm{P}_\ell^{R_{\mathfrak{p}}}(t) = 1/(1 - t)^2 = (1 + t)^2/(1 - 2t^2 + t^4)$, so $R_{\mathfrak{p}}$ is not Golod.

Golod rings are defined by an extremal property; this might be why they appear frequently as solutions to extremal problems.

Example 5.2.7. Let Q be a graded polynomial ring generated by Q_1 over k, and let S be a residue ring of Q, such that $\mathrm{rank}_k S_1 = \mathrm{rank}_k Q_1 = e$. A (now) famous theorem of Macaulay proves that there exists a *lex-segment* monomial ideal I, such that $R = Q/I$ and S $\mathrm{rank}_k S_n = \mathrm{rank}_k R_n$ for all n, and $\beta_{1j}^Q(S) \leq \beta_{1j}^Q(R)$ for the first graded Betti number, cf Remark 1.2.10.

Bigatti [43] and Hulett [90] in characteristic zero, and Pardue [128] in general, extend Macaulay's theorem to a coefficientwise inequality of Poincaré series in *two variables*

$$\mathrm{P}_S^Q(t, u) = \sum_{n,j} \beta_{nj}^Q(S)t^n u^j \preccurlyeq \sum_{n,j} \beta_{nj}^Q(R)t^n u^j = \mathrm{P}_R^Q(t, u).$$

As I is a *stable* monomial ideal, $\mathrm{P}_R^Q(t, u)$ is known explicitly from the minimal free resolution A of R over Q, given by Eliahou and Kervaire [60].

Peeva [129] constructs a DG algebra structure on A, such that $A_i A_j \subseteq \mathfrak{n}A_{i+j}$ when $i \geq 1$ and $j \geq 1$. By Proposition 5.2.4.4, this implies that R is Golod. The same conclusion is obtained by Aramova and Herzog [14], who verify that condition 5.2.4.1 holds. Thus, there are (in)equalities

$$\mathrm{P}_k^S(t, u) \preccurlyeq \frac{(1 + tu)^e}{1 + t - t\,\mathrm{P}_S^Q(t, u)} \preccurlyeq \frac{(1 + tu)^e}{1 + t - t\,\mathrm{P}_R^Q(t, u)} = \mathrm{P}_k^R(t, u).$$

Example 5.2.8. Let R be a Cohen-Macaulay ring of dimension d.

As the residue field of $R' = R[t]_{\mathfrak{m}[t]}$ is infinite, we can choose a regular sequence $\mathbf{g} = \{g_1, \ldots, g_d\}$ such that the length of $R'' = R'/(\mathbf{g})$ is equal to $\mathrm{mult}\,R$, the multiplicity of R, cf. Remark 1.2.9. Such a sequence is linearly independent modulo $\mathfrak{m}^2$, hence $\operatorname{edim} R'' = \operatorname{edim} R' - d = \operatorname{codim} R' = \operatorname{codim} R$. Thus, $1 + \operatorname{edim} R'' = 1 + \operatorname{length}(\mathfrak{m}''/\mathfrak{m}''^2) = \operatorname{length}(R''/\mathfrak{m}''^2) \leq \operatorname{length} R''$. This translates to an inequality $\operatorname{codim} R \leq \operatorname{mult} R - 1$, noted by Abhyankar [1].

If equality holds, then R is said to have *minimal multiplicity*. Such a ring is Golod. Indeed, it suffices to prove that for R'. Proposition 5.2.4.2 reduces the problem to R''. As $Z(K^{R''}) \subseteq \mathfrak{m}'' K^{R''}$ by Lemma 4.1.6.2 and $\mathfrak{m}''^2 = 0$, Proposition 5.2.4.1 shows that R'' is Golod. Another look at R'' shows that $\operatorname{mult} R = \operatorname{type} R + 1$, so for each M with $\operatorname{pd}_R M = \infty$ Lemma 4.2.7 yields

$$\mathrm{P}_M^R(t) = f(t) + \frac{at^{m+1}}{1 - (\operatorname{mult} R - 1)t}$$

with a positive $a \in \mathbb{Z}$, and $f(t) \in \mathbb{Z}[t]$ of degree $m = \dim R - \operatorname{depth} M$.

5.3. Golod modules. A *Golod module* is a finite R–module M, whose Poincaré series reaches the upper bound in the inequality of Proposition 4.1.4:

$$\mathrm{P}_M^R(t) = \frac{\sum_{i \geqslant 0} \operatorname{rank}_k \mathrm{H}_i(K^M) t^i}{1 - \sum_{j \geqslant 1} \operatorname{rank}_k \mathrm{H}_j(K^R) t^{j+1}} . \tag{5.3.1}$$

Thus, R is a Golod ring if and only if k is a Golod module. A result of Lescot [107] establishes a tight connection between Golod conditions on a ring and on its modules; part (1) is independently due to Levin [111].

Theorem 5.3.2. (1) *If R has a Golod module $M \neq 0$, then R is a Golod ring.*

(2) *If R is a Golod ring, and M is a finite R–module, then the module $\operatorname{Syz}_n^R(M)$ is Golod for $n \geq p = \operatorname{edim} R - \operatorname{depth} M$.*

Proof. (1) Referring successively to the definition, Proposition 3.3.3, and the inequality at the beginning of this section, we get a sequence of (in)equalities

$$\frac{\sum_{i \geqslant 0} \operatorname{rank}_k \mathrm{H}_i(K^M) t^i}{1 - \sum_{j \geqslant 1} \operatorname{rank}_k \mathrm{H}_j(K^R) t^{j+1}} \cdot \mathrm{P}_k^Q(t) = \mathrm{P}_M^R(t) \cdot \mathrm{P}_k^Q(t)$$

$$\preccurlyeq \mathrm{P}_M^Q(t) \cdot \mathrm{P}_k^R(t) \preccurlyeq \mathrm{P}_M^Q(t) \cdot \frac{\sum_{i \geqslant 0} \operatorname{rank}_k \mathrm{H}_i(K^R) t^i}{1 - \sum_{j \geqslant 1} \operatorname{rank}_k \mathrm{H}_j(K^R) t^{j+1}} .$$

Lemma 4.1.3 shows that the expressions at the ends are equal, so equalities hold throughout. Cancelling $\mathrm{P}_M^Q(t)$ from both sides of the last equality, we see that $\mathrm{P}_k^R(t)$ satisfies the defining formula for Golod rings.

(2) Set $M_i = \operatorname{Syz}_i^R(M)$. Tensoring the short exact sequence of complexes from the proof of Theorem 5.2.2 with M_i, we get exact sequences

$$0 \to K^{M_i} \to M_i \otimes_R G \to M_i \otimes_R (G \otimes_R V) \to 0 .$$

Setting $q = \operatorname{codepth} R$, note that

$$\mathrm{H}_n(M_i \otimes_R G) = \operatorname{Tor}_n^R(M_i, k) ;$$

$$\mathrm{H}_n((M_i \otimes_R G) \otimes_R V) = \bigoplus_{j=1}^{q} \operatorname{Tor}_{n-j-1}^R(M_i, k) \otimes_k \mathrm{H}_j(K^R) .$$

The short exact sequences produce commutative diagrams with exact rows

$$
\begin{array}{ccccc}
\operatorname{Tor}^R_{n+1}(M_i,k) & \xrightarrow{\iota^i_{n+1}} & \displaystyle\bigoplus_{j=1}^{q}\operatorname{Tor}^R_{n-j}(M_i,k)\otimes \mathrm{H}_j(K^R) & \longrightarrow & \mathrm{H}_n(K^{M_i}) \\
\big\downarrow{\scriptstyle\eth^i_n} & & \big\downarrow{\scriptstyle\oplus_j(\eth^i_j\otimes\mathrm{id})} & & \\
\operatorname{Tor}^R_n(M_{i+1},k) & \xrightarrow{\iota^{i+1}_n} & \displaystyle\bigoplus_{j=1}^{p}\operatorname{Tor}^R_{n-j-1}(M_{i+1},k)\otimes \mathrm{H}_j(K^R) & \longrightarrow & \mathrm{H}_{n-1}(K^{M_{i+1}})
\end{array}
$$

where the connecting homomorphisms $\eth^i_n$ come from the exact sequences

$$0 \to M_{i+1} \to R^{\beta_i} \to M_i \to 0\,.$$

Note that $\mathrm{H}_s(K^M) = 0$ for $s > p$, so ι^0_s is surjective for $s > p+1$. Assume by induction that ι^i_s is surjective for $s > p - i + 1$. Since $\eth^i_j$ is surjective for all j, both vertical arrows are onto, so the diagram shows that ι^{i+1}_s is surjective for $s > p - i$. We conclude that ι^p_s is surjective for $s > 1$. Since the complex $G \otimes_R V$ starts in degree 2, the map ι^p_1 is surjective as well. Thus,

$$\operatorname{Tor}^R_n(M_p,k) \cong \mathrm{H}_n(K^{M_p}) \oplus \bigoplus_{j=1}^{q}\operatorname{Tor}^R_{n-j-1}(M_p,k)\otimes_k \mathrm{H}_j(K^R) \qquad \text{for} \quad n \ge 0$$

and hence

$$\beta^R_n(M_p) = \operatorname{rank}_k \mathrm{H}_n(K^M) + \sum_{j=1}^{q}\beta^R_{n-j-1}(M_p)\operatorname{rank}_k \mathrm{H}_j(K^R) \qquad \text{for} \quad n \ge 0\,.$$

These numerical equalities add up to the defining equation (5.3.1).   □

Betti numbers of modules over Golod rings are well documented in the next theorem. It collects results from work of several authors: Ghione and Gulliksen [72] establish the rational expression in (1); the bound on its numerator, and a proof that $\beta^R_{n+1}(M) > \beta^R_n(M)$ for $n \ge \operatorname{edim} R$, are due to Lescot [107]; the exponential growth result in (5) is established by Peeva [130], while the asymptotic formula in (3) is a consequence of the result of Sun [150].

Hypersurfaces are excluded from the statement of the theorem, as they have been dealt with in Section 5.1.

Theorem 5.3.3. *Let R be a Golod ring with* $\operatorname{codepth} R = q \ge 2$, *and let M be a finite R-module with* $\operatorname{edim} R - \operatorname{depth} M = p$. *If* $\operatorname{pd}_R M = \infty$, *then:*

(1) *There exist polynomials with positive integer coefficients, $p(t)$ of degree $p-1$ and $q(t)$ of degree $\le q$, such that*

$$\mathrm{P}^R_M(t) = p(t) + \frac{q(t)}{1 - \sum_{j=1}^{q}\operatorname{rank}_k \mathrm{H}_j(K^R)t^{j+1}} \cdot t^p\,.$$

(2) $\operatorname{cx}_R M = \infty$ *and* $\operatorname{curv}_R M = \beta > 1\,.$

(3) $\beta_n^R(M) \sim \alpha\beta^n$ *for some real number* $\alpha > 0$.

(4) $\lim\limits_{n\to\infty} \dfrac{\beta_{n+1}^R(M)}{\beta_n^R(M)} = \beta$.

(5) $\dfrac{\beta_{n+1}^R(M)}{\beta_n^R(M)} \geq \min\left\{ \left.\dfrac{\beta_{i+1}^R(M)}{\beta_i^R(M)} \right| p \leq i < p+q \right\} = \gamma > 1$ *for* $n \geq p$.

Proof. The module $M_p = \mathrm{Syz}_p^R(M)$ is Golod by Theorem 5.3.2.2, and has

$$p = \mathrm{edim}\,R - \mathrm{depth}\,M_p = \mathrm{depth}\,R - \mathrm{depth}\,M_p + \mathrm{codepth}\,R \geq q + 2.$$

Proposition 1.2.8 shows that its depth is equal to depth R. Replacing M by M_p, we see that it suffices to prove the theorem when M is a Golod module, and $p = 0$. For simplicity, we write β_n instead of $\beta_n^R(M)$.

(1) then holds by definition.

(5) Let $\widehat{R} = Q/I$ be a minimal Cohen presentation. With $r_i = \mathrm{rank}_Q\,\mathrm{Syz}_i^Q(\widehat{M})$ and $s_j = \mathrm{rank}_Q\,\mathrm{Syz}_j^Q(\widehat{R})$, we have

$$\mathrm{P}_M^R(t) = \frac{\sum_i \mathrm{rank}_k\,\mathrm{H}_i(K^M)t^i}{1 - \sum_j \mathrm{rank}_k\,\mathrm{H}_j(K^R)t^{j+1}} = \frac{\mathrm{P}_{\widehat{M}}^Q(t)}{1 + t - t\,\mathrm{P}_{\widehat{R}}^Q(t)} = \frac{\sum_{i=1}^p r_i t^{i-1}}{1 - t\sum_{j=1}^q s_j t^{j-1}}$$

where the first equality holds because M is Golod, the second by Lemma 4.1.3.1, and the third by Proposition 3.3.5.2. This yields numerical relations

$$\beta_{n+1} = \beta_n + s_1\beta_{n-1} + \cdots + s_{q-1}\beta_{n-q+1} + r_{n+1} \quad \text{for} \quad n \geq 0.$$

Since $s_1 \geq 1$ because R is not a hypersurface, we get $\beta_{n+1} > \beta_n$ for $n \geq 0$. Thus, $\min\{\beta_{i+1}^R(M)/\beta_i^R(M) \mid 0 \leq i < q\} = \gamma > 1$, so assume by induction that $n \geq q-1$ and $\beta_{i+1}/\beta_i \geq \gamma$ for $i \leq n$. As $r_{n+1} = 0$ for $n \geq q - 1$, we have

$$\beta_{n+1} = \beta_n + s_1\beta_{n-1} + \cdots + s_{q-1}\beta_{n-q+1}$$
$$\geq \gamma\beta_{n-1} + s_1\gamma\beta_{n-2} + \cdots + s_{q-1}\gamma\beta_{n-q} = \gamma\beta_n.$$

(2) is a direct consequence of (5).

(3) Let ρ be the radius of convergence of $\mathrm{P}_M^R(t)$. Since this is a series with positive coefficients, ρ is a root of the polynomial $g(t) = 1 - \sum_{j=1}^q s_j t^j$. Because $g(0) = 1 > 0$ and $g(1) = 1 - \sum \mathrm{rank}_Q\,\mathrm{Syz}_{j+1}^Q(R) < 0$, we have $\rho < 1$. As $g'(\rho) = -\sum_{j=1}^q js_j\rho^{j-1} < 0$, the root ρ is simple. Let $\xi_1, \ldots, \xi_m$ be the remaining roots of $g(t)$; assuming that one of them is equal to $\zeta\rho$ for some $\zeta \in \mathbb{C}$ with $|\zeta| = 1$ and $\zeta \neq 1$, we get

$$1 = \left| \sum_{j=1}^q s_j(\zeta\rho)^j \right| < \sum_{j=1}^q \left| s_j(\zeta\rho)^j \right| = \sum_{j=1}^q s_j\rho^j = 1$$

which is absurd. Thus, $\rho < |\xi_i|$ for $i = 1, \ldots, m$. Set $\xi_0 = \rho$, and write $\mathrm{P}_M^R(t)$ as a sum of prime fractions $\alpha_{ih}/(1 - \xi_i^{-1}t)^h$ with $\alpha_{ih} \in \mathbb{C}$ for $i = 0, \ldots, m$. Expanding each one of them by the binomial formula, we see that

$$\beta_n = \alpha\rho^{-n} + \sum_{i=1}^{m}\sum_{h=1}^{n_i} \alpha_{ih}\binom{h+n-1}{h-1}\xi_i^{-n} \qquad \text{with} \quad \alpha \neq 0.$$

Thus, $\alpha = \lim_{n\to\infty} \beta_n\rho^n > 0$; as $\rho^{-1} = \text{curv}_R M = \beta$, this is what we want.
(4) is a direct consequence of (3). $\square$

The next result is due to Scheja [143]; the proof is from [21].

Proposition 5.3.4. *Let $e = \text{edim}\,R$ and $r = \text{rank}_k\,\mathrm{H}_1(K^R)$. If $\text{codepth}\,R = 2$, then either R is Golod, and then $\mathrm{P}_k^R(t) = (1+t)^{e-1}/(1-t-(r-1)t^2)$, or R is a complete intersection, and then $\mathrm{P}_k^R(t) = (1+t)^{e-2}/(1-2t+t^2)$.*

Proof. Completing if necessary, we may assume that $R \cong Q/I$, with $(Q, \mathfrak{n}, k)$ a regular local ring, and $I \subseteq \mathfrak{n}^2$. The Auslander-Buchsbaum Equality yields $\text{pd}_Q\,R = \text{depth}\,Q - \text{depth}\,R = \text{codepth}\,R = 2$. By Example 2.1.2, the minimal free resolution A of R over Q is a DG algebra, such that $A_1A_1 \subseteq \mathfrak{n}A_2$, unless I is generated by a regular sequence. In the latter case R is a complete intersection by definition; in the former, it is Golod by Proposition 5.2.4.4. The Poincaré series come from Proposition 3.3.5.2 and formula (5.0.1), respectively. $\square$

We complement the discussion in Section 1 with Iyengar's [92] construction of minimal resolutions over *Golod* rings of codepth 2; the very different case of complete intersection is treated by Avramov and Buchweitz [31].

Example 5.3.5. Let $R = Q/I$ be a Golod ring with codepth $R = 2$. If $\text{edim}\,R - \text{depth}\,M = p$, then $N = \text{Syz}_p^R(M)$ has $\text{pd}_Q\,N = 2$ by Proposition 1.2.8 and Lemma 1.2.6. By Proposition 2.2.5, the minimal free resolution U of N over Q is a DG module over the DG algebra A of Example 2.1.2. If $F(A, U) - F$ is the resolution of N over R given by Theorem 3.1.1, then

$$\sum_{n=0}^{\infty} \text{rank}_R\,F_n t^n = \frac{\mathrm{P}_N^Q(t)}{1+t-t\,\mathrm{P}_R^Q(t)} = \mathrm{P}_N^R(t)$$

because the R–module N is Golod by Theorem 5.3.2.2. Thus, F is minimal.

6. Tate Resolutions

A process of killing cycles of odd degree by adjunction of divided powers – rather than polynomial – variables, was systematically used by H. Cartan in his spectacular computation [50] of the homology of Eilenberg-MacLane spaces [56]. Their potential for commutative algebra was realized by Tate[24] [151].

Section 1 presents most of Tate's paper. Section 3 contains a major theorem of Gulliksen [77] and Schoeller [139]; it is proved essentially by Gulliksen's arguments,

---

[24]From the introduction of [151]: 'Our "adjunction of variables" is a naïve approach to the exterior algebras and twisted polynomial rings familiar to topologists, and the ideas involved were clarified in my mind by conversations with John Moore.'

but in a new framework motivated by Quillen's construction of the cotangent complex in characteristic 0 and developed in Section 2.

6.1. Construction. Let A be a DG algebra, and let $z \in A$ be a cycle.

Construction 6.1.1. Divided powers variable. The $\Bbbk$–algebra $\Bbbk\langle x \rangle$ on a *divided powers variable* x of positive even degree is the free $\Bbbk$–module with basis $\{x^{(i)} : |x^{(i)}| = i|x|\}_{i \geqslant 0}$ and multiplication table

$$x^{(i)}x^{(j)} = \binom{i+j}{i} x^{(i+j)} \quad \text{for} \quad i, j \geq 0;$$

it is customary to set $x^{(1)} = x$, $x^{(0)} = 1$, and $x^{(i)} = 0$ for $i < 0$.

Set $A\langle x \rangle^\natural = A^\natural \otimes_\Bbbk \Bbbk\langle x \rangle$. If $z \in A$ is cycle of *positive odd* degree, then

$$\partial\left(\sum_i a_i x^{(i)}\right) = \sum_i \partial(a_i)x^{(i)} + \sum_i (-1)^{|a_i|} a_i z x^{(i-1)}$$

is a differential on $A\langle x \rangle$, that extends that of A, and satisfies the Leibniz rule.

For uniformity of notation, when $|x|$ is odd $A\langle x \,|\, \partial(x) = z \rangle$ stands for the algebra $A[x \,|\, \partial(x) = z]$, described in Construction 2.1.7; when $|x| = 0$ we set $A\langle x \,|\, \partial(x) = 0 \rangle = A[x \,|\, \partial(x) = 0]$, cf. Construction 2.1.8.

Example 6.1.2. Let B be a strictly graded commutative algebra. An element $u \in B_d$ is *regular* on B if it is not invertible, and has the smallest possible annihilator: When d is even this means that $(0 :_B u) = 0$, the usual concept for commutative rings; when d is odd this means that $(0 :_B u) = Bu$, because $u^2 = 0$ implies $uB \subseteq (0 :_B u)$.

It is easy to see that u is regular if and only if $\pi \colon B\langle x \,|\, \partial(x) = u \rangle \to B/(u)$, $\pi\left(\sum b_i x^{(i)}\right) = b_0 + (u)$, is a quasi-isomorphism. Indeed, this is classical if d is even. When d is odd, $z = \sum b_i x^{(i)}$ is a cycle precisely when $ub_i = 0$ for $i > 0$; if u is regular, then $b_i = a_i u$ for each i, so $z = b_0 + \partial\left(\sum_i (-1)^{|a_i|} a_i x^{(i+1)}\right)$; else, there is a $v \notin (u)$ such that $uv = 0$, hence $\mathrm{Ker}\,\mathrm{H}(\pi) \ni \mathrm{cls}(ux) \neq 0$.

A *semi-free Γ-extension* $A\langle X \rangle$ of A is a DG algebra obtained by iterated (possibly, transfinite) sequence of adjunctions of the three types of variables described above; we say that the elements of X are *Γ-variables* over A.

Remark 6.1.3. Let $A \hookrightarrow A\langle X \rangle$ be a semi-free Γ-extension, with $\partial(x_\lambda) = z_\lambda \in A$. Consider the semi-free extension $A[Y \,|\, \partial(y_\lambda) = z_\lambda]$, and let $\alpha \colon A[Y] \to A\langle X \rangle$ be the morphism of DG algebras defined by $\alpha(y_\lambda) = x_\lambda$ for each λ. A simple induction yields $\alpha(y_\lambda^i) = (i!)x_\lambda^{(i)}$. Thus, if $n!$ is invertible in A, then

$$\beta\left(x_{\lambda_1}^{(i_1)} \cdots x_{\lambda_q}^{(i_q)}\right) = \frac{y_{\lambda_1}^{i_1} \cdots y_{\lambda_q}^{i_q}}{(i_1)! \cdots (i_q)!} \quad \text{for} \quad i_1 + \cdots + i_q \leq n$$

defines a morphism of complexes $(A\langle X \rangle)_{\leqslant n} \to (A[Y])_{\leqslant n}$, inverse to $\alpha_{\leqslant n}$. When A_0 is a $\mathbb{Q}$-algebra, α and β are inverse isomorphisms of DG algebras.

Let $\varphi\colon R \to S$ be a homomorphism of commutative rings. A factorization of φ through a semi-free Γ-extension $R \hookrightarrow R\langle X\rangle$ followed by a surjective quasi-isomorphism $R\langle X\rangle \to S$ is called a *Tate resolution* of S over R (or, of the R-algebra S). Revisiting the proof of Proposition 2.1.10, this time killing classes of odd degree by adjunction of divided powers variables, one gets

Proposition 6.1.4. *Each R-algebra S has a Tate resolution. If R is noetherian and S is a finitely generated R-algebra (and a residue ring of R), then such a resolution exists with all X_i finite (and $X_0 = \varnothing$).* $\square$

We take a close look at the effect that an adjunctions of Γ-variables has on homology. If $\iota\colon A \hookrightarrow A\langle X\rangle$ is a semi-free Γ-extension, and $\partial(X) \subseteq A$, then $W = \{\mathrm{cls}(\partial(x)) \in \mathrm{H}(A) \mid x \in X\}$ is contained in the kernel of the algebra homomorphism $\mathrm{H}(\iota)\colon \mathrm{H}(A) \to \mathrm{H}(A\langle X\rangle)$, hence there is an induced homomorphism of graded algebras $\bar\iota\colon \mathrm{H}(A)/(W)\,\mathrm{H}(A) \to \mathrm{H}(A\langle X\rangle)$.

Let $A \hookrightarrow A\langle x \mid \partial(x) = z\rangle$ be an extension with $|z| = d$ and $w = \mathrm{cls}(z)$.

Remark 6.1.5. When $d \geq 0$ is even, there is an exact sequence of chain maps

$$0 \to A \xrightarrow{\iota} A\langle x\rangle \xrightarrow{\vartheta} A \to 0 \qquad \text{where} \qquad \vartheta(a + xb) = b\,,$$

has degree $-d - 1$, and the homology exact sequence is

$$\cdots \to \mathrm{H}_{n-d}(A) \xrightarrow{w} \mathrm{H}_n(A) \xrightarrow{\mathrm{H}_n(\iota)} \mathrm{H}_n(A\langle x\rangle) \xrightarrow{\mathrm{H}_n(\vartheta)} \mathrm{H}_{n-d-1}(A) \to \cdots\,.$$

In a special case, it appears in most textbooks on commutative algebra: $A = K(\boldsymbol{f}; Q)$ is the Koszul complex on $\boldsymbol{f}$ and $A\langle x\rangle = K(\boldsymbol{f}, g; Q)$ is that on $\boldsymbol{f} \cup \{g\}$.

Remark 6.1.6. When $d > 0$ is odd, there is an exact sequence of chain maps

$$0 \to A \xrightarrow{\iota} A\langle x\rangle \xrightarrow{\vartheta} A\langle x\rangle \to 0 \qquad \text{where} \qquad \vartheta\left(\sum_i a_i x^{(i)}\right) = \sum_i a_i x^{(i-1)}$$

of degree $-d - 1$, and the homology exact sequence is

$$\cdots \to \mathrm{H}_{n-d}(A\langle x\rangle) \xrightarrow{\eth_{n+1}} \mathrm{H}_n(A) \xrightarrow{\mathrm{H}_n(\iota)} \mathrm{H}_n(A\langle x\rangle) \xrightarrow{\mathrm{H}_n(\vartheta)} \mathrm{H}_{n-d-1}(A\langle x\rangle) \to \cdots$$

Unlike the preceding case, multiplication by w does not appear as a map in this sequence. To analyze its impact, consider the spectral sequence of the filtration $\sum_{i \leqslant p} A x^{(i)}$. Its module ${}^2\mathrm{E}_{p,q}$ is the homology of the complex

$$\mathrm{H}_{q-d(p+1)}(A) \xrightarrow{w} \mathrm{H}_{q-dp}(A) \xrightarrow{w} \mathrm{H}_{q-d(p-1)}(A)\,,$$

so for all q there are equalities

$${}^2\mathrm{E}_{0,q} = \frac{\mathrm{H}_q(A)}{w\,\mathrm{H}_{q-d}(A)} \qquad \text{and} \qquad {}^2\mathrm{E}_{p,q} = \frac{(0 :_{\mathrm{H}(A)} w)_{q-pd}}{w\,\mathrm{H}_{q-(p+1)d}(A)} \qquad \text{when } p \geq 1\,.$$

Setting $s = \inf\{n \mid (0 :_{\mathrm{H}(A)} w)_n \neq w\,\mathrm{H}_{n-d}(A)\}$, we get ${}^2\mathrm{E}_{p,q} = 0$ for $p \geq 1$ and $q < dp + s$. Since ${}^r d_{p,q}$ maps ${}^r\mathrm{E}_{p,q}$ to ${}^r\mathrm{E}_{p-r,\,q+r-1}$, we have ${}^2\mathrm{E}_{0,q} = {}^\infty\mathrm{E}_{0,q}$ for

$q \leq 2d + s$ and ${}^2\mathrm{E}_{1,q} = {}^\infty\mathrm{E}_{1,q}$ for $q \leq 3d + s$. The spectral sequence converges to $\mathrm{H}(A\langle x \rangle)$, so for $n \leq s$ the inclusion $\iota \colon A \hookrightarrow A\langle X \rangle$ induces isomorphisms

$$\bar{\iota}_n \colon \frac{\mathrm{H}_n(A)}{w\,\mathrm{H}_{n-d}(A)} \cong \mathrm{H}_n(A\langle X \rangle) \quad \text{for } 0 \leq n \leq d + s,$$

and short exact sequences

$$0 \to \frac{\mathrm{H}_n(A)}{w\,\mathrm{H}_{n-d}(A)} \xrightarrow{\ \bar{\iota}_n\ } \mathrm{H}_n(A\langle x \rangle) \to \frac{(0 :_{\mathrm{H}(A)} w)_n}{w\,\mathrm{H}_{n-d}(A)} \to 0 \quad \text{for } d + s + 1 \leq n \leq 2d + s.$$

We extend to the graded commutative setup a basic tool of a commutative algebraist's trade: A sequence $\boldsymbol{u} = u_1, \ldots, u_i, \cdots \subseteq B$ is *regular* on B if $(\boldsymbol{u}) \neq B$ and the image of u_i is regular on $B/(u_1, \ldots, u_{i-1})B$ for each $i \geq 1$. The following result is from the source [151]. When $A = A_0$, the extension $A\langle X \rangle$ is an usual Koszul complex and condition (ii) means that $\mathrm{H}_n(A\langle X \rangle) = 0$ for $n > 0$, so we have an extension of the classical characterization of regular sequences. On the other hand, when $\partial^A = 0$, condition (iii) extends the homological description of regular elements in Example 6.1.2.

Proposition 6.1.7. *Let A be a DG algebra, let $\boldsymbol{w} = w_1, \ldots, w_j, \ldots$ be a sequence of classes in $\mathrm{H}(A)$, let $\boldsymbol{z}$ be a sequence of cycles such that $\mathrm{cls}(z_j) = w_j$ for each i, and let $\iota \colon A \hookrightarrow A\langle X \mid \partial(X) = z \rangle$ be a semi-free Γ-extension.*

Implications (i) $\implies$ (ii) $\implies$ (iii) *then hold among the conditions:*

(i) *The sequence $\boldsymbol{w}$ is regular in $\mathrm{H}(A)$.*

(ii) *The canonical map $\bar{\iota} \colon \dfrac{\mathrm{H}(A)}{(\boldsymbol{w})\,\mathrm{H}(A)} \to \mathrm{H}(A\langle X \rangle)$ is an isomorphism.*

(iii) *The canonical map $\mathrm{H}(\iota) \colon \mathrm{H}(A) \to \mathrm{H}(A\langle X \rangle)$ is surjective.*

The conditions are equivalent if $\boldsymbol{w} = w_1$, or $|w_j| > 0$ for all j, or each $\mathrm{H}_i(A)$ is noetherian over $S = \mathrm{H}_0(A)$ and all w_j of degree zero are in the Jacobson radical of S. In these cases, any permutation of a regular sequence is itself regular.

Proof. Homology commutes with direct limits, so we may assume that the sequence $\boldsymbol{z}$ is finite, say $\boldsymbol{z} = z_1, \ldots, z_i$, and induce on i. For $i = 1$, set $z = z_1$, $w = w_1$, and $d = |w|$. When d is even Remark 6.1.5 readily yields (iii) $\implies$ (i) $\implies$ (ii), so assume that d is odd.

(i) $\implies$ (ii). If w is regular on $\mathrm{H}(A)$, then ${}^2\mathrm{E}_{p,q} = 0$ for $p > 0$ in the spectral sequence in Remark 6.1.6, hence

$$\mathrm{H}_q(A)/w\,\mathrm{H}_{q-d}(A) = {}^2\mathrm{E}_{0,q} = {}^\infty\mathrm{E}_{0,q} = \mathrm{H}_q(A\langle X \rangle).$$

(iii) $\implies$ (i). If w is not regular, then by Remark 6.1.6 the sequence

$$\mathrm{H}_{s+d+1}(A) \xrightarrow{\ \mathrm{H}(\iota)\ } \mathrm{H}_{s+d+1}(A\langle x \rangle) \longrightarrow \frac{(0 :_{\mathrm{H}(A)} w)_s}{w\,\mathrm{H}_{s-d}(A)} \longrightarrow 0$$

is exact, with non-trivial quotient. This contradicts the surjectivity of $\bar{\iota}$.

Let $i > 1$, assume that the proposition has been proved for sequences of length $i - 1$, set $X' = x_1, \ldots, x_{i-1}$, $X = X' \cup \{x_i\}$, and consider the semi-free Γ-extensions $\iota' \colon A \to A\langle X' \rangle$ and $A\langle X' \rangle \to A\langle X' \rangle \langle x_i \rangle = A\langle X \rangle$.

(i) $\implies$ (ii). If $\boldsymbol{w} = w_1, \ldots, w_i$ is regular on $\mathrm{H}(A)$, then so is the sequence $\boldsymbol{w}' = w_1, \ldots, w_{i-1}$, hence $\mathrm{H}(A)/(\boldsymbol{w}')\,\mathrm{H}(A) \cong \mathrm{H}(A\langle X'\rangle)$ by the induction hypothesis. As w_i is regular on $\mathrm{H}(A)/(\boldsymbol{w}')\,\mathrm{H}(A)$, the basis of the induction yields an isomorphism $\mathrm{H}(A)/(\boldsymbol{w})\,\mathrm{H}(A) \cong \mathrm{H}(A\langle X\rangle)$.

(iii) $\implies$ (i). Since the homomorphism $\mathrm{H}(\iota)$ factors as

$$\mathrm{H}(A) \xrightarrow{\ \alpha\ } \mathrm{H}(A\langle X'\rangle) \xrightarrow{\ \beta\ } \frac{\mathrm{H}(A\langle X'\rangle)}{w_i\,\mathrm{H}(A\langle X'\rangle)} \xrightarrow{\ \gamma\ } \mathrm{H}(A\langle X\rangle),$$

where $\alpha = \mathrm{H}(\iota')$, we see that γ is onto. The basis of our induction shows that w_i is regular on $\mathrm{H}(A\langle X'\rangle)$, and that γ is bijective. Thus, $\beta\alpha$ is onto, so

$$\mathrm{H}_n(A\langle X'\rangle) = \alpha(\mathrm{H}_n(A)) + w_i\,\mathrm{H}_{n-d}(A\langle X'\rangle) \qquad \text{for} \quad n \in \mathbb{Z} \text{ and } d = |w_i|.$$

When $d > 0$, induction on n shows that α is surjective; when $d = 0$, the same conclusion comes from Nakayama's Lemma, which applies since $\mathrm{H}_n(A\langle X'\rangle)$ is a finite $\mathrm{H}_0(A\langle X'\rangle)$–module. The surjectivity of α and the induction hypothesis imply that $\boldsymbol{w}'$ is regular on $\mathrm{H}(A)$; thus, $\boldsymbol{w}$ is regular on $\mathrm{H}(A)$. $\qquad\square$

Theorem 6.1.8. *Let Q be a ring, let B be a DG algebra resolution of $S = Q/(s_1, \ldots, s_e)$, and choose $x_1, \ldots, x_e \in B_1$ such that $\partial(x_i) = s_i$ for $i = 1, \ldots, e$. For $\boldsymbol{f} = f_1, \ldots, f_r \in Q$ with $f_j = \sum_{i=1}^{e} a_{ij} s_i$ for $j = 1, \ldots, r$, set $z_j = \sum_{i=1}^{e} \bar{a}_{ij} x_i$, where overlines denote images in $R = Q/(\boldsymbol{f})$.*

If $\boldsymbol{f}$ is Q–regular, then $C = \bar{B}\langle y_1, \ldots, y_r \,|\, \partial(y_j) = z_j\rangle$ resolves S over R.

Proof. With $A = Q\langle Y \,|\, \partial(y_j) = f_j\rangle$, we have quasi-isomorphisms of DG algebras $\alpha\colon A \to R$ and $\beta\colon B \to S$, and hence induced quasi-isomorphisms

$$\bar{B} = B \otimes_Q R \xleftarrow{B \otimes_Q \alpha} B \otimes_Q A \xrightarrow{\beta \otimes_Q A} S \otimes_Q A = S\langle y_1, \ldots, y_r \,|\, \partial(y_j) = 0\rangle .$$

As $z_j = (B \otimes_Q \alpha)\big(\sum_{i=1}^{e} a_{ij} x_i - y_j\big)$ and $(\beta \otimes_Q A)\big(\sum_{i=1}^{e} a_{ij} x_i - y_j\big) = -y_j$, we see that $\mathrm{H}_1(\bar{B})$ is a free module on $w_1, \ldots, w_r$, where $w_j = \mathrm{cls}(z_j)$, and $\mathrm{H}(\bar{B})$ is the exterior algebra $\bigwedge \mathrm{H}_1(\bar{B})$. Thus, the sequence $w_1, \ldots, w_r$ is regular on $\mathrm{H}(\bar{B})$, so $\mathrm{H}(C) = \mathrm{H}(\bar{B})/(w_1, \ldots, w_r) = S$ by the preceding proposition. $\qquad\square$

For $B = Q\langle x_1, \ldots, x_e \,|\, \partial(x_i) = s_i\rangle$ and $t_i = \bar{s}_i \in R$, the theorem yields

Corollary 6.1.9. *If both sequences $\boldsymbol{f}$ and $\boldsymbol{s}$ are Q–regular, then the DG algebra $C = R\langle x_1, \ldots, x_e; y_1, \ldots, y_r \,|\, \partial(x_i) = t_i\ \partial(y_j) = z_j\rangle$ resolves S over R.* $\qquad\square$

Remark. A result of Blanco, Majadas, and Rodicio [41] rounds off this circle of ideas: $\mathrm{H}_n(R\langle X_1, X_2\rangle) = 0$ for $n \geq 1$ if and only if $\{\mathrm{cls}(\partial(x)) \mid x \in X_2\}$ is a basis of $H = \mathrm{H}_1(R\langle X_1\rangle)$ over $S = \mathrm{H}_0(R\langle X_1\rangle)$, and the canonical map of graded algebras $\bigwedge_S H \to \mathrm{H}(R\langle X_1\rangle)$ is bijective.

In characteristic 0, the theorem holds with $R\langle X\rangle$ replaced by $R[X]$, cf. Remark 6.1.3; in general, the use of divided powers is essential:

Example 6.1.10. Let $Q = k[[s]]$, where k is a field of characteristic $p > 0$, set $R = Q/(s^{m+1})$ for some $m \geq 1$, and let t be the image of s.

For each $i \geq 0$, the DG algebra $G = R[y_1, y_2 \mid \partial(y_1) = t ; \partial(y_2) = t^m y_1]$ has $G_{2i} = y_2^i$ and $G_{2i+1} = Ry_1 y_2^i$, with $\partial(y_2^i) = it^m y_1 y^{i-1}$ and $\partial(y_1 y_2^i) = ty_2^i$. Thus, $H_{2ip}(G) \cong H_{2ip-1}(G) \cong k$ for $i \geq 0$. If $R[Y]$ is a resolvent of k, then it contains a subalgebra isomorphic to G, and so cannot be minimal.

6.2. Derivations. Throughout this section, $A \hookrightarrow A\langle X \rangle$ is a semi-free Γ-extension. First, we describe a convenient basis.

Remark 6.2.1. The following conventions are in force: $x^{(i)} = 0$ and $x^{(0)} = 1$ for all $x \in X$ and all $i < 0$; when $|x|$ is odd, $x^{(i)}$ is defined only for $i \leq 1$, and $x^{(1)} = x$; when $|x| = 0$, $x^{(i)}$ stands for x^i.

Order $X = \{x_\lambda\}_{\lambda \in \Lambda}$, first by $x_\lambda < x_\mu$ if $|x_\lambda| < |x_\mu|$, then by well-ordering each X_n. For every sequence of Γ-variables $x_\mu < \cdots < x_\nu$, and every sequence of integers $i_\mu \geq 1, \ldots, i_\nu \geq 1$, the product $x_\mu^{(i_\mu)} \cdots x_\nu^{(i_\nu)}$ is called a *normal Γ-monomial* on X; its degree is $i_\mu|x_\mu| + \cdots + i_\nu|x_\nu|$; by convention, 1 is a normal monomial. The normal monomials form the *standard basis* of $A\langle X \rangle^\natural$ over $A^\natural$.

To contain a proliferation of signs, we use the *canonical bimodule structure* carried by each DG module V over a graded commutative DG algebra B. Namely, B operates on V on the right by $vb = (-1)^{|v||b|}bv$ for all $v \in V$ and $b \in B$. This operation is right associative (that is, $v(bb') = (vb)b'$), distributive, unitary, and commutes with the original action: $(bv)b' = b(vb')$.

Remark 6.2.2. Let U be a module over $A\langle X \rangle^\natural$.

A map of $\Bbbk$–modules $\vartheta \colon A\langle X \rangle \to U$, such that

$$\vartheta(a) = 0 \qquad \qquad \text{for all} \quad a \in A ;$$
$$\vartheta(bb') = \vartheta(b)b' + (-1)^{|b||\vartheta|}b\vartheta(b') \qquad \text{for all} \quad b, b' \in A\langle X \rangle ;$$
$$\vartheta(x^{(i)}) = \vartheta(x)x^{(i-1)} \qquad \text{for all} \quad x \in X_{\text{even}} \text{ and all } i \in \mathbb{N},$$

is called an *A–linear Γ-derivation*; it is a homomorphism of $A^\natural$–modules.

It is easy to see by induction that

$$\vartheta(x^{(i)}) = \begin{cases} ix^{(i-1)} & \text{if } |x| = 0 ; \\ x^{(i-1)} & \text{if } |x| \geq 1, \end{cases}$$

and

$$\vartheta\left(x_{\lambda_1}^{(i_{\lambda_1})} \cdots x_{\lambda_q}^{(i_{\lambda_q})}\right) = \sum_{j=1}^{q}(-1)^{s_{j-1}}x_{\lambda_1}^{(i_{\lambda_1})} \cdots \vartheta\left(x_{\lambda_j}^{(i_{\lambda_j})}\right) \cdots x_{\lambda_q}^{(i_{\lambda_q})}$$

where $s_j = |\vartheta|\left(i_{\lambda_1}|x_{\lambda_1}| + \cdots + i_{\lambda_j}|x_{\lambda_j}|\right)$.

In particular, ϑ is determined by its value on X. Conversely, each homogeneous map $X \to U$ extends to a (necessarily unique) A–linear Γ-derivation $\vartheta \colon A\langle X \rangle \to U$. Indeed, define the action of ϑ on the standard basis by the formulas above, and extend it by A–linearity; it suffices to check the Leibniz rule on products of normal monomials, and this is straightforward.

Let $\mathrm{Der}_A^\gamma\left(A\langle X\rangle, U\right)$ be the set of all A–linear Γ-derivations from $A\langle X\rangle$ to U. It is easy to see that $\mathrm{Der}_A^\gamma\left(A\langle X\rangle, U\right)$ is a submodule of $\mathrm{Hom}_A\left(A\langle X\rangle, U\right)$, for the operation of $A\langle X\rangle^\natural$ on the target: $(b\alpha)(b') = b\alpha(b')$.

If U is a DG module over $A\langle X\rangle$, and ϑ is an A–linear Γ-derivation, then so is $\partial\vartheta - (-1)^{|\vartheta|}\vartheta\partial$, hence $\mathrm{Hom}_A\left(A\langle X\rangle, U\right)$ contains $\mathrm{Der}_A^\gamma\left(A\langle X\rangle, U\right)$ as a DG submodule over $A\langle X\rangle$; we call it the *DG module of A–linear Γ-derivations*. If ϑ is a cycle in $\mathrm{Der}_A^\gamma\left(A\langle X\rangle, U\right)$, that is, if $\partial\vartheta = (-1)^{|\vartheta|}\vartheta\partial$, then we say that ϑ is a *chain Γ-derivation*. Clearly, if $\beta\colon U \to V$ is a homomorphism of DG modules over $A\langle X\rangle$, then $\vartheta \mapsto \beta\circ\vartheta$ is a natural homomorphism

$$\mathrm{Der}_A^\gamma\left(A\langle X\rangle, \beta\right) \colon \mathrm{Der}_A^\gamma\left(A\langle X\rangle, U\right) \to \mathrm{Der}_A^\gamma\left(A\langle X\rangle, V\right),$$

of DG modules over $A\langle X\rangle$, which is a chain map if β is one.

Over a commutative ring, the derivation functor is representable by module of Kähler differentials. The next proposition establishes the representability of the functor of Γ-derivations of semi-free Γ-extensions.

Proposition 6.2.3. *There exist a semi-free DG module* $\mathrm{Diff}_A^\gamma A\langle X\rangle$ *over* $A\langle X\rangle$ *and a degree zero chain Γ-derivation* $d\colon A\langle X\rangle \to \mathrm{Diff}_A^\gamma A\langle X\rangle$ *such that*

(1) $\left(\mathrm{Diff}_A^\gamma A\langle X\rangle\right)^\natural$ *has a basis* $dX = \{dx : |dx| = |x|\}_{x\in X}$ *over* $A\langle X\rangle^\natural$.

(2) $d(x) = dx$ *for all* $x \in X$.

(3) $\partial(b(dx)) = \partial(b)(dx) + (-1)^{|b|} b\, d(\partial(x))$ *for all* $b \in A\langle X\rangle$.

(4) *The map* $\beta \mapsto \beta\circ d$ *is a natural in U isomorphism*

$$\mathrm{Hom}_{A\langle X\rangle}\left(\mathrm{Diff}_A^\gamma A\langle X\rangle, U\right) \to \mathrm{Der}_A^\gamma\left(A\langle X\rangle, U\right)$$

of DG modules over $A\langle X\rangle$, with inverse given by

$$\widetilde{\vartheta}\left(\sum_{x\in X} a_x dx\right) = \sum_{x\in X}(-1)^{|\vartheta||a_x|} a_x \vartheta(x).$$

Remark. We call $\mathrm{Diff}_A^\gamma A\langle X\rangle$ the *DG module of Γ-differentials* of $A\langle X\rangle$ over A, and d the *universal chain Γ-derivation* of $A\langle X\rangle$ over A.

Proof. Let D be a module with basis $dX = \{dx : |dx| = |x|\}_{x\in X}$ over $A\langle X\rangle^\natural$. By Remark 6.2.2, there is a unique degree zero Γ-derivation $d\colon A\langle X\rangle \to D$, such that $d(x) = dx$ for all $x \in X$. A short computation shows that $\partial\circ d - d\circ\partial\colon A\langle X\rangle \to D$ is an A–linear Γ-derivation. It is trivial on X, hence $\partial\circ d = d\circ\partial$. In particular, $\partial^2(dx)) = d(\partial^2(x)) = 0$ for all $x \in X$, so $\partial^2 = 0$.

The DG module $\mathrm{Diff}_A^\gamma A\langle X\rangle = (D, \partial)$ has the first three properties by construction. The last one is verified by inspection. $\qquad\square$

In combination with Proposition 1.3.2, the preceding result yields:

Corollary 6.2.4. *If $U \to V$ is a (surjective) quasi-isomorphism of DG modules over $A\langle X \rangle$, then so is $\mathrm{Der}_A^\gamma (A\langle X \rangle, U) \to \mathrm{Der}_A^\gamma (A\langle X \rangle, V)$.*   □

Construction 6.2.5. Indecomposables. Let J denote the kernel of the morphism $A_0 \to S = \mathrm{H}_0(A\langle X \rangle)$, and let $X^{(\geqslant 2)}$ be the set of normal Γ-monomials $x_\mu^{(i_\mu)} \cdots x_\nu^{(i_\nu)}$ that are *decomposable*, that is, satisfy $i_\mu + \cdots + i_\nu \geq 2$. It is clear that $A + JX + AX^{(\geqslant 2)}$ is a DG submodule of $A\langle X \rangle$ over A, hence the projection $\pi \colon A \to A\langle X \rangle / (A + JX + AX^{(\geqslant 2)})$ defines a complex of free S–modules

$$\mathrm{Ind}_A^\gamma A\langle X \rangle : \quad \cdots \longrightarrow SX_{n+1} \xrightarrow{\ \delta_{n+1}\ } SX_n \longrightarrow \cdots .$$

We call it the *complex of Γ-indecomposables* of the extension $A \hookrightarrow A\langle X \rangle$. It is used to construct DG Γ-derivations, by means of the next lemma.

Lemma 6.2.6. *Let V be a complex of S–modules, let U a DG module over A with $U_i = 0$ for $i < 0$, and let $\beta \colon U \to V$ a surjective quasi-isomorphism.*

For each chain map $\xi \colon \mathrm{Ind}_A^\gamma A\langle X \rangle \to V$ of degree n there exists a degree n chain Γ-derivation $\vartheta \colon A\langle X \rangle \to U$ such that $\beta\vartheta = \xi\pi$; any two such derivations are homotopic by a homotopy that is itself an A–linear Γ-derivation.

Furthermore, for each family $\{u_x \in U_0 \mid \beta(u_x) = \xi(x) \text{ for } x \in X_n\} \subseteq U$ there is a chain Γ-derivation ϑ, satisfying $\vartheta(x) = u_x$ for all $x \in X_n$.

Proof. The canonical projection $A\langle X \rangle \to \mathrm{Ind}_A^\gamma A\langle X \rangle$ is an A–linear chain Γ-derivation, so Proposition 6.2.3.4 yields a morphism $D \to \mathrm{Ind}_A^\gamma A\langle X \rangle$ of DG modules over $A\langle X \rangle$, that maps dx to x for each $x \in X$. It induces a morphism of complexes of free S–modules $S \otimes_{A\langle X \rangle} D \to \mathrm{Ind}_A^\gamma A\langle X \rangle$ that is bijective on the bases, and hence is an isomorphism. Thus, we have isomorphisms

$$\mathrm{Hom}_{A\langle X \rangle} (D, V) \cong \mathrm{Hom}_S \left(S \otimes_{A\langle X \rangle} D, V \right) \cong \mathrm{Hom}_S \left(\mathrm{Ind}_A^\gamma A\langle X \rangle, V \right)$$

On the other hand, for the semi-free module $D = \mathrm{Diff}_A^\gamma A\langle X \rangle$ over $A\langle X \rangle$, Corollary 6.2.4 gives the surjective quasi-isomorphism below

$$\mathrm{Der}_A^\gamma (A\langle X \rangle, U) \xrightarrow{\ \simeq\ } \mathrm{Der}_A^\gamma (A\langle X \rangle, V) \cong \mathrm{Hom}_{A\langle X \rangle} (D, V)$$

while the isomorphism comes from Proposition 6.2.3.4.

Concatenating these two sequences of morphisms, we get a surjective quasi-isomorphism $\alpha \colon \mathrm{Der}_A^\gamma (A\langle X \rangle, U) \to \mathrm{Hom}_S (\mathrm{Ind}_A^\gamma A\langle X \rangle, V)$, so we can choose a cycle $\vartheta \in \mathrm{Der}_A^\gamma (A\langle X \rangle, U)$ with $\alpha(\vartheta) = \xi$: this is the desired chain Γ-derivation. $\vartheta(x) = u_x$ for all $x \in X_n$. Any two choices differ by the boundary of some $v \in \mathrm{Der}_A^\gamma (A\langle X \rangle, U)$, that is, of a Γ-derivation. Finally, observe that the $\vartheta_i = 0$ for $i > n$, and the choice of ϑ_n is only subject to the condition $\beta_0 \vartheta_n = \xi_0 \pi_n$, so $\vartheta_n(x) = u_x$ for $x \in X_n$ is a possible choice.   □

In these notes, applications of the lemma go through the following

Proposition 6.2.7. *Assume that $A\langle X \rangle \to S$ is a quasi-isomorphism.*

If $x \in X_n \subset \operatorname{Ind}_A^\gamma A\langle X \rangle$ is such that $\bar{x} = x \in X_n + \delta_{n+1}$ is a free direct summand of $\operatorname{Coker} \delta_{n+1}$, then there is an A-linear chain Γ-derivation $\vartheta \colon A\langle X \rangle \to A\langle X \rangle$ of degree $-n$, with $\vartheta(x) = 1$ and $\vartheta(X_n\{x\}) = 0$.

Proof. Since $Sx \cap \operatorname{Im} \delta_{n+1} = 0$, the homomorphism of S–modules $\xi_n \colon SX_n \to S$ defined by $\xi_n(x) = 1$ and $\xi_n(X_n \smallsetminus \{x\}) = 0$ extends to a morphisms of complexes $\operatorname{Ind}_A^\gamma A\langle X \rangle \to S$; apply the lemma with $U = A\langle X \rangle$ and $V = S$. $\qquad\square$

6.3. Acyclic closures. The notion is introduced by Gulliksen in [83], where the main results below may be found. Our approach is somewhat different, as it is based on techniques from the preceding section.

Construction 6.3.1. Acyclic closures. Let A be a DG algebra, such that A_0 is a local ring $(R, \mathfrak{m}, k)$, and each R–module $\operatorname{H}_n(A)$ is finitely generated, let $A \to S$ be a surjective augmentation, and set $J = \operatorname{Ker}(R \to S)$.

Successively adjoining finite packages of Γ-variables in degrees 1, 2, 3, etc., one arrives at a semi-free Γ-extension $A \hookrightarrow A\langle X \rangle$, such that $\operatorname{H}_0(A\langle X \rangle) = S$, and $\operatorname{H}_n(A\langle X \rangle) = 0$ for $n \neq 0$ (recall the argument for Proposition 2.1.10).

The Third Commandment imposes the following decisions: (1) $X = X_{\geqslant 1}$; (2) $\partial(X_1)$ minimally generates $J \mod \partial_1(A_1)$; (3) $\{\operatorname{cls}(\partial(x)) \mid x \in X_{n+1}\}$ minimally generates $\operatorname{H}_n(A\langle X_{\leqslant n} \rangle)$ for $n \geq 1$. Extensions obtained in that way are called *acyclic closures* of S over A.

The set X_n is finite for each n, so we number the Γ-variables X by the natural numbers, in such a way that $|x_i| \leq |x_j|$ for $i < j$. The standard basis of Remark 6.2.1 is then indexed by infinite sequences $I = (i_1, \ldots, i_j, \ldots)$, such that i_j is a non-negative integer, $i_j \leq 1$ when $|x_j|$ is odd, and $i_j = 0$ for $j > q = q(I)$; we call such an I an *indexing sequence*, and set $x^{(I)} = x_1^{(i_1)} \cdots x_q^{(i_q)}$.

We set $|I| = \sum_{j=0}^{\infty} i_j |x_j|$. For an indexing sequence $H = (h_1, \ldots, h_j, \ldots)$, we set $I > H$ if $|I| > |H|$; when $|I| = |H|$, we set $I > H$ if there is an $\ell \geq 0$, such that $i_\ell > h_\ell$, and $i_j = h_j$ for $j > \ell$. We now have a linear order on all indexing sequences, and we linearly order the basis accordingly. Since $|x^{(I)}| = |I|$, it refines the order on the variables, and is just (an extension of) the usual *degree-lexicographic order*.

To recognize an acyclic closure when we see one, we prove:

Lemma 6.3.2. *A semi-free Γ-extension $A \hookrightarrow A\langle X \rangle$ is an acyclic closure of S if and only if $X = X_{\geqslant 1}$, $\operatorname{H}_0(A\langle X \rangle) = S$, and the complex of free S–modules $\operatorname{Ind}_A^\gamma A\langle X \rangle$ of Construction 6.2.5 is minimal.*

Proof. As $X_0 = \varnothing$, the complex $\operatorname{Ind}_A^\gamma A\langle X \rangle$ is trivial in degrees ≤ 0.

Assume that $A\langle X \rangle$ is an acyclic closure. For $x' \in X_{n+1}$, write $z = \partial(x')$ as $\sum_x a_x x + w$ with $x \in X_n$, $a_x \in R$, and $w \in A\langle X_{<n} \rangle$. If $n = 1$, then $\partial(z) = 0$ means $\sum_x a_x \partial(x) \in \partial_1(A_1)$, so $a_x \in \mathfrak{m}$ by (2). If $n \geq 2$, then $\sum_x a_x \operatorname{cls}(\partial(x)) = 0$, so $a_x \in \mathfrak{m}$ by (3); thus, $\operatorname{Ind}_A^\gamma A\langle X \rangle$ is minimal.

Assume that $\mathrm{Ind}_A^\gamma A\langle X\rangle$ is minimal. If (2) or (3) fails, then there are $x_{i_1}, ..., x_{i_s}$ $\in X_{n+1}$; $a_{i_2}, ..., a_{i_s} \in R$; $y \in A\langle X_{\leqslant n}\rangle$, such that $x_{i_1} - a_{i_2}x_{i_2} - \cdots - a_{i_s}x_{i_s} - y$ is a cycle. In $A\langle X\rangle$, it is equal to $\partial(u + \sum_{v \in X_{n+1}} c_v v + w)$, with $u \in RX_{n+2}$, $c_v \in R$, and $w \in A\langle X_{\leqslant n}\rangle$. As $\partial(c_v v) = \partial(c_v)v - c_v\partial(v) \in JX_{n+1} + AX^{(\geqslant 2)}$ and $\partial(w) \in AX^{(\geqslant 2)}$, where $X^{(\geqslant 2)}$ is the set of decomposable Γ-monomials from Construction 6.2.5, we get $\partial(u) = x_{i_1} - a_2x_{i_2} - \cdots - a_sx_{i_s} \notin \mathfrak{m}\,\mathrm{Ind}_A^\gamma A\langle X\rangle$. This contradicts the minimality of $\mathrm{Ind}_A^\gamma A\langle X\rangle$. □

We are now ready to prove a key technical fact.

Lemma 6.3.3. *If $A\langle X\rangle$ is an acyclic closure of k over A, then there exist A–linear chain Γ-derivations $\vartheta_i \colon A\langle X\rangle \to A\langle X\rangle$ for $i \geq 1$, such that:*

(1)
$$\vartheta_i(x_h) = \begin{cases} 0 & \text{for } |x_h| \leq |x_i| \text{ and } h \neq i; \\ 1 & \text{for } h = i. \end{cases}$$

(2) *Each ϑ_i is unique up to an A–linear Γ-derivation homotopy.*

(3) *When I is an indexing sequence, q is such that $i_j = 0$ for $j > q$, $\vartheta^I = \vartheta_q^{i_q} \cdots \vartheta_1^{i_1}$, and H is an indexing sequence, then*

$$\vartheta^I\big(x^{(H)}\big) = \begin{cases} 0 & \text{for } H < I; \\ 1 & \text{for } H = I. \end{cases}$$

Remark. In the composition ϑ^I, the indices of ϑ_{i_j} appear in *decreasing* order.
Proof. By Lemma 6.3.2, $\mathrm{Ind}_A^\gamma A\langle X\rangle$ is a complex of k–vector spaces with trivial differential, so derivations ϑ_i satisfying (1) and (2) are provided by Proposition 6.2.7. As the ϑ_i are Γ-derivations, (3) follows by induction on $\sum_j i_j$. □

The next result is due to Gulliksen [83].

Theorem 6.3.4. *Let A be a DG algebra, such that A_0 is a local ring $(R, \mathfrak{m}, k)$, and each R–module $\mathrm{H}_n(A)$ is finitely generated. If $A\langle X\rangle$ is an acyclic closure of k over A, then $\partial(A\langle X\rangle) \subseteq JA\langle X\rangle$, where $J_0 = \mathfrak{m}$, and $J_n = A_n$ for $n > 0$.*

Proof. Take an arbitrary $b \in A\langle X\rangle$, and write its boundary in the standard basis: $\partial(b) = \sum_H a_H x^{(H)}$. We have to prove that if $|H| = |b|-1$, then $a_H \in \mathfrak{m}$. Assuming the contrary, we can find an indexing sequence I with $a_I \notin \mathfrak{m}$ and $a_H \in \mathfrak{m}$ for $H > I$. Using the preceding lemma, we get

$$\pm\partial_1(\vartheta^I(b)) = \vartheta^I(\partial(b)) = a_I + \sum_{H > I} a_H \vartheta^I\big(x^{(H)}\big) \equiv a_I \mod \mathfrak{m}A\langle X\rangle .$$

This is a contradiction, because $\partial_1(A\langle X\rangle_1) = \mathfrak{m}$. □

The important special case when $A = A_0$ is proved independently by Gulliksen [77] (using derivations) and Schoeller [139] (using Hopf algebras):

Theorem 6.3.5. *If $(R, \mathfrak{m}, k)$ is a local ring and $R\langle X\rangle$ is an acyclic closure of the R–algebra k, then $R\langle X\rangle$ is a minimal resolution of the R–module k.* □

As a first application we prove a result of Levin [111], where $\mathrm{H}_N^R(t)$ denotes the *Hilbert series* $\sum_{n=0}^{\infty} \mathrm{rank}_k(\mathfrak{m}^n N/\mathfrak{m}^{n+1} N) t^n$ of a finite R–module N.

Theorem 6.3.6. *For each finite R–module M there is an integer s such that*

$$\mathrm{P}_{\mathfrak{m}^i M}^R(t) = \mathrm{H}_{\mathfrak{m}^i M}^R(-t)\, \mathrm{P}_k^R(t) \qquad \text{for each } i \geq s.$$

Proof. By Theorem 6.3.5, there is a minimal resolution $U = R\langle X\rangle$ of k over R that is a semi-free DG module over the Koszul complex $K = R\langle X_1\rangle$. Choose s as in Lemma 4.1.6.3, so that for $i \geq s$ the complexes

$$C^i: \quad 0 \to \mathfrak{m}^{i-e} M \otimes_R K_e \to \cdots \to \mathfrak{m}^{i-1} M \otimes_R K_1 \to \mathfrak{m}^i M \otimes_R K_0 \to 0$$

are exact (here $e = \mathrm{edim}\, R$). Fix such an i, set $N = \mathfrak{m}^i M$, and for each $p \geq 0$ set $F^p = \bigoplus_{|e_\lambda| \leq p}(\mathfrak{m} N \otimes_R K) e_\lambda$.

Take $z \in Z_n(\mathfrak{m} N \otimes_R U) \cap F^p$. When $p = 0$ we have

$$Z_n(\mathfrak{m} N \otimes_R K) = Z_n(\mathfrak{m}^{i+1} M \otimes_R K) = \partial(\mathfrak{m}^i M \otimes_R K_{n+1}) = \partial(N \otimes_R K_{n+1})$$

with the second equality due to the exactness of C^{i+1+n}. When $p > 0$, assume by induction that $Z(\mathfrak{m} N \otimes_R U) \cap F^{p-1} \subseteq \partial(N \otimes_R U)$, and write $z = \sum_{\lambda \in \Lambda_p} a_\lambda e_\lambda + v$ with $v \in F^{p-1}$. and $\{e_\lambda\}_{\lambda \in \Lambda}$ a $K^\natural$–basis of $U^\natural$. Now

$$0 = \partial(z) = \sum_{\lambda \in \Lambda_p} \partial(a_\lambda) e_\lambda \pm \sum_{\lambda \in \Lambda_p} a_\lambda \partial(e_\lambda) + \partial(v)$$

implies $\partial(a_\lambda) = 0$ for $\lambda \in \Lambda_p$, hence $a_\lambda = \partial(b_\lambda)$ with $b_\lambda \in N \otimes_R K$, and

$$z = \sum_{\lambda \in \Lambda_p} \partial(b_\lambda) e_\lambda + v = \partial\left(\sum_{\lambda \in \Lambda_p} b_\lambda e_\lambda\right) \mp \sum_{\lambda \in \Lambda_p} b_\lambda \partial(e_\lambda) + v.$$

Since $u = \sum_{\lambda \in \Lambda_p} b_\lambda \partial(e_\lambda) \in N \otimes_R \mathfrak{m} U = \mathfrak{m} N \otimes_R U$, we see that $u + v$ lies in $Z(\mathfrak{m} N \otimes_R U) \cap F^{p-1} \subseteq \partial(N \otimes_R U)$, and hence $z \in \partial(N \otimes_R U)$.

We have $Z(\mathfrak{m} N \otimes_R U) \subseteq \partial(N \otimes_R U)$, so $\mathfrak{m} N \otimes_R U \subseteq N \otimes_R U$ induces the zero map in homology, that is, $\mathrm{Tor}^R(\mathfrak{m}^{i+1} M, k) \to \mathrm{Tor}^R(\mathfrak{m}^i M, k)$ is trivial. Thus, for each $n \in \mathbb{Z}$ we get an exact sequence

$$0 \to \mathrm{Tor}_n^R(\mathfrak{m}^i M, k) \to \mathrm{Tor}_n^R(\mathfrak{m}^i M/\mathfrak{m}^{i+1} M, k) \to \mathrm{Tor}_{n-1}^R(\mathfrak{m}^{i+1} M, k) \to 0$$

of k–vector spaces. They yield an equality of Poincaré series

$$\mathrm{P}_{\mathfrak{m}^j M}^R(t) + t\, \mathrm{P}_{\mathfrak{m}^{j+1} M}^R(t) = \mathrm{rank}_k(\mathfrak{m}^j M/\mathfrak{m}^{j+1} M)\, \mathrm{P}_k^R(t) \qquad (*_j)$$

for each $j \geq i$. Multiplying $(*_j)$ by $(-t)^{j-i}$, and summing the resulting equalities in $\mathbb{Z}[[t]]$ over $j \geq i$, we obtain $\mathrm{P}_{\mathfrak{m}^i M}^R(t) = \mathrm{H}_{\mathfrak{m}^i M}^R(-t)\, \mathrm{P}_k^R(t)$, as desired. $\square$

The[25] reader will note that the argument above may be used to yield a new proof of Theorem 4.1.8. By that result, $\mathrm{P}^R_M(t) = \mathrm{P}^R_{M/\mathfrak{m}^i M}(t) - t\,\mathrm{P}^R_{\mathfrak{m}^i M}(t)$; by Hilbert theory we know that $\mathrm{H}^R_{\mathfrak{m}^i M}(t)(1-t)^{\dim M} \in \mathbb{Z}[t]$ (for each i), hence

Corollary 6.3.7. *If all R–modules of finite length have rational Poincaré series, then $\mathrm{P}^R_M(t)$ is rational for all R–modules M.*   □

A second application, from Gulliksen [78], treats *partial acyclic closures*.

Proposition 6.3.8. *If $e = \operatorname{edim} R$, then $\mathrm{H}_{\geqslant 1}(R\langle X_{\leqslant n}\rangle)^{e+1} = 0$ for each $n \geq 1$.*

Proof. By Theorem 6.3.5, $Z_i(R\langle X\rangle) = (\partial(R\langle X\rangle))_i \subseteq \mathfrak{m}(R\langle X\rangle)_i$ for $i \geq 1$, so

$$Z_i(R\langle X_{\leqslant n}\rangle) = Z(R\langle X\rangle) \cap R\langle X_{\leqslant n}\rangle_i \subseteq \mathfrak{m}(R\langle X\rangle) \cap R\langle X_{\leqslant n}\rangle_i = \mathfrak{m}(R\langle X_{\leqslant n}\rangle_i)\,.$$

Thus, every cycle $z \in Z_i(R\langle X_{\leqslant n}\rangle)$ can be written in the form

$$z = \sum_{j=1}^e t_j v_j = \sum_{j=1}^e \partial(x_j)v_j = \sum_{j=1}^e x_j\partial(v_j) + \partial\left(\sum_{j=1}^e x_j v_j\right),$$

so each element of $\mathrm{H}_{\geqslant 1}(R\langle X_{\leqslant n}\rangle)$ is represented by a cycle in $X_1 R\langle X_{\leqslant n}\rangle$. As $(X_1 R\langle X_{\leqslant n}\rangle)^{e+1} = (X_1)^{e+1} R\langle X_{\leqslant n}\rangle = 0$, we get $\mathrm{H}_{\geqslant 1}(R\langle X_{\leqslant n}\rangle)^{e+1} = 0$.   □

Remark 6.3.9. To study 'uniqueness' of acyclic closures, one needs the category of *DG algebras with divided powers*: these are DG algebras, whose elements of positive even degree are equipped with a family of operations $\{a \mapsto a^{(i)}\}_{i \geqslant 0}$ that satisfy (among other things) the conditions imposed on the Γ-variables in Construction 6.1.1; they are also known as *DG Γ-algebras*[26], due to the use by Eilenberg-MacLane [56] and Cartan [50] of $\gamma_i(a)$ to denote $a^{(i)}$.

It is proved in [83] that $R\langle X\rangle$ has a unique structure of DG Γ-algebra, that extends the natural divided powers of the Γ-variables in X, and if $R\langle X'\rangle$ is an acyclic closure of S, then $R\langle X\rangle \cong R\langle X'\rangle$ as DG Γ-algebras over R.

7. Deviations of a Local Ring

In this chapter $(R, \mathfrak{m}, k)$ is a local ring.

We describe a sequence of homological invariants of the R–module k, that are 'logarithmically' related to its Betti numbers. They are introduced formalistically in Section 1, and shown to measure the deviation of R from being regular or a complete intersection in Section 3. In Section 2 we develop tools for their study, that reduce some problems over the singular ring R to problems over a regular ring, by means of *minimal models* for R.

---

[25]Attentive.
[26]This accounts for the Γ's appearing from Section 6.1 onward.

7.1. Deviations and Betti numbers.

We need an elementary observation.

Remark 7.1.1. For each formal power series $P(t) = 1 + \sum_{j=1}^{\infty} b_j t^j$ with $b_j \in \mathbb{Z}$, there exist uniquely defined $e_n \in \mathbb{Z}$, such that

$$P(t) = \frac{\prod_{i=1}^{\infty}(1 + t^{2i-1})^{e_{2i-1}}}{\prod_{i=1}^{\infty}(1 - t^{2i})^{e_{2i}}}$$

where the product converges in the (t)-adic topology of the ring $\mathbb{Z}[[t]]$.

Indeed, let $p_j(t) = (1 - (-t)^j)^{(-1)^{j+1}}$. Setting $P_0(t) = 1$, assume by induction that $P_{n-1}(t) = \prod_{h=1}^{n-1} p_h(t)^{e_h}$ satisfies $P(t) \equiv P_{n-1}(t) \pmod{t^n}$ with uniquely defined e_h. If $P(t) - P_{n-1}(t) \equiv e_n t^n \pmod{t^{n+1}}$, then set $P_n(t) = P_{n-1}(t) \cdot p_n(t)^{e_n}$. The binomial expansion of $p_n(t)^{e_n}$ shows that $P(t) \equiv P_n(t) \pmod{t^{n+1}}$, and that e_n is the only integer with that property.

The exponent e_n defined by the product decomposition of the Poincaré series $\mathrm{P}_k^R(t)$ is denoted[27] $\varepsilon_n(R)$ and called the n'th *deviation* of R (for reasons to be clarified in Section 3); we set $\varepsilon_n(R) = 0$ for $n \leq 0$. Here are the first few relations between Betti numbers $\beta_n = \beta_n^R(k)$ and deviations $\varepsilon_n = \varepsilon_n(R)$:

$$\beta_1 = \varepsilon_1; \qquad \beta_3 = \varepsilon_3 + \varepsilon_2\varepsilon_1 + \binom{\varepsilon_1}{3};$$

$$\beta_2 = \varepsilon_2 + \binom{\varepsilon_1}{2}; \qquad \beta_4 = \varepsilon_4 + \varepsilon_3\varepsilon_1 + \binom{\varepsilon_2}{2} + \varepsilon_2\binom{\varepsilon_1}{2} + \binom{\varepsilon_1}{4}.$$

Remark 7.1.2. The equality $\mathrm{P}_k^R(t) = \mathrm{P}_k^{\widehat{R}}(t)$ and the uniqueness of the product decomposition show that $\varepsilon_n(R) = \varepsilon_n(\widehat{R})$ for all n.

A first algebraic description of the deviations is given by

Theorem 7.1.3. *If* $R\langle X \rangle$ *is an acyclic closure of* k *over* R, *then*

$$\operatorname{card} X_n = \varepsilon_n(R) \qquad for \quad n \in \mathbb{Z}.$$

Proof. By the minimality of $R\langle X \rangle$ established in Theorem 6.3.5, we have equalities $\operatorname{Tor}^R(k,k) = \mathrm{H}(R\langle X \rangle \otimes_R k) = k\langle X \rangle = \bigotimes_{x \in X} k\langle x \rangle$. Furthermore,

$$\sum_{n=0}^{\infty} \operatorname{rank}_k(k\langle x \rangle_n)t^n = \begin{cases} 1 + t^{2i-1} & \text{if } x \in X_{2i-1}; \\ 1/(1 - t^{2i}) & \text{if } x \in X_{2i}, \end{cases}$$

---

[27]This numbering is at odds with [83], where ε_n stands for $\varepsilon_{n+1}(R)$.

by Constructions 2.1.7 and 6.1.1, so we get

$$P_k^R(t) = \frac{\displaystyle\prod_{i=1}^{\infty}(1 + t^{2i-1})^{\mathrm{card}(X_{2i-1})}}{\displaystyle\prod_{i=1}^{\infty}(1 - t^{2i})^{\mathrm{card}(X_{2i})}}.$$

The coefficient of t^n depends only on the first n factors, so the product converges in the t-adic topology, and the desired equalities follow from Remark 7.1.1.   □

We record a couple of easy consequences.

Corollary 7.1.4. $\varepsilon_n(R) \geq 0$ *for all* $n \in \mathbb{Z}$.   □

Corollary 7.1.5. *If* $\widehat{R} = Q/I$ *is a minimal regular presentation, then* $\varepsilon_1(R) = \nu_Q(\mathfrak{n})$ *and* $\varepsilon_2(R) = \nu_Q(I)$.

Proof. By Remark 7.1.2, we may assume that $R = Q/I$. Condition (6.3.1.0) for acyclic closures then yields the first equality, and implies that $R\langle X_1 \rangle$ is the Koszul complex K^R. The second equality now comes from Lemma 4.1.3.3.   □

A most important property of deviations is their behavior under change of rings: it is additive, as opposed to the multiplicative nature of Betti numbers. The logarithmic nature of the deviations will reappear in Theorem 7.4.2.

Proposition 7.1.6. *If* $g \in R$ *is regular, then*

$$\varepsilon_1(R/(g)) = \begin{cases} \varepsilon_1(R) - 1 & \text{if } g \notin \mathfrak{m}^2; \\ \varepsilon_1(R) & \text{if } g \in \mathfrak{m}^2; \end{cases}$$

$$\varepsilon_2(R/(g)) = \begin{cases} \varepsilon_2(R) & \text{if } g \notin \mathfrak{m}^2; \\ \varepsilon_2(R) + 1 & \text{if } g \in \mathfrak{m}^2; \end{cases}$$

$$\varepsilon_n(R/(g)) = \varepsilon_n(R) \quad \text{for} \quad n \geq 3.$$

Proof. For $R' = R/(g)$, Proposition 3.3.5 yields $P_k^R(t) = (1 - t^2) P_k^{R'}(t)$ if $g \in \mathfrak{m}^2$ and $P_k^R(t) = (1 + t) P_k^{R'}(t)$ if $g \notin \mathfrak{m}^2$; now apply Remark 7.1.1.   □

7.2. Minimal models. In this section $(Q, \mathfrak{n}, k)$ is a local ring.

A *minimal DG algebra* over Q is a semi-free extension $Q \hookrightarrow Q[Y]$ such that $Y = Y_{\geq 1}$ and the differential ∂ is *decomposable* in the sense that

$$\partial(Y_1) \subseteq QY_0 \quad \text{and} \quad \partial(Y_n) \subseteq \sum_{i=1}^{n-1} QY_{i-1}Y_{n-i} \quad \text{for} \quad n \geq 2$$

where Y_0 denotes a minimal set of generators of $\mathfrak{n}$; when $\partial(Y_1) \subseteq \mathfrak{n}^2$, the minimality condition can be rewritten in the handy format $\partial(Q[Y]) \subseteq (Y)^2 Q[Y]$.

Remark 7.2.1. Along with a minimal DG algebra $Q[Y]$, we consider the residue DG algebras $Q[Y]/(Y_{<n})Q[Y] = k[Y_{\geqslant n}]$ for $n \geq 1$. Their initial homology is easy to compute: for degree reasons, the decomposability of ∂ implies that $\partial(k[Y_{\geqslant n}]_i) = 0$ when $i \leq 2n$, hence

$$
H_i\left(k[Y_{\geqslant n}]\right) = \begin{cases} k & \text{if } i = 0; \\ 0 & \text{if } 0 < i < n; \\ kY_i & \text{if } n \leq i < 2n. \end{cases}
$$

Mimicking the proof of Lemma 6.3.2, we get a criterion for minimality:

Lemma 7.2.2. *A semi-free extension* $Q \hookrightarrow Q[Y]$ *with* $Y = Y_{\geqslant 1}$ *is minimal if and only if* $\partial(Y_1)$ *minimally generates* $\mathrm{Ker}\left(Q \to H_0(Q[Y])\right)$ *and* $\partial(Y_n)$ *minimally generates* $H_{n-1}(Q[Y_{\leqslant n-1}])$ *for* $n \geq 2$. $\square$

Minimal DG algebras are 'as unique as' minimal complexes.

Lemma 7.2.3. *Each quasi-isomorphism* $\phi\colon Q[Y] \to Q[Y']$ *of minimal DG algebras over* Q *is an isomorphism.*

Proof. Consider the restrictions $\phi^{\leqslant n}\colon Q[Y_{\leqslant n}] \to Q[Y'_{\leqslant n}]$ and the morphisms $\phi^{>n} = (k \otimes_{Q[Y_{\leqslant n}]} \phi)\colon k[Y_{>n}] \to k[Y'_{>n}]$ induced by ϕ. By the preceding lemma, $\partial(Y_1)$ and $\partial(Y'_1)$ are minimal sets of generators of $\mathrm{Ker}(Q \to H_0(Q[Y]))$, so by Nakayama $\phi_1\colon QY_1 \to QY'_1$ is an isomorphism of Q–modules, and hence $\phi^{\leqslant 1}$ is an isomorphism of DG algebras over Q.

Assume by induction that $\phi^{\leqslant n}$ is bijective for some $n \geq 1$. By Proposition 1.3.3 then so is $\phi^{>n}$, hence $H_{n+1}(\phi^{>n})$ is an isomorphism. By Remark 7.2.1, this is simply $\phi^{>n}_{n+1}\colon kY_{n+1} \to kY'_{n+1}$, hence $(k \otimes_{Q[Y_{\leqslant n}]} \phi^{\leqslant n+1})\colon k[Y_{\leqslant n+1}] \to k[Y_{\leqslant n+1}]$ is bijective; by Nakayama, so is $\phi^{\leqslant n+1}$. $\square$

If $R = Q/I$, then a *minimal model* of R over Q is a quasi-isomorphism $Q[Y] \to R$, where $Q[Y]$ is a minimal DG algebra.

Proposition 7.2.4. *Each residue ring* $R = Q/I$ *has a minimal model over* Q. *Any two minimal models are isomorphic DG algebras over* Q.

Proof. Going through the construction in 2.1.10 of a resolvent of R over Q and strictly observing the Third Commandment, one gets a quasi-isomorphism $Q[Y] \to R$; the DG algebra $Q[Y]$ is minimal by Lemma 7.2.2. If $Q[Y'] \to R$ is a quasi-isomorphism from a minimal DG algebra, then by the lifting property of Proposition 2.1.9 there is a quasi-isomorphism $Q[Y] \to Q[Y']$ of DG algebras over Q; it is an isomorphism by Lemma 7.2.3. $\square$

Remark. The proposition is from Wolffhardt [159], where minimal models are called 'special algebra resolutions'. The 'model' terminology is introduced in [23] so as to reflect the similarity with the DG algebras (over $\mathbb{Q}$) used by Sullivan [148] to encode the rational homotopy type of finite CW complexes. This parallel will bear fruits in Section 8.2.

Minimal models are not to be confused with acyclic closures: If $R = Q/I$, then the minimal model and acyclic closure of R over Q coincide in two cases only – when I is generated by a regular sequence, or when $R \supseteq \mathbb{Q}$.

It is proved in [159] when R contains a field (by using bar constructions) and in [18] in general (by using Hopf algebras) that the deviations of R can be read off a minimal model. More generally, we have:

Proposition 7.2.5. *Let $(Q, \mathfrak{n}, k)$ be a regular local ring.*

If $Q[Y]$ is a semi-free extension with $Y = Y_{\geqslant 1}$ and $\mathrm{H}(Q[Y]) = R$, then for each $n \in \mathbb{Z}$ there is an inequality $\mathrm{card}\,Y_n \geq \varepsilon_{n+1}(R)$; equalities hold for all n if and only if $\partial(Y_1) \subseteq \mathfrak{n}^2$ and $Q[Y]$ is a minimal DG algebra.

Recall that two DG algebras A and A' are *quasi-isomorphic* if there exists a sequence of quasi-isomorphisms of DG algebras $A \simeq A^1 \simeq \cdots \simeq A^m \simeq A'$, pointing in either direction. The next result is from Avramov [23].

Theorem 7.2.6. *Let $R\langle X \rangle$ be an acyclic closure of k over R, let $\widehat{R} \cong Q/I$ be a minimal Cohen presentation, and let $Q[Y]$ be a minimal model of $\widehat{R}$ over Q.*

For each $n \geq 1$ the DG algebras $R\langle X_{\leqslant n} \rangle$ and $k[Y_{\geqslant n}] = Q[Y]/(Y_{< n})$ are quasi-isomorphic, and $\mathrm{card}\,Y_n = \mathrm{card}\,X_{n+1} = \varepsilon_{n+1}(R)$ for $n \geq 0$.

Thus, a minimal model of R over Q contains essentially the same information as an acyclic closure of k over R; the model has an advantage: it is defined over a regular ring, where relations are easier to compute than over R.

In view of Proposition 6.3.8, minimal models are *homologically nilpotent*:

Corollary 7.2.7. *If $e = \mathrm{edim}\,R$ and $Q[Y]$ is a minimal model of $\widehat{R}$, then for each $n \geq 1$ the product of any $(e + 1)$ elements of $\mathrm{H}_{\geqslant 1}(k[Y_{\geqslant n}])$ is equal to 0.* $\square$

Wiebe [158] proves the next result through a lengthy computation.

Corollary 7.2.8. *If $\widehat{R} = Q/I$ is a minimal regular presentation and E is the Koszul complex on a minimal generating set of I, then $\varepsilon_3(R) = \nu_Q(\mathrm{H}_1(E))$.*

Proof. Remark 7.1.2, the theorem, and Lemma 7.2.2 yield $\varepsilon_3(R) = \varepsilon_3(\widehat{R}) = \mathrm{card}\,Y_2 = \nu_Q(\mathrm{H}_1(E))$. $\square$

Part of Theorem 7.2.6 is generalized in

Proposition 7.2.9. *If $Q[Y]$ is a minimal DG algebra over a regular local ring Q, then there exists an acyclic closure $Q[Y]\langle X \rangle$ of k over $Q[Y]$, such that $X = \{x_y : y \in Y, |x_y| = |y| + 1\}$ and for each $y \in Y$ there is an inclusion*

$$\partial(x_y) - y \in \sum_{j=0}^{n-1}(Y_j)Q[Y_{< n}]\langle X_{\leqslant n} \rangle \quad \text{where } n = |y|.$$

We start the proofs of the theorems with a couple of general lemmas.

Lemma 7.2.10. *For a (surjective) morphism of DG algebras $\phi\colon A \to A'$ and a set $\{z_\lambda\}_{\lambda \in \Lambda} \subseteq Z(A)$ there is a unique (surjective) morphism of DG algebras*

$$\phi\langle X\rangle\colon A\langle\{x_\lambda\}_{\lambda \in \Lambda} \mid \partial(x_\lambda) = z_\lambda\rangle \to A'\langle\{x_\lambda\}_{\lambda \in \Lambda} \mid \partial(x_\lambda) = \phi(z_\lambda)\rangle$$

$$\text{with} \quad \phi\langle X\rangle|_A = \phi \quad \text{and} \quad \phi\langle X\rangle(x_\lambda) = x_\lambda \quad \text{for all } \lambda \in \Lambda.$$

If ϕ is a quasi-isomorphism, then so is $\phi\langle X\rangle$.

Proof. The first assertion is clear. Since homology commutes with direct limits, for the second one we may assume that X is finite. By induction, it suffices to treat the case $X = \{x\}$. The result then follows from the homology exact sequences of Remarks 6.1.5 and 6.1.6, and the Five-Lemma. $\qquad\square$

Lemma 7.2.11. *Let $A = k[Y']$ be a minimal DG algebra over k, and let $kY' \subset A$ denote the span of the variables. For a linearly independent set $Z = \{z_\lambda\}_{\lambda \in \Lambda} \subset kY' \cap Z(A)$ the canonical morphism of DG algebras*

$$\xi\colon B = A\langle\{x_\lambda\}_{\lambda \in \Lambda} \mid \partial(x_\lambda) = z_\lambda\rangle \to A/(Z)$$

$$\text{with} \quad \xi(x_\lambda^{(i)}) = 0 \quad \text{for all } \lambda \in \Lambda \text{ and } i > 0$$

is a surjective quasi-isomorphism.

Proof. Consider the subalgebra $C = k[Z]\langle\{x_\lambda\}_{\lambda \in \Lambda} \mid \partial(x_\lambda) = z_\lambda\rangle \subseteq B$. By (an easy special case of) Proposition 6.1.7, the canonical projection $\epsilon\colon C \to k$ is a quasi-isomorphism. Since B is a semi-free DG module over C, Proposition 1.3.2 shows that $\pi = B \otimes_C \epsilon\colon B \to A/(Z)$ is a quasi-isomorphism. $\qquad\square$

Proof of Theorem 7.2.6. The canonical map $R\langle X\rangle \to \widehat{R}\otimes_R R\langle X\rangle = \widehat{R}\langle X\rangle$ is a quasi-isomorphism, and induces quasi-isomorphisms $R\langle X_{\leqslant n}\rangle \to \widehat{R}\langle X_{\leqslant n}\rangle$ for $n \geq 1$, so we assume that $R = Q/I$ and take a minimal model $\rho\colon Q[Y] \to R$.

Choose a minimal generating set $Y_0 = \{t_1, \ldots, t_e\}$ of $\mathfrak{m}$ and pick $s_1, \ldots, s_e \in Q$ with $\rho(s_i) = t_i$. Lemma 7.2.10 yields a quasi-isomorphism

$$R\langle X_1 \mid \partial(x_i) = t_i\rangle \xleftarrow{\ \rho\langle X_1\rangle\ } Q[Y]\langle X_1 \mid \partial(x_i) = s_i\rangle$$

for a set of Γ-variables $X_1 = \{x_1, \ldots, x_e\}$. On the other hand, as $s_1, \ldots, s_e$ is a Q-regular sequence, so $Q\langle X_1\rangle \to k$ is a quasi-isomorphism, and then Proposition 1.3.2 yields a quasi-isomorphism $\pi^1\colon Q[Y]\langle X_1\rangle \to k[Y]$.

Let $n \geq 1$, and assume by induction that we have constructed surjective quasi-isomorphisms of DG algebras

$$R\langle X_{\leqslant n}\rangle \xleftarrow{\ \rho^n\ } Q[Y]\langle X_{\leqslant n}\rangle \xrightarrow{\ \pi^n\ } k[Y_{\geqslant n}]$$

The classes of $\{\partial(x) \mid x \in X_{n+1}\}$ form a basis of $\mathrm{H}_n(R\langle X_{\leqslant n}\rangle)$ by (6.3.1.2). Since ρ^n is a surjective quasi-isomorphism, these cycles are images of cycles in $Q[Y]\langle X_{\leqslant n}\rangle$, so by Lemma 7.2.10 we get a surjective quasi-isomorphism

$$R\langle X_{\leqslant n+1}\rangle = R\langle X_{\leqslant n}\rangle\langle X_{n+1}\rangle \xleftarrow{\ \rho^n\langle X_{n+1}\rangle\ } Q[Y]\langle X_{\leqslant n}\rangle\langle X_{n+1}\rangle = Q[Y]\langle X_{\leqslant n+1}\rangle\,.$$

The same lemma yields a surjective quasi-isomorphism

$$Q[Y]\langle X_{\leqslant n+1}\rangle = Q[Y]\langle X_{\leqslant n}\rangle\langle X_{n+1}\rangle \xrightarrow{\pi^n\langle X_{n+1}\rangle} k[Y_{\geqslant n}]\langle X_{n+1}\rangle \;.$$

As $H_n(k[Y_{\geqslant n}]) = kY_n$ by Remark 7.2.1, the differential ∂_{n+1} induces an isomorphism $kX_{n+1} \to kY_n$. Lemma 7.2.11 yields a surjective quasi-isomorphism $\xi^{n+1}\colon k[Y_{\geqslant n}]\langle X_{n+1}\rangle \to k[Y_{\geqslant n+1}]$. We set $\pi^{n+1} = \xi^{n+1}\circ\pi\langle X_{n+1}\rangle$, and note that $\mathrm{card}(X_{n+1}) = \varepsilon_{n+1}(R)$ by Theorem 7.1.3. $\square$

Proof of Proposition 7.2.5. Choose $\boldsymbol{g} = g_1,\ldots,g_s \in I$ and $\boldsymbol{f} = f_1,\ldots,f_r \in I$ such that $\boldsymbol{f} \cup \boldsymbol{g}$ is a minimal set of generators of I, and $\boldsymbol{g}$ maps to a basis of $I/(I\cap\mathfrak{n}^2)$. The sequence $\boldsymbol{g}$ is then Q–regular, and $R = Q'/I'$ is a minimal regular presentation, with $Q' = Q/(\boldsymbol{g})$ and $I' = I/(\boldsymbol{g})$; set $\mathfrak{n}' = \mathfrak{n}/(\boldsymbol{g})$.

After a linear change of the variables of degree 1, we may assume that there is a sequence of variables $\boldsymbol{y} = y_1,\ldots,y_s$ in Y_1, such that $\partial(y_i) = g_i$ for $1 \leq i \leq s$ The Koszul complex $Q[\boldsymbol{y} \,|\, \partial(y_i) = g_i]$ is then a resolvent of Q', so $Q[Y] \to Q[Y]/(\boldsymbol{g},\boldsymbol{y}) = Q'[Y']$ is a quasi-isomorphism by Proposition 1.3.2. As $\mathrm{card}(Y'_n) = \mathrm{card}(Y_n) - s$ for $n = 0,1$, and $\mathrm{card}(Y'_n) = \mathrm{card}(Y_n)$ otherwise, we may assume that $R = Q/I$ is a minimal presentation, and $Q[Y]$ is a semi-free extension with $H(Q[Y]) = R$; we then have $\mathrm{card}(Y_0) = \varepsilon_1(R)$.

Let $Q[Y']$ be a minimal model of R over Q, so that $\mathrm{card}(Y'_n) = \varepsilon_{n+1}(R)$ by Theorem 7.2.6. Proposition 2.1.9 yields morphisms $\gamma\colon Q[Y'] \to Q[Y]$ and $\beta\colon Q[Y] \to Q[Y']$ of DG algebras over Q, such that $H(\beta) = \mathrm{id}^R$. By Lemma 7.2.3, $\beta\gamma$ is an automorphism of $Q[Y']$, so for each $n \geq 1$, the composition of $\gamma^{\geqslant n}\colon k[Y'_{\geqslant n}] \to k[Y_{\geqslant n}]$ with $\beta^{\geqslant n}\colon k[Y_{\geqslant n}] \to k[Y'_{\geqslant n}]$ is then bijective. In particular, $\beta^{\geqslant n}$ is onto, so $\mathrm{card}(Y_n) \geq \mathrm{card}(Y'_n) = \varepsilon_{n+1}(R)$ for $n \geq 1$.

Assume that $\mathrm{card}(Y_n) = \varepsilon_{n+1}(R)$ for all n. The equalities for $n \leq 1$ mean that $I \subseteq \mathfrak{n}^2$ and $\partial(Y_1)$ minimally generates I; equality for $n \geq 2$ implies that $H_n(\beta^{\geqslant n})$ is bijective, hence in $k[Y_{\geqslant n}]$ the differential ∂_{n+1} is trivial; this is equivalent to saying that in $Q[X]$ the differential ∂_{n+1} is decomposable. $\square$

Proof of Proposition 7.2.9. We first construct surjective quasi-isomorphisms

$$\pi^n\colon Q[Y]\langle X_{\leqslant n}\rangle \to k[Y_{\geqslant n}] \quad \text{with } X_n = \{x_y : y \in Y_{n-1}\}.$$

Let $Y_0 = \{s_1,\ldots,s_e\}$ be a system of generators of $\mathfrak{n}$, set $X_1 = \{x_1,\ldots,x_e\}$, and $Q[Y]\langle X_1 \,|\, \partial(x_i) = s_i\rangle$. As in the proof of Theorem 7.2.6 we have a quasi-isomorphism $\pi^1\colon Q[Y]\langle X_1\rangle \to k[Y]$.

Assume that π^n has been constructed for some $n \geq 1$. Now each $y \in Y_n$ is a cycle in $k[Y_{\geqslant n}]$, so $y = \pi^n(z_y)$ for some cycle z_y; write z_y in the form $y + \sum_{y' \in Y'_n} a_{y'}y' + v$, with $a_{y'} \in \mathfrak{n}$ for $y' \in Y'_n = Y_n\smallsetminus\{y\}$, and $v \in Q[Y_{<n}]\langle X_{\leqslant n}\rangle$. Since $a_{y'} = \partial(b_{y'})$ for appropriate $b_{y'} \in QX_1$, after replacing z_y with the homologous cycle $z_y - \partial\big(\sum_{y' \in Y'_n} b_{y'}y'\big)$, we can assume that $z_y - y \in Q[Y_{<n}]\langle X_{\leqslant n}\rangle$. Lemma 7.2.10 yields a surjective quasi-isomorphism

$$Q[Y]\langle X_{\leqslant n+1}\rangle = Q[Y]\langle X_{\leqslant n}\rangle\langle X_{n+1}\rangle \xrightarrow{\pi^n\langle X_{n+1}\rangle} k[Y_{\geqslant n}]\langle X_{n+1}\rangle \;.$$

Lemma 7.2.11 yields a surjective quism $\xi^n \colon k[Y_{\geqslant n}]\langle X_{n+1}\rangle \to k[Y_{\geqslant n+1}]$, so to complete the induction step set $\rho^{n+1} = \rho^n\langle X_{n+1}\rangle$ and $\pi^{n+1} = \xi^{n+1}\circ\pi^n\langle X_{n+1}\rangle$.

In the limit, we get a quasi-isomorphism $\pi \colon Q[Y]\langle X\rangle \to k$. As $\partial(X_1) = Y_0$ minimally generates $\mathfrak{n}$, and $\partial(X_n)$ minimally generates $H_n(Q[Y]\langle X_{<n}\rangle)$ for $n \geq 2$, the DG algebra $Q[Y]\langle X\rangle$ is an acyclic closure of k over $Q[Y]$. By the choice of z_y above, and Theorem 6.3.4, we see that $\partial(x_y) - y$ lies in

$$\left(Q[Y_{<n}]\langle X_{\leqslant n}\rangle\right)_n \cap \left((Y)Q[Y]\langle X\rangle\right)_n = \sum_{j=0}^{n-1} \left((Y_j)Q[Y_{<n}]\langle X_{\leqslant n}\rangle\right)_n.$$

This is the desired condition on the differential. $\qquad\Box$

7.3. Complete intersections. Now we can 'explain' the term *deviation*.

Remark 7.3.1. By Corollary 7.1.5 $\varepsilon_1(R) = \operatorname{edim} R$, so the following conditions are equivalent: (i) R is a field; (ii) $\varepsilon_1(R) = 0$; (iii) $\varepsilon_i(R) = 0$ for $i \geq 1$.

More generally, the regularity of a ring is detected by its deviations.

Theorem 7.3.2. *The following conditions are equivalent.*

(i) *R is regular.*
(ii) *$\varepsilon_2(R) = 0$.*
(iii) *$\varepsilon_n(R) = 0$ for $n \geq 2$.*

Proof. When R is regular, the Koszul complex K^R is exact, yielding $\mathrm{P}^R_k(t) = (1+t)^{\dim R}$; thus, (i) $\implies$ (iii) by the uniqueness of the product decomposition (7.1.1). On the other hand, (ii) $\implies$ (i) by Corollary 7.1.5. $\qquad\Box$

Recall that R is a *complete intersection* if $\widehat{R}$ has a minimal Cohen presentation Q/I, with I generated by a regular sequence; in that case, $\varepsilon_2(R) = \operatorname{codim} R$ by Corollary 7.2.8. Complete intersections of codimension 0 (respectively, ≤ 1) are precisely the regular (respectively, hypersurface) rings.

The next result gives characterizations of complete intersections in terms of vanishing of deviations, due to Assmus [15] for (iii) and to Gulliksen [78] for (iv) and [82] for (v). The use of minimal models in their proofs is new.

Theorem 7.3.3. *The following conditions are equivalent.*

(i) *R is a complete intersection.*
(ii) *$\varepsilon_3(R) = 0$.*
(iii) *$\varepsilon_n(R) = 0$ for $n \geq 3$.*
(iv) *$\varepsilon_n(R) = 0$ for $n \gg 0$.*
(v) *$\varepsilon_{2i}(R) = 0$ for $i \gg 0$.*

Proof. We may take $R \cong Q/I$ with $(Q, \mathfrak{n}, k)$ regular, $\boldsymbol{f} = f_1, \ldots, f_r$ minimally generating I, and $\boldsymbol{f} \subseteq \mathfrak{n}^2$, cf. Remark 7.1.2; let E be the Koszul complex on $\boldsymbol{f}$.

(i) $\implies$ (iii). If $\boldsymbol{f}$ is a Q–regular sequence, then $r = \operatorname{codim} R$, so the deviations of R are computed by Theorem 7.3.2 and Proposition 7.1.6.

(ii) $\implies$ (i). By Corollary 7.2.8 we have $H_1(E) = 0$, so $\boldsymbol{f}$ is Q–regular.

(v) $\implies$ (iv) Take n big enough, so that $\varepsilon_{2i}(R) = 0$ for $2i \geq n$. By Theorem 7.2.6, the DG algebra $k[Y_{\geqslant n}]$ is a polynomial ring with variables of even degree. Their boundaries have odd degree, so are trivial, and thus $H(k[Y_{\geqslant n}]) = k[Y_{\geqslant n}]$ is a polynomial ring. Each element of $H_{\geqslant 1}(k[Y_{\geqslant n}])$ is nilpotent by Corollary 7.2.7, so we conclude that $Y_{\geqslant n} = \varnothing$.

(iv) $\implies$ (i) Taking, as we may, $\boldsymbol{g} = f_1, \ldots, f_c$ to be a maximal regular sequence in I, we set $R' = Q/(\boldsymbol{g})$. By Corollary 6.1.9, the R'–algebra k has a minimal free resolution of the form $C' = R'\langle x_1, \ldots, x_{e+c+i} \rangle$. By the choice of $\boldsymbol{g}$, the DG algebra $C^0 = C' \otimes_{R'} R$ can be extended to an acyclic closure $R\langle X \rangle$ of k by adjunction of Γ-variables of degree ≥ 2. We order them in such a way that $|x_i| \leq |x_j|$ for $i < j$, set $C^i = C'\langle x_{e+c+1}, \ldots, x_{e+c+i} \rangle$, and $s_i = \sup\{n \mid H_n(C^i) \neq 0\}$. Assuming that R is not a complete intersection, we prove that X is infinite by showing that $s_i = \infty$ for each $i \geq 0$.

Each element of I is a zero-divisor modulo $(\boldsymbol{g})$, so $\mathrm{pd}_{R'}(R'/IR') = \infty$ by Proposition 1.2.7.2; as $H(C^0) = \mathrm{Tor}^{R'}(R, k)$, we get $s_0 = \infty$. For the induction step, set $A = C^{i-1}$, $x = x_{e+c+i}$, $z = \partial(x)$, and $u = \mathrm{cls}(z)$, and assume that $s_i = s < \infty$. When $|x|$ is even, the homology exact sequence in Remark 6.1.6 yields $s_{i-1} \leq s + |x|$, contradicting the induction hypothesis. When $|x|$ is odd the sequence in Remark 6.1.5 shows that $H(A) = H_{\leqslant s}(A) + u H(A)$. A simple iteration yields $H(A) = H_{\leqslant s + e|u|}(A) + u^{e+1} H(A)$. But $u^{e+1} = 0$ by Proposition 6.3.8, hence $s_{i-1} \leq s + e|u|$; this contradiction proves that $s_i = \infty$. $\square$

The last result is vastly generalized in Halperin's [84] rigidity theorem:

Theorem 7.3.4. *If $\varepsilon_n(R) = 0$ and $n > 0$ then R is a complete intersection.* ∎

A proof that uses techniques developed in Section 6.2, and extends the theorem to a relative situation, is given in [30].

7.4. Localization. The theme of the preceding section may be summarized as follows: The deviations of a local ring reflect the character of its singularity. Thus, one would expect that they do not go up under localization, and in particular that the complete intersection property localizes.

It is instructive to generalize the discussion by considering the number

$$\mathrm{cid}(R) = \varepsilon_2(R) - \varepsilon_1(R) + \dim R$$

that in view of the next lemma we[28] call the *complete intersection defect* of R. The lemma also shows that in the definition of complete intersection the restriction to minimal presentations is spurious.

Lemma 7.4.1. *If $\widehat{R} \cong Q/I$ is a regular presentation, then*

$$\mathrm{cid}\, R = \nu_Q(I) - \mathrm{height}(I) \geq 0.$$

Furthermore, the following conditions are equivalent: (i) $\mathrm{cid}\, R = 0$; (ii) I *is generated by a regular sequence;* (iii) R *is a complete intersection,*

---

[28]Kiehl and Kunz introduced it in [96] by the expression in Lemma 7.4.1, and called it the *deviation* of R; that was before an infinite supply of deviations appeared on the scene.

Proof. Choose a regular sequence $\boldsymbol{g} = g_1, \ldots, g_q$ as in the proof of Proposition 7.2.5. With $Q' = Q/(\boldsymbol{g})$ and $I' = I/(\boldsymbol{g})$ we have a minimal Cohen presentation $\widehat{R} \cong Q'/I'$. Corollary 7.1.5 provides the first equality below, the catenarity of Q' yields the last one, and Krull's Principal Ideal Theorem gives the inequality:

$$\operatorname{cid} R = \nu_{Q'}(I') - \nu_{Q'}(\mathfrak{n}') + \dim R = \nu_Q(I) - \nu_Q(\mathfrak{n}) + \dim R$$
$$= \nu_Q(I) - (\dim Q - \dim R) = \nu_Q(I) - \operatorname{height}(I) \geq 0.$$

The equivalence (i) $\Longleftrightarrow$ (ii) now follows from the Cohen-Macaulay Theorem. Applied to the minimal presentation $\widehat{R} \cong Q'/I'$, it yields (ii) $\Longleftrightarrow$ (iii). $\square$

For the study of complete intersection defects and even deviations, the first part of the next theorem[29] suffices; it is due to Avramov [20]. For odd deviations one needs the second part, due to André [9].

Theorem 7.4.2. *If $R \to S$ is a faithfully flat homomorphism of local rings, then for each $i \geq 1$ there is an integer $\delta_i \geq 0$, such that*

$$\varepsilon_{2i}(R) \leq \varepsilon_{2i}(S) = \varepsilon_{2i}(R) + \varepsilon_{2i}(S/\mathfrak{m}S) - \delta_i;$$
$$\varepsilon_{2i-1}(S/\mathfrak{m}S) \leq \varepsilon_{2i-1}(S) = \varepsilon_{2i-1}(R) + \varepsilon_{2i-1}(S/\mathfrak{m}S) - \delta_i.$$

Furthermore, $\delta_i = 0$ for $i \gg 0$, and $\sum_{i=0}^{\infty} \delta_i \leq \operatorname{codepth}(S/\mathfrak{m}S)$. ∎

As in [20], we deduce:

Theorem 7.4.3. *If $R \to S$ is a flat local homomorphism, then*

$$\operatorname{cid} S = \operatorname{cid} R + \operatorname{cid}(S/\mathfrak{m}S).$$

In particular, S is a complete intersection if and only if both R and $S/\mathfrak{m}S$ are.

Proof. The first two equalities of Theorem 7.4.2 yield

$$\varepsilon_2(S) - \varepsilon_1(S) = \varepsilon_2(R) - \varepsilon_1(R) + \varepsilon_2(S/\mathfrak{m}S) - \varepsilon_1(S/\mathfrak{m}S).$$

Classically, $\dim S = \dim R + \dim(S/\mathfrak{m}S)$, so we have the desired result. $\square$

Corollary 7.4.4. *For each prime ideal $\mathfrak{p}$ of R, $\operatorname{cid}(R_\mathfrak{p}) \leq \operatorname{cid} R$.*

Proof. As $\dim R = \dim \widehat{R}$, we have $\operatorname{cid} R = \operatorname{cid} \widehat{R}$ by Remark 7.1.2. Let $\widehat{R} \cong Q/I$ be a regular presentation, and pick prime ideals $\mathfrak{p}' \subseteq \widehat{R}$ and $\mathfrak{q} \subseteq Q$, such that $\mathfrak{p}' \cap R = \mathfrak{p}$ and $\mathfrak{p}' = \mathfrak{q}\widehat{R}$. As $\widehat{R}_{\mathfrak{p}'} \cong Q_\mathfrak{q}/I_\mathfrak{q}$ is a regular presentation,

$$\operatorname{cid} \widehat{R} = \nu_Q(I) - \operatorname{height}(I) \geq \nu_{Q_\mathfrak{q}}(I_\mathfrak{q}) - \operatorname{height}(I_\mathfrak{q}) = \operatorname{cid}(\widehat{R}_{\mathfrak{p}'}),$$

with equalities coming from Lemma 7.4.1, and inequality from the obvious relations $\nu_Q(I) \geq \nu_{Q_\mathfrak{q}}(I_\mathfrak{q})$ and $\operatorname{height}(I) \leq \operatorname{height}(I_\mathfrak{q})$. Finally, the theorem applied to the flat homomorphism $R_\mathfrak{p} \to \widehat{R}_{\mathfrak{p}'}$ yields $\operatorname{cid}(\widehat{R}_{\mathfrak{p}'}) \geq \operatorname{cid}(R_\mathfrak{p})$. $\square$

---

[29] For a more natural statement, cf. Remark 10.2.4.

It is now clear that complete intersections localize, a fact initially proved in [19]. This is sharpened in the corollary of the next theorem from [20], [9], which represents a quantitative extension to arbitrary local rings of the classical localization of regularity, cf. Corollary 4.1.2.

Theorem 7.4.5. *If $\mathfrak{p}$ is a prime ideal of R, then an inequality $\varepsilon_n(R_\mathfrak{p}) \leq \varepsilon_n(R)$ holds for all even n and for almost all odd n. When R is a residue ring of a regular local ring the inequalities hold for all n.*

Proof. If R is a residue ring of a regular local ring Q, let $Q[Y]$ be a minimal model of R. If $\mathfrak{q}$ is the inverse image of $\mathfrak{p}$ in Q, then $Q_\mathfrak{q}[Y]$ is a semi-free extension of $Q_\mathfrak{q}$ with $\mathrm{H}(Q_\mathfrak{q}[Y]) \cong R_\mathfrak{p}$. Applying Proposition 7.2.5 first to $Q[Y]$, then to $Q_\mathfrak{q}[Y]$, we get $\varepsilon_{n+1}(R) = \mathrm{card}(Y_n) \geq \varepsilon_{n+1}(R_\mathfrak{p})$.

In general, pick (by faithful flatness) a prime ideal $\mathfrak{p}'$ in $\widehat{R}$, such that $\mathfrak{p} = \mathfrak{p}' \cap R$. By Remark 7.1.2 and the preceding case we then have $\varepsilon_n(R) = \varepsilon_n(\widehat{R}) \geq \varepsilon_n(\widehat{R}_{\mathfrak{p}'})$ for all n. On the other hand, Theorem 7.4.2 yields an inequality $\varepsilon_n(\widehat{R}_{\mathfrak{p}'}) \geq \varepsilon_n(R_\mathfrak{p})$ for all even n, and almost all odd n. □

Corollary 7.4.6. *If R is a complete intersection, then for each prime ideal $\mathfrak{p}$ of R the ring $R_\mathfrak{p}$ is a complete intersection with $\mathrm{codim}(R_\mathfrak{p}) \leq \mathrm{codim}\, R$.* □

Proof. In view of Theorem 7.3.3, the inequalities of deviations for $n = 2i \gg 0$ prove that $R_\mathfrak{p}$ is a complete intersection. Since $\mathrm{codim}\, R = \varepsilon_2(R)$ by Corollary 7.1.5, the inequality for $n = 2$ shows that $\mathrm{codim}(R_\mathfrak{p}) \leq \mathrm{codim}\, R$. □

For the first deviation, there is a more precise result of Lech [103]; a simpler proof is given by Vasconcelos [154].

Theorem 7.4.7. *For each prime ideal $\mathfrak{p}$ of R, $\varepsilon_1(R_\mathfrak{p}) + \dim(R/\mathfrak{p}) \leq \varepsilon_1(R)$.* ■

We spell out the obvious remaining problems. The first one has a positive solution when $\mathrm{char}(k) = 2$, due to André [8].

Problem 7.4.8. Let $R \to S$ be a faithfully flat homomorphism of local rings. Does an equality $\varepsilon_n(S) = \varepsilon_n(R) + \varepsilon_n(S/\mathfrak{m}S)$ hold for each $n \geq 3$?

Note that by Corollary 7.1.5 we have

$$\varepsilon_1(R) - \varepsilon_1(S) + \varepsilon_1(S/\mathfrak{m}S) = \mathrm{edim}(R) - \mathrm{edim}(S) + \mathrm{edim}(S/\mathfrak{m}S) \geq 0.$$

The inequality is strict unless a minimal generating set of $\mathfrak{m}$ extends to one of $\mathfrak{n}$, so additivity may fail for $n = 1$ and hence, by Theorem 7.4.2, also for $n = 2$.

Problem 7.4.9. Does $\varepsilon_n(R_\mathfrak{p}) \leq \varepsilon_n(R)$ hold for all $\mathfrak{p} \in \mathrm{Spec}\, R$ and odd $n \geq 3$?

It is easily seen from the proof of Theorem 7.4.5 that a positive solution of the first problem implies one for the second. Larfeldt and Lech [104] prove that the two problems are, in fact, equivalent.

8. Test Modules

Ring are 'non-linear' objects, so some of their properties are easier to verify after translation into conditions on some canonically defined modules.

The Auslander-Buchsbaum-Serre Theorem provides a model: the regularity of a local ring $(R, \mathfrak{m}, k)$ is tested by checking the finiteness of the projective dimension of k. In terms of asymptotic invariants, this is stated as

$$\operatorname{cx}_R k = 0 \iff R \text{ is regular} \iff \operatorname{curv}_R k = 0.$$

The first two sections establish similar descriptions of complete intersections:

$$\operatorname{cx}_R k < \infty \iff R \text{ is a complete intersection} \iff \operatorname{curv}_R k \leq 1.$$

For algebras essentially of finite type, another classical test for regularity is given by the Jacobian criterion. Section 3 discusses extensions to complete intersections, in terms of the homology of Kähler differentials. The results there are partly motivated by (still open in general) conjectures of Vasconcelos.

8.1. Residue field. In this section $(R, \mathfrak{m}, k)$ is a local ring.

We start with a few general observations on the Betti numbers of k. They show that an extremal property of Poincaré series characterizes complete intersections – and places across the spectrum from Golod rings.

Remarks 8.1.1. Set $e = \operatorname{edim} R$, $r = \operatorname{rank}_k H_1(K^R)$, and $\varepsilon_n = \varepsilon_n(R)$.

(1) There is an inequality of formal power series

$$P_k^R(t) \succcurlyeq \frac{(1+t)^e}{(1-t^2)^r} .$$

Indeed, Corollary 7.1.5 yields $P_k^R(t) = (1+t)^e (1-t^2)^{-r} Q(t)$, with $Q(t) = \prod_{i=2}^{\infty} (1 + t^{2i-1})^{\varepsilon_{2i-1}} / \prod_{i=2}^{\infty} (1 - t^{2i})^{\varepsilon_{2i}}$; also, $Q(t) \succcurlyeq 1$ by Corollary 7.1.4.

(2) Theorem 7.3.3 shows that equality holds in (1) if and only if R is a complete intersection. In that case, $\operatorname{cx}_R k = r = \operatorname{codim} R$, and

$$\beta_n^R(k) = \sum_{i=0}^{e-r} \binom{e-r}{i} \binom{n+r-1-i}{r-1} \qquad \text{for} \quad n \geq 0.$$

(3) If R is not a hypersurface, then $\beta_n^R(k) > \beta_{n-1}^R(k)$ for $n \geq 1$.
Indeed, then $e \geq 1$ and $r \geq 2$, so we have coefficientwise (in)equalities

$$\sum_{n=0}^{\infty} (\beta_n^R(k) - \beta_{n-1}^R(k)) t^n = (1-t) P_k^R(t) = \frac{(1+t)^{e-1}}{(1-t^2)^{r-1}} Q(t) \succcurlyeq \frac{1}{(1-t)} .$$

Gulliksen [78], [82] extends the Auslander-Buchsbaum-Serre Theorem in

Theorem 8.1.2. *The following conditions are equivalent.*

(i) *R is a complete intersection (respectively, of codimension $\leq c$).*
(ii) *$\operatorname{cx}_R M < \infty$ (respectively, $\operatorname{cx}_R M \leq c$) for each finite R-module M.*
(iii) *$\operatorname{cx}_R k < \infty$ (respectively, $\operatorname{cx}_R k \leq c$).*

Proof. By Proposition 4.2.4 we have $\mathrm{cx}_R M \leq \mathrm{cx}_R k$, and the complexity of k is equal to $\mathrm{codim}\, R$ by Remark 8.1.1.2, so (i) implies (ii).

When R is not a complete intersection Theorem 7.3.3 gives infinitely many indices i_d with $\varepsilon_{2i_d}(R) > 0$. Remark 7.1.1 then yields an inequality

$$\mathrm{P}_k^R(t) \succcurlyeq \frac{1}{(1 - t^{2i_1}) \cdots (1 - t^{2i_d})} \succcurlyeq \frac{1}{(1 - t^{2i})^d} = \sum_{n=0}^{\infty} \binom{n + d - 1}{d - 1} t^{(2i)n}$$

with $i = i_1 \cdots i_d$. Thus, $\mathrm{cx}_R k \geq d$ for each $d \geq 1$, so (iii) implies (i). $\qquad\square$

Remark 8.1.3. We get a new proof of Corollary 7.4.6 by recycling the classical argument for regularity. Proposition 4.2.4.1 and Remark 8.1.1.2 yield

$$\mathrm{cx}_{R_\mathfrak{p}} \left(R_\mathfrak{p}/\mathfrak{p}R_\mathfrak{p} \right) \leq \mathrm{cx}_R(R/\mathfrak{p}) \leq \mathrm{cx}_R k = \mathrm{codim}\, R \,,$$

so $R_\mathfrak{p}$ is a complete intersection of codimension $\leq \mathrm{codim}\, R$ by the theorem.

We finish this section by a computation of the curvature of k in terms of the deviations of R; the purely analytical argument is from Babenko [38].

Proposition 8.1.4. *If R is not a complete intersection, then*

$$\mathrm{curv}_R k = \limsup_n \sqrt[n]{\varepsilon_n(R)} \,.$$

Proof. Note that $\limsup_n \sqrt[n]{\varepsilon_n(R)} = 1/\eta$, where η is the radius of convergence of the series $E(t) = \sum_{n=1}^{\infty} \varepsilon_n(R) t^n$. By the definition of $\mathrm{curv}_R k$, we have to show that $\eta = \rho$, where ρ is the radius of convergence of the Poincaré series $P(t) = \mathrm{P}_k^R(t)$; note that $\rho > 0$ by Remark 4.2.3.5. By Corollary 7.1.4, we have $\varepsilon_n(R) = \varepsilon_n \geq 0$, so by the product formula of Remark 7.1.1 we get a coefficientwise inequality $P(t) \succcurlyeq E(t)$, hence $\eta \geq \rho > 0$.

To prove that $\rho \geq \eta$, we show that if $0 < \gamma < \eta$, then $P(t)$ converges at $t = \gamma$. We have $\eta < 1$, because $E(t)$ has integer coefficients and $\varepsilon_n > 0$ for infinitely many n by Theorem 7.3.3. For $j \geq 1$ we then get

$$0 < -\ln(1 - \gamma^j) = \sum_{h=1}^{\infty} \frac{\gamma^{jh}}{h} < \sum_{h=1}^{\infty} \gamma^{jh} = \frac{\gamma^j}{1 - \gamma^j} < \frac{\gamma^j}{1 - \eta} \,.$$

By a similar computation, $0 < \ln(1 + \gamma^j) < (1 - \eta)^{-1} \gamma^j$, so

$$0 \leq L(\gamma) = \sum_{i=1}^{\infty} \left(\varepsilon_{2i-1} \ln(1 + \gamma^{2i-1}) - \varepsilon_{2i} \ln(1 - \gamma^{2i}) \right) \leq \frac{E(\gamma)}{(1 - \eta)} < \infty \,.$$

The numerical series with non-negative coefficients $L(\gamma)$ converges, so the product in 7.1.1 converges at $t = \gamma$, as desired. $\qquad\square$

8.2. Residue domains. In this section $(R, \mathfrak{m}, k)$ is a local ring. The results that follow are from Avramov [23].

Theorem 8.2.1. *If $\mathfrak{p}$ is a prime ideal of R such that $R_\mathfrak{p}$ is not a complete intersection, then there is a real number $\beta > 1$ with the property that*

$$\beta_n^R(R/\mathfrak{p}) \geq \beta^n \qquad for \quad n \geq 0.$$

The converse may fail: the ring $R_\mathfrak{p}$ in Example 5.2.6 is a complete intersection, but $\operatorname{curv}_R(R/\mathfrak{p}) > 1$ by Theorem 5.3.3.2. Finite modules over a complete intersection have curvature ≤ 1 by Proposition 4.2.4 and Remark 8.1.1.2, so

Corollary 8.2.2. *The following conditions are equivalent:*

(i) *R is a complete intersection.*
(ii) *$\operatorname{curv}_R M \leq 1$ for each finite R–module M.*
(iii) *$\operatorname{curv}_R k \leq 1$.*   □

The key to the proof of the theorem is to look at deviations.

Theorem 8.2.3. *When R is not a complete intersection there exist a sequence of integers $0 < s_1 < \cdots < s_j < \ldots$ and a real number $\gamma > 1$, such that*

$$\varepsilon_{s_j}(R) \geq \gamma^{s_j} \qquad for \quad j \geq 1$$

and $s_{j+1} = i_j(s_j - 1) + 2$ with integers $2 \leq i_j \leq \operatorname{edim} R + 1$.

Remark. The last result and its proof are 'looking glass images' – in the sense of [49], [33] – of a theorem of Félix, Halperin, and Thomas [66] on the rational homotopy groups $\pi_n(X) \otimes_{\mathbb{Z}} \mathbb{Q}$ of a finite CW complex X; it relies heavily on Félix and Halperin's [64] theory of rational Ljusternik-Schnirelmann category.

That theorem was used by Félix and Thomas [65] to prove Corollary 8.2.2 for graded rings over fields of characteristic 0, but the L.-S. category arguments do not extend to local rings or to positive characteristic. This is typical of a larger picture: a theorem in rational homotopy or local algebra raises a conjecture in the other field, but a proof usually requires new tools.

The arguments below use the properties of minimal models already established in Section 7.2, and the additional information contained in the next lemma, proved at the end of the section.

Lemma 8.2.4. *Let $k[Y]$ be a minimal DG algebra, such that $\operatorname{H}_n(k[Y]) = 0$ for $n \geq m$. If $\phi \colon k[Y] \to k[U]$ is a surjective morphism of DG algebras, such that U is a set of exterior variables and $\partial(U) = 0$, then $\operatorname{card}(U) < m$.*

As the proof of the theorem for rational homotopy groups, the one of the theorem on deviations proceeds in three steps. The lemma is needed for the first claim, which (now) can be obtained directly from Theorem 7.3.4, or from its precursor in [34]: *If R is not a complete intersection, then $\varepsilon_n(R) \neq 0$ for $n \gg 0$.* We present the original argument in order to keep the notes self-contained, and because of its intrinsic interest. The exposition of the arguments for Claims 2 and 3 follows [30].

Proof of Theorem 8.2.3. As may assume that R is complete, we take a minimal regular presentation $R \cong Q/I$ and a minimal model $Q[Y]$ of R over Q.

Note that the DG algebra $k[Y] = Q[Y]/\mathfrak{n}Q[Y]$ is minimal by Remark 7.2.1, that card $Y_n = \varepsilon_{n+1}(R)$ for $n \geq 1$ by Theorem 7.2.6, and that

$$H_n(k[Y]) \cong \operatorname{Tor}_n^Q (k, R) = 0 \qquad \text{for} \quad n \geq m = \operatorname{edim} R + 1. \tag{$*$}$$

Assuming that R is not a complete intersection, we show that the numbers

$$a(n) = \operatorname{card} Y_n \qquad \text{and} \qquad s(n) = \sum_{j=n}^{2n} a(j)$$

satisfy the following list of increasingly stronger properties:

Claim 1. The sequence $s(n)$ is unbounded.

Claim 2. The sequence $a(n)$ is unbounded.

Claim 3. There exist positive integers $r_1, r_2, \ldots$ with $r_{j+1} = i_j r_j + 1$ and $2 \leq i_j \leq m$, and a real number $v > 1$, such that $a(n) \geq v^{r_j}$ for each $j \geq 0$.

The last claim yields the theorem: with $\gamma = \sqrt{v}$ and $s_j = r_j + 1$ we have

$$\varepsilon_{s_j}(R) = a(r_j) > v^{r_j} = \gamma^{2r_j} \geq \gamma^{r_j+1} = \gamma^{s_j} \qquad \text{for} \quad j \geq 1.$$

For the rest of the proof, we write $Y_{[n]}$ for the span of $\bigcup_{j=n}^{2n} Y_j$, and abuse notation by letting Y_n stand also for the k–linear span of the variables $y \in Y_n$; thus, $Y_{[n]}^i$ is the k–linear span of all products involving i elements of $Y_{[n]}$.

Proof of Claim 1. Assume that there is a $c \in \mathbb{N}$ such that $s(n) \leq c$ for all $n \geq 1$.

We are going to construct for all $r \geq 1$ and $h \geq 0$ surjective morphisms of DG algebras $\phi_h^r \colon k[Y] \to k[U_h^r]$, where each U_h^r is a set $\{u_{hr+1}^r, \ldots, u_{hr+r}^r\}$ of exterior variables subject to the restrictions

$$|u_{n+1}^r| > |u_n^r| + 1 \qquad \text{for} \quad n \geq 1;$$

$$|u_{hr+i}^r| > |u_{hr+1}^r| + \cdots + |u_{hr+i-1}^r| + 1 \qquad \text{for} \quad i = 2, \ldots, r.$$

The second condition forces $\partial(U_h^r) = 0$, so in view of $(*)$ Lemma 8.2.4 implies that $r < m$. This contradiction establishes the unboundedness of $s(n)$.

By Theorems 7.2.6 and 7.3.3, there is an infinite sequence $y_1, y_2, \cdots \in Y_{\mathrm{odd}}$ with $|y_{h+1}| > |y_h| + 1$. Setting $n_h = |y_h|$, note that the compositions $k[Y] \to k[Y_{\geq n_h}] \to k[Y_{\geq n_h}]/(Y_{\geq n_h} \setminus \{y_h\})$ have the desired properties for $r = 1$.

Assume by induction that morphisms ϕ_h^r have been constructed for some $r \geq 1$. We fix $n \geq 0$, simplify the notation by setting $u_{ij} = u_{(n+i)r+j}^r$, $U_i = U_{n+i}^r$, and $\phi_i = \phi_{n+i}^r$, for $i = 0, \ldots, c$ and $j = 1, \ldots, r$, and embark on an auxiliary construction. Choose an index $q > |u_{c1}| + \cdots + |u_{cr}| + 1$ such that $Y_q \neq \varnothing$, and pick $y \in Y_q$. For $i = 0, \ldots, c$ the intervals

$$I_i = \left[(q + |u_{i1}| + 1), (q + |u_{i1}| + \cdots + |u_{ir}| + 1) \right]$$

are disjoint and contained in the interval $[q, 2q]$.

Since $s(q) \leq c$, we can choose an index i such that $Y_s = \varnothing$ for all $s \in I_i$. The restriction of ϕ_i to $B = k[Y_{<q}]$ yields a surjective morphism of DG algebras $B \to k[U_i]$. Tensoring it with $k[Y]$ over B, we get a surjective morphism $k[Y] \to C = k[U_i] \otimes_B k[Y]$. Note that $C^\natural = k[U_i] \otimes_k k[Y_{\geq q}]^\natural$ is equal to $k[U_i]$ in degrees $\leq q - 1$, to kY_q in degree q, and has no algebra generators in degrees from I_i. As the differential ∂^C is decomposable, the ideal of C generated by the variables $Y_{\geq q} \smallsetminus \{y\}$ is closed under the differential of C.

We have now constructed a surjective morphism of DG algebras

$$k[Y] \to C \to C/(Y_{\geq q} \smallsetminus \{y\})C = k[u_1 \ldots, u_{r+1}]$$

where $u_j = u_{ij}$ for $j = 1, \ldots, r$ and u_{r+1} is the image of y; clearly, the condition $|u_j| > |u_1| + \cdots + |u_{j-1}| + 1$ holds for $j = 2, \ldots, r + 1$. To end the auxiliary construction, choose an integer n' such that $rn' > (n+c+1)r$. Setting $n_1 = 1$ and $n_h = n'_{h-1}$ for $h \geq 2$, and applying the construction to n_h for $h = 1, 2, \ldots$, we get a sequence of surjective morphisms ϕ_{h+1}^{r+1} with the desired properties.

Proof of Claim 2. We assume that there exists a number c such that $\mathrm{rank}_k Y_n \leq c$ for all n, and work out a contradiction.

Fix for the moment an integer $n \geq 1$. For every $y \in Y_{\geq n}$ there are uniquely defined $\alpha_i(y) \in Y_{[n]}^i \subseteq k[Y_{\geq n}]$, such that

$$\partial(y) \equiv \sum_{i \geq 2} \alpha_i(y) \quad \mathrm{mod} \ \left((Y_{> 2n})k[Y_{\geq n}]\right).$$

Clearly, the maps $\alpha_i \colon Y_{\geq n} \to Y_{[n]}^i$, where $y \mapsto \alpha_i(y)$, are k-linear. The minimality of $k[Y]$ is inherited by $k[Y_{\geq n}]$, so there we have $\partial(Y_{[n]}^i) = 0$ for all i. Recalling from Corollary 7.2.7 that

$$\left(\mathrm{H}_{\geq 1}(k[Y_{\geq n}])\right)^m = 0 \qquad \text{for every } n \geq 1 \tag{†}$$

we see that $Y_{[n]}^m$ consists of boundaries, so we get an inclusion

$$\sum_{i=2}^m Y_{[n]}^{m-i} \alpha_i(Y_{\geq n}) \supseteq Y_{[n]}^m. \tag{‡}$$

For degree reasons, $\alpha_i(Y_j) = 0$ when $j < in+1$ or $j > i(2n) + 1$, so

$$s(in+1) = \sum_{j=in+1}^{2in+2} \mathrm{rank}_k Y_j \geq \sum_{j=in+1}^{i(2n)+1} \mathrm{rank}_k \alpha_i(Y_j) = \mathrm{rank}_k \alpha_i(Y_{\geq n}). \tag{§}$$

Set $d = (2m)^m$ and choose by Claim 1 an integer n_0 such that $s(n_0) > (md)^2$. Assume by induction on j that we have integers $n_0, n_1, \ldots, n_j$ such that

$$m(n_{h-1} + 1) \geq n_h + 1 \text{ and } s(n_h) \geq (md)s(n_{h-1}) \quad \text{for } 1 \leq h \leq j. \tag{¶}$$

Choose $n_{j+1} = ln_j + 1$ such that $s(n_{j+1}) = \max\{s(in_j + 1) \mid 2 \le i \le m\}$. It is then clear that $m(n_j + 1) \ge n_{j+1} + 1$. Using (§) and (‡), we get

$$(m-1)s(n_j)^{m-2}s(n_{j+1}) \ge \sum_{i=2}^{m} s(n_j)^{m-i}s(in_j + 1)$$

$$\ge \sum_{i=2}^{m} \left(\operatorname{rank}_k Y_{[n_j]}\right)^{m-i} \operatorname{rank}_k \alpha_i(Y_{\geqslant n_j})$$

$$\ge \operatorname{rank}_k \left(\sum_{i=2}^{m} Y_{[n_j]}^{m-i}\alpha_i(Y_{\geqslant n_j})\right) \ge \operatorname{rank}_k Y_{[n_j]}^m$$

$$\ge \binom{s(n_j)}{m} \ge \frac{s(n_j)^m}{(2m)^m} = \frac{s(n_j)}{d}s(n_j)^{m-1}$$

$$\ge (m^2d)s(n_j)^{m-1}$$

so $s(n_{j+1}) \ge (md)s(n_j)$, completing the induction step. Clearly, (¶) implies that $m^j(n_0+1) \ge n_j+1$ and $s(n_j) \ge m^j s(n_0)d^j$ hold for $j \ge 1$. Thus, we get $c(n_j+1) \ge s(n_j)$, and hence $c(n_0 + 1) \ge s(n_0)d^j$ for all j. This is absurd.

Proof of Claim 3. Set $b = (2m)^{m+1}$, choose r_1 so that $a(r_1) = a > b$, and assume by induction that $r_1, \ldots, r_j$ have been found with the property that

$$r_h = i_{h-1}r_{h-1} + 1 \quad \text{with } 2 \le i_{h-1} \le m \quad \text{and} \quad a(r_h) \ge \frac{a(r_{h-1})^{i_{h-1}}}{b}$$

for $1 \le h \le j$. The condition $\beta(y) \equiv \partial(y) \mod ((Y_{>r_j})k[Y_{\geqslant r_j}])$ defines a k-linear homomorphism $\beta: Y_{\geqslant r_j} \to \sum_{i \geqslant 2} Y_{r_j}^i$. Noting that $\beta(y) = 0$ unless $|y| \equiv 1 \pmod{r_j}$, and using the fact that $k[Y]$ is minimal and satisfies condition (†), we obtain $\sum_{i=2}^{m} Y_{r_j}^{m-i}\beta(Y_{ir_j+1}) \supseteq Y_{r_j}^m$ as in the the proof of the preceding claim. It follows that

$$\sum_{i=2}^{m} a(r_j)^{m-i}a(ir_j + 1) \ge \binom{a(r_j)}{m} \ge \frac{a(r_j)^m}{(2m)^m} = (2m)\frac{a(r_j)^m}{b}.$$

The assumption that $a(ir_j + 1) < a(r_j)^i/b$ for $2 \le i \le m$, leads to the impossible inequality $(m-1)a(r_j)^m > (2m)a(r_j)^m$. Thus, $a(ir_j+1) \ge (r_j)^i/b$ for some $i = i_j$, so the induction step is complete with $r_{j+1} = i_j r_j + 1$.

The quantities $P_j = (i_j \cdots i_1)$ and $S_j = \sum_{h=2}^{j}(i_j \cdots i_{j-h})$ satisfy

$$P_j > P_j\left(\frac{1}{2} + \cdots + \frac{1}{2^j}\right) \ge P_j\left(\frac{1}{i_1} + \cdots + \frac{1}{i_1 \cdots i_j}\right) = S_j.$$

Thus, $P_j r_1 > S_j r_1 + 1 = r_{j+1}$, and hence $P_j > r_{j+1}/r_1$. Since $a > b$, we have

$$a(r_{j+1}) \ge \frac{a(r_j)^{i_j}}{b} \ge \frac{a(r_{j-1})^{(i_j i_{j-1})}}{b^{1+i_j}} \ge \cdots \ge \frac{a^{P_j}}{b^{S_j}} > \left(\frac{a}{b}\right)^{P_j} > \left(\frac{a}{b}\right)^{\frac{r_{j+1}}{r_1}}.$$

To finish the proof of the claim, note that $v > 1$ and set $v = \sqrt[r_1]{a/b}$.   □

Proof of Theorem 8.2.1. Since $\beta_n^R(R/\mathfrak{p}) \geq \beta_n^{R_\mathfrak{p}}(R_\mathfrak{p}/\mathfrak{p}R_\mathfrak{p})$ for each n, we may assume that $\mathfrak{p} = \mathfrak{m}$; set $\beta_n = \beta_n^R(R/\mathfrak{m})$ and $e = \operatorname{edim} R$.

As R is not a complete intersection, Theorem 8.2.3 provides an infinite sequence $s_1, s_2, \ldots$ with $(e+1)s_j > s_{j+1}$, such that $\varepsilon_{s_j}(R) \geq \gamma^{s_j}$ for some real number $\gamma > 1$. For $n \geq 2$ we have $\beta_n > \beta_1 = e > 1$ by Remark 8.1.1.3, hence

$$\beta = \min\left\{ \sqrt[e+1]{\gamma}, \beta_1, \ldots, \sqrt[s_1]{\beta_{s_1}} \right\} > 1$$

and $\beta_n \geq \beta^n$ for $s_1 \geq n \geq 0$. If $s_{j+1} \geq n > s_j$ with $j \geq 1$, then

$$\beta_n > \beta_{s_j} \geq \varepsilon_{s_j}(R) \geq \gamma^{s_j} \geq \beta^{(e+1)s_j} > \beta^{s_{j+1}} > \beta^n$$

so the desired inequality $\beta_n \geq \beta^n$ holds for all $n \geq 0$. $\square$

Proof of Lemma 8.2.4. Since $\phi^\natural \colon k[Y]^\natural \to k[U]$ is a surjective homomorphism of graded free k–algebras, the ideal $\operatorname{Ker} \phi^\natural$ has a linearly independent generating set $Y' = \{y'_1, \ldots, y'_j, \ldots\} \subset kY$, which we can assume ordered in such a way that $|y'_{j+1}| \geq |y'_j|$ for $j \geq 1$. As $\operatorname{Ker} \phi$ is a DG ideal, we have $\partial(y'_1) = 0$ and $\partial(y'_{j+1}) \in (y'_1, \ldots, y'_j)$ for $j \geq 1$. Assume that for some $j \geq 1$ the morphism ϕ factors through a quasi-isomorphism

$$\pi^j \colon A^j = k[Y]\langle x'_1, \ldots, x'_j\rangle \xrightarrow{\pi^j} k[Y]/(y'_1, \ldots, y'_j) = B^j$$

that maps $x'^{(n)}_i$ to zero for $1 \leq i \leq j$ and $n \geq 1$. As the $\overline{y}'_{j+1} = \pi^j(y'_{j+1})$ is a cycle in B^j, there is a cycle $z_{j+1} \in A^j$ such that $\pi^j(z_{j+1}) = \pi^j(y'_{j+1})$. By Lemma 7.2.10, π^j extends to a quasi-isomorphism

$$\pi^j\langle x'_{j+1}\rangle \colon A^{j+1} = A^j\langle x'_{j+1} \mid \partial(x'_{j+1}) = z_{j+1}\rangle \longrightarrow B^j\langle x'_{j+1} \mid \partial(x'_{j+1}) = \overline{y}'_{j+1}\rangle \,.$$

Lemma 7.2.11 shows that the map $\xi^j \colon B^j\langle x'_{j+1}\rangle \longrightarrow B^j/(y'_{j+1})$ that sends $\sum_i b_i x'^{(i)}_{j+1}$ to $b_0 + (y'_{j+1})$ is a quasi-isomorphism; thus, so is

$$\pi^{j+1} = \xi^j \circ \pi^j\langle x'_{j+1}\rangle \colon A^{j+1} = B^j\langle x'_{j+1}\rangle \longrightarrow B^j/(y'_{j+1}) = B^{j+1} \,.$$

In the limit, we obtain a factorization of ϕ in the form

$$k[Y] \hookrightarrow k[Y]\langle X'\rangle \xrightarrow{\pi} k[U]$$

with a surjective quasi-isomorphism π, such that $\operatorname{Ker} \pi$ is generated by $\operatorname{Ker} \phi$ and $x'^{(i)}_j$ with $i, j \geq 1$. By Lemma 7.2.10, π extends to a quasi-isomorphism

$$\pi\langle X''\rangle \colon k[Y]\langle X'\rangle\langle X''\rangle \longrightarrow k[U]\langle X'' \mid \partial(x''_j) = u_j\rangle \,.$$

where $X'' = \{x''_1, \ldots, x''_m\}$. The DG algebra on the right is quasi-isomorphic to k, so we get a semi-free resolution $W' = k[Y]\langle X' \cup X''\rangle$ of k over $k[Y]$. Another semi free resolution $W = k[Y]\langle X\rangle$ of k over $k[Y]$, such that $X = \{x_y : |x_y| = |y|+1, y \in Y\}$, and $\partial(k[Y]\langle X\rangle) \subseteq (Y)W$, is given by Proposition 7.2.9 (applied with $Q = k$). By Propositions 1.3.1 and 1.3.2, the vector spaces $V' = k \otimes_{k[Y]} W' = k\langle X' \cup X''\rangle$ and

$V = k \otimes_{k[Y]} W = k\langle X \rangle$ are quasi-isomorphic. As $\partial^V = 0$, we get (in)equalities of formal power series

$$\frac{\displaystyle\prod_{i=1}^{\infty}(1 + t^{2i-1})^{\mathrm{card}(X_{2i-1})}}{\displaystyle\prod_{i=1}^{\infty}(1 - t^{2i})^{\mathrm{card}(X_{2i})}} = \sum_n \mathrm{rank}_k\, V_n t^n = \sum_n \mathrm{rank}_k\, \mathrm{H}_n(V) t^n$$

$$= \sum_n \mathrm{rank}_k\, \mathrm{H}_n(V') t^n \preccurlyeq \sum_n \mathrm{rank}_k\, V'_n t^n$$

$$= \frac{\displaystyle\prod_{i=1}^{\infty}(1 + t^{2i-1})^{\mathrm{card}(X'_{2i-1})}}{\displaystyle\prod_{i=1}^{\infty}(1 - t^{2i})^{\mathrm{card}(X'_{2i})}} \cdot \frac{\displaystyle\prod_{i=1}^{\infty}(1 + t^{2i-1})^{\mathrm{card}(X''_{2i-1})}}{\displaystyle\prod_{i=1}^{\infty}(1 - t^{2i})^{\mathrm{card}(X''_{2i})}} \cdot$$

On the other hand, by construction we have for each j an equality

$$\mathrm{card}\, X_{j+1} = \mathrm{card}\, Y_j = \mathrm{card}\, Y'_j + \mathrm{card}\, U_j = \mathrm{card}\, X'_{j+1} + \mathrm{card}\, X''_{j+1}\,.$$

It follows that $\mathrm{H}(V') = V'$, that is, that $\partial(W') \subseteq (Y)W'$.

As $W'^{\natural}$ is a free module over $k[Y]\langle X' \rangle^{\natural}$, and $\mathrm{H}_{\geqslant 1}(W) = 0$, we see that

$$Z_{\geqslant 1}(k[Y]\langle X' \rangle) = Z_{\geqslant 1}(W') \cap (k[Y]\langle X' \rangle) = \partial(W') \cap (k[Y]\langle X' \rangle)$$
$$\subseteq (Y)W' \cap k[Y]\langle X' \rangle = (Y)k[Y]\langle X' \rangle\,.$$

Since $\pi\colon k[Y]\langle X' \rangle \to k[U]$ is a surjective quasi-isomorphism, we can find $z_1, \ldots, z_m \in Z(k[Y]\langle X' \rangle)$ with $\pi(z_i) = u_i$. For them we have

$$z_1 \cdots z_m \in \bigl(Z_{\geqslant 1}(k[Y]\langle X' \rangle)\bigr)^m \subseteq Z\bigl((Y)^m k[Y]\langle X' \rangle\bigr) \subseteq Z(Jk[Y]\langle X' \rangle)$$

where $J \subset k[Y]$ is defined by

$$J_n = \begin{cases} 0 & \text{for } n < m\,; \\ \partial(k[Y]_{m+1}) & \text{for } n = m\,; \\ k[Y]_n & \text{for } n > m\,. \end{cases}$$

For degree reasons, J is a DG ideal of $k[Y]$. By hypothesis $\mathrm{H}_n(k[Y]) = 0$ for $n \geq m$, so $\mathrm{H}(J) \cong \mathrm{H}(k[Y]) = 0$. Thus, the projection $\tau\colon k[Y] \to k[Y]/J$ is a quasi-isomorphism; Proposition 1.3.2 then shows that the induced map $k[Y]\langle X' \rangle \to k[Y]\langle X' \rangle / Jk[Y]\langle X' \rangle$ is one, hence

$$Z(Jk[Y]\langle X' \rangle) = \partial(Jk[Y]\langle X' \rangle) \subseteq \partial(k[Y]\langle X' \rangle)$$

hence $\mathrm{cls}(z) = 0$. The computation $0 = \mathrm{H}(\pi)(\mathrm{cls}(z_1) \cdots \mathrm{cls}(z_m)) = u_1 \ldots u_m \neq 0$ now yield the desired contradiction. $\qquad\square$

8.3. Conormal modules. In this section we fix a presentation $R = Q/I$, where $(Q, \mathfrak{n}, k)$ is a local or graded ring, and $I \subseteq \mathfrak{n}^2$ is minimally generated by $\boldsymbol{f}$. The R–module I/I^2 is called the *conormal module* of the presentation[30].

If $\boldsymbol{f}$ is a regular sequence, then it is well known and easy to see that the image of $\boldsymbol{f}$ modulo I^2 is a basis of the conormal module, and the projective dimension $\operatorname{pd}_Q R$ is finite. The starting point of the present discussion is a well known converse, due to Ferrand [68] and Vasconcelos [154]:

Theorem 8.3.1. *If $\operatorname{pd}_Q R < \infty$ and the R–module I/I^2 is free, then $\boldsymbol{f}$ is a regular sequence.* ∎

Later, Vasconcelos [155] conjectured a considerably stronger statement: *If $\operatorname{pd}_Q R < \infty$ and $\operatorname{pd}_R(I/I^2) < \infty$, then $\boldsymbol{f}$ is a regular sequence.* Various known cases of small projective dimension are surveyed in [156]; the one below is proved by Vasconcelos and Gulliksen.

Theorem 8.3.2. *The conjecture holds if $\operatorname{pd}_R(I/I^2) \leq 1$.*

Proof. In view of the preceding theorem, it suffices to assume that $\operatorname{pd}_R(I/I^2) = 1$, and draw a contradiction.

For the Koszul complex $E = Q\langle X_1 \mid \partial(X_1) = \boldsymbol{f}\rangle$, set $Z = \operatorname{Z}_1(E)$ and $H = \operatorname{H}_1(E)$. Tensoring the exact sequence $0 \to Z \to Q^r \to I \to 0$ with R over Q, we get an exact sequence of R–modules $Z/IZ \to R^r \to I/I^2 \to 0$. As $\partial(E_2) \subseteq IE_1$, we have an induced exact sequence $H \to R^r \to I/I^2 \to 0$. The assumption $\operatorname{pd}_R(I/I^2) = 1$ then implies that H contains a free direct summand $R\operatorname{cls}(z) \cong R$; note that $z \in \mathfrak{n}E_1$, because $\boldsymbol{f}$ minimally generates I.

Let $E\langle X_{\geqslant 2}\rangle = Q\langle X\rangle$ be an acyclic closure of $R = \operatorname{H}_0(E)$ over E, such that $\partial(x) = z$ for some $x \in X_2$. The cokernel of the differential $\delta_2 \colon RX_2 \to RX_1$ of the complex of indecomposables $\operatorname{Ind}_Q^\gamma Q\langle X\rangle$ is equal to H, so Proposition 6.2.7 yields a Q–linear Γ–derivation $\vartheta \colon Q\langle X\rangle \to Q\langle X\rangle$ of degree -2, with $\vartheta(x) = 1$.

The Γ–derivation $\theta = \operatorname{H}(\vartheta \otimes_Q k)$ of $Q\langle X\rangle \otimes_Q k = k\langle X\rangle$ has $\theta(x) = 1$. As $\partial(x) = z \otimes 1 = 0 \in E_1/\mathfrak{n}E_1$, each $x^{(i)}$ is a cycle. Assuming that $x^{(i)} = \partial(v)$, we get $1 = \theta^i(x^{(i)}) = \theta^i\partial(v) = 0$, which is absurd. Thus, $0 \neq \operatorname{H}_{2i}(k\langle X\rangle) = \operatorname{Tor}_{2i}^Q(R, k)$ for all $i \geq 0$, contradicting the hypothesis that $\operatorname{pd}_Q R$ is finite. □

Next we present the results of Avramov and Herzog [35] on graded ring.

Theorem 8.3.3. *Let $Q = k[s_1, \ldots, s_e]$ be a graded polynomial ring over a field k of characteristic 0, with variables of positive degree, let I be a homogeneous ideal of Q, and set $R = Q/I$. The following conditions are equivalent.*

(i) *R is a complete intersection.*
(ii) *$\operatorname{pd}_R(I/I^2) < \infty$.*
(iii) *$\operatorname{cx}_R(I/I^2) < \infty$.*
(iv) *$\operatorname{curv}_R(I/I^2) \leq 1$.*

If R is not a complete intersection, then $\operatorname{curv}_R I/I^2 = \operatorname{curv}_R k$.

---

[30]Or: of the embedding $\operatorname{Spec}(R) \subseteq \operatorname{Spec}(Q)$.

The result is proved together with the next one:

Theorem 8.3.4. *If R is as in the preceding theorem, and $\Omega_{R|k}$ is its module of Kähler differentials over k, then the following conditions are equivalent.*

    (i) *R is a complete intersection.*

    (ii) $\mathrm{cx}_R(\Omega_{R|k}) < \infty$.

    (iii) $\mathrm{curv}_R(\Omega_{R|k}) \leq 1$.

If R is not a complete intersection, then $\mathrm{curv}_R \Omega_{R|k} = \mathrm{curv}_R k$.

Remark. If $\mathrm{pd}_R \Omega_{R|k} < \infty$, then the theorem implies that R is a complete intersection – another conjecture of Vasconcelos – but there is more.

If $\mathfrak{p}$ is a minimal prime ideal, then $\Omega_{R_\mathfrak{p}|k} \cong (\Omega_{R|k})_\mathfrak{p}$ has finite projective dimension over $R_\mathfrak{p}$. Thus, it is free, hence $R_\mathfrak{p}$ is regular by the Jacobian criterion, and so R is *reduced* by Serre's criterion. Conversely, if R is a reduced complete intersection, then Ferrand [68] and Vasconcelos [154] prove that $\mathrm{pd}_R \Omega_{R|k} \leq 1$.

The asymptotic results are easy consequences of more precise termwise inequalities[31] for the graded invariants described in Remark 1.2.10.

Theorem 8.3.5. *In the notation of Theorem 8.3.3, for all $n \geq 0$ and $j \in \mathbb{Z}$ there is an inequality between graded Betti numbers and deviations:*

$$\beta^R_{nj}(\Omega_{R|k}) \geq \varepsilon_{n+1,j}(R) \qquad and \qquad \beta^R_{nj}(I/I^2) \geq \varepsilon_{n+2,j}(R).$$

Our proof proceeds through a structural result on the resolution of I/I^2, that depends on the grading and on the characteristic; the following is open:

Problem 8.3.6. When R is a local ring and $R \cong Q/I$ is a regular presentation, does an inequality $\beta^R_n(I/I^2) \geq \varepsilon_{n+2}(R)$ hold for each $n \geq 0$?

In the arguments, we use graded versions of some basic constructions.

Remark 8.3.7. The first step in the construction of a minimal model of R over Q is a Koszul complex on the set $\boldsymbol{f}$ of minimal generators of I; we choose $\boldsymbol{f}$ to consist of homogeneous elements, so the first Koszul homology is a finite graded Q–module. Assume by induction that $\mathrm{H}_n(Q[Y_{\leqslant n}])$ has the same property for some $n \geq 1$; to kill it we adjoin a minimal set of homogeneous generators, and assign to each variable $y \in Y_{n+1}$ an internal degree, equal to that of $\partial(y)$.

Thus, we get a *graded minimal model* $Q[Y_{\geqslant 1}] = k[Y]$ of R over Q. Similar considerations yield a *graded acyclic closure* $R\langle X \rangle$ of k over R. The arguments in Sections 6.3 and 7.2 are compatible with the internal gradings, so the 'obvious' graded versions of the results proved there are available.

Remark 8.3.8. Proposition 6.2.3 can be repeated for ordinary (that is, not subject to a condition involving divided powers) k–linear derivations of the DG algebra $k[Y]$ over k, to produce a *DG module of differentials* $\mathrm{Diff}_k k[Y]$ over $k[Y]$. It is semi-free with basis $\{dy : |dy| = |y|;\ \deg(dy) = \deg(y)\}_{y \in Y}$, where $\deg(a)$

---

[31]Equalities hold for $n = 0$ by Corollary 7.1.5, but it appears that the other inequalities are strict unless R is a (reduced) complete intersection.

is the internal degree of a; the map $y \mapsto dy$ extends to a universal derivation $d \colon k[Y] \to \mathrm{Diff}_k\, k[Y]$; the differential is determined by $\partial(dy) = d(\partial(y))$; each k-linear derivation of $k[Y]$ into a DG module U over $k[Y]$ factors uniquely as the composition of d with a homomorphism of DG modules $\mathrm{Diff}_k\, k[Y] \to U$.

Consider the complex of free R–modules $L = R \otimes_{k[Y]} \mathrm{Diff}_k\, k[Y]$. (Using Lemma 7.2.3 on the uniqueness of minimal models and a functorial construction of $\mathrm{Diff}_k\, k[Y]$, it can be shown that this complex is defined uniquely up to isomorphism by the k–algebra R; we do not use that here, and refer to [35] for details.) For $g \in QY_1$, an easy computation shows that the differential

$$\partial_1 \colon L_1 \to L_0 \quad \text{acts by} \quad \partial_1(1 \otimes g) = 1 \otimes \sum_{i=1}^{e} \frac{\partial f}{\partial y_i} dy_i \quad \text{where} \quad \partial_1(g) = f \in Q\,.$$

On the other hand, the 'second fundamental exact sequence' for the module $\Omega_{R|k}$ of *Kähler differentials* of the k–algebra R has the form

$$I/I^2 \xrightarrow{\ \delta\ } R \otimes_Q \Omega_{Q|k} \to \Omega_{R|k} \to 0 \quad \text{with} \quad \delta(f + I^2) = 1 \otimes \sum_{i=1}^{e} \frac{\partial f}{\partial y_i} dy_i\,.$$

As $\Omega_{Q|k}$ is free with basis $\{dy_1, \ldots, dy_e\}$, we conclude that $\mathrm{H}_0(L) = \Omega_{R|k}$.

Recall from Remark 4.1.7, that an augmentation $\epsilon \colon F \to N$ of a complex of free R–modules F is essential, if for some lifting $\alpha \colon F \to G$ to a minimal resolution G of N, the map $k \otimes_R \alpha$ is injective. In that case, α maps F isomorphically onto a subcomplex of G, that splits off as a graded R–module.

Theorem 8.3.9. *The augmentation* $\epsilon^L \colon L \to \mathrm{H}_0(L) = \Omega_{R|k}$ *is essential.*

A special morphism is at the heart of the arguments to follow.

Construction 8.3.10. Euler morphisms. The graded algebra R has an *Euler derivation* $R \to \mathfrak{m}$, that multiplies each homogeneous element $a \in R$ by its (internal) degree. By Proposition 1.3.1, the R–linear map $\gamma \colon \Omega_{R|k} \to \mathfrak{m}$ that it defines lifts to a morphism $\omega \colon \mathrm{Diff}_k\, k[Y] \to V$ of DG modules over $k[Y]$, where $\epsilon^V \colon V \to \mathfrak{m}$ is a semi-free resolution of $\mathfrak{m}$ over $k[Y]$. We call such a lifting an *Euler morphism*; it is unique up to $k[Y]$–linear homotopy.

Lemma 8.3.11. *Let* $k[Y]$ *be a graded minimal model of* R *over* k, *and let* $U = k[Y]\langle X \rangle$ *be a graded acyclic closure of* k *over* $k[Y]$, *as in Remark 8.3.7.*

The DG module $V = \Sigma^{-1}(U/k[Y])$ *is a semi-free resolution of* $\mathfrak{m}$ *over* $k[Y]$, *and there is an Euler morphism* $\omega \colon \mathrm{Diff}_k\, k[Y] \to V$, *such that*

$$\omega(dy) \equiv -\deg(y) x_y \mod \mathfrak{n} X_{n+1} + \big(k[Y_{\leqslant n+1}]\langle X_{\leqslant n} \rangle\big)_{n+1} \quad \text{for} \quad y \in Y_n\,.$$

Proof. Set $D^n = \coprod_{|y| \leqslant n} k[Y] dy \subseteq \mathrm{Diff}_k\, k[Y] = D$.

The map $a \mapsto \deg(a)a$ is a k–linear chain Γ–derivation $k[Y] \to U$. In degree zero homology it induces the zero map $R \to k$, so it is homotopic to 0. If $\xi \colon D \to U$ is the $k[Y]$–linear morphism that corresponds to it by Proposition 6.2.3, then ξ is homotopic to 0. We set $\xi^n = \xi|_{D^n}$ and by induction on n construct $k[Y]$–linear homotopies $\sigma^n\colon D^n \to U$ between ξ^n and 0, such that

$$\sigma^n|_{D^{n-1}} = \sigma^{n-1}\,;$$
$$\sigma^n(dy) \equiv \deg(y)x_y \quad \mathrm{mod}\ \mathfrak{n}X_n + \big(k[Y_{\leqslant n}]\langle X_{<n}\rangle\big)_n\,. \tag{$*$}$$

If $|y| = 0$, then set $\sigma^0(dy) = \deg(y)x_y$: clearly, the formula above holds. Let $n \geq 1$, and assume by induction that σ^{n-1} has been found. It is easy to check that $\xi(dy) - \deg(y)\partial(x_y) - \sigma^{n-1}\partial(dy)$ is a cycle; as $n \geq 1$, it is a boundary, that we write as $\partial(u_y + v_y)$ with $u_y \in QX_{n+1}$, and $v_y \in k[Y_{\leqslant n+1}]\langle X_{\leqslant n}\rangle_{n+1}$. Because d is a derivation and $\partial(Y) \subseteq (Y)^2 k[Y]$, we get

$$d(\partial Y) \subseteq d\big((Y)^2 k[Y]\big) \subseteq (Y)d\big(k[Y]\big) = (Y)D$$

Since σ^{n-1} is $k[Y]$–linear, this implies:

$$\sigma^{n-1}\partial(dy) = \sigma^{n-1}d(\partial(y)) \in W_n = \mathfrak{n}Y_n + k[Y_{<n}]\langle X_{\leqslant n}\rangle_n\,.$$

By Proposition 7.2.9, we have $\partial(v_y) \in W_n$, hence

$$\partial(u_y) = \xi(dy) - \deg(y)\partial(x_y) - \sigma^{n-1}\partial(dy) - \partial(v_y) \in W_n\,.$$

By the same theorem, we conclude that $u_y \in \mathfrak{n}X_{n+1}$. The map $\sigma^n\colon D^n \to U$, $\sigma^n(dy) = \deg(y)x_y + u_y + v_y$, defines a homomorphism of DG modules over $k[Y]$ that satisfies ($*$). As for $|y| \leq n$ we have

$$\partial\sigma^n(dy) + \sigma^n\partial(dy) = \deg(y)\partial(x_y) + \partial(u_y + v_y) + \sigma^{n-1}\partial(dy) = \xi(dy)\,,$$

the induction step of the construction is complete.

In the limit, the maps σ^n define a homotopy $\sigma\colon D \to U$ between ξ and 0. Let $\omega\colon D \to V$ be the composition of σ with the canonical $k[Y]$–linear, degree -1 homomorphism $U \to U/k[Y] \to \Sigma^{-1}(U/k[Y]) = V$. As $\mathrm{Im}\,\xi \subseteq k[Y]$, the equality $\partial\sigma + \sigma\partial = \xi$ implies $\partial\omega = \omega\partial$, so ω is a chain map $D \to V$.

The homology exact sequence of $0 \to k[Y] \to U \to U/k[Y] \to 0$ yields $\mathrm{H}_n(U/k[Y]) = 0$ for $n \neq 1$ and $\mathrm{H}_1(V) = \mathfrak{m}$, so $V = \Sigma^{-1}(U/k[Y])$ is a semi-free resolution of $\mathfrak{m}$. For $n = 0$, formula ($*$) shows that $\mathrm{H}_0(\omega)\colon \Omega_{R|k} \to \mathfrak{m}$ is the homomorphism induced by the Euler derivation; for $n \geq 1$, the formula yields a congruence $\omega(dy) \equiv -\deg(y)x_y \quad \mathrm{mod}\ \mathfrak{n}X_{n+1} + \big(k[Y_{\leqslant n+1}]\langle X_{\leqslant n}\rangle\big)_{n+1}$. □

Proof of Theorem 8.3.9. Let $\omega\colon \mathrm{Diff}_k\,k[Y] \to V$ be the Euler morphism, constructed in the preceding lemma, and consider the induced morphism

$$\varpi\colon L = R \otimes_{k[Y]} \mathrm{Diff}_k\,k[Y] \xrightarrow{R \otimes \omega} R \otimes_{k[Y]} V = G$$

of complexes of graded R–modules. The lemma yields congruences

$$(k \otimes_R \varpi)(1 \otimes dy) \equiv 1 \otimes \deg(y)x_y \quad \mathrm{mod}\ (k\langle X_{\leqslant n}\rangle_{n+1}) \text{ for } y \in Y_n \text{ and } n \geq 0\,,$$

which show[32] that $k \otimes_R \varpi$ is injective. Furthermore, $\mathrm{H}_0(\varpi)$ is the homomorphism $\gamma\colon \Omega_{R|k} \to \mathfrak{m}$ defined by the Euler derivation.

---

[32]This is the only place where the hypothesis of characteristic 0 is used.

By Proposition 1.3.2, the quasi-isomorphism $\rho\colon k[Y] \to R$ induces a quasi-isomorphism $\rho \otimes V\colon V = k[Y] \otimes_{k[Y]} V \to R \otimes_{k[Y]} V = G$, so G is a minimal free resolution of $\mathfrak{m}$ over R. Let F be a minimal free resolution of $\Omega_{R|k}$ over R, let $\alpha\colon L \to F$ be a lifting of the identity map of $\Omega_{R|k}$, and let $\beta\colon F \to G$ be a lifting of γ. Since $H_0(\beta\alpha) = \gamma$, the morphisms ϖ and $\beta\alpha$ are homotopic. As noted in Remark 4.1.7, this yields

$$k \otimes_R \varpi = k \otimes_R (\beta\alpha) = (k \otimes_R \beta)(k \otimes_R \alpha),$$

so $k \otimes_R \alpha$ is injective. This is the desired assertion. $\qquad\square$

Proof of Theorem 8.3.5. By construction, L_n is a free R–module with basis Y_n, and $\mathrm{card}(Y_n) = \varepsilon_{n+1}(R)$ by Theorem 7.2.6. The inequalities for the Betti numbers of $\Omega_{R|k}$ follow from the result that we have just proved.

The morphism ϖ used in its proof induces a morphism $\varpi'\colon L' = \Sigma^{-1}L_{\geqslant 1} \to \Sigma^{-1}G_{\geqslant 1} = G'$, such that $\varpi'^{\natural}$ is a split injection of R–modules. An easy computation shows that $H_0(L') = I/I^2$, so replacing in the preceding argument F by a minimal resolution of I/I^2, we conclude that $\epsilon^{L'}\colon L' \to I/I^2$ is essential. That gives the second series of inequalities. $\qquad\square$

Proof of Theorem 8.3.3 and Theorem 8.3.4. In view of Corollary 8.2.2, in each case it suffices to prove the last assertion. Using Proposition 4.2.4.1, Theorem 8.3.5, and Proposition 8.1.4, we get

$$\mathrm{curv}_R\, k \geq \mathrm{curv}_R(I/I^2) = \limsup \sqrt[n]{\beta_n^R(I/I^2)} \geq \limsup \sqrt[n]{\varepsilon_n(R)} = \mathrm{curv}_R\, k\,.$$

We have Theorem 8.3.3. An identical argument yields Theorem 8.3.4. $\qquad\square$

9. Modules over Complete Intersections

Currently, homological algebra over complete intersections is an active area of research on infinite free resolutions. This chapter describes some basic techniques and results. Most proofs depend on a remarkable higher level structure on resolutions, introduced in Section 1 under more general hypotheses. It is then applied to modules over complete intersections, to study Betti numbers in Section 2, and other homological problems in Section 3.

9.1. Cohomology operators.
In this section $R = Q/(\boldsymbol{f})$, where $\boldsymbol{f} = f_1, \ldots, f_r$ is a regular sequence in a (not necessarily regular local) commutative ring Q. We denote $E = Q[y_1, \ldots, y_r \mid \partial(y_j) = f_j]$ the Koszul complex on $\boldsymbol{f}$, and let $\kappa\colon E \to R$ be its canonical augmentation.

Extending Shamash's [142] construction of resolutions over hypersurface sections, cf. Theorem 3.1.3, Eisenbud [57] produces (in a finite number of steps, if $\mathrm{pd}_Q M$ is finite). a free resolution of an R–module M starting from any free resolution of M over Q. Here we present a version of that construction, from Avramov and Buchweitz [31]; the result is somewhat weaker, but easier to prove and sufficient for our purposes.

Theorem 9.1.1. *Let M be a finite R–module, let $\epsilon^U : U \to M$ be a DG module resolution of M over E such that U_n is a free Q–module for each n.*

Let $G = Q\langle v_1, \dots, v_r \rangle$ be a Q–module with basis $\{ v^{(H)} = v_1{}^{(h_1)} \cdots v_r{}^{(h_r)} :$ $|v^{(H)}| = 2(h_1|v_1| + \cdots + h_r|v_r|),\ H = (h_1, \dots, h_r) \in \mathbb{N}^r \}$, and set

$$C_n(E, U) = \bigoplus_{i \geqslant 0} \overline{G}_i \otimes_R \overline{U}_{n-i,};$$

$$\partial(v^{(H)} \otimes u) = -\sum_{j=1}^{r} v^{(H_j)} \otimes y_j u + v^{(H)} \otimes \partial(u)$$

where $\overline{G}_i = R \otimes_Q G_i$, $\overline{U}_j = R \otimes_Q U_j$, and $H_j = (h_1, \dots, h_j - 1, \dots, h_r)$.
Then $(C(E, U), \partial)$ is a free resolution of M over R.

Remark. The Koszul complex K on a regular sequence s with $(s) \supseteq f$ is a DG module over E; by inspection, $C(E, K) = C$, the resolution of Corollary 6.1.9.
Proof. Let $\mu \colon E \otimes_Q E \to E$ be the morphism of DG algebras, given by the multiplication of the exterior algebra. An elementary computation shows that $\operatorname{Ker} \mu$ is generated by $y_j' = y_j \otimes 1 - 1 \otimes y_j$, for $j = 1, \dots, r$. Thus, μ is the composition of $(E \otimes_Q E) \hookrightarrow D = (E \otimes_Q E)\langle v_1, \dots, v_r \mid \partial(v_j) = y_j' \rangle$ with $\nu \colon D \to E$, where $\nu(v^{(H)}) = 0$ if $|H| > 0$. By Proposition 1.3.2, the map

$$E \otimes_Q \kappa \colon E \otimes_Q E \to E \otimes_Q R = R\langle y_1, \dots, y_r \mid \partial(y_j) = 0 \rangle$$

is a quasi-isomorphism. As $(E \otimes_Q \kappa)(y_j') = y_j$, we see that $H(E \otimes_Q E)$ is the exterior algebra on $H_1(E \otimes_Q E)$, itself a free R–module with basis $\operatorname{cls}(y_1'), \dots, \operatorname{cls}(y_r')$. Thus, Proposition 6.1.7 applied to the Γ-extension $E \otimes_Q E \hookrightarrow D$, shows that ν is a quasi-isomorphism of DG algebras.

Since ν is a morphism of semi-free DG modules over E for the action of E on the right, by Proposition 1.3.3 so is $\nu \otimes_E U \colon D \otimes_E U \to E \otimes_E U = U$, hence $H(D \otimes_E U) \cong M$. On the other hand, $(D \otimes_E U)^\natural \cong E^\natural \otimes_Q G \otimes_Q U^\natural$ is a semi-free DG module for the action of E on the left. Thus, by Proposition 1.3.2 the morphism $\kappa \otimes_E U \colon D \otimes_E U \to R \otimes_E D \otimes_E U$ is a quasi-isomorphism. Comparison shows that $R \otimes_E D \otimes_E U = C(E, U)$ as complexes of R–modules. $\square$

Construction 9.1.2. Cohomology operators. Let $\mathcal{S} = R[\chi_1, \dots, \chi_r]$ be a graded algebra with variables $\chi_1, \dots, \chi_r$ of degree[33] -2. In the notation of the preceding theorem, set $\chi_j \cdot v^{(H)} = v^{(H_j)}$ for $1 \leq j \leq r$. These are R–linear endomorphisms of degree -2 of $C(E, U)^\natural$. They clearly commute with each other, and a glance at the formula for the differential ∂ of the complex $C(E, U)$ shows that they are chain maps: $\chi_j \partial = \partial \chi_j$. Thus, $C(E, U)$ is a DG module over the graded *algebra*[34] *of cohomology operators* $\mathcal{S}$ of the presentation $R \cong Q/(f)$.

---

[33]This will not be surprising, once the χ_j's reveal their cohomological nature.
[34]The algebra $\mathcal{S}$ itself has a trivial differential; this nicely illustrates the fact that DG module structures are to be found in all walks of life.

The construction above, taken from [31], is a variant of that of Eisenbud [57], cf. Construction 9.1.5. The introduction of operators of degree -2 on (co)homology is due to Gulliksen [80]; other constructions have been given by Mehta [119] and Avramov [25]. For a long time, it had been held that they coincide, but a close reading of the published arguments has revealed serious flaws. In fact, they yield the same result, but only up to sign: this is proved in [37]; ironically, that proof introduces two new constructions.

Proposition 9.1.3. *For each R–module N there are $\mathcal{S}$–linear homomorphisms*

$$\chi_j \colon \operatorname{Tor}_n^R(M, N) \to \operatorname{Tor}_{n-2}^R(M, N)$$
$$\chi_j \colon \operatorname{Ext}_R^n(M, N) \to \operatorname{Ext}_R^{n+2}(M, N) \qquad \text{for} \quad 1 \le j \le c \quad \text{and all} \quad n,$$

which turn $\operatorname{Tor}^R(M, N)$ and $\operatorname{Ext}_R(M, N)$ into modules over $\mathcal{S}$.

These structures depend only on $\boldsymbol{f}$, are natural in both module arguments, and commute with the connecting maps induced by short exact sequences.

Proof. For the first statement, observe that for each R–module N, the complexes $\mathrm{C}(E, U) \otimes_R N$ and $\operatorname{Hom}_R(\mathrm{C}(E, U), N)$ have an induced structure of DG $\mathcal{S}$–module. Naturality in N is clear, as is linearity of the connecting homomorphisms induced by an exact sequence $0 \to N' \to N \to N'' \to 0$.

If $\beta \colon M' \to M$ is a homomorphism of R–modules, and U' is a resolution of M' given by Construction 2.2.7, then by the lifting property of Proposition 1.3.1 there is a morphism $\alpha \colon U' \to U$ of DG modules over E such that $\mathrm{H}(\alpha) = \beta$. The expressions for the differential in Theorem 9.1.1, and for the action of χ_j in Construction 9.1.2 show that $v^{(H)} \otimes u' \mapsto v^{(H)} \otimes \beta(u')$ defines a morphism of DG $\mathcal{S}$–modules $\mathrm{C}(E, \alpha) \colon \mathrm{C}(E, U') \to \mathrm{C}(E, U)$. All choices of α are homotopic, so the degree 0 maps of $\mathcal{S}$–modules $\mathrm{H}(\mathrm{C}(E, \alpha) \otimes_R N)$ and $\mathrm{H} \operatorname{Hom}_R(\mathrm{C}(E, \alpha), N)$ are uniquely defined, and equal respectively to $\operatorname{Tor}^R(\beta, N)$ and $\operatorname{Ext}_R(\beta, N)$. This proves naturality in M, and independence from the choice of U.

Let $0 \to M' \to M \to M'' \to 0$ be a short exact sequence of R–modules, and choose a semi-free resolution U'' of M'' over E, such that $U''^{\natural}$ is a free module over $E^{\natural}$. By the usual 'Horseshoe Lemma' argument, there exists a differential on $U^{\natural} = U'^{\natural} \oplus U''^{\natural}$, such that U becomes a DG module resolution of M over E, and the canonical exact sequence $0 \to U' \to U \to U'' \to 0$ is one of DG modules over E. Due to the expression for the differential in Theorem 9.1.1, it gives rise to an exact sequence of DG modules over $\mathcal{S}$:

$$0 \to \mathrm{C}(E, U') \to \mathrm{C}(E, U) \to \mathrm{C}(E, U'') \to 0$$

that splits over R. It induces short exact sequences of DG modules over $\mathcal{S}$

$$0 \to \mathrm{C}(E, U') \otimes_R N \to \mathrm{C}(E, U) \otimes_R N \to \mathrm{C}(E, U'') \otimes_R N \to 0$$
$$0 \to \operatorname{Hom}_R(\mathrm{C}(E, U''), N) \to \operatorname{Hom}_R(\mathrm{C}(E, U), N) \to \operatorname{Hom}_R(\mathrm{C}(E, U'), N) \to 0$$

Their connecting maps commute with the action of the operators χ_j. $\qquad\square$

The importance of the algebra of cohomology operators stems from

Theorem 9.1.4. *If M and N are finite modules over a noetherian ring R, such that $R = Q/(\boldsymbol{f})$ for some Q–regular sequence $\boldsymbol{f}$, then the $\mathcal{S}$–module $\mathrm{Ext}_R(M, N)$ is finite if and only if $\mathrm{Ext}_Q^n(M, N) = 0$ for $n \gg 0$.*

Remark. Most of the remaining results in this chapter are based on this theorem. Section 2 uses the 'if' part; different proofs for it are given in each one of the papers quoted in Construction 9.1.2; here we use an elementary argument to establish a special case, that suffices for many applications. Section 3 is based on the converse statement in the special case $N = k$, proved in [25]; the general result is established in [32].

Partial proof of Theorem 9.1.4. Assume that Q is noetherian, finite projective Q–modules are free, and $\mathrm{pd}_R M$ is finite. Proposition 2.2.8 then yields a DG module resolution U of M over E, which is a finite complex of free Q–modules. By the preceding result, we may use U to compute the action of $\mathcal{S}$. As $\mathrm{Hom}_R(C(E, U), R)$ is a semi-free DG module over $\mathcal{S}$ with underlying module $\mathcal{S} \otimes_R \mathrm{Hom}_Q(U, R)$, we see that it suffices to prove the

Claim. If $\mathcal{F}$ is a semi-free $\mathcal{S}$–module of finite rank, then for each finite R–module N the $\mathcal{S}$–module $\mathrm{H}(\mathcal{F} \otimes_R N)$ is noetherian.

The advantage is that now we can induce on $n = \mathrm{rank}_\mathcal{S} \mathcal{F}$. If $n = 1$, then $\mathcal{F}$ is a shift of $\mathcal{S}$, so $\mathrm{H}(\mathcal{F} \otimes_R N) \cong \Sigma^r \mathcal{S} \otimes_R N$ is a finite $\mathcal{S}$–module. If $n > 0$, then choose a basis element $u \in \mathcal{F}$ of minimal degree. As $\partial(u) = 0$ for degree reasons, $\mathcal{S}u$ is a DG submodule of $\mathcal{F}$, and $\mathcal{G} = \mathcal{F}/\mathcal{S}u$ is semi-free of rank $n - 1$. The homology exact sequence now yields an exact sequence of degree zero homomorphisms of $\mathcal{S}$–modules $\mathcal{S}u \otimes_R N \to \mathrm{H}(\mathcal{F} \otimes_R N) \to \mathrm{H}(\mathcal{G} \otimes_R N)$, where the two outer ones are noetherian by induction. The claim follows.   □

Eisenbud [57] shows how to compute the operators from any resolution.

Construction 9.1.5. Eisenbud operators. A *lifting* to Q of a free resolution (F, ∂) of M over R is a pair $(\widetilde{F}, \widetilde{\partial})$ consisting of a free Q–module $\widetilde{F}$ and a degree -1 endomorphism $\widetilde{\partial}$ of $\widetilde{F}$, such that $(F, \partial) = (\widetilde{F} \otimes_Q R, \widetilde{\partial} \otimes_Q R)$.

Liftings always exist – just take arbitrary inverse images in Q of the elements of the matrices of the differentials ∂_n. The relation $\partial^2 = 0$ yields $\widetilde{\partial}^2(\widetilde{F}) \subseteq (\boldsymbol{f})\widetilde{F}$, hence for $j = 1, \ldots, r$ there are degree -2 endomorphisms of Q–modules $\widetilde{\tau}^j \colon \widetilde{F} \to \widetilde{F}$, such that $\widetilde{\partial}^2 = \sum_{j=1}^r f_j \widetilde{\tau}^j$.

Each lifting produces a family of *Eisenbud operators*

$$\boldsymbol{\tau} = \{\tau^j = \widetilde{\tau}^j \otimes_Q R \colon F \to F\}_{1 \leqslant j \leqslant r}.$$

Proposition 9.1.6. *Let $\boldsymbol{\tau}$ be a family of Eisenbud operators defined by $\boldsymbol{f}$.*

For $1 \leq j \leq r$ the maps τ^j are chain maps of degree -2, that are defined uniquely up to homotopy, commute with each other up to homotopy, commute up to homotopy with any comparison of resolutions $F' \to F$ constructed over a *homomorphism of R–modules* $\beta\colon M' \to M$, *and satisfy*

$$\mathrm{H}(\mathrm{Hom}_R\left(\tau^j, N\right)) = -\chi_j\,.$$

Proof. Let $(\widetilde{F}', \widetilde{\partial}')$ be a lifting of a free resolution (F', ∂') of an R–module M', choose a family of maps $\widetilde{\tau}' = \{\widetilde{\tau}'^j\}\colon F' \to F'$ as above, and set $\tau' = \{\tau'^j = \widetilde{\tau}'^j \otimes_Q R\}$. If $\alpha\colon F' \to F$ is a chain map and $\widetilde{\alpha}\colon \widetilde{F}' \to \widetilde{F}$ is a map of Q–modules such that $\widetilde{\alpha} \otimes_Q R = \alpha$, then the equality $\partial\alpha = (-1)^{|\alpha|}\alpha\partial'$ implies that for $1 \leq j \leq r$ there exist Q–linear homomorphisms $\sigma^j\colon \widetilde{F}' \to \widetilde{F}$ with $|\sigma^j| = |\alpha| - 1$ and $\widetilde{\partial}\widetilde{\alpha} - (-1)^{|\alpha|}\widetilde{\alpha}\widetilde{\partial}' = \sum_{j=1}^r f_j\sigma^j$. Thus, we have

$$\sum_{j=1}^r f_j\big(\widetilde{\tau}^j\widetilde{\alpha} - \widetilde{\alpha}\widetilde{\tau}'^j\big) = \widetilde{\partial}^2\widetilde{\alpha} - \widetilde{\alpha}\widetilde{\partial}'^2$$

$$= \left(\sum_{j=1}^r \widetilde{\partial}f_j\sigma^j + (-1)^{|\alpha|}\widetilde{\partial}\widetilde{\alpha}\widetilde{\partial}'\right) + (-1)^{|\alpha|}\left(\sum_{j=1}^r f_j\sigma^j\widetilde{\partial}' - \widetilde{\partial}\widetilde{\alpha}\widetilde{\partial}'\right)$$

$$= \sum_{j=1}^r f_j\big(\widetilde{\partial}\sigma^j - (-1)^{|\sigma^j|}\sigma^j\widetilde{\partial}'\big)\,.$$

Since the elements of $\boldsymbol{f}$ are linearly independent modulo I^2, we get

$$\tau^j\alpha - \alpha\tau'^j = \partial(\sigma^j \otimes_Q R) - (-1)^{|\sigma^j|}(\sigma^j \otimes_Q R)\partial' \qquad \text{for} \quad 1 \leq j \leq r\,,$$

that is, $\sigma^j \otimes_Q R\colon F' \to F$ is a homotopy from $\tau^j\alpha$ to $\alpha\tau'^j$. We can now get most of the desired assertions by suitably specializing the maps chosen above.

First, letting $\alpha = \partial' = \partial$ and $\widetilde{\alpha} = \widetilde{\partial}' = \widetilde{\partial}$, we can set $\sigma^j = 0$ for $1 \leq j \leq r$, and so conclude that each τ^j is a chain map. Next, taking $\alpha = \mathrm{id}^F$ and varying τ', we see that $\tau^1, \ldots, \tau^r$ are defined uniquely up to homotopy. Then, keeping $\alpha = \tau^j$ and $\tau' = \tau$, we see that τ^j commutes up to homotopy with each τ^i. Finally, choosing α to be a lifting of a homomorphism of R–modules $\beta\colon M' \to M$, we obtain that $\tau^j\alpha$ and $\alpha\tau'^j$ are homotopic for each j.

The resolution $F = (\mathrm{C}(E, U), \partial)$ of Theorem 9.1.1 has an obvious lifting:

$$\widetilde{F} = U^{\natural} \otimes_Q G \qquad \text{with} \quad \widetilde{\partial}(v^{(H)} \otimes u) = -\sum_{j=1}^r v^{(H_j)} \otimes y_j u + v^{(H)} \otimes \partial(u)\,.$$

From it we get $\widetilde{\partial}^2(v^{(H)} \otimes u) = -\sum_{j=1}^r v^{(H_j)} \otimes f_j u = -\sum_{j=1}^r f_j\chi_j(v^{(H)} \otimes u)$ and hence $\mathrm{H}(\mathrm{Hom}_R\left(\tau^j, N\right)) = -\chi_j$ for $1 \leqslant j \leqslant r$. □

Remark 9.1.7. For any integer d with $1 \leq d \leq r$, the operators $\chi_1, \ldots, \chi_d$ act on $\mathrm{Ext}_R^*(M, k)$ in two ways: the initial one, from $R = Q/(\boldsymbol{f})$, and a new one, from the presentation $R = P/(f_1, \ldots, f_d)$ with $P = Q/(f_{d+1}, \ldots, f_r)$.

These actions coincide. Indeed, if $(\widetilde{F}, \widetilde{\partial})$ is a lifting to Q of a free resolution (F, ∂) of M over R, then it is clear that $(\widetilde{F} \otimes_Q P, \widetilde{\partial} \otimes_Q P)$ is a lifting of (F, ∂) to P. In this case we have $(\widetilde{\partial} \otimes_Q P)^2 = \sum_{j=1}^d f_j(\widetilde{\tau}^j \otimes_Q P)$. Thus, we may use $\widetilde{\tau}^j \otimes_Q P$

to compute the operation of χ_j coming from the new presentation. It remains to observe that $(\widetilde{\tau}^j \otimes_Q P) \otimes_P R = \tau^j \otimes_Q R$.

9.2. Betti numbers. Our method for studying homology over complete intersections is to use the action of the algebra of cohomology operators, in order to replace 'degree by degree' computations by 'global' considerations.

At that level, we are essentially dealing with finite graded modules over polynomial rings. This converts homological algebra back into commutative algebra, and opens the door to the use geometric methods to study cohomology. Such an approach was pioneered by Quillen [132] for cohomology of groups, and has evolved into a powerful tool of modular representation theory, cf. Benson [42] and Evens [62] for monographic expositions. Geometric methods are used in [25] to study resolutions over commutative rings.

For reference and comparison, the next theorem is presented along the lines of Theorem 5.3.3. It is compiled from four papers: the fact that $P_M^R(t)$ is rational with denominator $(1 - t^2)^{\text{codim } R}$ is from Gulliksen [80]; the comparison of the orders of the poles, and (3), are from Avramov [25]; the first part of (5) comes from Eisenbud [57], the second from Avramov, Gasharov, and Peeva [32].

Theorem 9.2.1. *Let R be a complete intersection with* $\text{edim } R = e$ *and* $\text{codim } R = r$. *For a finite R–module $M \neq 0$ with* $\text{depth } R - \text{depth } M = m$ *and* $\text{pd}_R M = \infty$, *the following hold.*

(1) *There is a polynomial $p(t) \in \mathbb{Z}[t]$ with $p(\pm 1) \neq 0$, such that*

$$P_M^R(t) = \frac{p(t)}{(1 + t)^c(1 - t)^d} \qquad \text{with} \quad c < d.$$

(2) $\text{cx}_R M = d \leq \text{codim } R$ *and* $\text{curv}_R M = 1$.

(3) $\beta_n^R(M) \sim \dfrac{b}{2^c(d - 1)!} \, n^{d-1}$ *where $b = p(1) > 0$.*

(4) $\displaystyle\lim_{n \to \infty} \frac{\beta_{n+1}^R(M)}{\beta_n^R(M)} = 1$.

(5.1) $\dfrac{\beta_{n+1}^R(M)}{\beta_n^R(M)} = 1$ *and* $\text{Syz}_{n+2}^R(M) \cong \text{Syz}_n^R(M)$ *for $n > m$ if $\text{cx}_R M = 1$.*

(5.2) $\dfrac{\beta_{n+1}^R(M)}{\beta_n^R(M)} > 1$ *and* $\text{Syz}_{n+2}^R(M) \twoheadrightarrow \text{Syz}_n^R(M)$ *for $n \gg 0$ if $\text{cx}_R M \geq 2$.*

Remark. A more precise version of the last inequality is proved in [32]: there are polynomials $h_\pm(t)$ of degree $d - 2$ with leading terms $a_\pm > 0$, such that for $n \gg 0$ the difference $\beta_{n+1}^R(M) - \beta_n^R(M)$ is equal to $h_+(n)$ if n is even, and to $h_-(n)$ if n is odd; however, it is possible that $a_+ \neq a_-$, cf. Example 9.2.4.

Example 9.2.2. By Remark 8.1.1.2, we have $\beta_n^R(k) \sim 2^{e-r}n^{r-1}/(r-1)!$, so $c = -\dim R$, $d = \text{codim } R$, $b = 1$, and $p_k(t) = 1$.

Recall that if R is a complete intersection, then $\text{mult } R \geq 2^{\text{codim } R}$.

Example 9.2.3. If $\text{mult}(R) = 2^r$, then for each M there is an integer valued polynomial $b(t) \in \mathbb{Q}[t]$ such that $\beta_n^R(M) = b(n)$ for $n \gg 0$, cf. [28]. This generalizes a well known property of complete intersections of *quadrics*.

Not all Betti sequences are eventually given by some polynomial in n.

Example 9.2.4. Let $q = \binom{e+1}{2} - \text{rank}_k \, \mathfrak{m}^2/\mathfrak{m}^3$ be the number of 'quadratic relations' of R. It is proved in [28] that

$$P_{R/\mathfrak{m}^2}^R(t) = \frac{(1-t)^q + (1+t)^{e-q-1} \cdot (et-1)}{(1-t)^r \cdot (1+t)^{r-q-1} \cdot t} .$$

Thus, when $q \leq r - 2$ the Poincaré series has poles at $t = 1$ *and* at $t = -1$, so the even and odd Betti numbers are each given by a different polynomial. For instance, if $R = k[s_1, s_2]/(s_1^{a_1}, s_2^{a_2})$, with $a_i \geq 3$, then $\beta_n^R(R/\mathfrak{m}^2)$ is equal to $\frac{3}{2}n+1$ if n is even, and to $\frac{3}{2}n + \frac{3}{2}$ if n is odd.

As (5.1) shows, if a Betti sequence is bounded, then it stabilizes after at most depth R steps. However, if $\text{cx}_R M \geq 2$, then there exist modules whose Betti sequence *strictly decreases* over an initial interval of any given length. This shows that no bound on the degree of the polynomial $p(t)$ can be expressed as a function only of invariants of the ring R:

Example 9.2.5. Let R be a complete intersection of codimension $c \geq 2$. Fix $N = \text{Syz}_n^R(k)$, with $n > \dim R$, and let F be its minimal free resolution. The module N is maximal Cohen-Macaulay, cf. 1.2.8, hence the complex

$$0 \to N^* \to F_0^* \xrightarrow{\partial_0^*} F_1^* \xrightarrow{\partial_1^*} F_2^* \to \cdots ,$$

where $-^* = \text{Hom}_R(-, R)$, is exact and minimal. Splice it to the right of a minimal free resolution of N^*: now you are holding a 'doubly infinite' exact complex of finite free R-modules, that you can truncate at will. The cokernel of ∂_s^* is guaranteed to have $s + 1$ strictly decreasing Betti numbers at the beginning of its resolution, cf. Remark 8.1.1.3.

Before starting on the proof, we make a general observation.

Remark 9.2.6. Let Q be a regular local ring, $\boldsymbol{f}$ be a Q-regular sequence, and set $R = Q/(\boldsymbol{f})$. If M is a finite R-module, then $\text{Ext}_R(M, k)$ is a finite module over $R[\chi_1, \ldots, \chi_r]$ by Theorem 9.1.4.

Since $\mathfrak{m}$ annihilates $\text{Ext}_R(M, k) = 0$, we see that $\mathcal{M} = \text{Ext}_R(M, k)$ is a finite module over the graded polynomial ring $\mathcal{P} = k[\chi_1, \ldots, \chi_r]$. In particular, the Hilbert-Serre Theorem applies to the graded $\mathcal{P}$-module $\mathcal{M}$, and shows that $P_M^R(t) = q(t)/(1 - t^2)^r$ for some polynomial $q(t) \in \mathbb{Z}[t]$.

Proof of Theorem 9.2.1. The hypotheses of the theorem and its conclusions do not change if one replaces (R, M) by $(R', M \otimes_R R')$, where R' is the completion of the local ring $R[u]_{\mathfrak{m}[u]}$. Thus, we assume that $R = Q/(\boldsymbol{f})$, where Q is regular with infinite residue field k, and $\boldsymbol{f}$ is a regular sequence.

Let F be a minimal free resolution of M over R, and set $\beta_n = \beta_n^R(M)$.

(1) Due to Remark 9.2.6, $P_M^R(t)$ can be written in the form

$$P_M^R(t) = \sum_{j=0}^{d-1} \frac{m_j}{(1-t)^{d-j}} + \sum_{i=0}^{c-1} \frac{\ell_i}{(1+t)^{c-i}} + f(t),$$

with $\max\{c,d\} \le r$ and $f(t) \in \mathbb{Q}[t]$. Thus, for $n \gg 0$, there are equalities

$$\beta_n = \begin{cases} \dfrac{m_0}{(d-1)!} \cdot n^{d-1} + \dfrac{\ell_0}{(c-1)!} \cdot n^{c-1} + g_+(n) & \text{for even } n; \\[2ex] \dfrac{m_0}{(d-1)!} \cdot n^{d-1} - \dfrac{\ell_0}{(c-1)!} \cdot n^{c-1} + g_-(n) & \text{for odd } n; \end{cases} \tag{$*$}$$

with $m_0 \ne 0$, and polynomials $g_\pm(t)$ of degree $< \max\{c,d\} - 1$. As the Betti numbers of M are positive, we have $d \ge c$, and $d > 0$.

Assume next that $d = c$, so that $\ell_0 \ne 0$. The positivity of Betti numbers implies that $m_0 \pm \ell_0 > 0$, hence $m_0 > 0$.

Set $\gamma(j,2s) = \sum_{i=-j}^{j} (-1)^i \beta_{2s-i}$. Formula $(*)$ shows that for all $s, h \gg 0$ the function $2s \mapsto \gamma(2h, 2s)$ is given by a polynomial in $2s$ of degree d with leading coefficient $a_0 = \big((4h+1)\ell_0 + m_0\big)/(d-1)!$, and the function $2s \mapsto \gamma(2h+1, 2s)$, by a polynomial of the same degree with leading coefficient $a_1 = \big((4h+3)\ell_0 - m_0\big)/(d-1)!$. Thus: if $\ell_0 < 0$, then $a_0 < 0$ for $h \gg 0$, so $\gamma(2h, 2s) < 0$; if $\ell_0 > 0$, then $a_1 > 0$ for $h \gg 0$, so $\gamma(2h+1, 2s) > 0$.

Localization of F at a minimal prime ideal $\mathfrak{p}$ of R yields an exact sequence

$$0 \to L_{s,j} \to (F_{2s+j})_{\mathfrak{p}} \to \cdots \to (F_{2s})_{\mathfrak{p}} \to \cdots \to (F_{2s-j})_{\mathfrak{p}} \to N_{s,j} \to 0.$$

Counting lengths over $R_{\mathfrak{p}}$, we get an equality

$$\gamma(j,2s) \cdot \text{length}(R_{\mathfrak{p}}) = (-1)^j \big(\text{length}(L_j) + \text{length}(N_j) \big)$$

which shows that $\gamma(2h, 2s) > 0$ and $\gamma(2h+1, 2s) < 0$, regardless of the sign of ℓ_0. We have a contradiction, so we conclude that $d > c$.

(3) Since $d > c$, formula $(*)$ yields $\lim_{n \to \infty} \beta_n / n^{d-1} = m_0/(d-1)!$. On the other hand, $m_0 = \lim_{t \to 1} (1-t)^d P_M^R(t) = p(1)/2^c$.

(2) and (4) are trivial consequences of (1) and (3).

(5) By Theorem 9.1.4, $\text{Ext}_R(M,k)$ is a finite graded module over the polynomial ring $k[\chi_1, \ldots, \chi_r]$. Thus, its graded submodule

$$\{\mu \in \text{Ext}_R(M,k) \mid (\chi_1, \ldots, \chi_r)^m \mu = 0 \text{ for some } m\}$$

is finite-dimensional, and hence is trivial, say, in degrees $> s$. Since k is infinite, we can find a linear combination χ of $\chi_1, \ldots, \chi_r$, that is a non-zero-divisor on $\text{Ext}_R^{>s}(M,k)$. Thus, the operator χ is injective on $\text{Ext}_R^{>s}(M,k)$. Dualizing, we see that $\chi\colon \text{Tor}_{n+2}^R(M,k) \to \text{Tor}_n^R(M,k)$ is surjective when $n > s$.

Changing bases, we may assume that $\chi = \chi_1$, and switch attention to the presentation $R = P/(f)$, where $P = Q/(f_2, \ldots, f_r)$ and f is the image of f_1; note that f is P–regular. Let $(\widetilde{F}, \widetilde{\partial})$ be a lifting of the complex (F, ∂) to P, and let $\widetilde{\tau}\colon \widetilde{F} \to \widetilde{F}$ and $\tau = \widetilde{\tau} \otimes_P R\colon F \to F$ be the degree -2 endomorphisms from Construction 9.1.5. By Remark 9.1.7, we have $\chi = \text{Hom}_R(\tau, k)$.

Since χ_n is surjective for $n > s$, so are the maps $\tilde{\tau}_{n+2} \colon \tilde{F}_{n+2} \to \tilde{F}_n$ and $\tau_{n+2} \colon F_{n+2} \to F_n$ by Nakayama. The chain map τ induces surjections $\mathrm{Syz}^R_{n+2}(M) \to \mathrm{Syz}^R_n(M)$ for $n > s$. Localize the defining exact sequence

$$0 \longrightarrow \mathrm{Syz}^R_{n+2}(M) \longrightarrow F_{n+1} \xrightarrow{\ \partial_{n+1}\ } F_n \longrightarrow \mathrm{Syz}^R_n(M) \longrightarrow 0$$

at a minimal prime $\mathfrak{p}$ of R. For $n > s$ a lengths count over $R_\mathfrak{p}$ yields

$$\beta_{n+1} - \beta_n = \big(\mathrm{length}\,\mathrm{Syz}^{R_\mathfrak{p}}_{n+2}(M_\mathfrak{p}) - \mathrm{length}\,\mathrm{Syz}^{R_\mathfrak{p}}_n(M_\mathfrak{p})\big)/\mathrm{length}(R_\mathfrak{p}) \geq 0,$$

Next we assume that $\beta_n = \beta_{n+1} = b \neq 0$ for some $n > s$, and show that $\beta_{n+2} = b$. Since $\tilde{\tau}_{n+2}$ is surjective for $n > s$, we have $\tilde{F}_{n+2} = E \oplus G$ with $E = \mathrm{Ker}\,\tilde{\tau}_{n+2}$, and the restriction θ of $\tilde{\tau}_{n+2}$ to G is an isomorphism with $\tilde{F}_n$. Let $\zeta \colon G \to \tilde{F}_{n+1}$ be the restriction of $\tilde{\partial}_{n+2}$. As $\tilde{\partial}_{n+1}\zeta$ is the restriction to G of $\tilde{\partial}_{n+1}\tilde{\partial}_{n+2} = f\tilde{\tau}_{n+2}$, where $P/(f) = R$, we have $\tilde{\partial}_{n+1}\zeta = f\theta$, and hence

$$\textstyle\bigwedge^b \tilde{\partial}_{n+1} \bigwedge^b \zeta = \bigwedge^b(\tilde{\partial}_{n+1}\zeta) = \bigwedge^b(f\theta) = f^b \bigwedge^b \theta.$$

Note that G, $\tilde{F}_{n+1}$, and $\tilde{F}_n$ have rank b and fix isomorphisms of P with $\bigwedge^b(G)$, $\bigwedge^b(\tilde{F}_{n+1})$, and $\bigwedge^b(\tilde{F}_n)$. The maps $\bigwedge^b \tilde{\partial}_{n+1}$, $\bigwedge^b \zeta$, and $\bigwedge^b \theta$ are then given by multiplication with elements of P, say y, z, and u, respectively. The equality above becomes $yz = f^b u$. As θ is bijective so is $\bigwedge^b(\theta)$, hence u is a unit in P. As f is P-regular, so is y, hence $\tilde{\partial}_{n+1}$ is injective. From $\tilde{\partial}_{n+1}\tilde{\partial}_{n+2}(E) = f\tilde{\tau}_{n+2}(E) = 0$ we now see that $E \subseteq \mathrm{Ker}\,\tilde{\partial}_{n+2}$, so $\mathrm{Im}\,\tilde{\partial}_{n+2}$ is a homomorphic image of $\tilde{F}_{n+2}/E \cong G$. Remarking that

$$(\mathrm{Coker}\,\tilde{\partial}_{n+3}) \otimes_P R \cong \mathrm{Coker}\,\partial_{n+3} = \mathrm{Syz}^R_{n+2}(M)$$

we conclude that $\mathrm{Syz}^R_{n+2}(M)$ is a homomorphic image of the free R–module $G \otimes_P R \cong R^b$. It follows that $\beta_{n+2} \leq b = \beta_n$. On the other hand, we already know that $\beta_{n+2} \geq \beta_{n+1} \geq \beta_n$, hence all three are equal to b. Thus, the sequence $\{\beta_n\}_{n>s}$ is either strictly increasing or constant: we have proved (5.2).

If $\beta_{n+2} = \beta_n$, then $\mathrm{rank}_P \tilde{F}_{n+2} = \mathrm{rank}_P \tilde{F}_n$, so $E = 0$ and the surjective homomorphism τ_{n+2} is bijective. To finish the proof of (5.1), we show that for $m = \mathrm{depth}\,R - \mathrm{depth}\,M$ the complex $F_{> m}$ is periodic of period 2. It is the minimal free resolution of $N = \mathrm{Syz}^R_{m+1}(M)$, and N is a maximal Cohen-Macaulay module by Proposition 1.2.8. Thus, $F^*_{> m} = \mathrm{Hom}_R(F_{> m}, R)$ is exact except in degree 0, and $\mathrm{H}_0(F^*_{> m}) = N^*$. Since $F^*_{> m}$ is minimal, N^* is a syzygy of $C_n = \mathrm{Coker}\,\partial^*_n$ for each $n \geq m$. For $n \gg 0$ the minimal resolution of C_n is periodic of period 2, hence so is $F^*_{> m}$. $\qquad\square$

9.3. Complexity and Tor.
Let $(R, \mathfrak{m}, k)$ be a local ring.

If $R = Q/I$ is a complete intersection and Q is regular, then the finite global dimension of Q implies that *all* R–modules have finite complexity. However, to study a *specific* R–module, it often pays off to use an intermediate (singular) complete intersection P, that retains the crucial property $\mathrm{pd}_P M < \infty$. With this approach, the following *factorization theorem* is proved in [25].

Theorem 9.3.1. *Let $R \cong Q/I$ be a regular presentation with I generated by a regular sequence. If k is infinite, then for each finite R–module M the surjection $Q \to R$ factors as $Q \to P \to R$, with the kernels of both maps generated by regular sequences, $\mathrm{pd}_P M < \infty$, and $\mathrm{cx}_R M = \mathrm{pd}_P R$.*

Proof. As in Remark 9.2.6, consider the finite graded module $\mathcal{M} = \mathrm{Ext}_R^*(M, k)$ over the ring $\mathcal{P}$ defined by the presentation $R = Q/I$. Elementary dimension theory shows that the Krull dimension of $\mathcal{M}$ over $\mathcal{P}$ is equal to $\mathrm{cx}_R M = d$. As k is infinite, we may choose a homogeneous system of parameters $\chi_1, \ldots, \chi_d$ for $\mathcal{M}$, and extend it to a basis $\chi_1, \ldots, \chi_r$ of $\mathcal{P}^2$, the degree 2 component of $\mathcal{P}$.

It is not hard to see that I can be generated by a Q–regular sequence $\boldsymbol{f} = f_1 \ldots, f_r$ that defines the operators $\chi_1, \ldots, \chi_r$. Remark 9.1.7 identifies $k[\chi_1, \ldots, \chi_d] \subseteq \mathcal{P}$ with the ring $\mathcal{P}'$ of cohomology operators of a presentation $R = P/(f_1, \ldots, f_d)$, where $P = Q/(f_{d+1}, \ldots, f_r)$. As $\mathcal{M}$ is finite over $\mathcal{P}'$, Theorem 9.1.4 shows that $\mathrm{Ext}_P^n(M, k) = 0$ for $n \gg 0$, that is, $\mathrm{pd}_P M < \infty$. $\square$

To deal with intrinsic properties of the R–module M, a concept of *virtual projective dimension* is introduced in [25] by the formula[35]

$$\mathrm{vpd}_R M = \inf \left\{ \mathrm{pd}_{Q'} \widehat{M} \, \middle| \, \begin{array}{l} Q' \text{ is a local ring such that } \widehat{R} \cong Q'/(\boldsymbol{f}') \\ \text{for some } Q'\text{–regular sequence } \boldsymbol{f}' \end{array} \right\}.$$

Clearly, $\mathrm{vpd}_R M < \infty$ whenever R is a complete intersection.

Recall that $\mathrm{pd}_R M$ is finite if and only if $\mathrm{cx}_R M = 0$. It is easy to see that in that case, $\mathrm{vpd}_R M = \mathrm{pd}_R M$. Thus, the following result extends of the Auslander-Buchsbaum Equality.

Theorem 9.3.2. *If M is a finite R–module and $\mathrm{vpd}_R M$ is finite, then*

$$\mathrm{vpd}_R M = \mathrm{depth}\, R - \mathrm{depth}_R M + \mathrm{cx}_R M \,.$$

Proof. We may assume that R is complete with infinite residue field.

Choosing Q' with $\mathrm{pd}_{Q'} M = \mathrm{vpd}_R M$, we have

$$\begin{aligned}
\mathrm{vpd}_R M = \mathrm{pd}_{Q'} M &= \mathrm{depth}\, Q' - \mathrm{depth}\, M \\
&= \mathrm{pd}_{Q'} R + \mathrm{depth}\, R - \mathrm{depth}\, M \geq \mathrm{cx}_R M + \mathrm{depth}\, R - \mathrm{depth}\, M
\end{aligned}$$

where the inequality comes from Corollary 4.2.5.4. On the other hand, the preceding theorem provides a ring P from which R is obtained by factoring out a regular sequence, and that satisfies $\mathrm{pd}_P R = \mathrm{cx}_R M$, so we get

$$\begin{aligned}
\mathrm{vpd}_R M \leq \mathrm{pd}_P M &= \mathrm{depth}\, P - \mathrm{depth}\, M \\
&= \mathrm{pd}_P R + \mathrm{depth}\, R - \mathrm{depth}\, M = \mathrm{cx}_R M + \mathrm{depth}\, R - \mathrm{depth}\, M
\end{aligned}$$

where the inequality holds by definition. $\square$

---

[35]If k is infinite; otherwise, R and M are replaced by $\widetilde{R} = R[u]_{\mathfrak{m}[u]}$ and $\widetilde{M} = M \otimes_R \widetilde{R}$.

Next we study the vanishing of Tor functors over a local ring. The subject starts with a famous rigidity theorem of Auslander [16] and Lichtenbaum [112]:

Theorem 9.3.3. *If M and N are finite modules over a regular ring R and*

$$\operatorname{Tor}_i^R(M,N) = 0 \quad \text{for some} \quad i > 0,$$

then $\operatorname{Tor}_n^R(M,N) = 0$ for all $n \geq i$.   ∎

Heitmann [85] proves that rigidity may fail, even with R Cohen-Macaulay and $\operatorname{pd}_R M$ finite. On the other hand, there are partial extensions of the theorem to complete intersections. The first one is due to Murthy [123].

Theorem 9.3.4. *If M and N are finite modules over a complete intersection R of codimension r, and for some $i > 0$ there are equalities*

$$\operatorname{Tor}_i^R(M,N) = \cdots = \operatorname{Tor}_{i+r}^R(M,N) = 0$$

then $\operatorname{Tor}_n^R(M,N) = 0$ for all $n \geq i$.

In codimension 1, this is complemented by Huneke and Wiegand [91]:

Theorem 9.3.5. *If M and N are finite modules over a hypersurface R, and*

$$\operatorname{Tor}_i^R(M,N) = \operatorname{Tor}_{i+1}^R(M,N) = 0 \quad \text{for some} \quad i > 0,$$

then either M or N has finite projective dimension.

When the vanishing occurs outside of an initial interval, Jorgensen [95] draws the conclusion from the vanishing of fewer Tor's.

Theorem 9.3.6. *Let M be a finite module over a complete intersection R, such that $\operatorname{cx}_R M = d$ and $\operatorname{depth} R - \operatorname{depth} M = m$. For a finite R–module N, the following are equivalent.*

(i) $\operatorname{Tor}_n^R(M,N) = 0$ *for $n > m$.*

(ii) $\operatorname{Tor}_n^R(M,N) = 0$ *for $n \gg 0$.*

(iii) $\operatorname{Tor}_i^R(M,N) = \cdots = \operatorname{Tor}_{i+d}^R(M,N) = 0$ *for some $i > m$.*

The number of vanishing Tor's in (iii) cannot be reduced further; the next example elaborates on a construction from [95].

Example 9.3.7. For $i \geq 1$, $R = k[[s_1, \ldots, s_{2r}]]/(s_1 s_{r+1}, \ldots, s_r s_{2r})$, and $N = R/(\bar{s}_{r+1}, \ldots, \bar{s}_{2r})$ there is a module M_i, such that $\operatorname{cx}_R M_i = r$, $\operatorname{Tor}_n^R(M_i, N) = 0$ for $i < n \leq i + r$, but $\operatorname{Tor}_n^R(M_i, N) \neq 0$ for infinitely many n.

Corollary 6.1.9 yields a minimal resolution

$$F = R\langle x_1, \ldots, x_{2r} \mid \partial(x_j) = \bar{s}_j, \partial(x_{r+j}) = \bar{s}_{r+j} x_j \text{ for } 1 \leq j \leq r \rangle$$

of $M = R/(\bar{s}_1, \ldots, \bar{s}_r) \cong k[[s_{r+1}, \ldots, s_{2r}]]$. As M is maximal Cohen-Macaulay, $F^* = \operatorname{Hom}_R(F, R)$ is exact in degrees $\neq 0$. It is easy to see that the sequence

$$F_1 \xrightarrow{\partial_1} F_0 \xrightarrow{\sigma} F_0^* \xrightarrow{-\partial_1^*} F_1^* \quad \text{with} \quad \sigma(x) = s_{r+1} \cdots s_{2r} x$$

is exact. The splice of F with $\Sigma^{-1}F^*$ along σ is a doubly infinite complex (G, ∂) of free R–modules. By construction, $\mathrm{H}_n(G \otimes_R N) = \mathrm{Tor}^R_n(M, N)$ for $n \geq 1$ and $\mathrm{H}_n(G \otimes_R N) = \mathrm{Ext}^{-n}_R(M, N)$ for $n \leq -2$; as $\sigma \otimes_R N = 0$ trivial, these equalities extend to $n = 0$ and $n = -1$, respectively.

For $1 \leq j \leq r$, consider $N_j = N/(\bar{s}_{j+1}, \ldots, \bar{s}_r)N \cong k[[s_1, \ldots, s_j]]$, note that $\mathrm{Tor}^R(M, N_j)$ and $\mathrm{Ext}_R(M, N_j)$ are annihilated by $(\bar{s}_1, \ldots, \bar{s}_{2r})$, and set

$$T_j(t) = \sum_{n=0}^{\infty} \mathrm{rank}_k \, \mathrm{Tor}^R_n(M, N_j)\, t^n \quad \text{and} \quad E_j(t) = \sum_{n=0}^{\infty} \mathrm{rank}_k \, \mathrm{Ext}^n_R(M, N_j)\, t^n\,.$$

The exact sequences $0 \to N_j \xrightarrow{s_j} N_j \to N_{j-1} \to 0$ induce (co)homology sequences in which multiplication by s_j is the zero map, so

$$T_{j-1}(t) = T_j(t) + tT_j(t) \qquad \text{and} \qquad E_{j-1}(t) = E_j(t) + \frac{1}{t}E_j(t)\,.$$

Since and $N_0 \cong k$, we have $T_0(t) = E_0(t) = \mathrm{P}^R_M(t) = 1/(1-t)^r$, and hence

$$T_r(t) = \frac{T_0(t)}{(1+t)^r} = \frac{1}{(1-t^2)^r} \qquad \text{and} \qquad E_r(t) = \frac{t^r E_0(t)}{(1+t)^r} = \frac{t^r}{(1-t^2)^r}\,.$$

Now set $M_i = \mathrm{Im}\,\partial_{-r-i-1}$; as $\mathrm{Tor}^R_n(M_i, N) \cong \mathrm{H}_{n-r-i}(G \otimes_R N)$ for $n \geq 1$, these equalities establish the desired property.

We start the proofs with a couple of easy lemmas.

Lemma 9.3.8. *Let $f_1, \ldots, f_d$ be a regular sequence in a commutative ring Q, and set $R = Q/(f_1, \ldots, f_d)$. If*

$$\mathrm{Tor}^R_s(M, N) = \cdots = \mathrm{Tor}^R_t(M, N) = 0$$

for integers s and t with $s + d \leq t$, then there are isomorphisms

$$\mathrm{Tor}^Q_{s+d-1}(M, N) \cong \mathrm{Tor}^R_{s-1}(M, N)\,;$$
$$\mathrm{Tor}^Q_{s+d}(M, N) = \cdots = \mathrm{Tor}^Q_t(M, N) = 0\,;$$
$$\mathrm{Tor}^Q_{t+1}(M, N) \cong \mathrm{Tor}^R_{t+1}(M, N)\,.$$

Proof. The Cartan-Eilenberg change of rings spectral sequence 3.2.1 has

$$^2\mathrm{E}_{p,q} = \mathrm{Tor}^R_p\left(\mathrm{Tor}^Q_q(M, R), N\right) \implies \mathrm{Tor}^Q_{p+q}(M, N)\,.$$

If E is the Koszul complex resolving R over Q, then

$$\mathrm{Tor}^Q_q(M, R) = \mathrm{H}_q(M \otimes_Q E) = M \otimes_Q E_q = M^{\binom{d}{q}}\,,$$

hence $^2\mathrm{E}_{p,q} = \mathrm{Tor}^R_p(M, N)^{\binom{d}{q}}$. Thus, $^2\mathrm{E}_{p,q} = 0$ for $s \leq p \leq t$. It follows that the only possibly non-zero module in total degree $s+d-1$ is $^2\mathrm{E}_{s-1,d} = \mathrm{Tor}^R_{s+d-1}(M, N)$, that all modules in total degree n for $s + d \leq n \leq t$ are trivial, and that the only possibly non-zero module in total degree $t + 1$ is $^2\mathrm{E}_{t+1,0} = \mathrm{Tor}^R_{t+1}(M, N)$. For degree reasons, no non-trivial differential can enter or quit these modules. This gives the desired isomorphisms. $\square$

Proof of Theorem 9.3.4. Since $\mathrm{Tor}_{i+r}^{Q}(M,N) = 0$ and $\mathrm{Tor}_{i+r+1}^{Q}(M,N) \cong \mathrm{Tor}_{i+r+1}^{R}(M,N)$ by the lemma, we get $\mathrm{Tor}_{i+r+1}^{Q}(M,N) = 0$ by Theorem 9.3.3, hence $\mathrm{Tor}_{i+r+1}^{R}(M,N) = 0$. Iteration yields $\mathrm{Tor}_{n}^{R}(M,N) = 0$ for $n > i+r$. $\square$

C. Miller [120] provides a simple proof of Theorem 9.3.5, based on

Lemma 9.3.9. *Let M, N be finite modules over a complete intersection R.*

If $\mathrm{Tor}_{n}^{R}(M,N) = 0$ for $n \geq 1$, then $\mathrm{cx}_R M + \mathrm{cx}_R N = \mathrm{cx}_R(M \otimes_R N)$.

If $\mathrm{Tor}_{n}^{R}(M,N) = 0$ for $n \gg 0$, then $\mathrm{cx}_R M + \mathrm{cx}_R N \leq \mathrm{codim}\, R$.

Proof. As $\mathrm{P}_{M \otimes_Q N}^{R}(t) = \mathrm{P}_{M}^{R}(t) \cdot \mathrm{P}_{N}^{R}(t)$, cf. the proof of Proposition 4.2.4.6, comparison of orders of poles at $t = 1$ and Theorem 9.2.1.2 yield the first assertion. For the second one, replace M by a high syzygy M'; then $\mathrm{cx}_R M' + \mathrm{cx}_R N = \mathrm{cx}_R(M' \otimes_R N) \leq \mathrm{codim}\, R$, the inequality coming from *loc. cit.* $\square$

Proof of Theorem 9.3.5. By Theorem 9.3.4, $\mathrm{Tor}_{n}^{R}(M,N) = 0$ for $n \geq i$; thus, $\mathrm{cx}_R M + \mathrm{cx}_R N \leq 1$ by lemma 9.3.9, so $\mathrm{pd}_R M$ or $\mathrm{pd}_R N$ is finite. $\square$

We use the factorization theorem to give a short
Proof of Theorem 9.3.6. Only (iii) $\implies$ (i) needs a proof. We may assume that R is complete with infinite residue field. By hypothesis, there are $s, t \in \mathbb{N}$, such that $m < s < s + d \leq t$ and $\mathrm{Tor}_{j}^{R}(M,N) = 0$ for $s \leq j \leq t$. For the smallest such s, choose P as in Theorem 9.3.1; as $\mathrm{pd}_P M$ is finite,

$$\mathrm{pd}_P M = \mathrm{depth}\, P - \mathrm{depth}\, M = \mathrm{depth}\, R + d - \mathrm{depth}\, M = m + d.$$

We see that if $s > m + 1$, then $\mathrm{Tor}_{s+d-1}^{P}(M,N) = 0$; the first isomorphism in Lemma 9.3.8 yields $\mathrm{Tor}_{s-1}^{R}(M,N) = 0$, contradicting the minimality of s. Thus, $s = m + 1$; it follows that $t + 1 > \mathrm{pd}_P M$, and so $\mathrm{Tor}_{t+1}^{P}(M,N) = 0$. The last isomorphism of the lemma yields $\mathrm{Tor}_{t+1}^{R}(M,N) = 0$. Iterate. . . $\square$

10. Homotopy Lie Algebra of a Local Ring

It is a remarkable phenomenon that very sensitive homological information on a local ring is encrypted in a *non-commutative* object – a graded Lie algebra. We construct it, and show that its very existence affects the size of free resolutions, while its structure influences their form.

This chapter provides a short introduction to a huge area of research: the use of non-commutative algebra for the construction and study of free resolutions. We start by providing a self-contained construction of a graded Lie algebra, whose universal enveloping algebra is the Ext-algebra of the local ring.

10.1. Products in cohomology. We revert to a commutative ring $\Bbbk$, and consider *graded associative*[36] algebras over $\Bbbk$. The primitive example of an associative algebra is a matrix ring. The graded version is the $\Bbbk$–module of homogeneous homomorphisms $\operatorname{Hom}_{\Bbbk}(C,C)$, with composition as product and the identity map as unit. If C is a complex, then the derivation on $\operatorname{Hom}_{\Bbbk}(C,C)$, used since Section 1.1, turns it into an associative DG algebra. It appears in

Construction 10.1.1. Ext algebras. When $\epsilon\colon F \to L$ is a free resolution of a $\Bbbk$–module L, Proposition 1.3.2 yields an isomorphism

$$\operatorname{H}\operatorname{Hom}_{\Bbbk}(F,\epsilon) : \ \operatorname{H}\operatorname{Hom}_{\Bbbk}(F,F) \cong \operatorname{H}\operatorname{Hom}_{\Bbbk}(F,L) = \operatorname{Ext}_{\Bbbk}(L,L)\,.$$

Thus, F defines a structure of graded $\Bbbk$–algebra on $\operatorname{Ext}_{\Bbbk}(L,L)$.

In degree zero, it is the usual product of $\operatorname{Hom}_{\Bbbk}(L,L)$, an invariant of the $\Bbbk$–module L. To see that all of it is invariant, take a resolution F' of L, and choose morphisms $\alpha\colon F \to F'$ and $\alpha'\colon F' \to F$, lifting the identity of L. As $\alpha'\alpha$ also is such a morphism, there is a homotopy σ with $\alpha'\alpha = \operatorname{id}^{F} + \partial\sigma + \sigma\partial$. Define $\phi\colon \operatorname{Hom}_{\Bbbk}(F,F) \to \operatorname{Hom}_{\Bbbk}(F',F')$ by $\phi(\beta) = \alpha\beta\alpha'$. If β and $\gamma\colon F \to F$ are chain maps, then

$$\phi(\beta\gamma) = \alpha\beta\gamma\alpha' = \alpha\beta(\alpha'\alpha)\gamma\alpha' - \alpha\beta(\partial\sigma)\gamma\alpha' - \alpha\beta(\sigma\partial)\gamma\alpha'$$

$$= (\alpha\beta\alpha')(\alpha\gamma\alpha') - (-1)^{|\beta|}\partial(\alpha\beta\sigma\gamma\alpha') - (-1)^{|\gamma|}(\alpha\beta\sigma\gamma\alpha')\partial$$

$$= \phi(\beta)\phi(\gamma) + \partial\tau + (-1)^{|\tau|}\tau\partial$$

with $\tau = -(-1)^{|\beta|}\phi(\beta\sigma\gamma)$. In homology, this shows that $\operatorname{H}(\phi)$ is an homomorphism of algebras. As ϕ is a quasi-isomorphism by Propositions 1.3.2 and 1.3.3, $\operatorname{H}(\phi)$ is an isomorphism. It is also unique: all choices for α and α' are homotopic to the original ones, producing homotopic maps ϕ, and hence the same $\operatorname{H}(\phi)$.

We have finished the construction of the Ext *algebra* of the $\Bbbk$–module L, with the *composition product*[37].

The next structure[38] might at first seem complicated.

Remark 10.1.2. A *graded Lie algebra* [39] over $\Bbbk$ is a $\Bbbk$–module $\mathfrak{g} = \{\mathfrak{g}^n\}_{n\in\mathbb{Z}}$ equipped with a $\Bbbk$–bilinear pairing, called the *Lie bracket*

$$[\ ,\ ]\colon \mathfrak{g}^i \times \mathfrak{g}^j \to \mathfrak{g}^{i+j} \qquad \text{for}\quad i,j\in\mathbb{Z}\,, \qquad (\vartheta,\xi) \mapsto [\vartheta,\xi]\,,$$

such that for all $\vartheta,\,\xi,\,\zeta \in \mathfrak{g}$ signed versions of the classical conditions hold:

(1) $[\vartheta,\xi] = -(-1)^{|\vartheta||\xi|}[\xi,\vartheta]$ (*anti-commutativity*)

(2) $[\vartheta,[\xi,\zeta]] = [[\vartheta,\xi],\zeta] + (-1)^{|\vartheta||\xi|}[\xi,[\vartheta,\zeta]]$ (*Jacobi identity*)

---

[36] That is, not assumed positively graded or graded commutative.

[37] Another pairing is the *Yoneda product*, that splices exact sequences representing elements of Ext; they differ by a subtle sign, treated with care by Bourbaki [45].

[38] An early appearance is in the form $\mathfrak{g}^{n+1} = \pi_n(X)$, the n'th homotopy group of a topological space X, with bracket given by the *Whitehead product*; the proof by Uehara and Massey [152] of the Jacobi identity for the Whitehead product was the first major application of the (then) newly discovered Massey triple product.

[39] In postmodern parlance, a *super Lie algebra*.

To deal with deviant behavior over rings without $\frac{1}{6}$, we extend the definition by requiring, in addition[40], that

($1\frac{1}{2}$) $[\vartheta, \vartheta] = 0$ for $\vartheta \in \mathfrak{g}^{\text{even}}$.

($2\frac{1}{3}$) $[v, [v, v]] = 0$ for $v \in \mathfrak{g}^{\text{odd}}$.

and that $\mathfrak{g}$ be endowed with a *square*[41]

$$\mathfrak{g}^{2h+1} \to \mathfrak{g}^{4h+2} \qquad \text{for} \quad h \in \mathbb{Z}, \qquad v \mapsto v^{[2]},$$

such that the following conditions are satisfied:

(3) $(v + w)^{[2]} = [v, w] - v^{[2]} - w^{[2]}$ for $v, w \in \mathfrak{g}^{\text{odd}}$ with $|v| = |w|$;

(4) $(av)^{[2]} = a^2 v^{[2]}$ for $a \in \mathbb{k}$ and $v \in \mathfrak{g}^{\text{odd}}$;

(5) $[v^{[2]}, \vartheta] = [v, [v, \vartheta]]$ for $v \in \mathfrak{g}^{\text{odd}}$ and $\vartheta \in \mathfrak{g}$.

A *Lie subalgebra* is a subset of $\mathfrak{g}$ closed under brackets and squares; with the induced operations, it is a graded Lie algebra in its own right. A homomorphism $\beta \colon \mathfrak{h} \to \mathfrak{g}$ of graded Lie algebras is a degree zero $\mathbb{k}$–linear map of the underlying graded $\mathbb{k}$–modules, such that $\beta[\vartheta, \xi] = [\beta(\vartheta), \beta(\xi)]$ and $\beta(\vartheta^{[2]}) = \beta(\vartheta)^{[2]}$.

One way to get a Lie structure is to partly forget an associative one. Let B be a graded associative algebra over $\mathbb{k}$. The underlying module of B, with bracket $[x, y] = xy - (-1)^{|x||y|} yx$ (the *graded commutator*) and square $v^{[2]} = v^2$ for $v \in B^{\text{odd}}$, is a graded Lie algebra, denoted $\text{Lie}(B)$: the axioms are readily verified by direct computations. The non-triviality of the operations measures how far the algebra B is from being graded commutative.

There is also a vehicle to go from Lie to associative algebras. A *universal enveloping algebra* of $\mathfrak{g}$ is a graded associative $\mathbb{k}$–algebra U together with a degree 0 homomorphism of graded Lie algebras $\iota \colon \mathfrak{g} \to \text{Lie}(U)$ with the following property: for each associative algebra B and each Lie algebra homomorphism $\beta \colon \mathfrak{g} \to \text{Lie}(B)$, there is a unique homomorphism of associative algebras $\beta' \colon U \to B$, such that $\beta = \beta' \iota$; we call β' the *universal extension* of β.

Remark 10.1.3. For the first few statements on enveloping algebras, one just needs to exercise plain abstract nonsense:

(1) Any two universal enveloping algebras of $\mathfrak{g}$ are isomorphic by a unique isomorphism, hence a notation $U_{\mathbb{k}}(\mathfrak{g})$ is warranted.

(2) Each homomorphism $\mathfrak{h} \to \mathfrak{g}$ of graded Lie algebras induces a natural homomorphism $U_{\mathbb{k}}(\mathfrak{h}) \to U_{\mathbb{k}}(\mathfrak{g})$ of graded (associative) algebras.

---

[40] Anticommutativity implies $2[\vartheta, \vartheta] = 0$ for $\vartheta \in \mathfrak{g}^{\text{even}}$, so ($1\frac{1}{2}$) is superfluous when $\mathbb{k} \ni \frac{1}{2}$. Jacobi yields $3[\vartheta, [\vartheta, \vartheta]] = 0$ for all ϑ, so ($2\frac{1}{3}$) is redundant when $\mathbb{k} \ni \frac{1}{3}$.

[41] Only needed if $\frac{1}{2} \notin \mathbb{k}$: conditions (3) and (4) imply that $2v^{[2]} = [v, v]$; when 2 is invertible, $v^{[2]} = \frac{1}{2}[v, v]$ satisfies condition (5), by the Jacobi identity.

(3) The graded $\Bbbk$–algebra $U_{\Bbbk}(\mathfrak{g})$ is isomorphic to the residue of the tensor algebra $T_k(\mathfrak{g})$ modulo the two-sided ideal generated by

$$\vartheta \otimes \xi - (-1)^{|\vartheta||\xi|}\xi \otimes \vartheta - [\vartheta,\xi] \qquad \text{for all} \quad \vartheta,\xi \in \mathfrak{g};$$

$$\upsilon \otimes \upsilon - \upsilon^{[2]} \qquad \text{for all} \quad \upsilon \in \mathfrak{g}^{\mathrm{odd}};$$

the map ι is the composition of $\mathfrak{g} \subseteq T(\mathfrak{g})$ with the projection $T(\mathfrak{g}) \to U$.

(4) Assume that $\mathfrak{g}^n = 0$ for $n \leq 0$, and that $\boldsymbol{\vartheta} = \{\vartheta_i\}_{i \geqslant 1}$ is a set of generators of $\mathfrak{g}$, linearly ordered so that $|\vartheta_i| \leq |\vartheta_j|$ for $i < j$. We consider *indexing sequences* $I = (i_1, i_2, \dots)$ of integers $i_j \geq 0$, such that $i_j \leq 1$ if $|\vartheta_j|$ is odd and $i_j = 0$ for $j \gg 0$. For each I, pick any q such that $i_j = 0$ for $j > q$, and form the (well defined) *normal monomial* $\vartheta^I = \vartheta_q^{i_q} \cdots \vartheta_1^{i_1} \in U_{\Bbbk}(\mathfrak{g})$.

The normal monomials span[42] $U_{\Bbbk}(\mathfrak{g})$. Indeed, (3) shows that $U_{\Bbbk}(\mathfrak{g})$ is spanned by *all* product of elements of $\boldsymbol{\vartheta}$. If such a product contains ϑ_i^2 with $|\vartheta_i|$ odd, then replace ϑ_i^2 by $\vartheta_i^{[2]}$; if it contains $\vartheta_i\vartheta_j$ with $i < j$, then replace it by $\vartheta_j\vartheta_i \pm [\vartheta_i, \vartheta_j]$; express each ϑ_i^2 and $[\vartheta_i, \vartheta_j]$ as a linear combination of generators. Applying the procedure to each of the new monomials, after a finite number of steps one ends up with a linear combination of normal monomials.

Returning to homological algebra, we show how basic constructions of Lie algebras create *co*homological structures. A *DG Lie algebra* over $\Bbbk$ is a graded Lie algebra $\mathfrak{g}$ with a degree -1 $\Bbbk$-linear map $\partial \colon \mathfrak{g} \to \mathfrak{g}$, such that $\partial^2 = 0$,

$$\partial[\vartheta,\xi] = [\partial(\vartheta),\xi] + (-1)^{|\vartheta|}[\vartheta,\partial(\xi)], \text{ and } \partial(\vartheta^{[2]}) = [\partial(\vartheta),\vartheta] \quad \text{for } \vartheta \in \mathfrak{g}^{\mathrm{odd}}.$$

A morphism of DG Lie algebras is a homomorphism of the underlying graded Lie algebras, that is also a morphism of complexes. Homology is a functor from DG Lie algebras to graded Lie algebras.

Lie algebras of derivations are paradigmatic throughout Lie theory. Here is a DG version, based on the Γ-free extensions of Chapter 6.

Lemma 10.1.4. *Let* $\Bbbk \hookrightarrow \Bbbk\langle X\rangle$ *be a semi-free Γ-extension. The inclusion*

$$\mathrm{Der}_{\Bbbk}^{\gamma}(\Bbbk\langle X\rangle, \Bbbk\langle X\rangle) \subseteq \mathrm{Lie}(\mathrm{Hom}_{\Bbbk}(\Bbbk\langle X\rangle, \Bbbk\langle X\rangle))$$

is one of DG Lie algebras.

Proof. The proof is a series of exercises on the Sign Rule.

If b, c, are elements of $\Bbbk\langle X\rangle$, and υ is a derivation of odd degree, then

$$
\begin{aligned}
\upsilon^2(bc) &= \upsilon\left(\upsilon(b)c + (-1)^{|b|}b\upsilon(c)\right)\\
&= \upsilon^2(b)c + (-1)^{|\vartheta|(|\vartheta|+|b|)}\upsilon(b)\upsilon(c) + (-1)^{|b|}\upsilon(b)\upsilon(c) + (-1)^{|b|+|b|}b\upsilon^2(c)\\
&= \upsilon^2(b)c + b\upsilon^2(c).
\end{aligned}
$$

---

[42]In fact, if $\boldsymbol{\vartheta}$ is a basis of $\mathfrak{g}$, then the normal monomials form a *basis* of $U_{\Bbbk}(\mathfrak{g})$: this is the contents of the celebrated Poincaré-Birkhoff-Witt Theorem. The original proof(s) provide one of the first applications of 'standard basis' techniques; for an argument in the graded framework, cf. Milnor and Moore [121]; for the case needed here, cf. Theorem 10.2.1.

If x is a Γ-variable of even degree, and ϑ, ξ are Γ-derivations, then

$$v^2(x^{(i)}) = v(v(x)x^{(i-1)}) = v^2(x)x^{(i-1)} - (v(x))^2 x^{(i-2)} = v^2(x)x^{(i-1)}$$

and

$$
\begin{aligned}
[\vartheta, \xi](x^{(i)}) &= \vartheta\xi(x^{(i)}) - (-1)^{|\vartheta||\xi|}\xi\vartheta(x^{(i)}) \\
&= \vartheta\left(\xi(x)x^{(i-1)}\right) - (-1)^{|\vartheta||\xi|}\xi\left(\vartheta(x)x^{(i-1)}\right) \\
&= \vartheta\xi(x)x^{(i-1)} + (-1)^{|\vartheta||\xi|}\xi(x)\vartheta(x)x^{(i-2)} \\
&\quad - (-1)^{|\vartheta||\xi|}\xi\vartheta(x)x^{(i-1)} - (-1)^{|\vartheta||\xi|+|\xi||\vartheta|}\vartheta(x)\xi(x)x^{(i-2)} \\
&= \left([\vartheta, \xi](x)\right)x^{(i-1)}
\end{aligned}
$$

A lengthier computation shows that $[\vartheta, \xi]$ is a derivation, completing the verification that $\mathrm{Der}_{\Bbbk}^{\gamma}\left(\Bbbk\langle X\rangle, \Bbbk\langle X\rangle\right)$ is a Lie subalgebra of $\mathrm{Lie}(\mathrm{Hom}_{\Bbbk}\left(\Bbbk\langle X\rangle, \Bbbk\langle X\rangle\right))$.

It remains to prove that the differential of the derivation complex satisfies the requirements for a graded Lie algebra. This is best done by 'interiorizing' it: as ∂ is a DG Γ-derivation of $\Bbbk\langle X\rangle$, it may be viewed as an element $\delta \in \mathrm{Der}_{\Bbbk}^{\gamma}\left(\Bbbk\langle X\rangle, \Bbbk\langle X\rangle\right)$, and then $\partial(\vartheta) = [\delta, \vartheta]$. The conditions on $\partial[\vartheta, \xi]$ and $\partial(v^{[2]})$ are now seen to be transcriptions of the Jacobi identity and its complement. $\qquad\square$

It is tempting to mimic the construction of Ext algebras: Choose a Tate resolution $\Bbbk\langle X\rangle$ of a commutative $\Bbbk$–algebra P, and associate with P the graded Lie algebra $\mathrm{H\,Der}_{\Bbbk}^{\gamma}\left(\Bbbk\langle X\rangle, \Bbbk\langle X\rangle\right)$. If $\mathbb{Q} \subseteq P$, then it is an invariant of P: this is proved by Quillen [133], as an outgrowth of his investigation of rational homotopy theory [131]. The general case is very different:

Example 10.1.5. The Tate resolution $A = \Bbbk\langle X\rangle$ of $\Bbbk$ over itself, with $X = \varnothing$, yields $\mathrm{H\,Der}_{\Bbbk}^{\gamma}(A, A) = 0$. If $\mathbb{F}_2 \subseteq \Bbbk$, then another Tate resolution of $\Bbbk$ is

$$B = \Bbbk\langle u, \{x_i, x_i'\}_{i \geqslant 0} \,|\, \partial(u) = 0, \partial(x_i) = u^{(2^i)}, \partial(x_i') = u^{(2^i)}x_i\rangle$$

with $|u| = 2$ (hence $|x_i| = 2^{i+1} + 1$ and $|x_i'| = 2^{i+2} + 2$). Using Corollary 6.2.4 and Construction 6.2.5, one gets $\mathrm{H\,Der}_{\Bbbk}^{\gamma}(B, B) \cong \mathrm{Hom}_{\Bbbk}\left(\mathrm{Ind}_{\Bbbk}^{\gamma} B, \Bbbk\right)$, and $\mathrm{Ind}_{\Bbbk}^{\gamma} B$ is the free $\Bbbk$–module with basis $\{x_i\}_{i \geqslant 1} \cup \{x_i'\}_{i \geqslant 0}$.

10.2. Homotopy Lie algebra. In this section $(R, \mathfrak{m}, k)$ is a local ring.

Some of the problems occurring in the last example may be circumvented, by using acyclic closures. We follows the ideas of Sjödin [144], and simplify the exposition by using complexes of derivations from Section 6.2.

Theorem 10.2.1. Let $R\langle X\rangle$ be an acyclic closure of k over R, where $X = \{x_i\}_{i \geqslant 1}$ and $|x_i| \leq |x_j|$ for $i < j$, and set $\pi(R) = \mathrm{H\,Der}_R^{\gamma}\left(R\langle X\rangle, R\langle X\rangle\right)$.

(1) $\pi(R)$ is a graded Lie algebra over k.

(2) $\mathrm{rank}_k \pi^n(R) = \varepsilon_n(R)$ for $n \in \mathbb{Z}$.

(3) $\pi(R)$ has a k-basis

$$\Theta = \{\theta_i = \mathrm{cls}(\vartheta_i) \mid \vartheta_i \in \mathrm{Der}_R^{\gamma}\left(R\langle X\rangle, R\langle X\rangle\right), \vartheta_i(x_j) = \delta_{ij} \text{ for } j \leq i\}_{i \geqslant 1}.$$

(4) *The normal monomials on Θ form a k–basis of $U_k(\pi(R))$.*

(5) $\mathrm{Der}_R^\gamma(R\langle X\rangle, R\langle X\rangle) \subseteq \mathrm{Hom}_R(R\langle X\rangle, R\langle X\rangle)$ *induces an injective homomorphism of graded Lie algebras* $\iota\colon \pi(R) \to \mathrm{Lie}(\mathrm{Ext}_R(k,k))$. *Its universal extension is an isomorphism of associative algebras*

$$\iota'\colon U_k(\pi(R)) \cong \mathrm{Ext}_R(k,k) \ .$$

Remark. By Remark 6.3.9, different choices of acyclic closures yield the same Lie algebra $\pi(R)$; it is called the *homotopy Lie algebra* of R.

Proof. (1) and (2). Let ϵ denote the quasi-isomorphism $R\langle X\rangle \to k$, and let $k \hookrightarrow k\langle X\rangle$ be the semi-free Γ-extension with trivial differential. Using Lemma 6.2.4, the minimality of $R\langle X\rangle$, and Proposition 6.2.3.4, we get

$$\pi(R) = \mathrm{H}\,\mathrm{Der}_R^\gamma(R\langle X\rangle, R\langle X\rangle) \cong \mathrm{H}\,\mathrm{Der}_R^\gamma(R\langle X\rangle, k)$$
$$\cong \mathrm{H}\,\mathrm{Der}_k^\gamma(k\langle X\rangle, k) = \mathrm{Der}_k^\gamma(k\langle X\rangle, k) \cong \mathrm{Hom}_k(kX, k) \ .$$

So $\pi(R)$ is a k–module; by Theorem 7.1.3, $\mathrm{rank}_k \pi^n(R) = \mathrm{card}(X_n) = \varepsilon_n(R)$.

(3) Lemma 6.3.3.1 provides a set of R–linear Γ-derivations $\{\vartheta_i\}_{i\geqslant 1}$, with $\vartheta_i(x_j) = \delta_{ij}$ for $j \leq i$. As $\{x_i\}_{i\geqslant 1}$ is a basis of kX, the isomorphisms above imply that Θ is linearly independent in $\pi(R)$. Since Θ_n has $\varepsilon_n(R)$ elements for each n, it is a basis by (2).

(4) and (5). In the commutative diagram

$$\begin{array}{ccc} \mathrm{Der}_R^\gamma(R\langle X\rangle, R\langle X\rangle) & \longrightarrow & \mathrm{Hom}_R(R\langle X\rangle, R\langle X\rangle) \\ {\scriptstyle \mathrm{Der}_R^\gamma(R\langle X\rangle, \epsilon)} \Big\downarrow {\scriptstyle \simeq} & & {\scriptstyle \simeq} \Big\downarrow {\scriptstyle \mathrm{Hom}_R(R\langle X\rangle, \epsilon)} \\ \mathrm{Der}_R^\gamma(R\langle X\rangle, k) & \overset{\iota}{\longrightarrow} & \mathrm{Hom}_R(R\langle X\rangle, k) \end{array}$$

the left hand arrow is a quasi-isomorphism, as noted for (1); the right hand one is a quasi-isomorphism by Proposition 1.3.2.

Let $\theta^I \in U_k(\pi(R))$ be a normal monomial on Θ. For a normal Γ-monomial $x^{(H)}$ on X, by Lemma 6.3.3.3 we see that $\iota'(\theta^I)(x^{(H)}) = \mathrm{cls}(\vartheta^I(x^{(H)}))$ is equal to 0 if $H < I$, and to 1 if $H = I$. As the normal Γ-monomials on X form a basis of $k\langle X\rangle$, the triangular form of the matrix implies that the images of the normal monomials on Θ form a basis of $\mathrm{Hom}_R(R\langle X\rangle, k) = \mathrm{Ext}_R(k,k)$. Thus, the homomorphism ι' is surjective, and the images of the normal monomials are linearly independent. By (4) and Remark 10.1.3.4, these monomials generate $U_k(\pi(R))$, so we conclude that ι' is an isomorphism, as desired. $\qquad\qquad\square$

Sjödin [144] shows how to compute the Lie operations on $\pi^1(R)$.

Example 10.2.2. Let $R = Q/I$, where $(Q, \mathfrak{n}, k)$ is regular, $\mathfrak{n}$ is minimally generated by $s_1, \ldots, s_e$, and I is minimally generated by $f_1, \ldots, f_r$, with

$$f_j = \sum_{1 \leqslant h \leqslant i \leqslant e} a_{hi,j} s_h s_i \quad \text{with} \quad a_{hi,j} \in Q \quad \text{for } 1 \leq j \leq r \ .$$

Using overbars to denote images in R, and setting $z_j = \sum_{h \leqslant i} \bar{a}_{hi,j} \bar{s}_h x_i$, we see that the acyclic closure $R\langle X \rangle$ of k over R is then obtained from

$$R\langle x_1, \ldots, x_{e+r} \mid \partial(x_i) = \bar{s}_i \text{ for } 1 \leq i \leq e \,; \partial(x_{e+j}) = z_j \text{ for } 1 \leq j \leq r \rangle$$

by adjunction of Γ-variables of degree ≥ 3. Let $\vartheta_1, \ldots \vartheta_e$ be the Γ-derivations of $R\langle x_1, \ldots, x_e \rangle$, defined by $\vartheta_i(x_h) = \delta_{ih}$. To extend them to Γ-derivations of $R\langle x_1, \ldots, x_{e+r} \rangle$, such that $\partial \vartheta_i = -\vartheta_i \partial$, note that

$$\vartheta_i \partial(x_{e+j}) = \vartheta_i \left(\sum_{h \leqslant i} \bar{a}_{hi,j} \bar{s}_h x_i \right) = \sum_{h=1}^{i} \bar{a}_{hi,j} \bar{s}_h = \partial \left(\sum_{h=1}^{i} \bar{a}_{hi,j} x_h \right),$$

and set $\vartheta_i(x_{e+j}) = - \sum_h \bar{a}_{hi,j} x_h$ for $1 \leq i \leq e$ and $1 \leq j \leq r$. This yields

$$[\vartheta_h, \vartheta_i](x_{e+j}) = -\bar{a}_{hi,j} \quad \text{for } h < i \qquad \text{and} \qquad \vartheta_i^{[2]}(x_{e+j}) = -\bar{a}_{ii,j}.$$

Thus, on the basis elements of Theorem 10.2.1, the Lie bracket $\pi^1(R) \times \pi^1(R) \to \pi^2(R)$ and square $\pi^1(R) \to \pi^2(R)$ are given by

$$[\theta_h, \theta_i] = - \sum_{j=1}^{r} a'_{hi,j} \theta_{e+j} \quad \text{for } h < i \qquad \text{and} \qquad (\theta_i)^{[2]} = - \sum_{j=1}^{r} a'_{ii,j} \theta_{e+j}$$

where a' denotes the image in k of $a \in R$.

Consider the k–subspace of $\pi^2(R)$, spanned by the commutators and squares of all elements of $\pi^1(R)$ (in fact, the squares suffice, cf. 10.1.2). By the preceding computation, its rank q is equal to that of the $\binom{e+1}{2} \times r$ matrix $(a_{hi,j})$ reduced modulo $\mathfrak{n}$, that is $q = \binom{e+1}{2} - \mathrm{rank}_k(I / I \cap \mathfrak{m}^3)$. In particular, $\pi^1(R)$ generates $\pi^2(R)$ if and only if the r quadratic forms $f_j = \sum_{h \leqslant i} a_{hi,j} s_h s_i$ are linearly independent in $\mathrm{gr}_{\mathfrak{n}}(Q)$. At the other extreme, the Lie subalgebra of $\pi(R)$ generated by $\pi^1(R)$ is is reduced to $\pi(R)$ itself if and only if $I \subseteq \mathfrak{m}^3$.

Example 10.2.3. By Theorems 10.2.1.2 and 7.3.3, if $\pi(R)$ is finite dimensional, then R is a complete intersection and $\pi(R)$ is concentrated in degrees 1 and 2, so the preceding example determines its structure. The Lie subalgebra $\pi^{\geqslant 2}(R)$ is central in $\pi(R)$, and its universal enveloping algebra is the polynomial ring $\mathcal{P} = k[\theta_1, \ldots, \theta_r]$. An isomorphism of $\mathcal{P}$–modules $\mathrm{Ext}_R(k, k) \cong \mathcal{P} \otimes_k \mathcal{E}$, where $\mathcal{E}$ is the vector space underlying the exterior algebra on $\pi^1(R) \cong \mathrm{Hom}_k(\mathfrak{m}/\mathfrak{m}^2, k)$, refines the equality $\mathrm{P}_k^R(t) = (1 + t)^e / (1 - t^2)^r$.

Conversely, each graded Lie algebra $\mathfrak{g}$ with $\mathrm{rank}_k \mathfrak{g}^1 \geq \mathrm{rank}_k \mathfrak{g}^2$ and $\mathfrak{g}^n = 0$ for $n \neq 1, 2$ is of the form $\pi(R)$ for an appropriate complete intersection: one starts by fixing the desired quadratic parts $g_j = \sum_{h \leqslant i} a_{hi,j} s_h s_i$ of the relations, and uses a 'prime avoidance' argument to find elements p_j in a high power of $\mathfrak{m}$, such that the sequence $g_1 + p_1, \ldots, g_r + p_r$ is regular, cf. [144].

We conclude with some general remarks on the homotopy Lie algebra. A detailed study belongs to a different exposition.

Remark 10.2.4. A local homomorphism of local rings $\varphi \colon R \to S$ induces a homo-morphism of graded Lie algebra $\pi(\varphi) \colon \pi(S) \to \pi(R) \otimes_k \ell$, where ℓ is the residue field of S. This yields a contravariant functor with remarkable properties. For example, if φ is flat, then for each i there is an exact sequence

$$0 \longrightarrow \pi^{2i-1}(S/\mathfrak{m}S) \longrightarrow \pi^{2i-1}(S) \longrightarrow \pi^{2i-1}(R) \otimes_k \ell$$

$$\xrightarrow{\eth^{2i-1}} \pi^{2i}(S/\mathfrak{m}S) \longrightarrow \pi^{2i}(S) \longrightarrow \pi^{2i}(R) \otimes_k \ell \longrightarrow 0$$

where $\eth^{2i-1} = 0$ for almost all i, and $\sum_{i=0}^{\infty} \operatorname{rank} \eth^{2i-1} \leq \operatorname{codepth}(S/\mathfrak{m}S)$: this is proved (in dual form) in [20] and [9]. The Lie algebra $\pi(R)$ is a looking glass version of the Lie algebra of rational homotopy groups in algebraic topology, cf. [23] and [33] for a systematic discussion.

Remark. The original construction of $\pi(R)$ proceeded in two steps.

The first, initiated by Assmus [15], and completed by Levin [108] and Schoeller [139], constructs a homomorphism $\Delta \colon \operatorname{Tor}^R(k,k) \to \operatorname{Tor}^R(k,k) \otimes_k \operatorname{Tor}^R(k,k)$ of Γ-algebras, giving $\operatorname{Tor}^R(k,k)$ a structure of Hopf algebra. The second identifies the composition product as the dual of Δ under the isomorphism of Hopf algebras $\operatorname{Ext}_R(k,k) = \operatorname{Hom}_k\left(\operatorname{Tor}^R(k,k),k\right)$.

At that point, a structure theorem due to Milnor and Moore [121] in charac-teristic 0 and to André in characteristic $p > 0$ (adjusted by Sjödin [145] for $p = 2$) shows that such a Hopf algebra is the universal enveloping algebra of graded Lie algebra. In fact, these results prove much more: namely, an equivalence (of certain subcategories) of the categories of Γ-Hopf algebras and graded Lie algebras, given in one direction by the universal enveloping algebra functor.

Remark. A graded Lie algebra $\operatorname{H}^*(R,k,k)$ is attached to R by the simplicially defined tangent cohomology of André and Quillen. There is a homomorphism of graded Lie algebras $\operatorname{H}^*(R,k,k) \to \pi(R)$, cf. [4]. It is bijective for complete intersec-tions, or when $\operatorname{char}(k) = 0$, cf. [133]; when $\operatorname{char}(k) = p > 0$, this holds in degrees $\leq 2p$, but not always in degree $2p+1$, cf. [7]. The computation of $\operatorname{H}^*(R,k,k)$ is very difficult in positive characteristic: for the small ring $R = \mathbb{F}_2[s_1,s_2]/(s_1^2,s_1s_2,s_2^2)$ it requires the book of Goerss [73]; for comparison, $\pi(R)$ is the free Lie algebra on the 2-dimensional vector space $\pi^1(R)$.

10.3. Applications.
Once again, $(R,\mathfrak{m},k)$ denotes a local ring.

We relate this chapter to the bulk of the notes by discussing two kinds of applications of $\pi(R)$ to the study of resolutions. The structure of $\pi(R)$ is reflected in the Poincaré series of finite R-modules. To illustrate the point, we characterize Golod rings in terms of their homotopy Lie algebras.

Example 10.3.1. Let V be a vector space over k. A graded Lie algebra $\mathfrak{g}$ is *free* on V, if $V \subseteq \mathfrak{g}$ and each degree zero k-linear map from V to a graded Lie algebra $\mathfrak{h}$ extends uniquely to a homomorphism of Lie algebras $\mathfrak{g} \to \mathfrak{h}$.

It is easy to see that free Lie algebras exist on any V: just take $\mathfrak{g}$ to be the subspace of the tensor algebra $\operatorname{T}_k(V)$, spanned by all commutators of elements of

V, and all squares of elements of V_{odd}. Using the universal property of $\mathrm{T}_k(V)$ one sees that $\mathfrak{g}$ is a free Lie algebra on V, and comparing it with that of $\mathrm{U}_k(\mathfrak{g})$ one concludes that these two algebras coincide.

Avramov [21] and Löfwall [113] prove that R is Golod if and only if $\pi^{\geqslant 2}(R)$ is free, and then it is the free Lie algebra on $V = \mathrm{Hom}_k\left(\Sigma\, \mathrm{H}_{\geqslant 1}(K^R), k\right)$.

This nicely 'explains' the formula (5.0.1) for the Poincaré series of k over a Golod ring: $f(t) = (1+t)^e$ is the Hilbert series of the vector space $\mathcal{E}$ underlying the exterior algebra on $\pi^1(R) \cong \mathrm{Hom}_k\left(\mathfrak{m}/\mathfrak{m}^2, k\right)$; the expression $g(t) = 1/(1 - \sum_i \mathrm{rank}\, \mathrm{H}_i(K^R)t^{i+1})$ is the Hilbert series of $\mathcal{T}$, the tensor algebra on V, and $\mathrm{P}_k^R(t) = f(t)g(t)$ reflects an isomorphism $\mathrm{Ext}_R(k,k) \cong \mathcal{T} \otimes_k \mathcal{E}$ of $\mathcal{T}$–modules.

The bracket and the square in a free Lie algebra are as non-trivial as possible, so the cohomological descriptions of Golod rings above and of complete intersections in Example 10.2.3 put a maximal distance between them (compare Remark 8.1.1.3). However, there exists a level at which these descriptions coalesce: the Lie algebra $\pi^{\geqslant 3}(R)$ is free, because it is trivial in the first case, and because freeness is inherited by Lie subalgebras, cf. [105], in the second.

Remark 10.3.2. It is proved in [28], using results from [24], that the following conditions on a local ring R are equivalent:

(i) $\pi(R)$ contains a free Lie subalgebra of finite codimension;

(ii) for some $r \in \mathbb{Z}$, the Lie algebra $\pi^{\geqslant r}(R)$ is free;

(iii) for some $s \in \mathbb{Z}$, the DG algebra $R\langle X_{< s}\rangle$ admits a trivial Massey operation, cf. Remark 5.2.1.

When they hold, the ring R is called *generalized Golod (of level $\leq s$)*. Such rings abound in small codepth: this is the case when $\mathrm{edim}\, R - \mathrm{depth}\, R \leq 3$ (Avramov, Kustin, and Miller [36]), or when $\mathrm{edim}\, R - \mathrm{depth}\, R = 4$ and R is Gorenstein (Jacobsson, Kustin, and Miller [94]) or an almost complete intersection (Kustin and Palmer [102]). Kustin [99], [100] proves that certain determinantal relations define generalized Golod rings.

Theorem 10.3.3. *Let R be a generalized Golod ring of level $\leq s$.*

(1) *There is a polynomial $\mathrm{den}(t) \in \mathbb{Z}[t]$, and for each finite R–module M there is a polynomial $q(t) \in \mathbb{Z}[t]$, such that $\mathrm{P}_M^R(t) = q(t)/\mathrm{den}(t)$; the numerator for $M = k$ divides $\prod_{2i+1<s}(1 + t^{2i+1})^{\varepsilon_{2i+1}(R)}$.*

(2) *If $\mathrm{cx}_R\, M = \infty$, then $\mathrm{curv}_R\, M = \beta > 1$ and there is a real number α such that $\beta_n^R(M) \sim \alpha\beta^n$.* ∎

The first part is proved by Avramov [28], the second by Sun [150]. Both use, among other things, a theorem of Gulliksen [81] that extends Remark 9.2.6.

Theorem 10.3.4. *Let $R\langle X\rangle$ be an acyclic closure of k over R. If M is a finite R–module, then for each $n \geq 1$ there is a polynomial $h_n(t) \in \mathbb{Z}[t]$, such that*

$$\sum_{i=0}^{\infty} \mathrm{rank}_k\, \mathrm{H}_i(M \otimes_R R\langle X_{\leqslant n}\rangle)t^i = \frac{h_n(t)}{\prod_{2j\leq n}(1 - t^{2j})^{\varepsilon_{2j}(R)}}\,. \qquad\blacksquare$$

Theorem 10.3.3, or results that it generalizes, has been used in essentially all cases when the Poincaré series is known to be rational for all finite modules. It is difficult to resist asking the next question; a positive answer would be unexpected and very useful; a negative one might be equally interesting, since it would most likely involve unusual constructions.

Problem 10.3.5. If $P^R_M(t)$ is rational for each M, is then R generalized Golod?

Next we describe applications of $\pi(R)$ that do not make specific assumptions on its form. They use the following easy consequence of Theorem 10.2.1.

Remark 10.3.6. Let $\mathfrak{h}$ be a graded Lie subalgebra of $\pi(R)$. By completing a basis of $\mathfrak{h}$ to one of $\pi(R)$, and considering the corresponding basis of normal monomials of $U_k(\pi(R)) = \mathrm{Ext}_R(k,k)$, cf. Theorem 10.2.1, one easily sees that there is an isomorphism $U_k(\mathfrak{g}) \cong U_k(\mathfrak{h}) \otimes_k V$ of left modules over $U_k(\mathfrak{h})$, where V is the tensor product of the exterior algebra on $(\mathfrak{g}/\mathfrak{h})^{\mathrm{odd}}$ with the symmetric algebra on $(\mathfrak{g}/\mathfrak{h})^{\mathrm{even}}$. By a simple count of basis elements,

$$H^k_V(t) = \frac{\prod_{i=1}^{\infty}(1+t^{2i-1})^{\ell^{2i-1}}}{\prod_{i=1}^{\infty}(1-t^{2i})^{\ell^{2i}}} \qquad \text{with} \quad \ell^n = \mathrm{rank}_k(\mathfrak{g}/\mathfrak{h})^n.$$

For each R–module M, the vector space $\mathrm{Ext}_R(M,k)$ is a left module over the universal enveloping algebra $\mathrm{Ext}_R(k,k)$: it suffices to make the obvious changes in the construction of cohomology products in Section 10.1. This module structure is the essential tool in the proof of the next result. In fact, it has already been used throughout Chapter 9, in a different guise: over a complete intersection, the actions on $\mathrm{Ext}_R(M,k)$ of the graded algebras denoted $\mathcal{P}$ in Remark 9.2.6 and in Remark 10.2.3 are the same, cf. [37].

Motivated by Proposition 4.2.4.1, we say that a module L over a local ring $(R,\mathfrak{m},k)$ is *extremal*, if $\mathrm{cx}_R L = \mathrm{cx}_R k$ and $\mathrm{curv}_R L = \mathrm{curv}_R k$. For instance, Theorems 8.3.3 and 8.3.4 show that (in the graded characteristic zero case) the conormal module and the module of differentials are extremal when R is not a complete intersection. Results from [29] add more instances, among them:

Theorem 10.3.7. *If R is a local ring of embedding dimension e, M is a finite R–module and L is a submodule such that $L \supseteq \mathfrak{m}M$, then*

$$P^R_L(t) \cdot (1+t)^e \succcurlyeq P^R_k(t) \cdot \mathrm{rank}_k\left(\frac{\mathfrak{m}M}{\mathfrak{m}L}\right).$$

For $\mathfrak{m}^iM \subseteq \mathfrak{m}^{i-1}M$, we get a quantitative version of Levin's characterization of regularity [108]: If $\mathfrak{m}M \neq 0$ and $\mathrm{pd}_R(\mathfrak{m}M) < \infty$, then R is regular:

Corollary 10.3.8. *If $\mathfrak{m}^iM \neq 0$ for some $i \geq 1$, then $\mathfrak{m}^iM$ is extremal.* □

Corollary 10.3.9. *Each non-zero R–module $M \neq k$ may be obtained as an extension of an extremal R–module by another such module. In particular, the extremal R–modules generate the Grothendieck group of R.* □

Here is another class of extremal modules from [29].

Remark 10.3.10. If $N \neq 0$ is a homomorphic image of a finite direct sum of syzygies of k, then $\mathrm{cx}_R N = \mathrm{cx}_R k$ and $\mathrm{curv}_R N = \mathrm{curv}_R k$. As a consequence, we get a result of Martsinkovsky [116]: $\mathrm{pd}_R N = \infty$, unless R is regular.

Proof of Theorem 10.3.7. The commutative diagram of R–modules

$$
\begin{array}{ccccccccc}
0 & \longrightarrow & L & \longrightarrow & M & \longrightarrow & N & \longrightarrow & 0 \\
& & \downarrow & & \downarrow & & \| & & \\
0 & \longrightarrow & \dfrac{L}{\mathfrak{m}L} & \longrightarrow & \dfrac{M}{\mathfrak{m}L} & \longrightarrow & N & \longrightarrow & 0
\end{array}
$$

induces a commutative square of homomorphisms of graded *left* modules

$$
\begin{array}{ccc}
\mathrm{Ext}_R(L,k) & \xrightarrow{\ \eth'\ } & \mathrm{Ext}_R(N,k) \\
\uparrow & & \| \\
\mathrm{Ext}_R\left(\dfrac{L}{\mathfrak{m}L},k\right) & \xrightarrow{\ \eth\ } & \mathrm{Ext}_R(N,k)
\end{array}
$$

over $E = \mathrm{Ext}_R(k,k)$, where $\eth'$ and $\eth$ are connecting maps of degree 1.

As $\mathfrak{m}$ annihilates $L/\mathfrak{m}L$ and N, we have isomorphisms

$$
\mathrm{Ext}_R\left(\frac{L}{\mathfrak{m}L},k\right) \cong E \otimes_k \mathrm{Hom}_k\left(\frac{L}{\mathfrak{m}L},k\right)
$$
$$
\mathrm{Ext}_R(N,k) \cong E \otimes_k \mathrm{Hom}_k(N,k)
$$

of graded E–modules. By Remark 10.3.6, $E \cong U \otimes_k \bigwedge(\pi^1(R))$ as graded left modules over the subalgebra $U = \mathrm{U}_k(\pi^{\geqslant 2}R)$. Noting that $\pi^1(R) = E^1$, we can rewrite $\eth$ as the top map of the commutative diagram

$$
\begin{array}{ccc}
U \otimes_k \bigwedge(E^1) \otimes_k \mathrm{Hom}_k\left(\dfrac{L}{\mathfrak{m}L},k\right) & \xrightarrow{\ \eth\ } & U \otimes_k \bigwedge(E^1) \otimes_k \mathrm{Hom}_k(N,k) \\
\cup \uparrow & & \uparrow \cup \\
U \otimes_k \mathrm{Hom}_k\left(\dfrac{L}{\mathfrak{m}L},k\right) & \longrightarrow & U \otimes_k E^1 \otimes_k \mathrm{Hom}_k(N,k) \\
\| & & \uparrow \cong \\
U \otimes_k \mathrm{Hom}_k\left(\dfrac{L}{\mathfrak{m}L},k\right) & \xrightarrow{\ U \otimes_k \eth^0\ } & U \otimes_k \mathrm{Ext}_R^1(N,k) \ .
\end{array}
$$

of homomorphisms of graded left U–modules.

The preceding information combines to yield a commutative square

$$\operatorname{Ext}_R(L,k) \quad \xrightarrow{\quad \eth' \quad} \quad \operatorname{Ext}_R(N,k)$$

$$\uparrow \qquad\qquad\qquad\qquad \uparrow$$

$$U \otimes_k \operatorname{Hom}_k\left(\frac{L}{\mathfrak{m}L},k\right) \xrightarrow{\; U \otimes_k \eth^0 \;} U \otimes_k \operatorname{Ext}_R^1(N,k)$$

of homomorphisms of graded U–modules, with injective right hand vertical arrow. Thus, $\operatorname{Ext}_R(N,k)$ contains a copy of the free module $\Sigma(U) \otimes_k \operatorname{Im}\eth_0$. By the commutativity of the square, $\operatorname{Ext}_R(L,k)$ contains a copy of $U \otimes_k \operatorname{Im}\eth^0$.

On the other hand, a length count in the cohomology exact sequence

$$0 \to \operatorname{Hom}_R(N,k) \to \operatorname{Hom}_R\left(\frac{M}{\mathfrak{m}L},k\right) \to \operatorname{Hom}_R\left(\frac{L}{\mathfrak{m}L},k\right) \xrightarrow{\eth^0} \operatorname{Ext}_R^1(N,k)$$

yields $\operatorname{rank}_k \eth^0 = \operatorname{rank}_k(\mathfrak{m}M/\mathfrak{m}L)$. Thus, $P_L^R(t) \succcurlyeq \operatorname{rank}_k(\mathfrak{m}M/\mathfrak{m}L) \cdot H_U^k(t)$. To finish the proof, multiply this inequality by $(1+t)^e$, then simplify the right hand side by using the equality $(1+t)^e \cdot H_U^k(t) = P_k^R(t)$ from 10.3.6.   □

References

[1] S. S. Abhyankar, *Local rings of high embedding dimension*, Amer. J. Math. **89** (1967), 1073–1077.

[2] J. Alperin, L. Evens, *Representations, resolutions, and Quillen's dimension theorem*, J. Pure Appl. Algebra **22** (1981), 1–9.

[3] M. André, *Méthode simpliciale en algèbre homologique et algèbre commutative*, Lecture Notes Math. **32**, Springer, Berlin, 1967.

[4] M. André, *L'algèbre de Lie d'un anneau local*, Symp. Math. **4**, (INDAM, Rome, 1968/69), Academic Press, London, 1970; pp. 337–375.

[5] M. André, *Hopf algebras with divided powers*, J. Algebra **18** (1971), 19–50.

[6] M. André, *Homologie des algèbres commutatives*, Grundlehren Math. Wiss. **204**, Springer, Berlin, 1974.

[7] M. André, *La $(2p+1)$ème déviation d'un anneau local*, Enseignement Math. (2) **23** (1977), 239–248.

[8] M. André, *Algèbre homologique des anneaux locaux à corps résiduels de caractéristique deux*, Sém. d'Algèbre P. Dubreil, Paris, 1979 (M.-P. Malliavin, ed.), Lecture Notes Math. **740**, Springer, Berlin, 1979; pp. 237–242.

[9] M. André, *Le caractère additif des déviations des anneaux locaux*, Comment. Math. Helv. **57** (1982), 648–675.

[10] D. Anick, *Constructions d'espaces de lacets et d'anneaux locaux à séries de Poincaré-Betti non rationnelles*, C. R. Acad. Sci. Paris Sér. A **290** (1980), 729–732.

[11] D. Anick, *Counterexample to a conjecture of Serre*, Ann. of Math. (2) **115** (1982), 1–33; *Comment*, ibid. **116** (1982), 661.

[12] D. Anick, *Recent progress in Hilbert and Poincaré series*, Algebraic topology. Rational homotopy, Louvain-la-Neuve, 1986 (Y. Félix, ed.), Lecture Notes Math. **1318**, Springer, Berlin, 1988; pp. 1–25.

[13] D. Anick, T. H. Gulliksen, *Rational dependence among Hilbert and Poincaré series*, J. Pure Appl. Algebra **38** (1985), 135–158.

[14] A. G. Aramova, J. Herzog, *Koszul cycles and Eliahou-Kervaire type resolutions*, J. Algebra **181** (1996), 347–370.

[15] E. F. Assmus, Jr., *On the homology of local rings*, Illinois J. Math. **3** (1959), 187–199.

[16] M. Auslander, *Modules over unramified regular local rings*, Illinois J. Math. **5** (1961), 631–647.

[17] M. Auslander, D. A. Buchsbaum, *Codimension and multiplicity*, Ann. of Math. (2) **68** (1958), 625–657; *Corrections*, ibid, **70** (1959), 395–397.

[18] L. L. Avramov, *On the Hopf algebra of a local ring*, Math USSR-Izv. **8** (1974), 259–284; [translated from:] Izv. Akad. Nauk. SSSR, Ser. Mat. **38** (1974), 253–277 [Russian].

[19] L. L. Avramov, *Flat morphisms of complete intersections*, Soviet Math. Dokl. **16** (1975), 1413–1417; [translated from:] Dokl. Akad. Nauk. SSSR, **225** (1975), 11–14 [Russian].

[20] L. L. Avramov, *Homology of local flat extensions and complete intersection defects*, Math. Ann. **228** (1977), 27–37.

[21] L. L. Avramov, *Small homomorphisms of local rings*, J. Algebra **50** (1978), 400–453.

[22] L. L. Avramov, *Obstructions to the existence of multiplicative structures on minimal free resolutions*, Amer. J. Math. **103** (1981), 1–31.

[23] L. L. Avramov, *Local algebra and rational homotopy*, Homotopie algébrique et algèbre locale; Luminy, 1982 (J.-M. Lemaire, J.-C. Thomas, eds.) Astérisque **113-114**, Soc. Math. France, Paris, 1984; pp. 15–43.

[24] L. L. Avramov, *Golod homomorphisms*, Algebra, algebraic topology, and their interactions; Stockholm, 1983 (J.-E. Roos, ed.), Lecture Notes Math. **1183**, Springer, Berlin, 1986; pp. 56–78.

[25] L. L. Avramov, *Modules of finite virtual projective dimension*, Invent. Math. **96** (1989), 71–101.

[26] L. L. Avramov, *Homological asymptotics of modules over local rings*, Commutative algebra; Berkeley, 1987 (M. Hochster, C. Huneke, J. Sally, eds.), MSRI Publ. **15**, Springer, New York 1989; pp. 33–62.

[27] L. L. Avramov, *Problems on infinite free resolutions*, Free resolutions in commutative algebra and algebraic geometry; Sundance, 1990 (D. Eisenbud, C. Huneke, eds.), Res. Notes Math. **2**, Jones and Bartlett, Boston 1992, pp. 3–23.

[28] L. L. Avramov, *Local rings over which all modules have rational Poincaré series*, J. Pure Appl. Algebra **91** (1994), 29–48.

[29] L. L. Avramov, *Modules with extremal resolutions*, Math. Res. Lett. **3** (1996), 319–328.

[30] L. L. Avramov, *Locally complete intersection homomorphisms, and a conjecture of Quillen on the vanishing of cotangent homology*, Preprint, 1997.

[31] L. L. Avramov, R.-O. Buchweitz, *Modules of complexity two over complete intersections*, Preprint, 1997.

[32] L. L. Avramov, V. N. Gasharov, I. V. Peeva, *Complete intersection dimension*, Publ. Math. I.H.E.S. (to appear).

[33] L. L. Avramov, S. Halperin, *Through the looking glass: A dictionary between rational homotopy theory and local algebra*, Algebra, algebraic topology, and their interactions; Stockholm, 1983 (J.-E. Roos, ed.), Lecture Notes Math. **1183**, Springer, Berlin, 1986; pp. 1–27.

[34] L. L. Avramov, S. Halperin, *On the non-vanishing of cotangent cohomology* Comment. Math. Helv. **62** (1987), 169–184.

[35] L. L. Avramov, J. Herzog, *Jacobian criteria for complete intersections. The graded case*, Invent. Math. **117** (1994), 75–88.

[36] L. L. Avramov, A. R. Kustin, and M. Miller, *Poincaré series of modules over local rings of small embedding codepth or small linking number*, J. Algebra **118** (1988), 162–204.

[37] L. L. Avramov, L.-C. Sun, *Cohomology operators defined by a deformation*, Preprint, 1996.

[38] I. K. Babenko, *On the analytic properties of Poincaré series of loop spaces*, Math. Notes **29** (1980), 359–366; [translated from:] Mat. Zametki **27** (1980), 751–765 [Russian].

[39] I. K. Babenko, *Problems of growth and rationality in algebra and topology*, Russian Math. Surv. **29** (1980), no. 2, 95–142; [translated from:] Uspekhi Mat. Nauk **41** (1986), no. 2, 95–142 [Russian].

[40] D. Bayer, M. E. Stillman, *Macaulay*, A computer algebra system for computing in Algebraic Geometry and Computer Algebra, 1990; available via anonymous ftp from `zariski.harvard.edu`.

[41] A. Blanco, J. Majadas, A. G. Rodicio, *On the acyclicity of the Tate complex*, J. Pure Appl. Algebra (to appear).

[42] D. Benson, *Representations and cohomology. I; II*, Cambridge Stud. Adv. Math. **31; 32**, Cambridge Univ. Press, Cambridge, 1991.

[43] A. Bigatti, *Upper bounds for the Betti numbers of a given Hilbert function*, Comm. Algebra **21** (1993), 2317–2334.

[44] R. Bøgvad, *Gorenstein rings with transcendental Poincaré series*, Math. Scand. **53** (1983), 5–15.

[45] N. Bourbaki, *Algèbre, X. Algèbre homologique*, Masson, Paris, 1980.

[46] W. Bruns, J. Herzog, *Cohen-Macaulay rings*, Cambridge Stud. Adv. Math. **39**, Cambridge Univ. Press, Cambridge, 1993.

[47] D. A. Buchsbaum, D. Eisenbud, *Algebra structures for finite free resolutions, and some structure theorems for ideals of codimension 3*, Amer. J. Math. **99** (1977), 447–485.

[48] R.-O. Buchweitz, G.-M. Greuel, F.-O. Schreyer, *Cohen-Macaulay modules on hypersurface singularities. II*, Invent. Math. **88** (1987), 165–182.

[49] L. Carroll, *Through the looking glass and what Alice found there*, Macmillan, London, 1871.

[50] H. Cartan, *Algèbres d'Eilenberg-MacLane*, Exposés 2 à 11, Sém. H. Cartan, Éc. Normale Sup. (1954–1955), Secrétariat Math., Paris, 1956; [reprinted in:] Œvres, vol. III, Springer, Berlin, 1979; pp. 1309–1394.

[51] H. Cartan, S. Eilenberg, *Homological Algebra*, Princeton Univ. Press, Princeton, NJ, 1956.

[52] S. Choi, *Betti numbers and the integral closure of ideals*, Math. Scand. **66** (1990), 173–184.

[53] S. Choi, *Exponential growth of Betti numbers*, J. Algebra **152** (1992), 20–29.

[54] J. A. Eagon, M. Fraser, *A note on the Koszul complex*, Proc. Amer. Math. Soc. **19** (1968), 251–252.

[55] S. Eilenberg, *Homological dimension and syzygies*, Ann. of Math. (2) **64** (1956), 328–336.

[56] S. Eilenberg, S. MacLane, *On the groups* $H(\Pi, n)$. *I*, Ann. of Math. (2) **58** (1953), 55–106.

[57] D. Eisenbud, *Homological algebra on a complete intersection, with an application to group representations*, Trans. Amer. Math. Soc. **260** (1980), 35–64.

[58] D. Eisenbud, *Commutative algebra, with a view towards algebraic geometry*, Graduate Texts Math. **150**, Springer, Berlin, 1995.

[59] D. Eisenbud, S. Goto, *Linear free resolutions and minimal multiplicity*, J. Algebra **88** (1984), 89–133.

[60] S. Eliahou, M. Kervaire, *Minimal resolutions of some monomial ideals*, J. Algebra **129** (1990), 1–25.

[61] E. G. Evans, P. Griffith, *Syzygies*, London Math. Soc. Lecture Notes Ser. **106**, Cambridge Univ. Press, Cambridge, 1985.

[62] L. Evens, *The cohomology of groups*, Oxford Math. Monographs, Clarendon Press, Oxford, 1991.

[63] C. T. Fan, *Growth of Betti numbers over noetherian local rings*, Math. Scand. **75** (1994), 161–168.

[64] Y. Félix, S. Halperin *Rational L.-S. category and its applications*, Trans. Amer. Math. Soc. **273** (1982), 1–37.

[65] Y. Félix, J.-C. Thomas, *The radius of convergence of the Poincaré series of loop spaces*, Invent. Math. **68** (1982), 257–274.

[66] Y. Félix, S. *The homotopy Lie algebra for finite complexes*, Publ. Math. I.H.E.S. **56** (1982), 179–202.

[67] Y. Félix, S. Halperin, C. Jacobsson, C. Löfwall, J.-C. Thomas, *The radical of the homotopy Lie algebra*, Amer. J. Math., **110** (1988), 301–322.

[68] D. Ferrand, *Suite régulière et intersection complète*, C. R. Acad. Sci. Paris Sér. A **264** (1967), 427–428.

[69] R. Fröberg, T. H. Gulliksen, C. Löfwall, *Flat families of local, artinian algebras, with an infinite number of Poincaré series*, Algebra, algebraic topology, and their interactions; Stockholm, 1983 (J.-E. Roos, ed.), Lecture Notes Math. **1183**, Springer, Berlin, 1986; pp. 56–78.

[70] V. N. Gasharov, I. V. Peeva, *Boundedness versus periodicity over commutative local rings*, Trans. Amer. Math. Soc, **320** (1990), 569–580.

[71] I. M. Gelfand, A. A. Kirillov, *Sur les corps liés aux algèbres enveloppantes des algèbres de Lie*, Publ. Math. I.H.E.S., **31** (1966), 509–523.

[72] F. Ghione, T. H. Gulliksen, *Some reduction formulas for Poincaré series of modules*, Atti Accad. Naz. Lincei Mem. Cl. Sci. Fis. Mat. Nat. (8) **58** (1975), 82–91.

[73] P. G. Goerss, *On the André-Quillen cohomology of commutative* $\mathbb{F}_2$*-algebras*, Astérisque **186**, Soc. Math. France, Paris, 1990.

[74] E. S. Golod, *On the homologies of certain local rings*, Soviet Math. Dokl. **3** (1962), 745–748; [translated from:] Dokl. Akad. Nauk. SSSR, **144** (1962), 479–482 [Russian].

[75] E. H. Gover, M. Ramras, *Increasing sequences of Betti numbers*, Pacific J. Math. **87** (1980), 65–68.

[76] V. K. A. M. Gugenheim, J. P. May, *On the theory and applications of differential torsion products*, Mem. Amer. Math. Soc, **142**, Amer. Math. Soc., Providence, RI, 1974.

[77] T. H. Gulliksen, *A proof of the existence of minimal algebra resolutions*, Acta Math. **120** (1968), 53–58.

[78] T. H. Gulliksen, *A homological characterization of local complete intersections*, Compositio Math. **23** (1971), 251–255.

[79] T. H. Gulliksen, *Massey operations and the Poincaré series of certain local rings*, J. Algebra **22** (1972), 223–232.

[80] T. H. Gulliksen, *A change of ring theorem with applications to Poincaré series and intersection multiplicity*, Math. Scand. **34** (1974), 167–183.

[81] T. H. Gulliksen, *On the Hilbert series of the homology of differential graded algebras*, Math. Scand. **46** (1980), 15–22.

[82] T. H. Gulliksen, *On the deviations of a local ring*, Math. Scand. **47** (1980), 5–20.

[83] T. H. Gulliksen, G. Levin, *Homology of local rings*, Queen's Papers Pure Appl. Math. **20**, Queen's Univ., Kingston, ON, 1969

[84] S. Halperin, *On the non-vanishing of the deviations of a local ring*, Comment. Math. Helv. **62** (1987), 646–653.

[85] R. Heitmann, *A counterexample to the rigidity conjecture for rings*, Bull. Amer. Math. Soc. (New Ser.) **29** (1993), 94–97.

[86] J. Herzog, *Komplexe, Auflösungen, und Dualität in der lokalen Algebra*, Habilitationsschrift, Regensburg, 1973.

[87] J. Herzog, B. Ulrich, J. Backelin, *Linear maximal Cohen-Macaulay modules over strict complete intersections*, J. Pure Appl. Algebra **71** (1991), 187–201.

[88] D. Hilbert, *Über die Theorie der algebraischen Formen*, Math. Ann. **36** (1890) 473–534; [reprinted in:] Gesammelte Abhandlungen, Band II: Algebra, Invariantentheorie, Geometrie, Springer, Berlin, 1970; pp. 199–257.

[89] M. Hochster, *Topics in the homological study of modules over commutative rings*, CBMS Regional Conf. Ser. in Math. **24**, Amer. Math. Soc., Providence, RI, 1975.

[90] H. Hulett, *Maximum Betti numbers of homogeneous ideals with a given Hilbert function*, Comm. Algebra **21** (1993), 2335–2350.

[91] C. Huneke, R. Wiegand, *Tensor products of modules, rigidity and local cohomology*, Math. Scand. (to appear).

[92] S. Iyengar, *Free resolutions and change of rings*, J. Algebra, **190** (1997), 195–213.

[93] C. Jacobsson, *Finitely presented graded Lie algebras and homomorphisms of local rings*, J. Pure Appl. Algebra **38** (1985), 243–253.

[94] C. Jacobsson, A. R. Kustin, and M. Miller, *The Poincaré series of a codimension four Gorenstein ideal is rational*, J. Pure Appl. Algebra **38** (1985), 255–275.

[95] D. A. Jorgensen, *Complexity and Tor on a complete intersection*, J. Algebra (to appear).

[96] R. Kiehl, E. Kunz, *Vollständige Durchschnitte und p-Basen*, Arch. Math. (Basel), **16** (1965), 348–362.

[97] A. R. Kustin, *Gorenstein algebras of codimension four and characteristic two*, Comm. Algebra **15** (1987), 2417-2429.

[98] A. R. Kustin, *The minimal resolution of a codimension four almost complete intersection is a DG algebra*, J. Algebra **168** (1994), 371–399.

[99] A. R. Kustin, *The deviation two Gorenstein rings of Huneke and Ulrich*, Commutative algebra, Trieste, 1994 (A. Simis, N. V. Trung, G. Valla, eds.), World Scientific, Singapore, 1994; pp. 140-163.

[100] A. R. Kustin, *Huneke-Ulrich almost complete intersections of Cohen-Macaulay type two*, J. Algebra **174** (1995), 373–429.

[101] A. R. Kustin, M. Miller, *Algebra structures on minimal resolutions of Gorenstein rings of embedding codimension four*, Math. Z. **173** (1980), 171–184.

[102] A. R. Kustin, S. Palmer Slattery, *The Poincaré series of every finitely generated module over a codimension four almost complete intersection is a rational function*, J. Pure Appl. Algebra **95** (1994), 271–295.

[103] C. Lech, *Inequalities related to certain couples of local rings*, Acta Math. **112** (1964), 69–89.

[104] T. Larfeldt, C. Lech, *Analytic ramification and flat couples of local rings*, Acta Math. **146** (1981), 201–208.

[105] J.-M. Lemaire, *Algèbres connexes et homologie des espaces de lacets*, Lecture Notes Math. **422**, Springer, Berlin, 1974.

[106] J. Lescot, *Asymptotic properties of Betti numbers of modules over certain rings*, J. Pure Appl. Algebra **38** (1985), 287–298.

[107] J. Lescot, *Séries de Poincaré et modules inertes*, J. Algebra **132** (1990), 22–49.

[108] G. Levin, *Homology of local rings*, Ph. D. Thesis, Univ. of Chicago, Chicago, IL, 1965.

[109] G. Levin, *Local rings and Golod homomorphisms*, J. Algebra **37** (1975), 266–289.

[110] G. Levin, *Lectures on Golod homomorphisms*, Matematiska Istitutionen, Stockholms Universitet, Preprint **15**, 1975.

[111] G. Levin, *Modules and Golod homomorphisms*, J. Pure Appl. Algebra **38** (1985), 299–304.

[112] S. Lichtenbaum, *On the vanishing of Tor in regular local rings*, Illinois J. Math. **10** (1966), 220–226.

[113] C. Löfwall, *On the subalgebra generated by the one-dimensional elements in the Yoneda Ext-algebra*, Algebra, algebraic topology, and their interactions; Stockholm, 1983 (J.-E. Roos, ed.), Lecture Notes Math. **1183**, Springer, Berlin, 1986; pp. 291–338.

[114] C. Löfwall, J.-E. Roos, *Cohomologie des algèbres de Lie graduées et séries de Poincaré-Betti non-rationnelles*, C. R. Acad. Sci. Paris Sér. A **290** (1980), 733–736.

[115] S. MacLane, *Homology*, Grundlehren Math. Wiss. **114** Springer, Berlin, 1967.

[116] A. Martsinkovsky, *A remarkable property of the (co)syzygy modules of the residue field of a non-regular local ring*, J. Pure Appl. Algebra **111** (1996), 9–13.

[117] H. Matsumura, *Commutative ring theory*, Cambridge Stud. Adv. Math. **8**, Cambridge Univ. Press, Cambridge, 1986.

[118] J. P. May *Matric Massey products*, J. Algebra **12** (1969), 533–568.

[119] V. Mehta, *Endomorphisms of complexes and modules over Golod rings*, Ph. D. Thesis, Univ. of California, Berkeley, CA, 1976.

[120] C. Miller, *Complexity of tensor products of modules and a theorem of Huneke-Wiegand*, Proc. Amer. Math. Soc. (to appear).

[121] J. W. Milnor, J. C. Moore, *On the structure of Hopf algebras*, Ann. of Math. (2) **81** (1965), 211–264.

[122] J. C. Moore *Algèbre homologique et homologie des espaces classifiants*, Exposé 7, Sém. H. Cartan, Éc. Normale Sup. (1959–1960), Sectétariat Math., Paris, 1957.

[123] M. P. Murthy, *Modules over regular local rings*, Illinois J. Math. **7** (1963), 558–565.

[124] M. Nagata, *Local rings*, Wiley, New York, 1962.

[125] D. G. Northcott, *Finite free resolutions*, Tracts in Pure Math., **71**, Cambridge Univ. Press, Cambridge, 1976.

[126] S. Okiyama, *A local ring is CM if and only if its residue field has a CM syzygy*, Tokyo J. Math. **14** (1991), 489–500.

[127] S. Palmer Slattery, *Algebra structures on resolutions of rings defined by grade four almost complete intersections*, J. Algebra **168** (1994), 371–399.

[128] K. Pardue, *Deformation classes of graded modules and maximal Betti numbers*, Illinois J. Math. **40** (1996), 564–585.

[129] I. Peeva, *0-Borel fixed ideals*, J. Algebra **184** (1996), 945–984.

[130] I. Peeva, *Exponential growth of Betti numbers*, J. Pure. Appl. Algebra (to appear).

[131] D. Quillen, *Rational homotopy theory*, Ann. of Math. (2) **90** (1969), 205–295.

[132] D. Quillen, *The spectrum of an equivariant cohomology ring I; II*, Ann. of Math. (2) **94** (1971), 549–572; 573–602.

[133] D. Quillen, *On the (co-)homology of commutative rings*, Applications of categorical algebra; New York, 1968 (A. Heller, ed.), Proc. Symp. Pure Math. **17**, Amer. Math. Soc., Providence, RI, 1970; pp. 65–87.

[134] M. Ramras, *Sequences of Betti numbers*, J. Algebra **66** (1980), 193–204.

[135] M. Ramras, *Bounds on Betti numbers*, Can. J. Math. **34** (1982), 589–592.

[136] P. Roberts, *Homological invariants of modules over commutative rings*, Sém. Math. Sup., **72**, Presses Univ. Montréal, Montréal, 1980.

[137] J.-E. Roos, *Relations between the Poincaré-Betti series of loop spaces and local rings*, Sém. d'Algèbre P. Dubreil; Paris, 1977-78 (M.-P. Malliavin, ed.), Lecture Notes Math. **740**, Springer, Berlin, 1979; pp. 285–322.

[138] J.-E. Roos, *Homology of loop spaces and of local rings*, Proc. 18th Scand. Congr. Math. Århus, 1980 (E. Balslev, ed.), Progress Math. **11**, Birkhäuser, Basel, 1982; pp. 441–468.

[139] C. Schoeller, *Homologie des anneaux locaux noethériens*, C. R. Acad. Sci. Paris Sér. A **265** (1967), 768–771.

[140] J.-P. Serre, *Sur la dimension homologique des anneaux et des modules noethériens*, Proc. Int. Symp., Tokyo-Nikko (1956), pp. 175-189.

[141] J.-P. Serre, *Algèbre locale. Multiplicités*, Lecture Notes Math. **11** Springer, Berlin, 1965.

[142] J. Shamash, *The Poincaré series of a local ring*, J. Algebra **12** (1969), 453–470.

[143] G. Scheja, *Über die Bettizahlen lokaler Ringe*, Math. Ann. **155** (1964), 155–172.

[144] G. Sjödin, *A set of generators for $\mathrm{Ext}_R(k,k)$*, Math. Scand. **38** (1976), 1–12.

[145] G. Sjödin, *Hopf algebras and derivations*, J. Algebra **64** (1980), 218–229.

[146] H. Srinivasan, *The non-existence of a minimal algebra resolutions despite the vanishing of Avramov obstructions*, J. Algebra **146** (1992), 251–266.

[147] H. Srinivasan, *A grade five Gorenstein algebra with no minimal algebra resolutions*, J. Algebra **179** (1996), 362–379.

[148] D. Sullivan, *Infinitesimal computations in topology*, Publ. Math. I.H.E.S. **47** (1978), 269–331.

[149] L.-C. Sun, *Growth of Betti numbers of modules over rings of small embedding codimension or small linkage number*, J. Pure Appl. Algebra, **96** (1994), 57–71.

[150] L.-C. Sun, *Growth of Betti numbers of modules over generalized Golod rings*, Preprint, 1996.

[151] J. Tate, *Homology of noetherian rings and local rings*, Illinois J. Math. **1** (1957), 14–25.

[152] H. Uehara, W. S. Massey, *The Jacobi identity for Whitehead products*, Algebraic geometry and topology. A symposium in honor of S. Lefschetz, Princeton Univ. Press, Princeton, NJ, 1957; pp. 361–377.

[153] V. A. Ufnarovskij, *Combinatorial and asymptotic methods in algebra*, Encyclopaedia of Math. Sci. **57**, Springer, Berlin, 1995; pp. 1–196; [translated from:] Current problems in mathematics. Fundamental directions, **57**, Akad. Nauk SSSR, VINITI, Moscow, 1990. pp. 5–177 [Russian].

[154] W. V. Vasconcelos, *Ideals generated by R–sequences*, J. Algebra **6** (1970), 309–316.

[155] W. V. Vasconcelos, *On the homology of I/I^2*, Comm. Algebra **6** (1978), 1801–1809.

[156] W. V. Vasconcelos, *The complete intersection locus of certain ideals*, J. Pure Appl. Algebra, **38** (1986), 367–378.

[157] J. Watanabe, *A note on Gorenstein rings of embedding codimension three*, Nagoya Math. J. **50** (1973), 227–232.

[158] H. Wiebe, *Über homologische Invarianten lokaler Ringe*, Math. Ann. **179** (1969), 257–274.

[159] K. Wolffhardt, *Die Betti-Reihe und die Abweichungen eines lokalen Rings*, Math. Z. **114** (1970), 66–78.

Department of Mathematics
Purdue University
West Lafayette, Indiana 47907, U.S.A.
E-mail address: avramov@math.purdue.edu

Generic Initial Ideals

Mark L. Green

Introduction

A very powerful technique in commutative algebra was introduced by Macaulay, who realized that studying the initial terms of elements of an ideal gives one great insight into the algebra and combinatorics of the ideal. The initial ideal depends on the choice of coordinates, but there is an object, the initial ideal in generic coordinates, which is coordinate-independent. Generic initial ideals appeared in the work of Grauert and Hironaka.

The present set of notes began as course notes for a course I gave at UCLA. My interest in the subject rekindled when I discovered the material in Chapters 5 and 6, and I am extremely grateful to the organizers – Joan Elias, José Giral, Rosa Miró-Roig and Santiago Zarzuela – of the Centre de Recerca Matemàtica Summer School in Commutative Algebra for giving me the opportunity to give a series of lectures on this subject in the summer of 1996 in Barcelona. I am especially honored to have been included as a geometric outlier among a cluster of distinguished algebraists – Luchezar Avramov, Craig Huneke, Peter Schenzel, Giuseppe Valla, and Wolmer Vasconcelos.

The table of contents will probably strike the reader as a strange brew of commutative algebra, geometry, and combinatorics, with a little bit of non-commutative algebra thrown in for good measure. These reflect to some extent the quirks of my mathematical personality, but I hope that the patient reader will come away convinced that there is indeed an interesting field of study here where ideas from these different areas meet as equals and work together in harmony. Readers coming to these notes from the geometric side will retrace my own struggle in coming to grips with the algebraic side of the subject, while those approaching from the algebraic side will encounter the geometric language of sheaves and varieties in what I hope is a friendly environment.

There are several general ideas that weave their way through this manuscript. The first is that by using gins, one can separate the geometry and the combinatorics, and indeed certain statements about, for example, curves in $\mathbf{P}^3$ are best understood as statements about their gins – an example is the proof of Laudal's Lemma in Chapter 4. The second, somewhat hidden, is that there is a "differential geometry of ideals," as exemplified by Lemma 2.16 in Chapter 2 and by my proof of Strano's theorem. A third idea is that some purely algebraic results can be elucidated if one thinks about them geometrically – see for example the proof of the Macaulay-Gotzmann estimates in Chapter 3.

These notes contain original results of mine, not published elsewhere, and of course many results by others, hopefully attributed correctly. Almost all of the proofs are new – indeed, my hope is to provide a unifying framework and a new perspective.

I would like to thank a number of mathematicians who influenced my thinking on this subject: D. Bayer, E. Bierstone, R. Braun, M. Cook, D. Eisenbud, Ph. Ellia, G. Floystad, G. Gotzmann, A. Iarrobino, J.M. Landsberg, R. Lazarsfeld, P. Milman, P. Pedersen, C. Peskine, C. Rippel, M. Stillman, R. Strano, B. Sturmfels. I also want to acknowledge my heavy reliance on the computer program Macaulay, without which these notes would not exist, and to extend my thanks to its creators, Dave Bayer and Mike Stillman, and my favorite scriptwriter, David Eisenbud. Michele Cook, Kristina Crona, Bo Ilic, Rich Liebling, and Mihnea Popa caught mistakes in an earlier version and made helpful suggestions for revisions, for which they have my gratitude and the readers' as well.

1. The Initial Ideal

Let V be a vector space over $\mathbf{C}$ with basis $x_1, x_2, \ldots, x_n$, and let $S = \oplus_k S^k V$ denote the symmetric algebra on V, or equivalently $\mathbf{C}[x_1, \ldots, x_n]$. We will use multi-index notation, so that $x^I = x_1^{i_1} x_2^{i_2} \cdots x_n^{i_n}$, where $I = (i_1, i_2, \ldots, i_n)$. We let $|I| = \sum_{j=1}^n i_j$.

We want to impose a total order on the monomials in a reasonable way. It turns out that there is more than one reasonable way to do this.

Definition 1.1. *A total order on the monomials of each degree is a* **multiplicative order** *if*
(1) $x_1 > x_2 > \cdots > x_n$, and
(2) If $x^I > x^J$, then $x^K x^I > x^K x^J$ for all multi-indices K.

Definition 1.2. *Given a monomial order, we extend it to monomials of different degrees by $I > J$ if $|I| < |J|$.*

Rremark.   Definition 1.2 is rather non-standard, but is the right thing for the purpose of ordering initial terms of syzygies. The more normal $I > J$ if $|I| > |J|$ is better in most other contexts.

For $n = 2$, Definition 1.1 forces in each degree d the order $x_1^d > x_1^{d-1} x_2 > \cdots > x_2^d$. However, for $n = 3$, the only information about the monomials of degree 2 we obtain is
$$x_1^2 > x_1 x_2 > x_1 x_3, x_2^2 > x_2 x_3 > x_3^2.$$
Either inequality between $x_1 x_3$ and x_2^2 is permitted.

Given a multiplicative order, we will use interchangeably the notations $x^I > x^J$ and $I > J$.

The two most famous multiplicative orders are:

Definition 1.3. *The **lexicographic order** is defined by saying that, if* $|I| = |J|$*, then* $x^I > x^J$ *if for some* k*,* $i_m = j_m$ *for* $m < k$ *and* $i_k > j_k$*. The **reverse lexicographic order** is defined by saying that, if* $|I| = |J|$*, then* $x^I > x^J$ *if for some* k*,* $i_m = j_m$ *for* $m > k$ *and* $i_k < j_k$*.*

Example 1.4. If $n = 3$ and $|I| = 2$, the lexicographic order is

$$x_1^2 > x_1 x_2 > x_1 x_3 > x_2^2 > x_2 x_3 > x_3^2.$$

The reverse lexicographic order is

$$x_1^2 > x_1 x_2 > x_2^2 > x_1 x_3 > x_2 x_3 > x_3^2.$$

Notice that these are the only possible multiplicative orders in this case. In general, there are other multiplicative orders. The lexicographic order is just the order words would appear in a dictionary if x_1 is the first letter of the alphabet and $>$ means "comes first in alphabetical order." In reverse lexicographic order, a monomial is dragged down by having a high x_n term, much like a mathematician being judged on his or her worst paper.

There is a classification of monomial orders that is useful for some later results.

Example 1.5. *(General Multiplicative Orders)* Let A be a finitely generated subgroup of $\mathbf{R}^N$ of rank n. If we identify $A \cong \mathbf{Z}^n$ with multi-indices in n variables, and put the lexicographic order on $\mathbf{R}^N$, then this induces an order on monomials of any given degree by $I > J$ if and only if $I - J$ maps to a positive element of $\mathbf{R}^N$ under the lexicographic order. This is easily seen to be a multiplicative order, since $I > J$ implies $I + K > J + K$. The Theorem below says that these in fact give all possible multiplicative orders.

I learned of this result from Christoph Rippel, who also supplied the proof. See [Ro1], [Ro2] for related results.

Theorem 1.6 (Robbiano). *Given any multiplicative order on the monomials in* n *variables, there exists an injective group homomorphism*

$$\alpha \colon \mathbf{Z}^n \xrightarrow{(H_0, H_1, \cdots, H_{N-1})} \mathbf{R}^N$$

for some $N \le n$ *such that the lexicographic order on* $\mathbf{R}^N$ *pulls back to the given monomial order, i.e. if* $|I| = |J|$ *and* $I > J$*, then* $\alpha(I) > \alpha(J)$*.*

Proof. We extend the monomial order from positive multi-indices to all integer multi-indices by $I - J > K - L$ if and only if $I + L > J + K$ for positive multi-indices I, J, K, L; one easily checks that the multiplicative property ensures that this is well-defined. We now extend this to rational multi-indices by $\frac{1}{k}I > \frac{1}{k}J$ if and only if $I > J$ for any positive integer k. This allows us to clear denominators. This procedure gives a well-defined answer because the multiplicative property ensures that for ordinary multi-indices I, J of the same degree, $I > J$ implies $kI > kJ$ for any positive integer k (by $kI > (k-1)I + J > (k-2)I + 2J > \cdots > kJ$). Let $H_0(q) = q_1 + \cdots + q_n$; we will also denote by H_0 the corresponding hyperplane in $\mathbf{Q}^n$. We thus have a total ordering on H_0 given by our monomial order. Let Δ be the set of strictly positive elements of H_0. Under the standard inclusion $\mathbf{Q}^n \subseteq \mathbf{R}^n$, let $\bar{\Delta}$ be the topological closure of Δ. Let $-\Delta = \{q \mid -q \in \Delta\}$. Note that $\bar{\Delta}$ is closed under positive scalar multiplication and that $\bar{\Delta} \cup -\bar{\Delta} = H_0$. Thus $\bar{\Delta}$ is a half-space, defined by a linear form $H_1 \in \mathbf{R}^{n*}$. Adjust the sign of H_1 so that $H_1 > 0$ on Δ. Now $H_1 \cap H_0 \cap \mathbf{Q}^n$ is a linear subspace of $\mathbf{Q}^n$ which has a total order, and therefore if it is not empty, has the order given by an element $H_2 \in \mathbf{R}^{n*}$. We may continue this procedure, decreasing the dimension of $H_0 \cap H_1 \cap \cdots \cap H_k$ as a real vector space, so that after at most $n-1$ steps, the procedure terminates. Now, on H_0, if $H_1(q) > 0$, then $q > 0$, while if $H_1(q) = 0$ and $H_2(q) > 0$, then $q > 0$, etc. By construction, the lexicographic order on $\mathbf{R}^N$ gives the order on H_0, and hence the monomial order. The map α is just $(H_0, H_1, \ldots, H_{n-1})$. $\qquad\square$

Corollary 1.7. *Given any multiplicative order and any finite set of monomials S in n variables, there exist positive integers $d_1, \ldots, d_n$ such that for any $I, J \in S$, $I > J$ if and only if $\sum_k d_k i_k < \sum_k d_k j_k$.*

Proof. Since S is finite, there exists a constant $D > 0$ such that for all i, $|H_i| < D$ on S, and such that for $I \in S$, $H_i(I) \neq 0$ implies that $|H_i(I)| > 1/D$. Now on elements of S of the same degree, the function

$$f = -H_{N-1} - D^2 H_{N-2} - \cdots - D^{2N-4} H_1$$

has the property that $I > J$ if and only if $f(I) < f(J)$. We may approximate f by a function with rational coefficients without changing this property, and can then clear denominators to replace f by a linear function e with integral coefficients. If we now let $d = e + BH_0$ for some large number B chosen so that the coefficients of d are positive integers, then we have the $d = (d_1, \ldots, d_n)$ we want. $\qquad\square$

Lemma 1.8. *Given a homogeneous ideal I, TFAE:*
(1) I is generated by monomials;
(2) If $f \in I$ and $f = \sum_J a_J x^J$, then $x^J \in I$ whenever $a_J \neq 0$.

Definition 1.9. *A homogeneous ideal satisfying either of the two equivalent properties of the preceding lemma is called a* **monomial ideal***.*

If $f \in S^k V$ is a homogeneous polynomial, write $f = \sum_I a_I x^I$. Let

$$I_m = \max(\{I \mid a_I \neq 0\}).$$

Then the **initial monomial** of f is

$$\mathrm{in}(f) = x^{I_m}.$$

We note the formulas

$$\mathrm{in}(fg) = \mathrm{in}(f)\,\mathrm{in}(g),$$
$$\mathrm{in}(f+g) \leq \max(\mathrm{in}(f), \mathrm{in}(g)).$$

If $I \subset S$ is a homogeneous ideal, then the **initial ideal** of I is the ideal $\mathrm{in}(I)$ generated by

$$\{\mathrm{in}(f) \mid f \in I\};$$

note that the set above is closed under multiplication. Clearly, $\mathrm{in}(I)$ is a monomial ideal.

Example. $I = (x_1^2 + 3x_1x_2, 2x_1^2 + x_2^2)$. The initial monomial of both generators is x_1^2. However, twice the first generator minus the second generator has initial monomial x_1x_2. No linear combination with constant coefficients of the generators has smaller initial monomial, so $\mathrm{in}(I)_2 = (x_1^2, x_1x_2)$. It is automatic that $\mathrm{in}(I)_3$ contains $(x_1^3, x_1^2x_2, x_1x_2^2)$, but by direct computation we can find an element of I which is equal to x_2^3, and thus has initial monomial x_2^3. Thus $\mathrm{in}(I)_3 = (x_1, x_2)^3$, and $\mathrm{in}(I) = (x_1^2, x_1x_2, x_2^3)$.

Definition 1.10. *For each monomial $x^J \in \mathrm{in}(I)$, there is an element $f_J \in I$ with $\mathrm{in}(f_J) = x^J$. A choice of elements f_J as x^J ranges over a basis for $\mathrm{in}(I)_d$ are called a **standard basis** for I_d.*

Remark. A standard basis is not unique. It is possible to make f_J unique by requiring that the coefficient in f_J of x^K is zero for every $x^K \in \mathrm{in}(I)_d$ with $K \neq J$. We will call this a **reduced standard basis** for I_d. This bears the same relation to a standard basis as reduced row echelon form does to row echelon form in Gaussian elimination.

Example. $I = (x_1^2 + 3x_1x_2, 2x_1^2 + x_2^2)$ revisited. A standard basis for I is $(x_1^2 + 3x_1x_2, x_1x_2 - (1/6)x_2^2, x_2^3)$. A reduced standard basis would be $(x_1^2 + (1/2)x_2^2, x_1x_2 - (1/6)x_2^2, x_2^3)$.

Proposition 1.11. *For any homogeneous ideal I, a standard basis $\{f_J\}$ for I_d is in fact a basis for I_d. In particular, I_d and $\mathrm{in}(I)_d$ have the same dimension.*

Proof. It is clear that the f_J are linearly independent, since if $f = \sum_J c_J f_J$ is a linear combination of them, then $\mathrm{in}(f) = x^{J_m}$, where $J_m = \max\{J \mid c_J \neq 0\}$. To see that they span, let K be the smallest multi-index such that there is an $f \in I_d$ not in the span of the f_J having $\mathrm{in}(f) = x^K$. If the coefficient of x^K in f is c, then $f - cf_K$ is an element of I_d not in the span of the f_J and having lower initial term than f. $\qquad\square$

We note that I and $\mathrm{in}(I)$ have the same Hilbert function, i.e. the same dimension in every degree.

Corollary 1.12. *If $I \subseteq J$, then $I = J$ if and only if $\mathrm{in}(I) = \mathrm{in}(J)$.*

Example 1.13. *Symmetric polynomials* Let $\mathcal{S}_n$ be the symmetric group on n letters, which acts on $k[x_1, \dots, x_n]$ by permuting the variables. The **symmetric polynomials** are the polynomials left invariant by this action, denoted $k[x_1, \dots, x_n]^{\mathcal{S}_n}$. The **elementary symmetric functions** σ_k are defined by

$$\prod_{i=1}^n (x + x_i) = x^n + \sum_{k=1}^n \sigma_k x^{n-k}.$$

The basic fact is that:

Theorem 1.14. *The natural map*

$$k[\sigma_1, \dots, \sigma_n] \to k[x_1, \dots, x_n]^{\mathcal{S}_n}$$

is an isomorphism.

There is a famous proof that appears in Artin's book on Galois theory that proves this using Galois theory. However, there is an easier proof using initial ideals, which is well-known – I am not sure who found it originally.

Proof. One must check both that the map is surjective and injective. If p is a symmetric polynomial, and if we use the lex order, then

$$\mathrm{in}(p) = x_1^{i_1} \cdots x_n^{i_n},$$

where $i_1 \geq i_2 \geq \cdots \geq i_n$ by symmetry. Now

$$\mathrm{in}(\sigma_1^{j_1} \cdots \sigma_n^{j_n}) = x_1^{j_1 + j_2 + \cdots + j_n} x_2^{j_2 + \cdots + j_n} \cdots x_n^{j_n}.$$

we see that these initial terms are all different, which proves injectivity of the map, and that any weakly decreasing sequence of exponents occurs, which proves surjectivity. $\square$

Proposition 1.15. *Let I be a monomial ideal. Then*

$$\mathrm{in}(I) = I.$$

Proof. Let $f_1, f_2, \dots, f_k$ be a set of monomials generating I. Let $f = \sum_{i=1}^k a_i f_i$, where $a_i \in S$. We do an induction on $\max\{\mathrm{in}(a_i f_i) \mid 1 \leq i \leq k\}$. If $\mathrm{in}(f)$ equals this, then it is of the form $\mathrm{in}(a_i)\mathrm{in}(f_i) = (\mathrm{in}(a_i))f_i \in \mathrm{in}(I)$. If not, then two or more of the leading terms of the $a_i f_i$ cancel; in this case, deleting the leading terms of those a_i achieving the maximum gives a new expression for f which decreases the quantity on which we are doing the induction. $\square$

Proposition 1.16. *Let I be a monomial ideal. Then for any p, there is a basis for the p'th syzygies of I of the form $\sum_i a_i s_i$, where s_i is a basis for the $(p-1)$'st syzygies of I and the a_i are monomials.*

Proof. If $M_1, \dots, M_N$ is a set of monomial generators for I and $\sum_i a_i M_i = 0$ is a syzygy, then choose i so that $\mathrm{in}(a_i M_i)$ is as large as possible. Then we may throw away all monomials appearing in any $a_j M_j$ which are not equal to $\mathrm{in}(a_i M_i)$. Now we rewrite this syzygy as a linear combination of syzygies of the form $x^J M_i - x^K M_j = 0$. Continuing inductively, we assume that there exists a basis for the

$(p-1)$'st syzygies with monomial coefficients and with the total monomial (defined inductively by multiplying the monomial M_i by the successive coefficients which occur) constant for all terms which occur. Now if we have a p'th syzygy $\sum_i a_i s_i$, we may break it down into a sum of terms all of which have the same total monomial, and thus with monomial coefficients.   □

Definition 1.17. *Let $\tilde{S} = S[T]$. We can now make $\tilde{S}$ into a graded ring by setting* $\deg(x_1^{i_1} \cdots x_n^{i_n} T^j) = i_1 + \cdots + i_n$; *in other words, we treat T as having degree 0. By a* **family of homogeneous** *S-modules we mean a finitely-generated homogeneous $\tilde{S}$-module M.*

We introduce the notation $S_t = \tilde{S}/(T-t)$, and thus $S_t \cong S$ as a graded ring. If M is a family of homogeneous S-modules, we let $M_t = M \otimes_{\tilde{S}} S_t$.

Theorem 1.18. *Let M be a family of homogeneous S-modules. TFAE:*
(1) The Hilbert function of M_t is constant;
(2) For any t, any resolution $0 \to E_\bullet \to M \to 0$ of M by free $\tilde{S}$-modules has the property that $0 \to E_\bullet \otimes S_t \to M_t \to 0$ is exact.

Proof. (2) → (1): Since the Hilbert function of M_t is the alternating sum of the Hilbert functions of the $E_i \otimes S_t$, and these are independent of t, the Hilbert function of M_t is constant.

(1) → (2): Let $E_\bullet$ be a resolution of M by free $\tilde{S}$ modules. It is of course enough to check that the sequence remains exact when tensored by S_t in each degree d. However, any graded homogeneous $\tilde{S}$-module K has the property that K_d is naturally an $R = \tilde{S}/(x_1, \ldots, x_n) \cong \mathbf{C}[T]$-module. By the classification of modules over a PID,
$$M_d \cong (\oplus_j R(b_j)) \oplus (\oplus_k R(c_k)/(p_k(T)))$$
for some polynomials $p_k(T)$. Since $R/(p(T)) \otimes R/(T-t)$ is zero if $T-t$ does not divide $p(T)$ and $R/(T-t)$ if $(T-t)$ does divide $p(T)$, we see that $(M_t)_d$ has constant dimension in t for all d if and only if the torsion part vanishes, i.e. M_d is free. In this case, all Tors of M_d vanish, so the sequence $(E_\bullet)_d \otimes R/(T-t)$ is a resolution of $(M_t)_d$ for all t. This is equivalent to what we want.   □

Definition 1.19. *We will call a family of homogeneous S-modules satisfying either of the equivalent conditions of the preceding theorem a* **flat family** *of S-modules.*

Remark.   We do not need here the full power of flatness, so we will not tie this in with the usual definition of flatness (see [E]).

Theorem 1.20 (Bayer). *Let I be an ideal. Then there is a flat family of ideals I_t with $I_0 = \mathrm{in}(I)$ and I_t canonically isomorphic to I for all $t \neq 0$.*

Proof. Choose positive integers d_i for $i = 1, \ldots, n$. If $J = (j_1, j_2, \ldots, j_n)$, let
$$d(J) = \sum_{i=1}^{n} d_i j_i.$$

Given I, we now choose the d_i so that $d(J) > d(K)$ if and only if $J < K$ whenever $|J| = |K|$ is the same as the degree of any generator of $\mathrm{in}(I)$ and or of any term appearing in the syzygies of $\mathrm{in}(I)$; we can do this for any multiplicative monomial order by Corollary 1.7. If $f \in S$, let

$$f = \sum_J a_J x^J.$$

Let

$$d(f) = \min_{a_J \neq 0} d(J).$$

If $\mathrm{in}(f) = x^{J_0}$, then $d(f) = d(J_0)$. We now let

$$f(t) = \sum_J a_J t^{d(J)-d(f)} x^J.$$

Note that $f(0) = \mathrm{in}(f)$ and $f(1) = f$. Also, for $t \neq 0$, the substitution $x_j \mapsto t^{d_j} x_j$ carries f to $t^{d(f)} f(t)$. This gives a 1-parameter group acting on S. We note that $d(fg) = d(f) + d(g)$, and thus

$$(fg)(t) = f(t)g(t).$$

However, $(f + g)(t)$ need not equal $f(t) + g(t)$.

Now let $\{x^J\}_{J \in A}$ be a set of generators for $\mathrm{in}(I)$, and let $f_J \in I$ be the element of the standard basis for I with initial term x^J. Let

$$I_t = (f_J(t))_{J \in A}.$$

Thus $I_0 = \mathrm{in}(I)$, and under the substitution $x_j \mapsto t^{d_j} x_j$, I is carried to I_t for all $t \neq 0$.

It remains to check flatness. Given any syzygy

$$\sum_{J \in A} u_J f_J = 0,$$

let

$$d = \min_{u_J \neq 0}(d(u_J) + d(J)).$$

Then

$$\sum t^{(d(u_J)+d(J)-d)} u_J(t) f_J(t) = 0.$$

Evaluated at 0, this gives

$$\sum_{d(u_J)+d(J)=d} \mathrm{in}(u_J) x^J = 0,$$

a non-trivial syzygy of $\mathrm{in}(I)$. Conversely, given any syzygy of the generators of $\mathrm{in}(I)$, we can break it up into pieces of the form

$$\sum_{J \in A} m_J x^J = 0,$$

where m_J is a monomial and

$$d(m_J) + d(J) = d$$

for all J appearing in the sum. Now $f = \sum_J m_J f_J \in I$, and $d(\mathrm{in}(f)) > d$. We may therefore write

$$f = \sum_{K \in A} h_K f_K,$$

where $d(h_K) + d(K) > d$. Now

$$t^{-d} \sum_J m_J(t) f_J(t) - \sum_K t^{(d(h_K) + d(K) - d)} h_K(t) f_K(t) = 0$$

deforms the syzygy $\sum_J m_J x^J$ into a syzygy of I_t. This completes the proof. □

Corollary 1.21. *(The Cancellation Principle) For any ideal I and any i and d, there is a complex of $S/\mathfrak{m}$-modules $V_\bullet^d$ such that*

$$V_i^d \cong \mathrm{Tor}_i^S(\mathrm{in}(I), S/\mathfrak{m})_d$$

and

$$H_i(V_\bullet^d) \cong \mathrm{Tor}_i^S(I, S/\mathfrak{m})_d.$$

Proof. Let I_t be the flat family constructed above and $E_\bullet$ a free $\tilde{S}$ resolution of the family. Let $R = \tilde{S}/(x_1, \dots, x_n) \cong \mathbf{C}[T]$ and $R_t = R/(T - t) \cong \tilde{S}/(x_1, \dots, x_n, T - t)$. Let $\bar{E}_\bullet = E_\bullet \otimes R$, and let $\bar{\phi}_i \colon \bar{E}_i \to \bar{E}_{i-1}$ be the maps. Let $\bar{E}_{i,t} = \bar{E}_i \otimes R_t$ and $\phi_i(t) \colon \bar{E}_{i,t} \to \bar{E}_{i-1,t}$ the induced maps. Since $\bar{E}_{\bullet,t} \cong E_\bullet \otimes (S_t/\mathfrak{m}_t)$, under the identification $S_t \cong S$ we have that $\mathrm{Tor}_i^S(I_t, S/\mathfrak{m}) \cong H_i(\bar{E}_{\bullet,t})$. Choose for each i a maximal minor of $\bar{\phi}_i(0)$ with non-vanishing determinant, and let $U \subseteq \mathbf{C}$ denote a Zariski open subset containing 0 for which all of these minors have non-vanishing determinant. If $R(U)$ is the regular functions on U, we may use these minors to find a resolution of our family of ideals restricted to U by free $R(U)$ modules $F_\bullet$ such that, using analogous notation to that used for the E's, if $\psi_i \colon F_i \to F_{i-1}$ are the maps, $\bar{\psi}_i(0) = 0$ for all i. Now we have that $\bar{F}_i \otimes R_0 \cong \mathrm{Tor}_i^S(I_0, S/\mathfrak{m})$ for all i and $H_i(\bar{F}_{\bullet,t}) \cong \mathrm{Tor}_i^S(I_t, S/\mathfrak{m})$ for all i and all $t \in U$. Now in our case, $I_0 = \mathrm{in}(I)$ and, for all t, $I_t \cong I$ for $t \neq 0$, and hence the Tor's of I_t are the same for all $t \neq 0$. Taking any $t \in U$ with $t \neq 0$, the complex $\bar{F}_{\bullet,t}$ satisfies the conclusion of this corollary. □

Remark. One way to paraphrase this Corollary is to say that the minimal free resolution of I is obtained from that of $\mathrm{in}(I)$ by cancelling some adjacent terms of the same degree. There is no *a priori* way to tell which potential cancellations occur without knowing more about I than just its initial ideal.

Definition 1.22. *If $g = (g_{ij}) \in GL(V)$ and $f \in S^d V$, we will denote by $g(f)$ the standard action of $GL(V)$ on $S^d V$ under the substitution*

$$x_i \mapsto \sum_j g_{ij} x_j.$$

We let $g(I) = \{g(f) \mid f \in I\}$. The **Borel subgroup** $B = \{(g_{ij}) \in GL(V) \mid g_{ij} = 0 \text{ for } j > i\}$; these are the lower triangular matrices. Let $T = \{(g_{ij}) \in GL(V) \mid g_{ij} = 0 \text{ for } j < i\}$; these are the upper-triangular matrices.

Lemma 1.23. *If $g \in T$ and $f \in S^d V$, then $\operatorname{in}(g(f)) = \operatorname{in}(f)$.*

Proof. It is enough to show that for $g \in T$, $\operatorname{in}(g(x^J)) = x^J$ for all multi-indices J. However,

$$g(x^J) = \prod_{i=1}^{n} (\sum_{k=1}^{n} g_{ik} x_k)^{j_i}$$
$$= \det(g) x^J + \text{lower terms.} \qquad \square$$

Definition 1.24. *An **elementary move** e_k for $1 \leq k \leq n-1$ is defined by*

$$e_k(x^J) = x^{\hat{J}}$$

where

$$\hat{J} = (j_1, \ldots, j_{k-1}, j_k + 1, j_{k+1} - 1, j_{k+2}, \ldots, j_n),$$

and where we adopt the convention that $x^J = 0$ if some $j_m < 0$.

Proposition 1.25. *Let I be a monomial ideal. Then TFAE:*
(1) If $x^J \in I$, then for every elementary move $e_k(x^J) \in I$;
(2) $g(I) = I$ for every g belonging to the Borel subgroup B;
(3) $\operatorname{in}(g(I)) = I$ for every g in some open neighborhood of the identity in B.

Proof. To see that (2) implies (1), we note that if $E_{i,j}$ denotes the matrix with a 1 in the (i,j) position and zeros elsewhere, then if $g = \operatorname{Id} + E_{k+1,k}$, we see that

$$g(x^J) = \sum_{m=0}^{j_{k+1}} \binom{j_{k+1}}{m} (e_k)^m (x^J).$$

If $g(I) = I$, then in particular it is a monomial ideal and thus every term of $g(x^J)$ belongs to I. In particular, $e_k(x^J) \in I$.

To see that (1) implies (2), we note that it is enough to check that $g(I) = I$ when g is of the form $g = m\operatorname{Id} + tE_{k+1,k}$, since these generate B. Now

$$g(x^J) = \sum_{m=0}^{j_{k+1}} \binom{j_{k+1}}{m} t^m (e_k)^m (x^J).$$

If $x^J \in I$ and I satisfies (1), then $g(x^J) \in I$ for all t, which completes the proof.

It is automatic that (2) implies (3). Assuming (3), and letting $g = \operatorname{Id} + E_{k+1,k}$, we have that

$$g(x^J) = \sum_{m=0}^{j_{k+1}} \binom{j_{k+1}}{m} (e_k)^m (x^J).$$

We see that its leading term $e_k^{j_{k+1}}(x^J) \in I$, and note that $g(e_k^{j_{k+1}}(x^J)) = e_k^{j_{k+1}}(x^J)$, so that this term is also in $g(I)$. Inductively, $g(e_k^m(x^J)) = e_k^m(x^J) + \sum_{r>k} b_r e_k^r(x^J)$ for some coefficients b_r, and thus inductively (on decreasing m) $e_k^m(x^J) \in I \cap g(I)$ for all m. Thus $x^J \in g(I)$, so $g(I) = I$, proving (2). $\qquad \square$

Definition 1.26. *A monomial ideal I is said to be* **Borel-fixed** *if any of the equivalent conditions of the preceding proposition holds.*

Example. *Some monomial ideals.* The ideal (x_1^2, x_2^2) is not Borel-fixed, because the elementary move $e_1(x_2^2) = x_1 x_2$ does not belong to I. The monomial ideal $(x_1^3, x_1^2 x_2, x_1 x_2^2, x_2^3, x_1^2 x_3^2)$ is Borel-fixed.

Theorem 1.27 (Galligo's Theorem). *For any multiplicative monomial order and any homogeneous ideal I, there is a Zariski open subset $U \subseteq GL(V)$ such that $\mathrm{in}(g(I))$ is constant and Borel-fixed for $g \in U$.*

Remark. We will call $\mathrm{in}(g(I))$ for $g \in U$ the **generic initial ideal** of I and denote it $\mathrm{gin}(I)$. Grauert studied this invariant for ideals in the ring of germs of analytic functions at a point of $\mathbf{C}^n$. A property which holds for $g(I)$ for all g belonging to a Zariski open subset of $GL(V)$ will be said to hold for **general coordinates**.

It is also worth remarking that although $\mathrm{gin}(I)$ is well-defined for any multiplicative order, it definitely depends on which order you are using. For example, if I is three general conics in x_1, x_2, x_3, then $\mathrm{gin}(I)_2$ is the highest three monomials in the order being used, and these are $x_1^2, x_1 x_2, x_1 x_3$ for lexicographic order and $x_1^2, x_1 x_2, x_2^2$ for reverse lexicographic order.

Proof. If I_d has dimension N, consider the $N \times \binom{n-1+d}{d}$ matrix $M(I_d)$ given by writing out a basis for I_d in terms of a decreasing basis for the monomials of degree d. Then the dimension of the intersection of $\mathrm{in}(I)_d$ with the highest k monomials of degree d is the rank of the submatrix $M_k(I_d)$ consisting of the first k columns of $M(I_d)$. The rank of $M_k(g(I)_d)$ is constant for g in a Zariski open subset of $GL(V)$, since the rank is the size of the largest minor having non-zero determinant. We thus see that for a given d, $\mathrm{in}(g(I))_d$ is constant on a Zariski open subset U_d of $GL(V)$. We may inductively choose the sets U_d so that $U_{d+1} \subseteq U_d$ for all d. Let us now define an ideal $\mathrm{gin}(I)$ by $\mathrm{gin}(I)_d = \mathrm{in}(g(I))_d$ for (any) $g \in U_d$. Now the ideal $\mathrm{gin}(I)$ is finitely generated, so by some degree d_0 we have all the generators. The Zariski open set we want is U_{d_0}.

We now wish to see that $\mathrm{in}(g(I))_d$ is Borel-fixed for g in this Zariski open set U. We may change coordinates so that the identity belongs to U. By Theorem 1.20, there is a flat family of ideals I_t with $I_0 = \mathrm{in}(I)$ and, for $t \neq 0$, $I_t = \delta_t(I)$ for some diagonal matrix δ_t. In particular, $\mathrm{in}(I_t) = I_0$ for all t. We now introduce the numbers $N_k(I)$ to be $\dim(\mathrm{in}(I) \cap J_k)$, where J_k denotes the span of the k highest monomials of degree d. For any flat family of ideals K_t, since $N_k(I)$ is the rank of the matrix $M_k(I)$, we have that $N_k(K_0) \leq N_k(K_t)$ for small values of t. Applying this to the family $g(I_t)$, we see that for any $g \in GL(V)$, $N_k(I_0) \leq N_k(g(I_0)) \leq N_k(g(I_t))$ for small values of t. For g near the identity, we have that $N_k(g(I_t)) = N_k(I_t) = N_k(I_0)$ for small values of t, and thus the inequalities are all equalities. It thus follows that for small values of t and g near the identity, $N_k(g(I_0)) = N_k(I_0)$. Since in general the numbers $N_k(I)$ determine $\mathrm{in}(I)_d$, this implies that $\mathrm{in}(g(I_0)) = I_0$ for g near the identity. However, by Proposition 1.25, this implies that I_0 is Borel-fixed. However, I_0 was $\mathrm{in}(g(I))$ for a general g. $\qquad \square$

Example 1.28. $\mathrm{in}(I)$ *Borel-fixed but* $\mathrm{in}(I) \neq \mathrm{gin}(I)$ If we take $I = (x_1^2, x_1 x_2, x_1 x_3 + x_3^2)$, then for the rlex order, $\mathrm{in}(I) = (x_1^2, x_1 x_2, x_1 x_3)$, which is Borel-fixed. However, $\mathrm{gin}(I) = (x_1^2, x_1 x_2, x_2^2)$. Thus although $\mathrm{gin}(I)$ is Borel-fixed, some non-generic initial ideals of I may also be Borel-fixed.

We now introduce the **diagram** of a monomial ideal. We may envision the monomials of degree d in n variables as an $(n-1)$-simplex whose vertices correspond to $x_1^d, \dots, x_n^d$, and where x^I corresponds to the point with barycentric coordinates $(i_1/d, \dots, i_n/d)$. We insert a 0 in a point if the corresponding monomial does not belong to the initial ideal of I. The **generic diagram** of I is the diagram of the generic initial ideal of I.

It is possible to define the initial term of syzygies of an ideal as well. There is not universal agreement on the best way to do this; personally, I incline to think that the following scheme, suggested by Frank Schreyer, is the best. At this point it is convenient to introduce the following notations:

Definition 1.29. *The notation* $x^{K_p} \otimes x^{K_{p-1}} \otimes \cdots \otimes x^{K_0} < x^{L_p} \otimes x^{L_{p-1}} \otimes \cdots \otimes x^{L_0}$ *is defined inductively in p to mean that either* $K_0 + \cdots + K_p < L_0 + \cdots + L_p$ *or* $K_0 + \cdots + K_p = L_0 + \cdots + L_p$ *and* $x^{K_{p-1}} \otimes \cdots \otimes x^{K_0} > x^{L_{p-1}} \otimes \cdots \otimes x^{L_0}$. *The case $p = 0$ is covered by Definitions 1.1 and 1.2.*

Remark. Here, $\otimes$ means $\otimes_{\mathbb{C}}$; it is really just a placeholder. This definition takes some getting used to – for example, $x_2^2 \otimes x_1 < x_2 \otimes x_1 x_2 < x_1 \otimes x_2^2$. On the other hand, $x_1 \otimes x_2^2 < x_2 \otimes x_1^2$ because $x_1 x_2^2 < x_1^2 x_2$.

Definition 1.30. *Let I be a homogeneous ideal with a minimal set of generators* $f_1, \dots, f_N$ *chosen so that* $\mathrm{in}(f_1) = x^{J_1}, \dots, \mathrm{in}(f_N) = x^{J_N}$ *are distinct. To a syzygy* $\sum_i a_i f_i = 0$ *we associate the vector* $\sum_i a_i \otimes x^{J_i}$. *The **highest term** of a syzygy $s = \sum_i a_i f_i$ is* $\max\{\mathrm{in}(a_i f_i) \mid i = 1, \dots, N\}$. *The **initial term** of $\sum_i a_i f_i$ is the maximal element among $\mathrm{in}(a_i) \otimes \mathrm{in}(f_i)$; it is denoted $\mathrm{in}(s)$. Choose a basis $s_1, \dots, s_{N_2}$ for the syzygies, each having distinct initial terms. The initial term of a second syzygy $\sum b_i s_i = 0$ is the maximal element among $\mathrm{in}(b_i) \otimes \mathrm{in}(s_i)$. Inductively, choose bases for the syzygies having distinct initial terms, and define the **initial term of a p'th syzygy** $\sum_i a_i s_i$ as the maximal element among $\mathrm{in}(a_i) \otimes \mathrm{in}(s_i)$; it will be of the form* $x^{K_p} \otimes x^{K_{p-1}} \otimes \cdots \otimes x^{K_0}$. *The **initial module of p'th syzygies** of I is the submodule of $S \otimes_{\mathbb{C}} \cdots \otimes_{\mathbb{C}} S$, with $p + 1$ tensor products, consisting of initial terms of p'th syzygies of I; it is an S-module under multiplication of the leftmost factor.*

We want to compute the minimal free resolution of a Borel fixed monomial ideal. We need the notation that for a multi-index $J = (j_1, \dots, j_n)$, $\max(J) = \max\{i \mid j_i > 0\}$. Similarly, $\min(J) = \min\{i \mid j_i > 0\}$.

Theorem 1.31 (Eliahou-Kervaire). *(cf. [E-K]) Let I be a Borel fixed monomial ideal with generators $x^{J_1}, \dots, x^{J_N}$. Then the initial module of first syzygies of*

I is minimally generated by $x_{i_1} \otimes x^{J_j}$ as $j = 1, \ldots N$ and $1 \leq i_1 < \max(J_j)$. More generally, the initial module of p'th syzygies of J is minimally generated by $x_{i_p} \otimes x_{i_{p-1}} \otimes \cdots \otimes x_{i_1} \otimes x^{J_j}$ where $1 \leq j \leq N$ and $i_p < i_{p-1} < \cdots < i_1 < \max(J_j)$.

Proof. We begin with the first syzygies of I. If $i < m = \max(J_j)$, then because I is Borel-fixed, by a series of elementary moves we see that $x^{J_j - m + i} \in I$. Thus there is a syzygy of the form $x_i x^{J_j} - x^{m+K} x^{J_k}$, where K is some multi-index in which i does not appear. Further, since $i + J_j = m + K + J_k$, we see that $x_i \otimes x^{J_j} > x^{m+K} \otimes x^{J_k}$, since either $|J_j| > |J_k|$ or $|J_j| = |J_k|$ and then, as $i < m$, $J_j < J_k$. Thus $x_i \otimes x^{J_j}$ is the leading term of this syzygy. Now, if we look at the second term $x^{m+K} \otimes x^{J_k}$, if $\min(m + K) < \max(J_k)$, then we may apply further syzygies of the same type and replace the second term by a term which is lower in our ordering. Since this procedure must terminate after a finite number of steps, as we are decreasing the term at each step, we may assume that we have a minimal set of generators for the first syzygies of I of the form above with $\min(m + K) \geq \max(J_k)$.

Now if $x^K \otimes x^{J_k} - x^L \otimes x^{J_l}$ is any element of a minimal set of generators of the first syzygies of I, we may assume that x^K and x^L have no common factors. By using the syzygies we already have, we can arrange that $\min(x^K) \geq \max(x^{J_k})$ and $\min(x^L) \geq \max(x^{J_l})$. Since K and L have no common factor, it follows from $K + J_k = L + J_l$ that $\max(x^{J_k}) \geq \max(x^L)$ and $\max(x^{J_l}) \geq \max(x^K)$. So $\min(x^K) \geq \max(x^L)$ and $\min(x^L) \geq \max(x^K)$. However, this forces both inequalities to be equalities, which in turn forces x^K and x^L to have a common factor unless they are both empty. Thus, after using the syzygies we already have, we can kill off every other syzygy. This proves that syzygies with initial term $x_{i_1} \otimes x^{J_j}$ with $i_1 < \max(J_j)$ give a minimal set of generators of the first syzygies of I.

Let us assume inductively that the result holds for e'th syzygies for $e \leq p - 1$ and try to prove it for p. We denote for $i_e < i_{e-1} \cdots < i_1 < \max(J_j)$ by $s^{J_j}_{i_e i_{e-1} \cdots i_1}$ the syzygy with leading term $x_{i_e} \otimes x_{i_{e-1}} \otimes \cdots \otimes x_{i_1} \otimes x^{J_j}$, where inductively we may assume that all other terms $x^L \otimes s^{J_k}_{l_e l_{e-1} \cdots l_1}$ which occur in these syzygies have $\min(L) \geq l_e$. We note that for $i_p < i_{p-1}$, $x_{i_p} s^{J_j}_{i_{p-1} \cdots i_1} - x_{i_{p-1}} s^{J_j}_{i_p i_{p-2} \cdots i_1}$ is a syzygy whose leading term is strictly lower than $x_{i_p} \otimes \cdots \otimes x_{i_1} \otimes x^{J_j}$, since tensoring with any variable preserves order of terms. We thus obtain syzygies $s^{J_j}_{i_p i_{p-1} \cdots i_1}$ with leading term $x_{i_p} \otimes \cdots \otimes x_{i_1} \otimes x^{J_j}$ and, by using these syzygies, we may make all other terms $x^L \otimes s^{J_k}_{l_p l_{p-1} \cdots l_1}$ which occur in these syzygies have $\min(L) \geq l_p$.

It remains to prove inductively that these syzygies span the p'th syzygies. Given any generator for the p'th syzygies of I, let $x^K \otimes s^{J_j}_{i_p i_{p-1} \cdots i_1}$ be the leading term, which we may assume by using the syzygies we already have that $\min(K) \geq i_p$. The term $x^{K+i_p} \otimes s^{J_j}_{i_{p-1} \cdots i_1}$ must be cancelled by a term appearing in one of the form $x^L \otimes s^{J_k}_{l_p \cdots l_1}$, and since $x^L \otimes s^{J_k}_{l_p \cdots l_1}$ is not the leading term, $|L| \geq |K|$. Note that $K + J_j + i_1 + \cdots + i_p = L + J_k + l_1 + \cdots + l_p$. Now if $s^{J_k}_{l_p \cdots l_1}$ contains a term of the form $x^M \otimes s^{J_j}_{i_{p-1} \cdots i_1}$, then $J_k + l_1 + \cdots + l_p = M + J_j + i_1 + \cdots + i_{p-1}$, and thus $K + i_p = L + M$. Taking into account the inequality on the degrees of L and K, and the fact $|M| > 0$, we have that $|M| = 1$ and $|L| = |K|$. Thus $x^M = x_m$ and $x^K \geq x^L$. Since $K + i_p = L + m$, we conclude that $i_p \geq m$. If $m \neq i_p$, then m appears in K, and thus $m \geq \min(K) \geq i_p$, so $m = i_p$ in any case, and hence $K = L$. If $x_m \otimes s^{J_j}_{i_{p-1}\cdots i_1}$ is not the leading term of $s^{J_k}_{l_p\cdots l_1}$, then we know $m \geq i_{p-1} > i_p$, which would be a contradiction. If it is the leading term, then $x_m \otimes s^{J_j}_{i_{p-1}\cdots i_1} = x_{l_p} \otimes s^{J_k}_{l_{p-1}\cdots l_1}$. But now $x^K \otimes s^{J_j}_{i_p i_{p-1}\cdots i_1} = x^L \otimes s^{J_k}_{l_p\cdots l_1}$, and these were supposed to be distinct terms. This is a contradiction. This shows that the p'th syzygies of I are spanned by syzygies of the form $s^{J_j}_{i_p\cdots i_1}$ with $i_p < \cdots < i_1 < \max(J_j)$. $\qquad\square$

Example. *Minimal free resolution of* $I = (x_1^2, x_1 x_2, x_1 x_3, x_2^3)$. The initial terms of possible first syzygies are $x_1 \otimes x_1 x_2, x_1 \otimes x_1 x_3, x_2 \otimes x_1 x_3, x_1 \otimes x_2^3$. The only second syzygy has leading term $x_1 \otimes x_2 \otimes x_1 x_3$. There are no third syzygies. The minimal free resolution thus has the form $S(-4) \to S^3(-3) \oplus S(-4) \to S^3(-2) \oplus S(-3) \to I \to 0$.

There is a **picture** that I find useful for using the Eliahou-Kervaire theorem. For the ideal $I = (x_1^3, x_1^2 x_2, x_1 x_2^2, x_2^3, x_1^2 x_3)$, the picture is as follows:

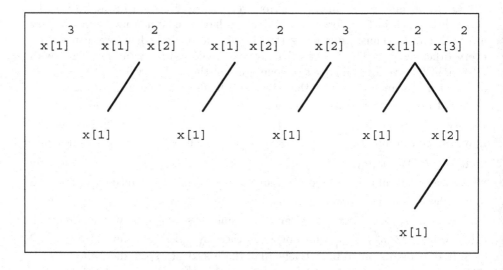

From each generator x^J of the monomial ideal, draw a line downwards for each variable x_i with $i < \max(J)$. The second row of the diagram corresponds to the first syzygies of the monomial ideal. Now from each x_i in the second row, draw a line downwards for each variable x_j with $j < i$. The third row represents the second syzygies of the monomial ideal. Continue this procedure until the process terminates. In the picture, from the fact that the generators have (from left to right) degrees 3,3,3,3,4 respectively, we get that the first syzygies have degrees 4,4,4,5,5 and the second syzygy has degree 6.

Corollary 1.32. *Let I be a Borel-fixed monomial ideal with $x^{J_1}, \ldots, x^{J_N}$ as generators. Let $c_k^d(I) = \text{card}\{J_j \mid |J_j| = d \text{ and } \max(J_j) = k\}$. Then*

$$\dim(\text{Tor}_p^S(I, S/\mathfrak{m})_d) = \sum_k c_k^{d-p}(I) \binom{k-1}{p}.$$

Corollary 1.33. *Let I be any homogeneous ideal. Then using the notations of the preceding corollary,*

$$\dim(\text{Tor}_p^S(I, S/\mathfrak{m})_d) \leq \sum_k c_k^{d-p}(\text{gin}(I)) \binom{k-1}{p}.$$

Proof. This follows from the preceding Corollary by the Cancellation Principle.

Corollary 1.34. *(Hilbert's Syzygy Theorem) The minimal free resolution $E_\bullet$ of any homogeneous ideal I has $E_p = 0$ for $p \geq n$.*

Proof. By Galligo's Theorem and the Cancellation Principle, it is enough to check this for Borel-fixed monomial ideals. However, if $p \geq n$, the sequence of inequalities $i_p < i_{p-1} < \cdots < i_1 < \max(J_j)$ cannot be satisfied.   □

Example 1.35. *(due to G. Evans and H. Hullett)* In three variables, let

$$I = (x_1 x_3, x_2 x_3, x_1^3, x_2^3, x_3^3)$$
$$J = (x_1^5, x_2^2, x_3^2, x_1^2 x_2, x_1^2 x_3)$$

In reverse lexicographic order,

$$\text{gin}(I) = \text{gin}(J) = (x_1^2, x_1 x_2, x_1 x_3^2, x_2^3, x_2^2 x_3, x_2 x_3^3, x_3^5).$$

In terms of the notation above, $c_1^2 = 1, c_2^2 = 1, c_2^3 = 1, c_3^3 = 2, c_3^4 = 1, c_3^5 = 1$ and all other $c_k^d = 0$. The minimal free resolution $E_\bullet$ of $\text{gin}(I) = \text{gin}(J)$ is therefore

$$E_0 = S^2(-2) \oplus S^3(-3) \oplus S(-4) \oplus S(-5)$$
$$E_1 = S(-3) \oplus S^5(-4) \oplus S^2(-5) \oplus S^2(-6)$$
$$E_2 = S^2(-5) \oplus S(-6) \oplus S(-7).$$

For I, the minimal free resolution has

$$E_0 = S^2(-2) \oplus S^3(-3)$$
$$E_1 = S(-3) \oplus S^4(-4) \oplus S(-6)$$
$$E_2 = S(-5) \oplus S(-7).$$

Notice that an $S(-5)$ and an $S(-6)$ have cancelled between E_2 and E_1, and that an $S(-4)$ and an $S(-5)$ have cancelled between E_1 and E_0. It is even more interesting to notice that there is an $S(-3)$ in E_1 and E_0 that did not cancel. The minimal free resolution of J has

$$E_0 = S^2(-2) \oplus S^2(-3) \oplus S(-5)$$
$$E_1 = S^4(-4) \oplus S^2(-6)$$
$$E_2 = S(-6) \oplus S(-7).$$

Here, different cancellations took place, and an $S(-6)$ is present in both E_2 and E_1 that did not cancel. One might legitimately wonder if there is an ideal K whose minimal free resolution has the one of the maximal cancellations permitted by the Cancellation Theorem coming from cancelling all cancellable terms of the resolution of J, namely

$$E_0 = S^2(-2) \oplus S^2(-3) \oplus S(-5)$$
$$E_1 = S^4(-4) \oplus S(-6)$$
$$E_2 = S(-7).$$

They point out that this is impossible, since the E_2 term forces S/K to be Gorenstein, and therefore the E_1 and E_0 terms of the resolution of K would be dual to each other after an appropriate twist, and they are not.

Alternatively, one might wonder whether it is possible to cancel all cancellable terms of the minimal free resolution of I to obtain an ideal K with minimal free resolution

$$E_0 = S^2(-2) \oplus S^2(-3)$$
$$E_1 = S^4(-4) \oplus S(-6)$$
$$E_2 = S(-5) \oplus S(-7).$$

This also can be ruled out. Since the Hilbert function of K is the same as that of J, which contains a regular sequence, then K is forced to contain a regular sequence, of the form Q_1, Q_2, C_1 consisting of 2 conics and a cubic. Let L be the ideal spanned by these three polynomials, and thus $K = L + (C_2)$ for some cubic C_2. From the exact sequence

$$0 \to K/L \to S/L \to S/K \to 0,$$

we have by the exact sequence for Tor that $\mathrm{Tor}_1^S(K/L, S/\mathfrak{m}) \cong (S/\mathfrak{m})^3(-4) \oplus (S/\mathfrak{m})(-6)$. This means that the three linear forms x_1, x_2, x_3 take C_2 to an element of L under multiplication, and this rules out there being a cubic independent of these which also does so. This contradiction rules out this possibility.

The moral content of this example is that neither of the "smallest" minimal free resolutions having the same Hilbert function as those of I and J can indeed be the minimal free resolution of an ideal. However, the minimal free resolutions of I and J do incorporate all possible cancellations between them, so that whatever rules forbid certain cancellations must be fairly complicated, since if certain cancellations take place that forbids other cancellations which would otherwise be permitted. This example highlights the mysterious aspects of which terms can and do cancel as we move from the minimal free resolution of $\text{gin}(I)$ to that of I.

A useful remark is:

Proposition 1.36. *If $x^J, x^K \in \text{in}(I)$, and either $|J| > |K|$ or $|J| = |K|$ and $x^J < x^K$, then $x^K \otimes x^J$ belongs to the module of initial terms of syzygies of I.*

Proof. The leading term of the Koszul syzygy $x^J \otimes x^K - x^K \otimes x^J$ is $x^K \otimes x^J$. $\square$

The rest of this section will give another way of looking at the initial ideal. Let V be a vector space of dimension n and choose an ordered basis $e_1, \ldots e_n$ for V.

Definition 1.37. *If $v = \sum_i v_i e_i$ is a non-zero element of V and $j = \min\{i \mid v_i \neq 0\}$, then we define the **initial term** of v to be $\text{in}(v) = e_j$. If $W \subseteq V$ is a linear subspace of V, then the **initial subspace** of W is $\text{in}(W) = \text{span}\{\text{in}(v) \mid v \in W\}$.*

Definition 1.38. *Let $I \subseteq \{1, 2, \ldots, n\}$, we denote by $V_I = \text{span}\{e_i \mid i \in I\}$; we will call these **standard flags** of V.*

Remark. For a subspace $W \subseteq V$, $\text{in}(W) = V_I$ for some set of indices I with $|I| = \dim(W)$.

Definition 1.39. *If $W \subseteq V$ is a linear subspace, then if $w_1, \ldots, w_d$ is a basis for W and $w_i = \sum_j w_{ij} e_j$, the $d \times n$ matrix $M_W = (w_{ij})$ will be called the **Plücker matrix** of W with respect to the basis $w_1, \ldots, w_d$. If $I \subseteq \{1, 2, \ldots, n\}$ is a set of indices with $|I| = d$, let $p_I(W)$ denote the determinant of the $d \times d$ minor of M_W taking the columns indexed by I; these are the **Plücker coordinates** of W.*

Remark. Whether or not $p_I(W)$ vanishes depends only on W and I, and not on the choice of basis of W.

Proposition 1.40. *Let $W \subseteq V$ be a linear subspace with $\text{in}(W) = V_I$. If we order the sets of indices of size $d = \dim(W)$ lexicographically (with $1 > 2 > \cdots > n$), then*

$$I = \max\{J \mid |J| = d \text{ and } p_J(W) \neq 0\}.$$

Proof. Choose a basis $w_1, \ldots, w_d$ for W such that $\text{in}(w_j) = e_{i_j}$, where $I = \{i_1, \ldots, i_d\}$ with $i_1 < i_2 < \cdots < i_d$. For this basis, it is clear that $p_I(W) \neq 0$, as the minor we are taking the determinant of is upper-triangular with all diagonal entries non-zero. If $J > I$ in lexicographic order, then the row-rank of the minor corresponding to J is equal to $\text{card}(I \cap J)$, which is $< d$. $\square$

Definition 1.41. *Let $F^i V = V_{\{i+1, i+2, \ldots, n\}} = \text{span}\{e_{i+1}, \ldots, e_n\}$, this is the **filtration on V associated to the ordered basis** $e_1, \ldots, e_n$.*

Proposition 1.42. *Let $\text{in}(W) = V_I$. Then $\dim(W \cap F^i V) = \text{card}(I \cap \{i + 1, i + 2, \ldots, n\})$.*

Proof. We have that for $v \neq 0$, $v \in F^i V$ if and only if $\text{in}(v) = e_j$ with $j > i$. The proposition follows immediately. $\square$

Definition 1.43. *Let $G(d, n)$ denote the* **Grassmannian** *of linear subspaces of V of dimension d. Then*

$$\text{Schub}_I = \{W \in G(d, n) \mid \text{in}(W) = V_I\}$$

is the **Schubert cell associated to the index** *I.*

Remark. We note that Schub_I is a quasi-projective variety, being the Zariski open subset $p_I \neq 0$ of the projective variety $\{p_J = 0 \text{ for } J > I\}$. We note that the Zariski closure

$$\overline{\text{Schub}_I} = \{p_J = 0 \text{ for all } J > I \text{ with } |J| = d\}.$$

We now let $G \subseteq GL(n, \mathbf{C})$ be a connected algebraic group with Lie algebra $\mathcal{G} \subseteq \mathcal{G}L(n, \mathbf{C})$.

Definition 1.44. *If $g \in G$ and $W \subseteq V$ a linear subspace of dimension d, let $gW = \{g(v) \mid v \in W\}$. This induces a map $\phi_W \colon G \to G(d, n)$.*

Proposition 1.45. *There is a non-empty Zariski open subset $U \subseteq G$ and a unique set of indices I such that $\text{in}(gW) = V_I$ for all $g \in U$.*

Proof. It follows from the definition that $G(d, n)$ is the disjoint union of the Schub_I as I ranges over subsets of size d of $\{1, 2, \ldots, n\}$. Choose $I = \min\{J \mid \phi_W(G) \subseteq \overline{\text{Schub}_J}\}$. Take $U = \phi_W^{-1}(\text{Schub}_I)$. This is non-empty by the minimality of I and Zariski open since it is the preimage of a Zariski open subset of $\overline{\text{Schub}_I}$ under the morphism ϕ_W. $\square$

Definition 1.46. *In the preceding Proposition, V_I is the* **generic initial subspace** *of W for G, denoted $\text{gin}(W)$.*

Example 1.47. If we look at the action of $GL(n, \mathbf{C})$ on $S^d V$, taking as basis the monomials x^J ordered by some monomial order and let $W = I_d \subseteq S^d V$, then $\text{gin}(W)$ is what we have denoted $\text{gin}(I)_d$.

Definition 1.48. *A subgroup $B \subseteq G$ with Lie algebra $\mathcal{B}$ has the* **infinitesimal property** *if, for all $\gamma \in \mathcal{B}$ and all $i < j$, either both $\gamma(e_i) = 0$ and $\gamma(e_j) = 0$ or $\text{in}(\gamma(e_i)) > \max(e_i, \text{in}(\gamma(e_j)))$.*

Example 1.49. If we have the action of $G = GL(n, \mathbf{C})$ on $S^d V$, with basis x^J ordered by a monomial order, and B is the Borel subgroup, then the infinitesimal property is satisfied.

Example 1.50.   If we let $G = GL(n, \mathbf{C})$ act on $V^{2,2}$ (using the language of Young diagrams), then if we take a basis of standard tableaux, then an acceptable order must respect the order on the total weight of the tableaux. The problem is that there exist tableaux having the same weight, e.g.

$$\begin{array}{cc} 1 & 2 \\ 3 & 4 \end{array} \qquad \begin{array}{cc} 1 & 3 \\ 2 & 4 \end{array}.$$

These do not have the infinitesimal property for the Borel subgroup of G–see the thesis of Christoph Rippel [Ri].

Proposition 1.51. *If W is a linear subspace of V with $\mathrm{gin}(W) = V_I$, and $B \subseteq G$ has the infinitesimal property, then for any $i \in I$ and $\gamma \in B$, if $\mathrm{in}(\gamma(e_i)) = e_j$, then $j \in I$.*

Proof. Letting i, j be as in the Proposition, filter V by $V^0 = V$, $V^1 = F^j V$, $V^2 = F^i V$. By the infinitesimal property, $\gamma(V^2) \subseteq V^1$. We now invoke (somewhat prematurely) Lemma 2.16 – if $w = e_i +$ lower terms $\in W$, then $\gamma(w) \in W + V^2$, from which we conclude that $\gamma(e_i) \in \mathrm{in}(W)$. $\qquad\qquad\square$

Remark.   For the case of ideals, the preceding proposition is the statement that the generic initial ideal is Borel-fixed.

Comments for students:

Another source for much of this material is Ch. 15 of Eisenbud's book [E]. He has a nice history of the origins of the subject, in which the names of Macaulay, Gröbner, Buchberger, Hironaka, Grauert, and Schreyer figure prominently. From my perspective, a real watershed in the subject was Bayer's thesis and his subsequent joint work with Mike Stillman, which begins relating rlex gins to geometry by discovering the link with saturation and regularity, featured in Chapter 2.

Bayer and Stillman also inaugurated the computational phase of the subject with their implementation of various algorithms in the computer program, Macaulay. Eisenbud has a nice series of computer projects at the end of Chapter 15 of [E] for those who want to get their feet wet computationally. It is a good idea to learn to use Macaulay or a similar language if you want to work in this subject; however, it is not at all necessary to become a wizard or guru – as I am living proof.

2. Regularity and Saturation

If $I, J \subseteq S$ are homogeneous ideals, we have the ideal quotient

$$(I : J) = \{P \in S \mid PQ \in I \text{ for all } Q \in J\}.$$

We let $\mathfrak{m}$ denote the maximal ideal $(x_1, \dots, x_n)$.

Definition 2.1. *A homogeneous ideal I is* **saturated** *if $(I : \mathfrak{m}) = I$. The* **saturation** I^{sat} *of I is*

$$I^{\text{sat}} = \cup_{k \geq 0}(I : \mathfrak{m}^k).$$

A homogeneous ideal I is **$\mathfrak{m}$-saturated** *if $I_d = I_d^{\text{sat}}$ for all $d \geq m$.*

Example. $I = (x_1^2, x_1 x_2, x_2^3)$ *in 2 variables. Since $(I : \mathfrak{m})_2$ contains x_2^2, I is not 2-saturated. However, I is 3-saturated. If we took the same ideal in 3 or more variables, it would be saturated.*

Associated to a homogeneous ideal I we have its **sheafification**, which is a coherent sheaf on $\mathbf{P}^{n-1}$. For d sufficiently large, the image of the sheaf map $e_d \colon I_d \otimes \mathcal{O}_{\mathbf{P}^{n-1}}(-d) \to \mathcal{O}_{\mathbf{P}^{n-1}}$ is constant, and this image is $\mathcal{I}$, the sheafification of I. The map e_d induces an injection $I_d \to H^0(\mathcal{I}(d))$. We repeat the well-known basic result:

Proposition 2.2. *A homogeneous ideal I is saturated if and only if the map $e_d \colon I_d \to H^0(\mathcal{I}(d))$ is an isomorphism for all $d \geq 0$, and $\mathfrak{m}$-saturated if and only if e_d is an isomorphism for $d \geq m$. I is $\mathfrak{m}$-saturated for $m \gg 0$. For any I, $I^{\text{sat}} = \oplus_d H^0(\mathcal{I}(d))$.*

Definition 2.3. *The* **satiety** *of I is the smallest m for which I is $\mathfrak{m}$-saturated. We denote it* sat(I).

Let

$$0 \to E_{n-1} \to \cdots \to E_1 \to E_0 \to I \to 0$$

be a minimal free resolution of I, where

$$E_p = \oplus_j S(-a_{pj}).$$

Then I is **$\mathfrak{m}$-regular** if $a_{pj} - p \leq m$ for all p, j. The **regularity** of I is the smallest m for which I is $\mathfrak{m}$-regular; the regularity of I thus equals $\max\{a_{pj} - p\}$. Alternatively, the regularity of I is the largest q for which $\text{Tor}_p^S(I, S/\mathfrak{m})_{p+q} \neq 0$ for some p.

Definition 2.4. *A coherent sheaf $\mathcal{F}$ on $\mathbf{P}^r$ is* **$\mathfrak{m}$-regular** *if $H^q(\mathcal{F}(m-q)) = 0$ for all $q > 0$.*

Remark. If $\mathcal{F}$ is $\mathfrak{m}$-regular, then it is $(m+1)$-regular, by an argument of Castelnuovo (see [G1] or see below for the case we need.) By the Serre vanishing theorem, every coherent sheaf $\mathcal{F}$ is $\mathfrak{m}$-regular for $m \gg 0$.

Definition 2.5. *The* **regularity** *of $\mathcal{F}$ is the smallest m for which $\mathcal{F}$ is $\mathfrak{m}$-regular. We denote it* reg$(\mathcal{F})$.

Example. $I = (x_1^2, x_1 x_2, x_2^3)$ *in 2 variables. The minimal free resolution of I is $0 \to S(-3) \oplus S(-4) \to S^2(-2) \oplus S(-3) \to I \to 0$. The maximum of $3-1, 4-1, 2-0, 3-0$ is 3, so* reg$(I) = 3$. *We note that the saturation of I is S, and that the sheafification of I is $\mathcal{O}_{\mathbf{P}^1}$, which is 0-regular.*

We state the well-known fact:

Proposition 2.6. *An ideal I is m-regular if and only if I is m-saturated and its sheafification $\mathcal{I}$ is m-regular. For a saturated homogeneous ideal I, the regularity of I equals the regularity of its sheafification.*

Proof. Assume I is m-regular. Let $\mathcal{I}$ be the ideal sheaf on $\mathbf{P}^{n-1}$ which is the sheafification of I. If $E_\bullet$ is the minimal free resolution of I, let $\mathcal{E}_\bullet$ be the sheafification of $E_\bullet$, so that $\mathcal{E}_\bullet$ is a resolution of $\mathcal{I}$. Note that

$$\mathcal{E}_p \cong \oplus_j \mathcal{O}_{\mathbf{P}^{n-1}}(-a_{pj}).$$

The image of

$$\alpha_k \colon H^0(\mathcal{E}_0(k)) \to H^0(\mathcal{I}(k))$$

is I_k, and by the hypercohomology spectral sequence α_k is surjective provided that

$$H^q(\mathcal{E}_q(k)) = 0$$

for all $q > 0$. The only case to check is $q = n - 1$, and here the vanishing is guaranteed if $a_{n-1j} < n + k$ for all j, and this follows from m-regularity if $k \geq m$.

The only differentials which can come into $H^q(\mathcal{I}(m-q))$ originate from terms of the form $H^{q+p}(E_p(m-q)) \cong \oplus_j \mathcal{O}_{\mathbf{P}^{n-1}}(m - q - a_{pj})$. For $q > 0$, these vanish by the Bott Vanishing Theorem unless $p + q = n - 1$ and $m - q - a_{pj} \leq -n$. However, the last inequality is equivalent to $a_{pj} \geq m + p + 1$, which contradicts the hypothesis that I is m-regular.

Conversely, if $\mathcal{I}$ is m-regular, then let $E_\bullet$ be the minimal free resolution of I, with $E_p = \oplus_j S(-a_{pj})$. If I has regularity k, with $k > m$, choose the largest p such that $a_{pj} = k + p$ for some j. If we sheafify $E_\bullet$, we have a resolution $\mathcal{E}_\bullet$ of $\mathcal{I}$. The E_1 term of the hypercohomology spectral sequence for $0 \to \mathcal{E}_\bullet(k + p - n) \to \mathcal{I}(k+p-n) \to 0$ has one term equal to $H^{n-1}(\mathcal{E}_p(k+p-n)) \cong H^{n-1}(\oplus_j \mathcal{O}_{\mathbf{P}^{n-1}}(k + p - n - a_{pj}))$. This contains a non-zero term that survives to the E_{p+1} term, when, unless $n - 1 = p = 0$, it must map isomorphically to $H^{n-1-p}(\mathcal{I}(k + p - n))$, which therefore is non-zero. So $\mathcal{I}$ is not $(k - 1)$-regular, unless $n - 1 - p = 0$, i.e. $p = n - 1$. However, in this case, since I is m-saturated and $m \leq k - 1$, $H^0(\mathcal{E}_0(k - 1)) \to H^0(\mathcal{I}(k - 1))$ is surjective, so this cannot occur. Hence $k \leq m$. This proves the second part. The third part follows from the first two. $\square$

We have the following result of Bayer-Stillman [B-S]:

Proposition 2.7. *I is m-saturated if and only if, for a general linear form $h \in V$, $(I : h)_d = I_d$ for all $d \geq m$.*

Proof. Since $I_d \subseteq (I : \mathfrak{m})_d \subseteq (I : h)_d$, the if direction is automatic for any h. So assume I is m saturated. For a finitely generated homogeneous S-module M, if $u \in M$, then $\mathrm{Ann}(u) = \{f \in S \mid fu = 0\}$. A prime ideal $\mathfrak{p}$ which is $\mathrm{Ann}(u)$ for some $u \in M$ is called an **associated prime** of M. It is a theorem that $\mathrm{Ass}(M)$, the set of associated primes of M, is finite ([M], [E]). A non-zero element $f \in S$ is a zero-divisor if and only if $f \in \cup_{\mathfrak{p} \in \mathrm{Ass}(M)} \mathfrak{p}$. Now let $M = \sum_{d \geq m} (S/I)_d$. Now either $\mathfrak{m} \in \mathrm{Ass}(M)$, or else a general linear form h does not belong to $\cup_{\mathfrak{p} \in \mathrm{Ass}(M)} \mathfrak{p}$, in which case h is not a zero-divisor for M. This is equivalent to $(I : h)_d = I_d$ for all $d \geq m$. If $\mathfrak{m} \in \mathrm{Ass}(M)$, then if $\mathfrak{m} = \mathrm{Ann}(u)$ for some non-zero $u \in M$, then

$u \in (I : \mathfrak{m})_d$ for some $d \geq m$, but $u \notin I_d$, so I is not m-saturated, which is a contradiction. $\square$

Corollary 2.8. *For a general linear form h, $(I : h)_d = I_d$ for $d \geq \operatorname{reg}(I)$.*

Proof. Since m-regular implies m-saturated, this follows directly from the Proposition. $\square$

Proposition 2.9. *For a Borel-fixed monomial ideal I, $(I : x_n) = (I : \mathfrak{m})$.*

Proof. Automatically, $(I : \mathfrak{m}) \subseteq (I : x_n)$. Conversely, if $f \in (I : x_n)$, then $x_n f \in I$. Now, applying a series of elementary moves to every monomial appearing in $x_n f$, we see that $x_i f \in I$ since I is Borel-fixed. So $f \in (I : \mathfrak{m})$, which proves the other inclusion. $\square$

Corollary 2.10. *For a Borel-fixed monomial ideal I, $\operatorname{sat}(I)$ is the degree of the largest generator of I involving x_n. In particular, I is saturated if no generator of I involves x_n.*

Proof. By the proposition above, I is m-saturated if and only if $(I : x_n)_d = I_d$ for all $d \geq m$. If I has a generator of degree $d \geq m + 1$ which involves x_n, then $(I : x_n)_{d-1} \neq I_{d-1}$. If I has no generator of degree $\geq m + 1$ which involves x_n, then if $x^J \in (I : x_n)_d$, but $x^J \notin I_d$ for some $d \geq m$, then $x_n x^J = x^K x^L$ where x^L is a generator of I and $|K| > 0$. By using the syzygies of a Borel-fixed ideal, we may arrange that $\min(K) \geq \max(L)$. Thus x_n must divide x^K, and hence $x^J \in I$, which is a contradiction. $\square$

Proposition 2.11. *For a Borel-fixed monomial ideal I, $\operatorname{reg}(I)$ is the maximal degree of a generator of I.*

Proof. If x^J is a generator of I, then by Corollary 1.32, if $k = \max\{i \mid x_i \text{ appears in } J\}$, then x^J contributes $\binom{k}{p}$ elements to the p'th syzygies of I in degree $|J| + p$. The regularity is thus the degree of the highest generator. $\square$

Example. $I = (x_1^2, x_1 x_2, x_2^3)$ in 2 variables. This is a Borel-fixed monomial ideal. It is not saturated because some of its generators involve x_2. The degree of the highest generator is 3, which we already saw computes $\operatorname{reg}(I)$.

Corollary 2.12. *For a homogeneous ideal I, using any order, $\operatorname{reg}(I) \leq \operatorname{reg}(\operatorname{in}(I))$. In particular, in any order, $\operatorname{reg}(I)$ is bounded by the degree of the highest generator of $\operatorname{gin}(I)$.*

Proof. Since I is a flat deformation of $\operatorname{in}(I)$, by the Cancellation Principle 1.21, the minimal free resolution of I is derived from that of $\operatorname{in}(I)$ by cancelling certain adjacent terms. Since the regularity is the maximal value of $a_{pj} - p$ for the minimal free resolution, this cannot go up when we cancel terms. The second statement now follows from Proposition 2.11 and the fact the $\operatorname{gin}(I)$ is Borel-fixed. $\square$

At this point, we introduce a somewhat finer invariant than the initial ideal. We could get by without these, but I am introducing them out of the conviction that they are useful in their own right.

Definition 2.13. *Let I be a graded homogeneous ideal over $k[x_1, \ldots, x_n]$ and h a linear form. For $f \in I_d$, we say that $f \in \tilde{L}_k(I, h)$ if $f = h^k g$ for some polynomial g, i.e. $g \in (I : h^k)$. If we write $f = \sum_{i \geq k} h^i f_{d-i}$, where $f_i \in k[x_1, \ldots, x_{n-1}]_i$, then we let $L_k(I, h)_d = \{f_{d-k} \mid f \in \tilde{L}_k(I, h)\}$, the $\mathbf{k}$'th ideal of leading terms of $\mathbf{I}$ with respect to $\mathbf{h}$. For $h = x_n$, we denote $L_k(I, x_n)$ and $\tilde{L}_k(I, x_n)$ by $L_k(I)$ and $\tilde{L}_k(I)$ respectively.*

For $h \in V$, let H be the corresponding hyperplane and I_H the restriction of I to H. Let $V_H = V/(h)$. If h is general, the coefficient of x_n is non-zero, so that we may identify $S^k V_H$ with $S^k \bar{V}$ in a natural way, where $\bar{V}$ is the vector space spanned by $x_1, \ldots, x_{n-1}$. We will consider I_H as a graded ideal in the symmetric algebra on $\bar{V}$.

Proposition 2.14. *For a general linear form h and a general choice of coordinates, then using the rlex order, $\mathrm{gin}(L_k(I)) = \mathrm{gin}((I : h^k)_H) = (\mathrm{gin}(I) : x_n^k)_{x_n} = L_k(\mathrm{gin}(I))$ holds for all $k \geq 0$.*

Proof. For any given degree, $\mathrm{in}(I : h^k)_H$ is constant on a Zariski open set of linear forms. We will make a general choice of coordinates by choosing which linear forms will be $x_1, \ldots, x_n$. For a general linear form h, as a first step we choose coordinates such that $h = x_n$. Thus $(I : h^k)h^k = \tilde{L}_k(I)$ for this choice of coordinates, and $(I : h^k)_H = L_k(I)$. Now $(I : h^k)_H$ and $L_k(I)$ both involve only $x_1, \ldots, x_{n-1}$, and hence a general choice of coordinates does not involve x_n for these, so we get the first equality of gins. Now we note that in reverse lexicographic order, $x_n^k | \mathrm{in}(p) \leftrightarrow x_n^k | p$, as if the leading term of p has a factor of x_n^k, then every term of p has such a factor. From this, we see the second equality. $\qquad \square$

Corollary 2.15. *For a general linear form, in rlex order, $\mathrm{gin}(I_H) = (\mathrm{gin}(I))_{x_n}$.*

The following general fact is useful in proving the next proposition, and is also used in Chapter 6:

Lemma 2.16. *Let W be a finite-dimensional vector space with a fixed descending flag $W = W^0 \supseteq W^1 \supseteq W^2 \supseteq \cdots$. Let $M \in \mathrm{End}(W)$ be such that, for all p, $M(W^p) \subseteq W^{p-1}$, and consider $g(t) = \mathrm{Exp}(tM) \in GL(W)$. Let $V \subseteq W$ be a fixed subspace and let $V^p(t) = W^p \cap g(t)(V)$. If the dimension of $V^p(t)$ is locally constant at $t = 0$, then*

$$M(V^p(0)) \subseteq V^{p-1}(0) + W^p.$$

Proof. If the dimension of $V^p(t)$ is locally constant at 0 and $v^0 \in V^p(0)$, then there is a power series $v(t) = v^0 + tv^1 + t^2 v^2 + \cdots$, where $v^k \in V$ for all k and $g(t)(v(t)) \in W^p$ for all t. Expanding out $g(t) = I + tM + (t^2/2)M^2 + \cdots$, we have, looking at the coefficient of t, that $Mv^0 + v^1 \in W^p$. Since $v^0 \in W^p$, by hypothesis we know that $Mv^0 \in W^{p-1}$, and thus $v^1 \in W^{p-1}$. Thus $v^1 \in V^{p-1}(0)$, and now $Mv^0 \in V^{p-1}(0) + W^p$, completing the proof. $\qquad \square$

Proposition 2.17. *For a general choice of coordinates, $\mathfrak{m} L_k(I) \subseteq L_{k-1}(I)$.*

Proof. We use Lemma 2.16. Filter $W = S^d V$ by $W^k = \{x_n^k p \mid p \in S^{d-k}V\}$. Take M to be the linear map $p \mapsto x_i \partial p / \partial x_n$. Note that $M(W^k) \subseteq W^{k-1}$. Now $g(t) = e^{tM}$ corresponds to the substitution $x_n \mapsto x_n + tx_i = h(t)$. If we have chosen general coordinates, then the dimension of $W^k \cap g(t)(I) = \tilde{L}_k(I, h(t))$ is locally constant, and thus we conclude that $x_i L_k(I) \subseteq L_{k-1}(I)$ for all i, proving the desired result. □

Proposition 2.18. *For a general h, and all $k \geq 0$, $(I : h^k) \subseteq I^{sat}$.*

Proof. If $h^k p \in I$, write $p = p_0 + hq$, where $p_0 \in L_k(I, h)$ and q is a polynomial. Then by Proposition 2.17, $x^J p_0 \in L_0(I, h)$ if $|J| = k$. It follows that for any L with $|L| \geq k$, that $x^L p$ is an element of I modulo h, so we have $x^L p + h U_L \in I$. Now multiplying by h^k, we see that $x^L h^k p + h^{k+1} U_L \in I$, and hence $h^{k+1} U_L \in I$. By Corollary 2.8, we have for a general h that $(I : h^{k+1})_m = I_m$ for $m >> 0$, and thus by taking $|L|$ sufficiently large, $U_L \in I$ and therefore $X^L p \in I$. It follows that $p \in I^{sat}$. □

Corollary 2.19. *For a general linear form h, $I^{sat} = \cup_k (I : h^k)$.*

Proof. This follows from $\cup_k (I : h^k) \subseteq I^{sat} = \cup_k (I : \mathfrak{m}^k) \subseteq \cup_k (I : h^k)$. □

Proposition 2.20. *For a general choice of coordinates,*

$$L_0(I^{sat}) = \sum_k L_k(I).$$

Proof. The inclusion $\subseteq$ is easy, since if $p \in I^{sat}$, then $p \in (I : \mathfrak{m}^k) \subseteq (I : x_n^k)$ for some k. So $p|_{x_n} \in (I : x_n^k)_{x_n}$. For a general choice of coordinates, this is equivalent to saying $p|_{x_n} \in L_k(I)$. The set of all possible $p|_{x_n}$ for $p \in I^{sat}$ is $L_0(I^{sat})$.

The more difficult inclusion $\supseteq$ is just a reinterpretation of the preceding proposition. □

Proposition 2.21. *For a general choice of coordinates, for the rlex order,*
(1) $\mathrm{gin}(I) = \sum_k \sum_{l \geq k} x_n^l \mathrm{gin}(L_k(I))$;
(2) $\mathrm{gin}(I^{sat}) = \sum_k \sum_l x_n^l \mathrm{gin}(L_k(I)) = \cup_k (\mathrm{gin}(I) : x_n^k)$.

Proof. The first statement is just that the initial term of any element of I is just the initial term of its leading term if we are using rlex. The second statement follows from the preceding Proposition, and the fact from Proposition 2.14 that $\mathrm{gin}(L_k(I)) = (\mathrm{gin}(I) : x_n^k)_{x_n}$. □

Example 2.22. *I is saturated but $\mathrm{in}(I)$ is not.* In 3 variables, let $I = (x_1^2, x_1 x_2 + x_3^2)$. This ideal is saturated. On the other hand, $\mathrm{in}(I) = (x_1^2, x_1 x_2, x_1 x_3^2, x_3^4)$, which is not saturated, since $x_1 x_3 \in (\mathrm{in}(I) : \mathfrak{m})$. However, if we use generic coordinates, then this kind of problem evaporates.

Lemma 2.23. *A homogeneous ideal I is $\mathfrak{m}$-saturated if and only if, in generic coordinates, for all $d \geq m$ and all $k \geq 0$, $L_k(I)_{d-k} = L_{k+1}(I)_{d-k}$.*

Proof. After a general change of coordinates, I is m-saturated if and only if $(I : x_n)_d = I_d$ for all $d \geq m$. If $f \in (I : x_n)_d$, then for some $k \geq 0$, $x_n^k | f$ but not $x_n^{k+1} | f$. We may write $f = x_n^k g$, where g is not divisible by x_n. Since $x_n f \in I$, we have that $x_n^{k+1} g \in I$, and thus $g \in \tilde{L}_{k+1}(I)_{d-k}$, and thus g determines a non-zero element $\bar{g} \in L_{k+1}(I)_{d-k}$. Now $f \in I$ iff $g \in \tilde{L}_k(I)_{d-k}$ iff $\bar{g} \in L_k(I)_{d-k}$. Thus $(I : x_n)_d \neq I_d$ iff for some $k \geq 0$, $L_k(I)_{d-k} \neq L_{k+1}(I)_{d-k}$. $\qquad\square$

Theorem 2.24 (Bayer-Stillman). *For a homogeneous ideal I, using rlex order,* $\mathrm{sat}(I) = \mathrm{sat}(\mathrm{gin}(I))$, *i.e. the maximal degree of a generator of* $\mathrm{gin}(I)$ *involving* x_n. *In particular, I is saturated iff no generator of rlex* $\mathrm{gin}(I)$ *involves* x_n.

Proof. In view of the preceding Lemma, after a general change of coordinates, I is m-saturated iff for all $d \geq m$, $L_k(I)_{d-k} = L_{k+1}(I)_{d-k}$. However, since $L_k(I) \subseteq L_{k+1}(I)$, this equality is equivalent to equality of their Hilbert functions, and hence to $\mathrm{gin}(L_k(I))_{d-k} = \mathrm{gin}(L_{k+1}(I))_{d-k}$ for all $k \geq 0$ and all $d \geq m$. However, by Proposition 2.14, in rlex order, $\mathrm{gin}(L_k(I)) = L_k(\mathrm{gin}(I))$, and now using Lemma 2.23 again, we see that the preceding condition is equivalent to $\mathrm{gin}(I)$ being m-saturated. This says that $\mathrm{sat}(I) = \mathrm{sat}(\mathrm{gin}(I))$. However, Corollary 2.10 says that for a Borel-fixed monomial ideal such as $\mathrm{gin}(I)$, its satiety is the degree of the largest generator involving x_n. $\qquad\square$

Proposition 2.25. *If M, N are graded homogeneous S-modules with $M \subseteq N$. For any p, q,*
(1) If $M_k = N_k$ for $q - p - 1 \leq k \leq q - p$, then $\mathrm{Tor}_p^S(M, S/\mathfrak{m})_q \to \mathrm{Tor}_p^S(N, S/\mathfrak{m})_q$ is an isomorphism;
(2) If $M_k = N_k$ for $k = q - p$, then $\mathrm{Tor}_p^S(M, S/\mathfrak{m})_q \to \mathrm{Tor}_p^S(N, S/\mathfrak{m})_q$ is surjective.

Proof. Consider the Koszul resolution $K_\bullet$ of $S/\mathfrak{m}$, so $K_p = \wedge^p V \otimes S(-p)$. By the symmetry of Tor, $\mathrm{Tor}_p^S(N/M, S/\mathfrak{m})_q \cong H_p(K_\bullet \otimes (N/M))_q$. The latter is zero if $(N/M)_{q-p} = 0$. The result now follows from the exact sequence for Tor. $\qquad\square$

In the following theorem's proof, we let $\bar{S} = S/(x_n)$ and $\bar{\mathfrak{m}}$ its maximal ideal.

Theorem 2.26. *In generic coordinates, if*

$$\Lambda_i = \{x_{n+1-i} = 0, x_{n+2-i} = 0, \dots, x_n = 0\},$$

then I is m-regular if and only if I is m-saturated and I_{Λ_i} is m-saturated for all $i = 1, \dots, n$.

Proof. Since by Proposition 2.6, m-regular implies m-saturated, neither condition can hold if I is not m-saturated. So assume that I is m-saturated. From the exact sequence

$$0 \to (I : x_n)(-1) \to I \to I_{x_n} \to 0,$$

we see that we have an exact sequence

$$\to \mathrm{Tor}_p^S(I, S/\mathfrak{m})_q \to \mathrm{Tor}_p^S(I_{x_n}, S/\mathfrak{m})_q \to$$
$$\mathrm{Tor}_{p-1}^S((I : x_n), S/\mathfrak{m})_{q-1} \to \mathrm{Tor}_{p-1}^S(I, S/\mathfrak{m})_q \to$$

for all p, q. We also have, because $\wedge^p V = \wedge^p \bar{V} \oplus \wedge^{p-1} \bar{V}$, that

$$\mathrm{Tor}_p^S(I_{x_n}, S/\mathfrak{m})_q \cong \mathrm{Tor}_p^{\bar{S}}(I_{x_n}, \bar{S}/\bar{\mathfrak{m}})_q \oplus \mathrm{Tor}_{p-1}^{\bar{S}}(I_{x_n}, \bar{S}/\bar{\mathfrak{m}})_{q-1}.$$

Now if I is m-regular, then $\mathrm{Tor}_p^S(I, S/\mathfrak{m})_q = 0$ for $q > p+m$ and by m-saturation and the Proposition above, $\mathrm{Tor}_{p-1}^S((I : x_n), S/\mathfrak{m})_{q-1} = \mathrm{Tor}_{p-1}^S(I, S/\mathfrak{m})_{q-1} = 0$ for $q > p+m$, so $\mathrm{Tor}_p^S(I_{x_n}, S/\mathfrak{m})_q = 0$ for $q > m+p$, and this, using a formula above, forces I_{x_n} to be m-regular, and hence by induction, all of the I_{Λ_i} are m-regular and hence, by Proposition 2.6, m-saturated.

Conversely, if all of the I_{Λ_i} are m-saturated, then inductively I_{x_n} is m-regular. Assume that I is not m-regular, so that $\mathrm{Tor}_p^S(I, S/\mathfrak{m})_q \neq 0$ for some $q > m+p$, and we may choose the largest $q - p$ for which this happens. By the argument above, in this range $\mathrm{Tor}_p^S(I, S/\mathfrak{m})_q = \mathrm{Tor}_p^S((I : x_n), S/\mathfrak{m})_q$. From the exact sequence for Tor, either

$$\mathrm{Tor}_{p+1}^S(I_{x_n}, S/\mathfrak{m})_{q+1} \neq 0,$$

which is forbidden by I_{x_n} being m-regular, or

$$\mathrm{Tor}_p^S(I, S/\mathfrak{m})_{q+1} \neq 0,$$

which is forbidden by the maximality of $q - p$. This is a contradiction, so I is m-regular. $\qquad\square$

Theorem 2.27 (Bayer-Stillman). *The regularity of I is equal to the regularity of the rlex gin(I), or equivalently by 2.11 to the degree of the largest generator of the rlex gin(I).*

Remark. If one does not work in generic coordinates, one only has that the regularity of in(I) is $\geq$ the regularity of I. An example is the ideal generated by $x_1^2, x_1 x_2 + x_3^2$.

Proof. We first show that I and gin(I) have the same regularity. It is logically equivalent to show that I is m-regular if and only if in(I) is m-regular. By Theorem 2.26 above, it is enough to show that I_{Λ_i} is m-saturated if and only if in$(I)_{\Lambda_i}$ is m-saturated. By Corollary 2.15,

$$\mathrm{in}(I)_{\Lambda_i} = \mathrm{in}(I_{\Lambda_i}),$$

and now we are done by Theorem 2.24.

By definition, the regularity of gin(I) is $\geq$ the largest degree of a generator of gin(I). Now by the Theorem, gin(I) is m-regular if and only if all of the I_{Λ_i} are m-saturated. If $d < m$ is the largest degree of a generator of gin(I), then $d \geq$ the largest degree of a generator of I_{Λ_i}, and hence I_{Λ_i} is d-saturated for all i, and hence I is $d < m$-regular, which is a contradiction. So the regularity of gin(I) is the largest degree of a generator of gin(I) (Of course, we could use Theorem 1.31 on the resolution of a Borel-fixed ideal to prove this, but this is a much simpler direct argument.) $\qquad\square$

Example. *Two quadrics in the plane.* If Q_1, Q_2 is a regular sequence of two quadrics in 3 variables, then the rlex gin of the ideal I they generate is $x_1^2, x_1 x_2, x_2^3$. The minimal free resolution of I is $0 \rightarrow S(-4) \rightarrow S^2(-2) \rightarrow I \rightarrow 0$. From Definition 2.3, we see that the maximum of $a_{pj} - p$ is 3, achieved at the left of the resolution, so $\text{reg}(I) = 3$. The maximal degree of a generator of the rlex gin of I is 3, as Theorem 2.27 claims. No generator of the rlex gin involves x_3, which agrees with the fact that this is the saturated ideal of 4 points in the plane.

For a large class of examples, see Chapter 4.

Proposition 2.28 (Crystallization Principle). *Let I be a homogeneous ideal generated in degrees $\leq d$. Using the rlex order, assume that $\text{gin}(I)$ has no generator in degree $d + 1$. Then $\text{gin}(I)$ is generated in degrees $\leq d$ and I is d-regular.*

Proof. By the Cancellation Principle (Corollary 1.21), for every k there is a complex $V_\bullet^k$ with

$$V_p^k = \text{Tor}_p^S(\text{gin}(I), S/\mathfrak{m})_k$$

for every p and with

$$H_p(V_\bullet^k) \cong \text{Tor}_p^S(I, S/\mathfrak{m})_k.$$

If $k > d$, $H_0(V_\bullet^k) = 0$, and thus $V_1^k \rightarrow V_0^k$ is surjective for $k > d$. So for $k > d$, if $V_1^k = 0$, then $V_0^k = 0$. By Theorem 1.31 on the structure of resolutions of Borel-fixed ideals, we know that for all k, if $V_0^k = 0$, then $V_1^{k+1} = 0$. So for $k > d$, $V_0^k = 0$ implies $V_0^{k+1} = 0$. By hypothesis, $V_0^{d+1} = 0$, so $V_0^k = 0$ for $k > d$. This says that all generators of $\text{gin}(I)$ have degree $\leq d$, and hence by Theorem 2.27 that I is d-regular. □

Corollary 2.29. *Let I be a homogeneous ideal. Assume that every generator of rlex $\text{gin}(I)$ in degree $d + 1$ is the initial term of a generator of I. Let J be the ideal generated by the generators of I having degree $\leq d$. Then J is d-regular.*

Proof. $\text{gin}(J) \subseteq \text{gin}(I)$, and $\text{gin}(J)_k = \text{gin}(I)_k$ for $k \leq d$. Thus any generator of $\text{gin}(J)$ in degree $d + 1$ would also be a generator of $\text{gin}(I)$, but could not be the initial term of a generator of I. Thus the proposition applies to J. □

Example. *Two quadrics in the plane with a common factor.* If $I = (Q_1, Q_2)$ is an ideal generated by two quadrics in the plane, then in rlex, $\text{gin}(I)_2$ is necessarily $x_1^2, x_1 x_2$, as this is the only Borel-fixed monomial ideal in degree 2 with 2 elements. If there is no new generator for the rlex gin in degree 3, then the crystallization principle proclaims that there are no further generators in any degree, so $\text{gin}(I) = (x_1^2, x_1 x_2)$. It follows that this is a saturated ideal whose Hilbert polynomial is $P_{S/I}(d) = d + 2$, from which it follows that I is the ideal of a line union a point. The line must be a common factor of Q_1, Q_2, and conversely two quadrics with a common factor have this gin. If there is a generator of $\text{gin}(I)$ in degree 3, by the Borel-fixed property it must be either x_2^3 or $x_1 x_3^2$. In the latter case, $\text{gin}(I^{\text{sat}}) = (x_1)$, so once again Q_1, Q_2 must have a line as common factor, and then there is also a point of intersection, ruling out this case. If the generator is x_2^3, then the quadrics form a regular sequence.

As may be seen from the foregoing results, the reverse lexicographic order is extremely well-suited to intersecting with a general hyperplane. The lexicographic order is analogously well-suited to projecting to a general subspace of lower dimension (I am grateful to David Eisenbud for pointing this out to me), as we will see in Chapter 6.

We collect together what we know in the following:

Theorem 2.30. *Using rlex order, for any homogeneous ideal I,*
(1) $\operatorname{sat}(I)$ is the degree of the largest generator of $\operatorname{gin}(I)$ involving x_n and $\operatorname{reg}(I)$ is the degree of the largest generator of $\operatorname{gin}(I)$.
(2) $\operatorname{gin}(I^{\mathrm{sat}}) = \cup_{k \geq 0}(\operatorname{gin}(I) : x_n^k)$
(3) I is saturated iff no generator of $\operatorname{gin}(I)$ involves x_n.
(4) For a general linear form h corresponding to a hyperplane H, $\operatorname{gin}(I_H) = (\operatorname{gin}(I))_{x_n}$. More generally, for any $k \geq 0$, $\operatorname{gin}((I : h^k)_H) = (\operatorname{gin}(I) : x_n^k)_{x_n}$.
(5) For a general hyperplane H, $\operatorname{reg}(I) = \max(\operatorname{sat}(I), \operatorname{reg}(I_H))$.
(6) For a general nested sequence of nested linear spaces $\Lambda_0 \supseteq \Lambda_1 \cdots \supseteq \Lambda_{n-1}$ where we have $\operatorname{codim}(\Lambda_i) = i$, $\operatorname{reg}(I) = \max_{i=0}^{n-1}\{\operatorname{sat}(I_{\Lambda_i})\}$.
(7) In any order, $\operatorname{reg}(I)$ is less than or equal to the degree of the highest generator of $\operatorname{gin}(I)$.

We now return to the study of the ideal of leading terms and relate the minimal free resolutions of the $L_k(I)$ to the minimal free resolution of I itself.

If V is a vector space and $M = \oplus_q M_q$ is a finitely-generated graded module over $S(V)$, we denote by $\mathcal{K}_m^\bullet(M, V)$ the Koszul complex

$$\cdots \to \wedge^{p+1}V \otimes M_{m-p-1} \to \wedge^p V \otimes M_{m-p} \to \wedge^{p-1}V \otimes M_{m-p+1} \to \cdots$$

with the indexes set up so that $\mathcal{K}_m^{-p} = \wedge^p V \otimes M_{m-p}$ and denote

$$\mathcal{K}_{p,q}(M, V) = H^{-p}(\mathcal{K}_{p+q}(M, V)).$$

Lemma 2.31. *If M is a finitely-generated graded $S(\bar{V})$-module, where $V \to \bar{V}$ is quotient with $\dim(\bar{V}) = \dim(V) - 1$, then*

$$\mathcal{K}_{p,q}(M, V) \cong \mathcal{K}_{p,q}(M, \bar{V}) \oplus \mathcal{K}_{p-1,q}(M, \bar{V}).$$

Proof. This follows easily from the decomposition $\wedge^p V \cong \wedge^p \bar{V} \oplus \wedge^{p-1} \bar{V}$, and the fact that the extra element of V acts trivially on M, which induces an isomorphism $\mathcal{K}_m^\bullet(M, V) \cong \mathcal{K}_m^\bullet(M, \bar{V}) \oplus \mathcal{K}_m^{\bullet-1}(M, \bar{V})$. $\square$

Proposition 2.32. *Let I be a graded homogeneous ideal in $k[x_1, \ldots, x_n]$ and let $V, \bar{V}$ be the vector space spanned by $x_1, \ldots, x_n$ and $x_1, \ldots, x_{n-1}$ respectively. There is a spectral sequence for any m with $E_1^{p,q} = \mathcal{K}_{-p-q,m+q}(L_p(I), \bar{V}) \oplus \mathcal{K}_{-p-q-1,m+q}(L_p(I), \bar{V})$ which abuts to $\mathcal{K}_{-p-q,m+p+q}(I, V)$.*

Proof. We filter I by $F^p I = \tilde{L}_p(I)$, and thus $\operatorname{Gr}^p I = L_p(I)$. The complex $\mathcal{K}_m^\bullet(I, V)$ is filtered by the F^p, and thus (see p. 440 of [G-H]) there is a spectral sequence with $E_1^{p,q} = H^{p+q}(\mathcal{K}_{m-p}^\bullet(L_p(I), V)) \cong \mathcal{K}_{-p-q,m+q}(L_p(I), V)$. Since x_n acts trivially on $L_k(I)$, by the preceding Lemma $E_1^{p,q} \cong \mathcal{K}_{-p-q,m+q}(L_p(I), \bar{V}) \oplus \mathcal{K}_{-p-q-1,m+q}(L_p(I), \bar{V})$. The spectral sequence abuts to $\mathcal{K}_{-p-q,m+p+q}(I, V)$. $\square$

Comments for students:

One of the most difficult things for algebraists is learning how to interpret algebraic results geometrically. One classic and encyclopaedic source is Hartshorne's book [H]; another quite beautiful treatment is Serre's classic paper [S], which is often used by students. Eisenbud's book [E] does a wonderful job on interpreting algebraic theorems geometrically, but he assumes that you already know some of the geometric language, e.g. sheaves.

Non-algebraists may be thrown by associated primes. These appear in Matsumura's book [M] or in Eisenbud's book [E], among many others.

The relation between the Koszul complex and Tor's, and a proof of the symmetry of Tor, appears in my notes on Koszul cohomology [G1], where there are also some geometric results relating to regularity, or alternatively in Eisenbud [E].

The regularity of an ideal is a crucial geometric property, and there are important conjectures on the regularity of projective varieties. For some results on specific varieties, one might consult [G1] while a powerful general result is that of [G-L-P].

3. The Macaulay-Gotzmann Estimates on the Growth of Ideals

Proposition 3.1. If $k_d > k_{d-1} > \cdots > k_1 \geq 0$ and $l_d > l_{d-1} > \cdots > l_1 \geq 0$, then

$$\binom{k_d}{d} + \binom{k_{d-1}}{d-1} + \cdots + \binom{k_1}{1} > \binom{l_d}{d} + \binom{l_{d-1}}{d-1} + \cdots + \binom{l_1}{1}$$

if and only if $(k_d, k_{d-1}, \ldots, k_1) > (l_d, l_{d-1}, \ldots, l_1)$ in the lexicographic order, i.e. $k_d > l_d$ or $k_d = l_d$ and $k_{d-1} > l_{d-1}$, etc. We use the convention $\binom{a}{b} = 0$ if $a < b$.

Proof. We note that the inequalities force $l_d \geq l_i + d - i$. If $k_d > l_d$, then we note

$$\binom{k_d}{d} \geq \binom{l_d + 1}{d}$$

$$= \binom{l_d}{d} + \binom{l_d - 1}{d - 1} + \cdots + \binom{l_d - (d-1)}{1} + 1$$

$$> \binom{l_d}{d} + \binom{l_{d-1}}{d-1} + \cdots + \binom{l_1}{1}.$$

If $k_d = l_d$, then we may subtract off these terms and proceed by induction on d, the case $d = 1$ being obvious. $\square$

Definition 3.2. Let $d > 0$ be an integer and $c > 0$ another. The **d'th Macaulay representation** of c is the unique way of writing

$$c = \binom{k_d}{d} + \binom{k_{d-1}}{d-1} + \cdots + \binom{k_\delta}{\delta},$$

where $k_d > k_{d-1} > \cdots > k_\delta \geq \delta > 0$. We will use the notations

$$c^{<d>} = \binom{k_d + 1}{d + 1} + \binom{k_{d-1} + 1}{d} + \cdots + \binom{k_\delta + 1}{\delta + 1},$$

$$c_{<d>} = \binom{k_d - 1}{d} + \binom{k_{d-1} - 1}{d - 1} + \cdots + \binom{k_\delta - 1}{\delta},$$

where by convention $\binom{a}{b} = 0$ if $a < b$.

Remark. Uniqueness of the d'th Macaulay representation is guaranteed by the proposition above. We remark that $k_d = \max\{k \mid c \geq \binom{k}{d}\}$. Inductively, all the k_i are chosen maximally so that $c - \sum_{j=i}^{d} \binom{k_j}{j} \geq 0$. It is also elementary from this Proposition to note that $c \mapsto c_{<d>}$ is weakly increasing in c and $c \mapsto c^{<d>}$ is increasing in c.

Example. The 3'rd Macaulay representation of 27. Since $\binom{6}{3} = 20 \leq 27 < \binom{7}{3} = 35$, we see that $27 = \binom{6}{3} + 7 = \binom{6}{3} + \binom{4}{2} + 1 = \binom{6}{3} + \binom{4}{2} + \binom{1}{1}$ is the Macaulay representation of 27 for $d = 3$. Now $27^{<3>} = \binom{7}{4} + \binom{5}{3} + \binom{2}{2} = 46$, while $27_{<3>} = \binom{5}{3} + \binom{3}{2} + \binom{0}{1} = 13$.

Theorem 3.3 (Macaulay's Estimate on the Growth of Ideals). Let I be a homogeneous ideal with the Hilbert function of S/I denoted $h(d)$. Then

$$h(d + 1) \leq h(d)^{<d>}.$$

Theorem 3.4 (Hyperplane Restriction Theorem). Let I be a homogeneous ideal and I_H its restriction to a general hyperplane. Let h, h_H be the Hilbert functions of $S/I, S_H/I_H$ respectively. Then

$$h_H(d) \leq h(d)_{<d>}.$$

Proposition 3.5. Let I be a homogeneous ideal with Hilbert function h for S/I. Then for all $d \geq 1$,

$$h(d) \leq (\sum_{j=0}^{d} h(j))_{<d>}.$$

Proof. It is easiest to prove all of these theorems and the proposition at once by doing induction on both d and the number of variables. In this proof, we use rlex order.

Proposition for $d, n - 1$ implies Hyperplane Restriction Theorem for d, n: If we work in general coordinates, we may take $H = x_n$. Furthermore, since $\mathrm{gin}(I_H) = \mathrm{gin}(I)_{x_n}$, we may replace I by $\mathrm{gin}(I)$ and thus reduce (since I and $\mathrm{gin}(I)$ have the same Hilbert functions) to the case where I is a Borel-fixed monomial ideal. We now write

$$I_d = J_d + x_n J_{d-1} + \cdots + x_n^d J_0,$$

where J_j is a Borel-fixed monomial ideal of degree j and involves only $x_1, \ldots, x_{n-1}$. Since I is Borel-fixed, using elementary moves we have that $\bar{\mathfrak{m}} J_j \subseteq J_{j+1}$. So let J

be the ideal in $n-1$ variables generated by the J_j. If the proposition is true in $n-1$ variables, then $h_{\bar{S}/J}(d) \leq (\sum_{j=0}^{d} h_{\bar{S}/J}(j))_{<d>}$. However, $h_{S/I}(d) = \sum_{j=0}^{d} h_{\bar{S}/J}(j)$ and $h_{S_H/I_H}(d) = h_{\bar{S}/J}(d)$, so the Proposition in degree d and $n-1$ variables implies the hyperplane restriction Theorem in n variables and degree d.

Hyperplane Restriction Theorem for d, n implies Macaulay's Estimate for $d-1, n$: There is an exact sequence

$$0 \to (I : H)_{d-1} \to I_d \to I_{H,d} \to 0.$$

From this, if h, h_H are the Hilbert functions for $S/I, S_H/I_H$ respectively, we obtain the inequality $h(d) \leq h(d-1)+h_H(d)$. Using the Hyperplane Restriction Theorem, $h_H(d) \leq h(d)_{<d>}$. Thus $h(d) - h(d)_{<d>} \leq h(d-1)$. By an easy binomial identity, for any c, $(c - c_{<d>})^{<d-1>} \geq c$. Taking upper $< d - 1 >$ of both sides of the inequality on h and noting that this is an increasing function, we get that $h(d) \leq h(d-1)^{<d-1>}$, as desired.

Macaulay's estimate for $d - 1, n$ and the Proposition for n and degrees $\leq d - 1$ implies the Proposition for d, n: It is convenient to introduce the notation, if c has d'th Macaulay representation $\binom{k_d}{d} + \cdots + \binom{k_\delta}{\delta}$, that $f_d(c) = \binom{k_d+1}{d} + \cdots + \binom{k_\delta+1}{\delta}$; note $(f_d(c))_{<d>} = c$. We also note that $f_d(c_{<d>}) \leq c$; we get an inequality because of terms with $k_m = m$. Let $h(d)$ denote the Hilbert function of I. Using the Proposition for $n, d - 1$, we may assume $h(d - 1) \leq (h(0) + \cdots + h(d - 1))_{<d-1>}$. Taking f_d of both sides and using $f_d(c_{<d>}) \leq c$, we have $f_{d-1}(h(d - 1)) \leq h(0) + \cdots + h(d-1)$. We need to show $f_d(h(d)) \leq h(0) + \cdots + h(d)$. By Macaulay's estimate, $h(d) \leq h(d-1)^{<d-1>}$. Now $h(d) + f_{d-1}(h(d-1)) \leq h(0) + \cdots + h(d)$. It is therefore enough to show that in general, if $c \leq b^{<d-1>}$, then $c + f_{d-1}(b) \geq f_d(c)$. If c has d'th Macaulay representation $\binom{k_d}{d} + \cdots + \binom{k_\delta}{\delta}$, then $b \geq \binom{k_d-1}{d-1} + \cdots + \binom{k_\delta-1}{\delta-1}$ if $\delta > 1$ and $b \geq \binom{k_d-1}{d-1} + \cdots + \binom{k_2-1}{1} + 1$ if $\delta = 1$. Now $f_{d-1}(b) \geq \binom{k_d}{d-1} + \cdots + \binom{k_\delta}{\delta-1}$ or $f_{d-1}(b) \geq \binom{k_d}{d-1} + \cdots + \binom{k_2}{1} + 1$, depending on δ. Adding, we see that $c + f_{d-1}(b) \geq \binom{k_d+1}{d} + \cdots + \binom{k_\delta+1}{\delta}$ in either case.

   To start the induction, we note that for $n = 1$ the only interesting case of the Proposition for any d is that $h(i) = 1$ for $i = 0, 1, \ldots, d$. Now $f_d(h(d)) = \binom{d+1}{d} = h(0) + \cdots + h(d)$. The other case we need to start the induction is $d = 1, n$ arbitrary, and here the only interesting case is $h(1) \neq 0$, which forces $h(0) = 1$, so the statement is $h(1) \leq (1 + h(1))_{<1>}$, but $(1 + h(1))_{<1>} = h(1)$, so this is clear. This completes the proof. $\qquad \square$

Definition 3.6. *A monomial ideal is said to be a* **lex segment ideal** *in degree d if I_d is spanned by the first $\dim(I_d)$ monomials in lexicographic order.*

Example. *The monomial ideal $(x_1, x_2^3, x_2^2 x_3, x_2 x_3^3, x_3^4)$ in 3 variables.*

This is a lex segment ideal in all degrees.

Example. *The lex gin of 2 general cubics in 4 variables.*

If C_1, C_2 are two general cubics in 4 variables and $I = (C_1, C_2)$, then using lex order, one may verify that $\text{gin}(I) = (x_1^2, x_1 x_2^2, x_1 x_2 x_3^2, x_1 x_2 x_3 x_4^2, x_1 x_2 x_4^4, x_1 x_3^6, x_2^6)$. It is Borel-fixed, as it must be since it is a gin. This is not a lex segment ideal in degree 6 – rather than x_2^6, the next term in lex order not in the ideal is $x_1 x_3^5$. This example was suggested by Bernd Sturmfels, and helped to motivate the work discussed in Chapter 6.

Proposition 3.7 (Macaulay). *If I is a lex segment ideal in degree d and has no generators in degree $d + 1$, then $h_{S/I}(d + 1) = h_{S/I}(d)^{<d>}$, i.e. I achieves the maximum for Macaulay's bound.*

Proof. In one variable, the result is obvious, since the only cases to consider are $h_{S/I}(d) = 0, 1$, and these are both easily verified. So we proceed by induction on the number of variables. We may write $I_d = J_d + x_1 J_{d-1} + \cdots + x_1^d J_0$, where the J_i involves only the variables $x_2, \ldots, x_n$. Because I is a lex segment ideal, for some i, $J_j = \mathbf{C}[x_2, \ldots, x_n]_j$ for $j < i$ and $J_j = 0$ for $j > i$, and J_i is a lex segment ideal in $n - 1$ variables. Thus $h_{S/I}(d) = h_{\bar{S}/J_i}(i) + \binom{n+i-1}{i+1} + \cdots + \binom{n-3+d}{d-1} + \binom{n-2+d}{d}$. Now $I_{d+1} = \bar{m}J_d + x_1(\bar{m}J_{d-1} + J_d) + \cdots + x_1^d(\bar{m}J_0 + J_1) + x_1^{d+1} J_0$, and $h_{S/I}(d + 1) = h_{\bar{S}/J_i}(i+1) + \binom{n+1}{i+2} + \cdots + \binom{n-1+d}{d+1}$. Since $h_{\bar{S}/J_i}(d) \leq \binom{n+i-2}{i}$, if may write the i'th Macaulay representation for $h_{\bar{S}/J_i}(d)$ as $\binom{k_i}{i} + \cdots + \binom{k_\delta}{\delta}$, then the d'th Macaulay representation for $h_{S/I}(d)$ is $\binom{n-2+d}{d} + \cdots + \binom{n-1+i}{i+1} + \binom{k_i}{i} + \cdots + \binom{k_\delta}{\delta}$. We now see that if the proposition is true for J_i, then it is true for I_d. $\square$

Theorem 3.8 (Gotzmann's Persistence Theorem). *Let I be a homogeneous ideal generated in degrees $\leq d + 1$. If in Macaulay's estimate,*

$$h(d + 1) = h(d)^{<d>},$$

then I is d-regular and

$$h(k + 1) = h(k)^{<k>}$$

for all $k \geq d$.

Remark. We could just as well assume I is generated in degrees $\leq d$, since we cannot achieve Macaulay's bound if there are any generators in degree $d + 1$.

Proof. By comparing I with $\text{gin}(I)$, we note that since the ideal generated by $\text{gin}(I)_d$ must satisfy Macaulay's bound, if $\text{gin}(I)$ had a generator in degree $d + 1$, then I would be off by at least that number of generators from achieving equality in Macaulay's bound. So $\text{gin}(I)$ has no new generators in degree $d + 1$, and since I is generated in degree $\leq d + 1$, we conclude by the Crystallization Principle (Proposition 2.28) that I is d regular and $\text{gin}(I)$ is generated in degrees $\leq d$. We may now replace I by $\text{gin}(I)$, so we may assume that I is a Borel-fixed monomial ideal. We inductively assume the theorem if either the degree or number of variables is smaller than d, n respectively, since the cases $n = 1$ and $d = 1$ are easy. Write

$$I_d = J_d + x_n J_{d-1} + \cdots + x_n^d J_0,$$

where J_j is homogeneous of degree j and involves only $x_1, \ldots, x_{n-1}$. The Borel-fixed property shows that $\mathfrak{m}J_i \subseteq J_{i+1}$. Denote by J the ideal generated by the J_i. If $\bar{S} = \mathbf{C}[x_1, \ldots, x_{n-1}]$, then $h_{S/I}(d) = h_{\bar{S}/J}(d) + \cdots + h_{\bar{S}/J}(0)$. We note that

$$I_{d+1} = J_{d+1} + x_n J_d + \cdots + x_n^{d+1} J_0,$$

and thus $h_{S/I}(d+1) = h_{\bar{S}/J}(d+1) + \cdots + h_{\bar{S}/J}(0)$. Let $h_{S/I}(d) = \binom{k_d}{d} + \cdots + \binom{k_\delta}{\delta}$ be the d'th Macaulay representation of $h_{S/I}(d)$. If equality holds in Macaulay's estimate, then $h_{S/I}(d+1) = \binom{k_d+1}{d+1} + \cdots + \binom{k_\delta+1}{\delta+1}$. Taking the difference, $h_{\bar{S}/J}(d+1) = \binom{k_d}{d+1} + \cdots + \binom{k_\delta}{\delta+1}$. By Proposition 3.5, $h_{\bar{S}/J}(d) \leq (h_{\bar{S}/J}(d) + \cdots + h_{\bar{S}/J}(0))_{<d>} = h_{S/I}(d)_{<d>}$. This latter is $\binom{k_d-1}{d} + \cdots + \binom{k_\delta-1}{\delta}$, and thus we see that equality holds for Macaulay's estimate for J in degree d, and also that $h_{\bar{S}/J}(d) = (h_{\bar{S}/J}(d) + \cdots + h_{\bar{S}/J}(0))_{<d>}$. By induction, this means that equality holds for Macaulay's estimate for J in all degrees $\geq d$. Now $I_{d+2} = J_{d+2} + x_n J_{d+1} + \cdots + x_n^{d+2} J_0$, so $h_{S/I}(d+2) = \sum_{j=0}^{d+2} h_{\bar{S}/J}(j) = h_{S/I}(d+1) + h_{\bar{S}/J}(d+2)$. However, the right hand side is the sum of $\binom{k_d+1}{d+1} + \cdots + \binom{k_\delta+1}{\delta+1}$ and $\binom{k_d+1}{d+2} + \cdots + \binom{k_\delta+1}{\delta+2}$, and hence is $\binom{k_d+2}{d+2} + \cdots + \binom{k_\delta+2}{\delta+2}$. Thus $h_{S/I}(d+2) = h_{S/I}(d+1)^{<d+1>}$, and thus equality in Macaulay's estimate holds in degree $d+1$ if it holds in degree d. Of course, this is formally equivalent to showing that it holds in all degrees $k \geq d$ if it holds in degree d. $\qquad\square$

Corollary 3.9. *If I is a homogeneous ideal generated in degrees $\leq d$ and equality holds for Macaulay's estimate for I in degree d, and if $h_{S/I}(d)$ has Macaulay representation in degree d given by $\binom{k_d}{d} + \cdots + \binom{k_\delta}{\delta}$, then*

$$h_{S/I}(k) = \binom{k_d + k - d}{k} + \cdots + \binom{k_\delta + k - d}{\delta + k - d}$$

for all $k \geq d$.

Remark. The decomposition of monomial ideals I_d into $\sum_i x_n^{d-i} J_i$ is a reflection of the filtration on any I given by $I_d \supset h(I : h)_{d-1} \supset h^2(I : h^2)_{d-2} \supset \cdots$ which has successive quotients $I_H, (I : h)_H, (I : h^2)_H, \ldots$. For H general, we are just decomposing $\mathrm{gin}(I)_d$ into the generic initial ideals of these quotients; this is just the decomposition for the rlex order $\mathrm{gin}(I)_d = \sum_k x_n^k \mathrm{gin}(L_k(I))_{d-k}$ using the ideals of leading terms.

Definition 3.10. For any I, the Hilbert function $h_{S/I}(d)$ is equal to a polynomial $P_{S/I}(d)$ for $d \gg 0$, called the **Hilbert polynomial** of S/I.

Theorem 3.11 (Gotzmann's Regularity Theorem). *Any graded ideal $I_\bullet$ has Hilbert polynomial of the form*

$$P_{S/I}(k) = \binom{k + a_1}{a_1} + \binom{k + a_2 - 1}{a_2} + \cdots + \binom{k + a_s - (s - 1)}{a_s}$$

where

$$a_1 \geq a_2 \geq \cdots \geq a_s \geq 0.$$

Furthermore, the associated ideal sheaf $\mathcal{I}$ is s-regular.

Remark.   We will prove a somewhat stronger result. Let

$$s_q = \mathrm{card}\{i \mid a_i \geq q - 1\}.$$

Note that $s_1 = s$. Then in fact for each $q > 0$,

$$H^q(\mathcal{I}(k - q)) = 0 \quad \text{for } k \geq s_q.$$

Proof. We denote $P(k) = P_{S/I}(k)$. Let H be a general hyperplane, and $\mathcal{I}' = \mathcal{I} \otimes \mathcal{O}_H$. We have the exact sequence

$$0 \to \mathcal{I}(k - 1) \to \mathcal{I}(k) \to \mathcal{I}'(k) \to 0$$

arising from restriction. If $P_H(k)$ is the Hilbert polynomial of $\mathcal{I}'$ viewed as a coherent sheaf on H, then we have by this sequence that

$$P_H(k) = P(k) - P(k - 1).$$

By induction on dimension of the ambient projective space, the case of dimension 0 being obvious, we may assume that $P_H(k)$ has the form

$$P_H(k) = \binom{k + b_1}{b_1} + \binom{k + b_2 - 1}{b_2} + \cdots + \binom{k + b_t - (t - 1)}{b_t}$$

where

$$b_1 \geq b_2 \geq \cdots b_t \geq 0.$$

We may further assume that if

$$t_q = \mathrm{card}\{i \mid b_i \geq q - 1\},$$

then for every $q > 0$,

$$H^q(\mathcal{I}'(k - q)) = 0 \quad \text{for } k \geq t_q.$$

We immediately conclude that

$$P(k) = \binom{k + a_1}{a_1} + \binom{k + a_2 - 1}{a_2} + \cdots + \binom{k + a_t - (t - 1)}{a_t} + e$$

where e is an unknown constant and $a_i = b_i + 1$. Thus $s_{q+1} = t_q$ for $q \geq 0$. We immediately see by the restriction sequence for $\mathcal{I}'$, together with Theorem B, that for all $q > 1$,

$$H^q(\mathcal{I}(k - q)) = 0 \quad \text{for } k \geq s_q.$$

It remains to show that this also holds for $q = 1$, and that $e \geq 0$.

Let $f_d = \mathrm{codim}(H^0(\mathcal{I}(d)), H^0(\mathcal{O}_{\mathbf{P}^r}(d)))$, and $f_{d,H}$ the analogous numbers for $\mathcal{I}'$. If $e < 0$, then for $k \gg 0$,

$$f_k < \binom{k + a_1}{a_1} + \binom{k + a_2 - 1}{a_2} + \cdots + \binom{k + a_t - (t - 1)}{a_t}.$$

By our result on codimensions for restriction to a general hyperplane, for $k \gg 0$,

$$f_{k,H} < \binom{k+b_1}{b_1} + \binom{k+b_2-1}{b_2} + \cdots + \binom{k+b_t-(t-1)}{b_t} = P_H(k),$$

which is a contradiction. So $e \geq 0$. Setting $a_{t+1} = a_{t+2} = \cdots = a_{t+e} = 0$, we get that $P(k)$ has the desired formula, where $s = t + e$.

By the vanishing of cohomology that we have so far, we have that

$$f_d \leq P(d), \quad \text{for } d \geq s_2 - 2,$$

with equality holding if and only if $H^1(\mathcal{I}(d)) = 0$. For $d = s - 1$, we may write

$$P(d) = \binom{d+a_1}{d} + \binom{d+a_2-1}{d-1} + \cdots + \binom{d+a_s-(s-1)}{d-(s-1)}.$$

By Macaulay's theorem, if $f_{s-1} < P(s-1)$, then it remains behind forever, contradicting the fact that $f_k = P(k)$ for $k \gg 0$. Thus $H^1(\mathcal{I}(s-1)) = 0$, which is the last thing we need to conclude $\mathcal{I}$ is s-regular. This completes the proof of Gotzmann's Regularity Theorem. $\qquad\square$

Remark. Using the notation of the theorem, for d sufficiently large, $h_{S/I}(d) = \binom{d+a_1}{d} + \binom{d-1+a_2}{d-1} + \cdots + \binom{d-(s-1)+a_s}{d-(s-1)}$, and thus every homogeneous ideal achieves Macaulay's bound for d sufficiently large.

This completes a survey of the fundamental results in this circle of ideas. We include some interesting but more specialized results.

Proposition 3.12. *Let I be a homogeneous ideal generated in degrees $\leq d$. If I_d is a lex segment ideal in degree d, then I_k is a lex segment ideal in degree k for all $k \geq d$.*

Proof. We need only check the case $k = d+1$, since then inductively we are done. If $x^J \in I_d$ and, in the lexicographic order, $x^K > x^{J+m}$, we need to show $x^K = x^{L+k}$ for some L satisfying $x^L \geq x^J$. If $k_m > 0$, we may take $k = m$. If $k_m = 0$, then choose p so that $k_i = j_i + \delta_{im}$ for $i < p$ and $k_p > j_p + \delta_{pm}$. Clearly $p < m$. If $k_p > j_p + 1$, we may take $k = p$. If $k_p = j_p + 1$, then if any $k_l > 0$ for some $l > p$, we can take $k = l$. Otherwise, $x^K = x_p x^J$, and we may take $k = p$. $\qquad\square$

Proposition 3.13. *Let I be a lex segment ideal in degree d. Then $(I : \mathfrak{m})$ is a lex segment ideal in degree $d - 1$.*

Proof. If $x^J \in (I : \mathfrak{m})_{d-1}$, assume $x^K > x^J$ in the lexicographic order. Then $x^{K+i} > x^{J+i}$ for all i. Since $x^{J+i} \in I_d$ for all i and I_d is a lex segment ideal, this implies $x^{K+i} \in I_d$ for all i, and thus $x^K \in (I : \mathfrak{m})_{d-1}$. $\qquad\square$

Proposition 3.14. *Let J_d be a lex segment ideal in degree d in $n - 1$ variables. Then $I_d = J_d + x_n(J_d : \bar{\mathfrak{m}}) + \cdots + x_n^d(J_d : \bar{\mathfrak{m}}^d)$ is a lex segment ideal in degree d in n variables.*

Proof. Given a monomial in I_d of the form $x^K x_n^k$ where K involves only variables $x_1, \ldots, x_{n-1}$, and $x^K \in (J_d : \bar{\mathfrak{m}}^k)$, assume that $x^L x_n^l > x^K x_n^k$, where L involves only $x_1, \ldots, x_n$. We need to show that $x^L \in (J : \bar{\mathfrak{m}}^l)$. If $l = k$, then $x^L > x^K$ and we are done, as $(J_d : \bar{\mathfrak{m}}^k)$ is a lex segment ideal by Proposition 3.13. If $k < l$, then $x^L x_n^{l-k} > x^K$, which implies that $x^L x^M > x^K$ for any M with $|M| = l - k$, and hence $x^L \in (J_d : \bar{\mathfrak{m}}^l)$. If $k > l$, then $x^L > x^K x_n^{k-l}$. Let L_0 consist of the last $k - l$ indices of x^L, then $x^{L-L_0} > x^K$. Hence $x^{L-L_0} \in (J : \bar{\mathfrak{m}}^k)$, and hence $x^L \in (J : \bar{\mathfrak{m}}^l)$.   □

Definition 3.15. *Given $c \geq 0$, $d \geq 1$, let $^{<d>}c$ be the smallest number e such that $e^{<d-1>} \geq c$.*

Remark. If the d'th Macaulay representation of $c = \binom{k_d}{d} + \cdots + \binom{k_\delta}{\delta}$, then if $\delta > 1$, $^{<d>}c = \binom{k_d-1}{d-1} + \cdots + \binom{k_\delta-1}{\delta-1}$, while if $\delta = 1$, then $^{<d>}c = \binom{k_d-1}{d-1} + \cdots + \binom{k_2-1}{1} + 1$.

Definition 3.16. *Given $c \geq 0$, $d \geq 2$, let $_{<d>}c$ denote the smallest integer e such that $e_{<d>} = c$.*

Remark. If the d'th Macaulay representation of c is $\binom{k_d}{d} + \cdots + \binom{k_\delta}{\delta}$, then $_{<d>}c = \binom{k_d+1}{d} + \cdots + \binom{k_\delta+1}{\delta}$.

Proposition 3.17. *Let I be a homogeneous ideal such that*

$$\sum_{j=0}^{d} h_{S/I}(j) = {}_{<d>}(h_{S/I}(d)).$$

Then $h_{S/I}(j-1) = {}^{<j>}h_{S/I}(j)$ for all $1 \leq j \leq d$. Furthermore, $I_j = (I_d : \mathfrak{m}^{d-j})$ for all $j \leq d$.

Proof. We proceed by induction on d, the case $d = 1$ being easy. Let the d'th Macaulay representation for $h_{S/I}(d) = \binom{k_d}{d} + \cdots + \binom{k_\delta}{\delta}$. Then $\sum_{j=0}^{d} h_{S/I}(j) = \binom{k_d+1}{d} + \cdots + \binom{k_\delta+1}{\delta}$. If $\delta > 1$, then by Macaulay's estimate, $h_{S/I}(d-1) \geq \binom{k_d-1}{d-1} + \cdots + \binom{k_\delta-1}{\delta-1}$. We also have $\sum_{j=0}^{d-1} h_{S/I}(j) = \binom{k_d}{d-1} + \cdots + \binom{k_\delta}{\delta-1}$, and therefore by Proposition 3.5, $h_{S/I}(d-1) \leq \binom{k_d-1}{d-1} + \cdots + \binom{k_\delta-1}{\delta-1}$. Combining the two inequalities, $h_{S/I}(d-1) = {}_{<d-1>}(\sum_{j=0}^{d-1} h_{S/I}(j))$, and so by induction on d we are done. If $\delta = 1$, then $\sum_{j=0}^{d-1} h_{S/I}(j) = \binom{k_d}{d-1} + \cdots + \binom{k_2}{1} + 1$. This implies by Proposition 3.5 that $h_{S/I}(d-1) \leq \binom{k_d-1}{d-1} + \cdots + \binom{k_2-1}{1} + 1 = {}^{<d>}h_{S/I}(d)$ On the other hand, by Macaulay's estimate, $h_{S/I}(d-1) \geq {}^{<d>}h_{S/I}(d)$, so we have equality, and $h_{S/I}(d-1) = (\sum_{j=0}^{d-1} h_{S/I}(j))_{<d-1>}$. We are now done with all but the last statement by induction on d.

If we replace I_j by $K_j = (I_d : \mathfrak{m}^{d-j})$, then $\sum_{j=0}^{d} h_{S/K}(j) \leq \sum_{j=0}^{d} h_{S/I}(j)$, while $h_{S/K}(d) = h_{S/I}(d)$, so by the inequality of Proposition 3.5, together with the fact that $\sum_{j=0}^{d} h_{S/I}(j) = {}_{<d>}h_{S/I}(d)$, we must have $\sum_{j=0}^{d} \dim(K_j) = \sum_{j=0}^{d} \dim(I_j)$ for all $j \leq d$, and hence, since $I_j \subseteq K_j$, we have $I_j = K_j$ for all $j \leq d$.   □

Lemma 3.18. *Let I be a homogeneous ideal. Then for a general hyperplane H,*
$(I : H^i)_H \subseteq (I_H : \mathfrak{m}_H^i)$.

Proof. If we write $H = \sum_{i=1}^n t_i x_i$, then any element of $(I : H^i)$ is represented by a polynomial P bi-homogeneous in x and t such that $H^i P \in I$. Differentiating with respect to t^J with $|J| = i$, we see that $x^J P + H U_J \in I$ for some U_J, and thus $x^J P|_H \in I_H$. Hence $P \in (I_H : \mathfrak{m}_H^i)$. $\square$

Theorem 3.19. *Let I be a homogeneous ideal such that $h_{S_H/I_H}(d) = c$ and $h_{S/I}(d) = {}_{<d>}c$ for a general hyperplane H. Then*

$$h_{S_H/(I_H:\mathfrak{m}_H^i)}(d - i) = {}^{<d-i+1>}h_{S_H/(I_H:\mathfrak{m}_H^{i-1})}(d - i + 1)$$

for all $0 < i \leq d$.

Proof. If we write

$$\mathrm{gin}(I)_d = J_d + x_n J_{d-1} + \cdots + x_n^d J_0,$$

where J_i are homogeneous of degree i and involve only $x_1, \ldots, x_{n-1}$, then the assumption on c implies that the ideal J in $n - 1$ variables generated by the J_j satisfies the hypotheses of Proposition 3.17. However, if we use reverse lexicographic order, $J_d = I_{H,d}$. The equality claimed in the theorem now follows from that of Proposition 3.17. $\square$

Remark. It is not true that if $h_{S_H/I_H}(d) = h_{S/I}(d)_{<d>}$, then $\mathrm{gin}(I)$ is a lex segment ideal in degree d. For example, if $n = 4$, the ideal $I = (x_1^2, x_1 x_2, x_2^2)$ has codimension 7, and $7_{<2>} = 3$, and indeed I_H has codimension 3. This is Borel-fixed, so $I = \mathrm{gin}(I)$, which is not a lex segment ideal. This does not violate the theorem because $_{<2>}3 = 6$, but it shows that there are ideals achieving Macaulay's bound whose gin's are not lex segment ideals.

Corollary 3.20. *Let I be a homogeneous ideal such that $h_{S/I}(d) = d + 1$ and $h_{S_H/I_H}(d) = 1$. Then $I_d = I(L)_d$ for some line L.*

Proof. This satisfies the hypotheses of the theorem, so using the notations of the proof, J_1 has codimension 1, so $(I_H : \mathfrak{m}_H^{d-1})_1$ has codimension 1, hence is the ideal of a point in H. Thus $I_{H,d} \supseteq I(P)_d$ for some point P, and by comparing codimensions we get equality. This implies that the base locus of I_d contains a line, so $I_d \subseteq I(L)_d$, but then equality holds since they have the same codimension. $\square$

Example 3.21. *6 cubics in 3 variables.* Given the ideal I generated by 6 cubics in 3 variables, $h_{S/I}(3) = 4 = \binom{4}{3}$, so $4^{<3>} = \binom{5}{4} = 5$. We conclude from Macaulay's bound that the ideal generated in degree 4 has codimension at most 5, so dimension at least 10. One way to achieve this is to take the six cubics to be the six cubics vanishing on a line L. By Gotzmann's persistence theorem, any set of 6 cubics achieving this bound must have $h_{S/I}(d) = d + 1$ for all $d \geq 3$. It follows that the base locus of the variety defined by the six cubics contains a curve of degree 1, and hence is a line, so this is thus the only possibility.

Alternatively, given 6 cubics in 3 variables, $4_{<3>} = \binom{3}{3} = 1$, so the restriction of the cubics to a general line has codimension at most 1 by the hyperplane restriction theorem. Once again, if the bound is achieved, then by Corollary 3.20, the six cubics are the ideal of a line, and then the restriction to a general line is the ideal of the intersection of the two lines.

The lex segment for six cubics in 3 variables is x_1^3, $x_1^2 x_2$, $x_1^2 x_3$, $x_1 x_2^2$, $x_1 x_2 x_3$, $x_1 x_3^2$, which is the ideal of the line x_1.

Example 3.22. *The case* $h_{S/I}(d) \leq d$.

If $h_{S/I}(d) = c \leq d$, then the Macaulay representation is $k_d = d, k_{d-1} = d - 1, \ldots, k_{d-c+1} = d - c + 1$. It follows that $c^{<d>} = c$. Assume I is generated in degrees $\leq d$. If $h_{S/I}(d+1) = c$, then it remains at this value for arbitrarily large degree, and thus I_d must be the ideal of a set of points of length c. If we achieve Macaulay's bound c and the variety of I_d is empty, i.e. I_d is base-point free, then by applying Gotzmann's theorem at each step, necessarily I_{d+c} is the full set of polynomials of degree $d + c$. In any case, the Hilbert function of I must stabilize after $\leq c$ steps.

Comments for students:

There is a very beautiful combinatorial side to this subject; see Richard Stanley's book [St] for a survey of some results.

Gotzmann's Persistence Theorem is from [Go], with an alternate proof of mine in [G2], where the Hyperplane Restriction Theorem first appears. For a perspective on Macaulay's theorem in the context of commutative algebra, one might consult the book of Bruns and Herzog [B-H]. Gotzmann originally proved his theorem as a tool for constructing moduli spaces – one wants to know that once the Hilbert function is specified, one has a bound on the regularity of the ideal. My interest in his result, and indeed in commutative algebra, grew out of my work on the explicit Noether-Lefschetz problem in Hodge theory; this work is described in my notes on Koszul cohomology [G1] and in Lecture 7 of my CIME lectures [G3]. Most of the proofs given here are new.

The question of how Hilbert functions of ideals can grow is quite interesting, and there are many results, and lots of open problems. If we are allowed to insert new generators at will, then using lex segment ideals, one can get any Hilbert function satisfying Macaulay's bound at each stage. One expects to improve on Macaulay's bound if one knows some additional geometric information, or if one knows that the ideal contains a certain kind of subideal, or if one knows more than one step of the history of the growth of the Hilbert function and one assumes that there are no new generators. For a conjecture about ideals that contain a regular sequence of quadrics, see [E-G-H]. For a fascinating table of some possible Hilbert functions, see [R]. Chapter 4 will deal with one example of how to use additional geometric information.

4. Points in $\mathbf{P}^2$ and Curves in $\mathbf{P}^3$

In this section, we will use rlex order. The reason is that this is the order adapted to hyperplane sections.

Let Γ be a collection of d points in the plane, and I the graded ideal of Γ. For generic coordinates, $\mathrm{gin}(I)$ is a Borel-fixed saturated ideal, and thus (by Theorem 2.30) x_3 does not appear in any generator of $\mathrm{gin}(I)$. It thus has generators of the form

$$x_1^k, x_1^{k-1}x_2^{\lambda_{k-1}}, \ldots, x_1 x_2^{\lambda_1}, x_2^{\lambda_0}.$$

Because $\mathrm{gin}(I)$ is Borel-fixed, $\lambda_0 > \lambda_1 > \cdots > \lambda_{k-1} > 0$.

Definition 4.1. *For a set Γ of d points, if*

$$\mathrm{gin}(I_\Gamma) = (x_1^k, x_1^{k-1}x_2^{\lambda_{k-1}}, \ldots, x_1 x_2^{\lambda_1}, x_2^{\lambda_0}),$$

*we will say that Γ has **invariants** $\lambda_0, \lambda_1, \ldots, \lambda_{k-1}$.*

We note that for any $m \geq 0$,

$$h_{S/I}(m) = \sum_{i=0}^{k-1} \min(m - i, \lambda_i).$$

Taking m large, we have that $d = \sum_i \lambda_i$. Notice that therefore $h^1(\mathcal{I}_\Gamma(m)) = d - \sum_{i=0}^{k-1} \min(m - i, \lambda_i)$ for any $m \geq 0$. Another consequence of the formula is:

Lemma 4.2. *Two set of points in the plane have the same invariants $\lambda_0, \ldots, \lambda_{k-1}$ if and only if they have the same Hilbert function.*

Example. *Four general points in the plane.* The ideal of 4 general points in $\mathbf{P}^2$ is $I = (Q_1, Q_2)$, where Q_1, Q_2 are general conics. In rlex order, we know that $\mathrm{gin}(I) = (x_1^2, x_1 x_2, x_2^3)$. Thus $\lambda_0 = 3$, $\lambda_1 = 1$, $k = 2$.

There is a nice way to draw a picture of the rlex gin of I in this situation. In the picture below, we show all monomials in x_1, x_2, x_3 of degree 5. x_1^5 is in the lower left corner, x_2^5 is in the lower right corner, and x_3^5 is at the top. The black dots (when I do not have PostScript graphics available, these are replaced by "X"s) denote monomials that belong to $\mathrm{gin}(I)$ and the empty circles ("zeros") denote monomials that are missing. In terms of this diagram, the fact that the ideal is saturated translates into the fact that the diagram in degree one lower is obtained by suppressing the bottom row of the diagram. It follows that if we draw a diagram of a set of points whose bottom row is all black dots, then we can reconstruct the diagram in any degree from the given one. Notice that in the diagram, $\lambda_0 = 3$ comes from the string of 3 zeros running from the top toward the bottom right corner, and $\lambda_1 = 1$ is the string of 1 zero parallel to it. This is a general fact.

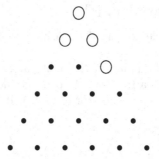

If we look in generic coordinates at $A = S/I$ as a module over $\bar{S} = \mathbf{C}[x_2, x_3]$, we see that $1, x_1, \ldots, x_1^{k-1}$ are a minimal set of generators. The relations have leading terms $x_2^{\lambda_j} \cdot x_1^j$. Thus a minimal free resolution of A over $\bar{S}$ is

$$0 \to \oplus_{j=0}^{k-1} \bar{S}(-\lambda_j - j) \to \oplus_{i=0}^{k-1} \bar{S}(-i) \to A \to 0.$$

Following Gruson-Peskine [G-P], we let

$$n_j = \lambda_j + j \qquad \text{for } j = 0, 1, \ldots, k-1.$$

Since $\lambda_{j-1} > \lambda_j$, it follows that $n_{j-1} = \lambda_{j-1} + j - 1 \geq \lambda_j + j = n_j$. The sequence $n_0 \geq n_1 \geq \cdots \geq n_{k-1} \geq 0$ is called the **numerical character** of Γ, and it is related to our invariants by

$$\lambda_j = n_j - j.$$

This invariant was introduced by Gruson and Peskine to obtain a necessary and sufficient condition for identifying Hilbert functions of general hyperplane sections of irreducible reduced non-degenerate curves in $\mathbf{P}^3$. An interesting version of their result is given below, but first we need a useful Proposition:

Proposition 4.3. *Let I, J be two homogeneous ideals. If $\text{in}(I) \cap \text{in}(J) \subseteq \text{in}(I \cap J)$, then $\text{in}(I + J) = \text{in}(I) + \text{in}(J)$.*

Proof. In the conclusion, it is obvious that the left-hand side contains the right. So assume $F \in I + J$, and write $F = P + Q$ where $P \in I$, $Q \in J$. We may assume by induction that we have chosen P, Q so that $\text{in}(P)$ is as small as possible. If $\text{in}(P) \neq \text{in}(Q)$, then either $\text{in}(F) = \text{in}(P)$ or $\text{in}(F) = \text{in}(Q)$, and therefore $\text{in}(F) \in \text{in}(I) + \text{in}(J)$. If $\text{in}(P) = \text{in}(Q)$, then by hypothesis there is an element $R \in I \cap J$ such that $\text{in}(R) = \text{in}(P) = \text{in}(Q)$. For some constant c, $\text{in}(P - cR) < \text{in}(P)$. Replace P, Q by $P - cR, Q + cR$. We have decreased $\text{in}(P)$, and this contradicts the assumption that it was already as small as possible. $\qquad\square$

Theorem 4.4 (Ellia-Peskine). *If $\lambda_i > \lambda_{i+1} + 2$ for some $i < k - 1$, then Γ contains a subset Γ_1 of $\lambda_0 + \cdots + \lambda_i$ points lying on a curve of degree $i + 1$. Furthermore, if $\Gamma = \Gamma_1 + \Gamma_2$, then Γ_2 has invariants $\lambda_{i+1}, \lambda_{i+2}, \ldots, \lambda_{k-1}$ and Γ_1 has invariants $\lambda_0, \lambda_1, \ldots, \lambda_i$. Furthermore, the decomposition of Γ into two subsets having invariants $\lambda_0, \ldots, \lambda_i$ and $\lambda_{i+1}, \ldots, \lambda_{k-1}$ is unique.*

Proof. There is no generator of the initial ideal of I in degree $m = \lambda_{i+1} + i + 2$. If J is the ideal generated by the elements of I of degree $\leq m - 1 = \lambda_{i+1} + i + 1$, then the initial ideal of J contains no generator of degree m. Thus by the crystallization principle (Proposition 2.28) every generator of $\mathrm{gin}(J)$ has degree $\leq m - 1$, and hence the initial ideal of J has generators $x_1^k, x_1^{k-1}x_2^{\lambda_{k-1}}, \ldots, x_1^{i+1}x_2^{\lambda_{i+1}}$. Now for a general hyperplane H, $\mathrm{gin}((J_H)^{\mathrm{sat}}) = x_1^{i+1}$ by Theorem 2.30, and hence is the ideal of $i + 1$ points on the line H. Thus the base locus of J contains a plane curve of degree $i+1$ given by a polynomial F. The elements of I corresponding to generators of $\mathrm{gin}(I)$ of degree $\leq m-1$ are of the form $g_0 F, g_1 F, \ldots, g_{k-i-1}F$, where $\mathrm{in}(g_j) = x_1^j x_2^{\lambda_{i+1}+j}$. The ideal generated by $g_0, \ldots, g_{k-i-1}$ is saturated, hence the ideal of a set of points Γ_2 with invariants $\lambda_{i+1}, \ldots, \lambda_{k-1}$; it is non-empty because $i < k - 1$. Note that $(I : F) = (g_0, \ldots, g_{k-i-1})$. We note that $\mathrm{gin}(F) \cap \mathrm{gin}(I) \subseteq \mathrm{gin}(J) = \mathrm{gin}((F) \cap I)$, so by Proposition 4.3, $\mathrm{gin}(I + (F)) = \mathrm{gin}(I) + \mathrm{gin}(F)$. Thus $\mathrm{gin}(I + (F)) = (x_1^{i+1}, x_1^i x_2^{\lambda_i}, \cdots, x_2^{\lambda_0})$, which is saturated and by counting degrees must be the ideal of a set of points Γ_1 which has invariants $\lambda_0, \ldots, \lambda_i$. Finally, note that $(I : F) \cdot (I + (F)) \subseteq I$, so $\Gamma_1 + \Gamma_2 \supseteq \Gamma$, which since both sides have the same number of points forces equality.

It remains to see uniqueness. Let $\Gamma = \hat{\Gamma}_1 + \hat{\Gamma}_2$ be another decomposition having the same invariants. Let $\hat{F} \in I_{\hat{\Gamma}_1, i+1}$, and note that $\hat{F} I_{\hat{\Gamma}_2, \lambda_{i+1}+1} \subseteq I_{\Gamma, \lambda_{i+1}+i+2}$, and by Hilbert functions we must have equality. Thus $\hat{F} I_{\hat{\Gamma}_2, \lambda_{i+1}+1} = F I_{\Gamma_2, \lambda_{i+1}+1}$. Let R be the greatest common divisor of $F, \hat{F}$ and $G = F/R$, $\hat{G} = \hat{F}/R$. Then $I_{\Gamma_2, \lambda_{i+1}+1} \subseteq \hat{G}$, and since gin of the left-hand side is not generated by single element if $i < k - 1$, we must have $\hat{G}$ constant, and hence $I_{\Gamma_2, \lambda_{i+1}+1} = I_{\hat{\Gamma}_2, \lambda_{i+1}+1}$, which forces $\Gamma_2 = \hat{\Gamma}_2$, which shows uniqueness. $\qquad\square$

Remark. Something a bit stronger that Theorem 4.4 is true. If $\lambda_i = \lambda_{i+1} + 2$, then one draws the same conclusion if the generator of $\mathrm{in}(I)$ of degree $\lambda_i + i$ represents a generator of I, because then Corollary 2.29 applies.

Example. *Five points in the plane, 4 of them on a line.*

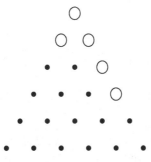

The ideal of 5 points in the plane, of which 4 are collinear, is $I = (LM, LM', F)$, where L is the line containing 4 of the points, M, M' are the linear forms vanishing at the fifth point, and F is a quartic cutting L at the 4 points and passing through the 5'th point. The rlex gin(I) $= (x_1^2, x_1x_2, x_2^4)$. So $\lambda_0 = 4$, $\lambda_1 = 1$, $k = 2$. Since $\lambda_0 > \lambda_1 + 2$, we conclude by Theorem 4.4 that any set of points with this rlex gin must consist of 5 points, 4 of them on a line.

Definition 4.5. *A set Γ of points is said to be in* **uniform position** *if every pair of subsets of Γ having the same number of points have the same Hilbert function.*

The relevance of this definition for us is:

Proposition 4.6 (Uniform Position Principle). Let X be an irreducible variety in $\mathbf{P}^r$ of dimension n. *Then for a general codimension-n linear space $\Lambda \subseteq \mathbf{P}^r$, $\Lambda \cap X$ is a set of points in uniform position.*

Proof. See pp. 109-113 of [A-C-G-H]. □

Definition 4.7. *A variety $X \subseteq \mathbf{P}^r$ is* **non-degenerate** *if X does not lie in any hyperplane.*

Corollary 4.8 (Gruson-Peskine). *If a set of points Γ in the plane with invariants $\lambda_0, \lambda_1, \ldots, \lambda_{k-1}$ is in uniform position, then*

$$\lambda_i - 1 \geq \lambda_{i+1} \geq \lambda_i - 2$$

for all $i = 0, 1, \ldots, k - 2$. In particular, this holds for a general hyperplane section of a reduced irreducible non-degenerate curve in $\mathbf{P}^3$.

Proof. If the inequalities are violated for i, we are in the situation of the theorem, so we have a decomposition $\Gamma = \Gamma_1 + \Gamma_2$. If the points Γ are in uniform position, then for any other decomposition $\Gamma = \hat{\Gamma}_1 + \hat{\Gamma}_2$ into subsets of the same sizes as Γ_1 and Γ_2, the Hilbert functions of $\hat{\Gamma}_1$ and $\hat{\Gamma}_2$ are the same as those as Γ_1, Γ_2 respectively. By Lemma 4.2, the same statement is true of the invariants, and hence by uniqueness of the decomposition $\Gamma = \Gamma_1 + \Gamma_2$, $\Gamma_1 = \hat{\Gamma}_1$. This is a contradiction to how $\hat{\Gamma}_1$ was chosen. □

Corollary 4.9. *If a set of points Γ in the plane with invariants $\lambda_0, \lambda_2, \ldots, \lambda_{k-1}$ is in uniform position, then*

$$d/k + (k - 1)/2 \leq \lambda_0 \leq d/k + k - 1,$$

with $\lambda_0 = d/k + k - 1$ holding if and only if Γ is a complete intersection of type $(k, d/k)$.

Proof. We have that $\lambda_0 \leq \lambda_1 + 2 \leq \lambda_2 + 4 \cdots \leq \lambda_{k-1} + 2(k - 1)$. Adding up all of these, we conclude that $k\lambda_0 \leq \sum_{i=0}^{k-1}(\lambda_i + 2i) = d + k(k - 1)$. So $\lambda_0 \leq \frac{d}{k} + k - 1$. We also have $\lambda_0 \geq \lambda_1 + 1 \geq \lambda_2 + 2 \geq \cdots \geq \lambda_{k-1} + k - 1$, from which it follows that $k\lambda_0 \geq d + k(k-1)/2$. Thus $\lambda_0 \geq \frac{d}{k} + \frac{k-1}{2}$. If equality holds in the upper inequality for λ_0, then $\lambda_i = d/k + k - 1 - 2i$. In particular, $\lambda_{k-1} = d/k - (k - 1)$. Thus I_Γ contains an element F of degree k and an element G of degree $d/k - (k - 1) + k - 1 = d/k$ which is not a multiple of F. By the remark following Theorem 4.4 (i.e. by the

Crystallization Principle), there can be no further generators of I_Γ, so $I_\Gamma = (F, G)$ and Γ is a complete intersection of type $(k, d/k)$.   □

Example 4.10. *8 points in $\mathbf{P}^2$ failing to impose independent conditions on cubics.* Given 8 points in the plane which fail to impose independent conditions on cubics with invariants $\lambda_0, \dots, \lambda_{k-1}$, we note that necessarily $\lambda_0 \geq 5$, as otherwise $h_{S/I}(3) = 8$. If $\lambda_1 \leq 2$, then λ_0 of the points lie on a line. If $\lambda_1 \geq 3$, then $\lambda_0 = 5, \lambda_1 = 3, k = 2$, and the points lie on a conic. Thus we obtain the classical result that either at least 5 of the points lie on a line or else all 8 lie on a conic. Note that in all of these cases, we can read off from the gin that the points impose independent conditions on cubics.

For our later considerations, certain notations estimates are helpful.

Definition 4.11. Let $i_m(\Gamma) = \dim(I_{\Gamma,m})$. For $m \geq 0$, let $d_m = i_m - i_{m-1}$. We will call $(d_k, d_{k+1}, \dots)$ the **difference sequence** of Γ. Here k is the degree of the lowest generator of I_Γ.

Example. *5 points in $\mathbf{P}^2$, 4 of them on a line.* For the example discussed earlier of 5 points in the plane, 4 of them on a line, we have $(d_2, d_3, d_4, \dots) = (2, 3, 5, 6, 7, \dots)$. These points are not in uniform position. Referring back to the diagram of this set of points, note that in general d_k is the number of black dots in the $k + 1$'st row of the diagram.

Proposition 4.12. Let Γ be d points in the plane in uniform position.
(1) $d_{m+1} \geq d_m + 2$ for all $\lambda_{k-1} + k - 1 \leq m < \lambda_0$;
(2) If $d_{m+1} = d_m + 2$ for some $\lambda_{k-1} + k - 1 \leq m < \lambda_0$, then I_Γ has no generators in degree $m + 1$;
(3) $d = \sum_{m=0}^{\lambda_0} (m + 1 - d_m)$.

Proof. (1) and (2) We note that $i_m = \binom{m+2}{2} - h_{S/I_\Gamma}(m)$, so $d_m = m + 1 - (h_{S/I_\Gamma}(m) - h_{S/I_\Gamma}(m-1))$. Using the formula for the Hilbert function at the beginning of this section, $h_{S/I_\Gamma}(m) - h_{S/I_\Gamma}(m-1) = \sum_{i=0}^{k-1} \min(m-i, \lambda_i) - \min(m-1-i, \lambda_i) = \text{card}\{i \mid m \leq \lambda_i + i\}$. Thus, $d_{m+1} - d_m = 1 + \text{card}\{i \mid m = \lambda_i + i\}$. If $d_{m+1} = d_m + 1$, then choose an i such that $\lambda_i + i > m > \lambda_{i+1} + i + 1$, this is possible as $\lambda_0 > m \geq \lambda_{k-1} + k - 1$. Then $\lambda_i \geq \lambda_{i+1} + 3$, violating the Ellia-Peskine Theorem. If $d_{m+1} = d_m + 2$, then as in the proof of the Ellia-Peskine Theorem, if there is a generator of I_Γ in degree $m + 1$, then the ideal generated by $I_{\Gamma,m}$ is m-regular, and we obtain that the points of Γ are not in uniform position.
(3) Since $d_m = i_m - i_{m-1}$ and $m + 1 = \binom{m+2}{2} - \binom{m+1}{2}$, we see that $\sum_{m=k}^{\lambda_0}(m + 1 - d_m) = h_{S/I_\Gamma}(\lambda_0) = d$.   □

Corollary 4.13. *Let Γ be d points and $\hat{\Gamma}$ be $\hat{d}$ points. If*

$$\sum_{m=0}^{k} d_m(\hat{\Gamma}) \geq \sum_{m=0}^{k} d_m(\Gamma)$$

for k sufficiently large, then $\hat{d} \leq d$.

Proof. This follows from (3). □

Proposition 4.14. *Let Γ be d points in the plane. Assume that I_Γ has g_m generators in degree m and s_m generators for the first syzygy module in degree m. Then $-d_{m-1} + 2d_m - d_{m+1} = s_{m+1} - g_{m+1}$.*

Proof. Let $0 \to \oplus_j S(-b_j) \to \oplus_i S(-a_i) \to I_\Gamma \to 0$ be the minimal free resolution of I_Γ. Now $i_m = \sum_i \binom{m-a_i+2}{2} - \sum_j \binom{m-b_j+2}{2}$. Taking differences, $d_m = \sum_i \max(m - a_i + 1, 0) - \sum_j \max(m - b_j + 1, 0)$. Now $d_m - d_{m-1} = \sum_{k \leq m} g_k - \sum_{k \leq m} s_k$. Finally, $(d_{m+1} - d_m) - (d_m - d_{m-1}) = g_{m+1} - s_{m+1}$, which is equivalent to the formula claimed.

An alternate proof: both sides of the formula are the same for I_Γ and $\mathrm{gin}(I_\Gamma)$. For the left-hand side, this is because they have the same Hilbert function, while for the right-hand side it follows from the Cancellation Theorem. For $\mathrm{gin}(I)$, $g_{m+1} = d_{m+1} - d_m - 1$ and $s_{m+1} = d_m - d_{m-1} - 1$, and the result follows by taking the difference of these two equations. □

Proposition 4.15 (Hilbert-Burch Theorem). *The minimal free resolution of $\mathrm{gin}(I_\Gamma)$ is*

$$0 \to \oplus_{i=0}^{k-1} S(-\lambda_i - i - 1) \to (\oplus_{i=0}^{k-1} S(-\lambda_i - i)) \oplus S(-k) \to \mathrm{gin}(I_\Gamma) \to 0.$$

The minimal free resolution of I_Γ is obtained from this by cancelling those $S(-\lambda_i - i)$ not corresponding to generators of I_Γ.

Proof. This is a direct consequence of Theorem 1.31 and the Cancellation Principle (Corollary 1.21). □

Remark. The preceding proposition illustrates the much more general Hilbert-Burch Theorem; see [E] for an algebraic treatment of this result.

Example 4.16. Γ be a set of d points in the plane in uniform position with $d_s \geq 1$ and $d_{s+1} \geq 4$. Then $d \leq s^2 + 1$.

For any $s \geq 1$ and any set of points Γ as above, we have that by (2) of Proposition 4.12, $d_{s+k} \geq \min(2k + 2, s + k + 1)$ for any $k \geq 1$. By (3), this implies that

$$d \leq 1 + 2 + \cdots + s + (s + 1 - 1) + (s + 2 - 4) + (s + 3 - 6) + \cdots + 1$$
$$= 1 + 2 + \cdots + s + s + (s - 2) + (s - 3) + \cdots + 1$$
$$= \frac{s(s+1)}{2} + s + \frac{(s-1)(s-2)}{2}$$
$$= s^2 + 1.$$

□

Proposition 4.17. *(Harris) Let Z be a set of points in the plane and $W \supset Z$ a complete intersection given by a regular sequence of degrees a, b. Then $m + 1 - d_m(W - Z) = d_{e+1-m}(W) - d_{e+1-m}(Z)$, where $e = a + b - 2$.*

Proof. Let H be a general hyperplane. The ring $R = S_H/I_W|_H$ is Gorenstein with socle in degree $a+b-2$, and thus the multiplication $R_m \times R_{a+b-2-m} \to R_{a+b-2} \cong \mathbf{C}$ is a perfect pairing. Now $I_{W-Z} = (I_W : I_Z)$, and thus if we let $\bar{I}_Z, \bar{I}_{W-Z}$ denote the images of I_Z, I_{W-Z} in $R = S/I_W$, we see that in R, $\bar{I}_{W-Z} = (0 : \bar{I}_Z)$. It follows that the codimension of $\bar{I}_{W-Z}$ in degree m equals the dimension of $\bar{I}_Z$ in degree $e - m$. Thus $i_{e-m}(Z) - i_{e-m}(W) = h_{S/IW-Z}(m)$. Taking differences, we see that $d_{e-m}(Z) - d_{e-m}(W) = h_{S/I_{W-z}}(m) - h_{S/IW-Z}(m+1) = m + 2 - d_{m+1}(W - Z)$. Replacing m by $m - 1$ gives the desired formula. $\qquad\square$

We now turn to the more challenging subject of curves in $\mathbf{P}^3$.

Let C be a reduced irreducible non-degenerate curve of degree d in $\mathbf{P}^3$, and let I_C be its homogeneous ideal. Let H be a hyperplane and $\Gamma = H \cap C$ the associated hyperplane section. By the exact sequence

$$0 \to \mathcal{I}_C(m - 1) \to \mathcal{I}_C(m) \to \mathcal{I}_\Gamma(m) \to 0,$$

we see that the restriction map $I_{C,m} \to I_{\Gamma,m}$ is surjective for $m >> 0$. Thus

$$I_\Gamma = (I_C|_H)^{\mathrm{sat}},$$

and the initial ideal of I_Γ in generic coordinates is constant on a Zariski open subset of hyperplanes. Let $\lambda_0, \lambda_1, \dots, \lambda_{s-1}$ be the invariants of Γ for a generic choice of H.

We now will look at the generic initial ideal $\mathrm{gin}(I_C)$ of I_C. Since I_C is saturated, no generator of $\mathrm{gin}(I_C)$ involves x_4. Let $\bar{h}$ be a general linear form on H, and L the corresponding line in H. Then we may write

$$\mathrm{gin}(I_C|_L) = (x_1^s, x_1^{s-1}x_2^{\mu_{s-1}}, \dots, x_1 x_2^{\mu_1}, x_2^{\mu_0}),$$

where $s > 0$ and $\mu_0 > \mu_1 > \cdots > \mu_{s-1} > 0$ because the ideal is Borel-fixed.

Definition 4.18. *For any pair of positive integers (i_1, i_2), let*

$$f_C(i_1, i_2) = \min\{i_3 \mid x_1^{i_1}x_2^{i_2}x_3^{i_3} \in \mathrm{gin}(I_C)\},$$

where we allow for the possibility $f_C(i_1, i_2) = \infty$.

Because $\mathrm{gin}(I_C)$ is Borel-fixed, if $f_C(i_1, i_2) < \infty$, then

$$f_C(i_1, i_2) > f_C(i_1, i_2 + 1) \geq f_C(i_1 + 1, i_2).$$

Thus if $0 < f_C(i_1, i_2) < \infty$, $x_1^{i_1}x_2^{i_2}x_3^{f_C(i_1,i_2)}$ is a generator of $\mathrm{gin}(I_C)$. We note that $f_C(i_1, i_2) = 0$ if and only if $x_1^{i_1}x_2^{i_2} \in \mathrm{gin}(I_C|_L)$. The generators of $\mathrm{gin}(I_C)$ are x_1^s, $x_1^{s-1}x_2^{\mu_{s-1}}$, ..., $x_1 x_2^{\mu_1}$, $x_2^{\mu_0}$ and $x_1^{i_1}x_2^{i_2}x_3^{f(i_1,i_2)}$ for $i_2 < \mu_{i_1}$ and $f_C(i_1, i_2) < \infty$. We note that if Γ has invariants $\lambda_0, \dots, \lambda_{k-1}$, then $s \geq k$ and $\mu_i \geq \lambda_i$ for all $0 \leq i < k$.

Recall that for $m >> 0$, $h_{S/I_C}(m) = 1 - g + dm$, where d is the **degree** of C and g is the **arithmetic genus** of C. For C reduced and irreducible, $g \geq 0$. For C

smooth, g is equal to the **geometric genus** of C, defined as $h^0(\Omega_C^1) = \frac{1}{2}h^1(C, \mathbf{R})$; if C is singular, we have only that g is $\geq$ the geometric genus, with the difference being computable from the singularities. In what follows, genus will always mean arithmetic genus.

Proposition 4.19. *Let C be a curve whose general hyperplane section has invariants $\lambda_0, \dots, \lambda_{s-1}$. The (arithmetic) genus of C is*

$$g(C) = 1 + \sum_{i=0}^{s-1}\left((i-1)\lambda_i + \binom{\lambda_i}{2}\right) - \sum_{f(i_1,i_2)<\infty} f(i_1, i_2).$$

Proof. For $m \gg 0$,

$$h_{S/\mathrm{gin}(I_C)}(m) = \sum_{i=0}^{s-1}\sum_{j=m+2-i-\lambda_i}^{m+1-i} j + \sum_{f(i_1,i_2)<\infty} f(i_1, i_2)$$

$$= \sum_{i=0}^{s-1}\left(\binom{m+2-i}{2} - \binom{m+2-i-\lambda_i}{2}\right) + \sum_{f(i_1,i_2)<\infty} f(i_1, i_2)$$

$$= \sum_{i=0}^{s-1}\left(\lambda_i(m+1-i) - \binom{\lambda_i}{2}\right) + \sum_{f(i_1,i_2)<\infty} f(i_1, i_2).$$

On the other hand, $h_{S/\mathrm{gin}(I_C)}(m) = 1 - g(C) + md$, where C has degree d. Since $\sum_i \lambda_i = d$, we may cancel and obtain

$$1 - g(C) = \sum_{i=0}^{s-1}\left(\lambda_i(1-i) - \binom{\lambda_i}{2}\right) + \sum_{f(i_1,i_2)<\infty} f(i_1, i_2),$$

from which the desired formula follows.   □

Example. *The general rational curve of degree 5 in $\mathbf{P}^3$.*

The rlex gin of the ideal I_C of the general rational quintic in $(x_1^3,\ x_1^2 x_2,\ x_1 x_2^2,\ x_2^3,\ x_1^2 x_3^2)$. For a general hyperplane H, by Theorem 2.30, $\mathrm{gin}(I_C|_H) = (x_1^3, x_1^2 x_2, x_1 x_2^2, x_2^3, x_1^2 x_3^2)$, and by the same theorem, if $\Gamma = C \cap H$, $\mathrm{gin}(I_\Gamma) = \mathrm{gin}((I_C|_H)^{\mathrm{sat}}) = (x_1^2, x_1 x_2^2, x_2^3)$, and thus the invariants are $\lambda_0 = 3$, $\lambda_1 = 2$, $k = 2$. The only finite but non-zero value of $f(i_1, i_2)$ is $f(2, 0) = 2$. The formula of Proposition 4.19 gives $g(C) = 0$.

If we want to draw a diagram representing the rlex gin of a curve in $\mathbf{P}^3$, the ideal way to do it would be to use a tetrahedron representing the monomials of some degree d. Unfortunately, 2-dimensional projections of these are hard to read, so I have come up with a schematic version.

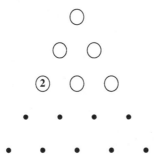

I am showing the monomials in x_1, x_2, x_3 in degree 4, with the same positions as for diagrams of ideals of points in $\mathbf{P}^2$. Referring to Definition 4.18, if $f_C(i, j) = 0$, we put a black dot ("X") in the spot corresponding to $x_1^i x_2^j$, i.e. in the $i + j$ row (counting the top as zero) and in the j'th spot from the left (counting the leftmost spot as 0). If $f_C(i, j) = k > 0$, we put a circle containing the number k in the spot i, j. If $f_C(i, j) = \infty$, we put a blank circle in the spot i, j. Once again, once we have a diagram whose bottom row is all black, saturation implies that we can reconstruct the rlex gin from the diagram.

Once we have this diagram, the diagram of the general hyperplane section Γ is obtained by replacing all circles with numbers in them with black dots.

Corollary 4.20. *Let C be a reduced irreducible non-degenerate curve whose generic hyperplane section has invariants $\lambda_0, \dots, \lambda_{s-1}$. Then*

$$g(C) \leq 1 + \sum_{i=0}^{s-1} \left((i-1)\lambda_i + \binom{\lambda_i}{2} \right).$$

Remark. In terms of diagrams, this says that the genus is obtained by counting 1 for each empty circle in the third row, 2 for each empty circle in the fourth row, etc.(ignore the first and second rows), and adding, and then subtracting the sum of the numbers inside circles. Try this for the diagram of the general rational curve of degree 5 and see that you get 0.

Example 4.21. *Constraints coming from $g \geq 0$.*

We cannot have a reduced irreducible curve C with $\mathrm{gin}(C) = (x_1^2, x_1x_2, x_2^3, x_2^2x_3)$. This is a Borel-fixed ideal, and $\mathrm{gin}(I_\Gamma)$ would be (x_1^2, x_1x_2, x_2^2), which corresponds to 3 points in uniform position. However, we would have $g = -1$, which is forbidden.

Definition 4.22. *We will call $\sum_{f_C(i_1,i_2)<\infty} f_C(i_1, i_2)$ the **number of sporadic zeros** of C. It is bounded above by $1 + \sum_{i=0}^{s-1}((i-1)\lambda_i + \binom{\lambda_i}{2})$. If it is zero, we will say that C has **no sporadic zeros**. If $x_1^{i_1} x_2^{i_2} x_3^{i_3} \notin \mathrm{gin}(I_C)$ but $f(i_1, i_2) < \infty$, we will say that $x_1^{i_1} x_2^{i_2} x_3^{i_3}$ is a **sporadic zero** of degree $i_1 + i_2 + i_3$.*

Example. *The general rational quintic.* Revisiting the general rational quintic in $\mathbf{P}^3$, we see that it has 2 sporadic zeros, at x_1^2 and $x_1^2 x_3$, so of degrees 2 and 3.

The name "sporadic zero" is best understood by referring to the diagram of the ideal of this curve. A circle with a number inside it represents that number of sporadic zeros.

Proposition 4.23. *The number of sporadic zeros in degree m is the dimension of the cokernel of the restriction map $I_{C,m}|_H \to I_{\Gamma,m}$, or equivalently the dimension of the kernel of multiplication by H, $H\colon H^1(\mathcal{I}_C(m-1)) \to H^1(\mathcal{I}_C(m))$.*

Proof. If $x_1^{i_1} x_2^{i_2} x_3^{i_3}$ is a sporadic zero of degree m, then it belongs to $\mathrm{gin}(I_\Gamma)$, but not to $\mathrm{gin}(I_C)$, and conversely, any such monomial is a sporadic zero. Noting that the dimension of the cokernel of the restriction map $\mathrm{gin}(I_C)|_{x_4=0} \to \mathrm{gin}(I_\Gamma)$ is the same as that of the ideals themselves, we obtain the first part. From the exact sequence

$$0 \to \mathcal{I}_C(m-1) \to \mathcal{I}_C(m) \to \mathcal{I}_\Gamma(m) \to 0,$$

we see that

$$I_{\Gamma,m}/\mathrm{im}(I_{C.m}) \cong \ker(H\colon H^1(\mathcal{I}_C(m-1)) \to H^1(\mathcal{I}_C(m))),$$

which implies the second part.

Remark. The relevance of the map $H\colon H^1(\mathcal{I}_C(m-1)) \to H^1(\mathcal{I}_C(m))$ to our considerations is inspired by the work of Strano, although of course he did not phrase things in terms of sporadic zeros.

Proposition 4.24. *Let C be a reduced irreducible curve in $\mathbf{P}^3$ with I_C having minimal free resolution $0 \to E_2 \to E_1 \to E_0 \to I_C \to 0$. Then $E_2 = 0$ if and only if C has no sporadic zeros.*

Proof. If $0 \to \mathcal{E}_2 \to \mathcal{E}_1 \to \mathcal{E}_0 \to \mathcal{I}_C \to 0$ is the sheafification of the exact sequence above, then for any m, we have that

$$h^1(\mathcal{I}_C(m)) \cong \ker(H^3(\mathcal{E}_2(m)) \to H^3(\mathcal{E}_1(m))).$$

If $\mathcal{E}_2 = \oplus_j \mathcal{O}_{\mathbf{P}^3}(-a_j)$, since the matrix defining the map from E_2 to E_1 has no constant term, we see that $H^1(\mathcal{I}_C(a_j - 4)) \neq 0$ for all j. Thus $E_2 = 0$ if and only if $H^1(\mathcal{I}_C(m)) = 0$ for all $m \geq 0$.

By the preceding Proposition, C has no sporadic zeros if and only if the map

$$H\colon H^1(\mathcal{I}_C(m-1)) \to H^1(\mathcal{I}_C(m))$$

is injective for all $m \geq 1$. Since $H^1(\mathcal{I}_C(m)) = 0$ for $m >> 0$, this implies that $H^1(\mathcal{I}_C(m)) = 0$ for all $m \geq 0$. $\qquad\square$

Corollary 4.25. *A reduced irreducible curve C has no sporadic zeros if and only if I_C is equal to the ideal of $m \times m$ minors of an $m \times (m+1)$ polynomial matrix which drops rank in codimension 2.*

Proof. If $E_2 = 0$, then the matrix giving the map $E_1 \to E_0$ is $m \times (m+1)$ for some m, and by Fitting ideals the ideal sheaf where this matrix drops rank is $\mathcal{I}_C$. Conversely, if such a matrix drops rank in the correct codimension (2 in this case), then its minimal free resolution is an Eagon-Northcott complex, which is just the resolution $\mathcal{E}_1 \to \mathcal{E}_0 \to \mathcal{I}_C \to 0$. Finally, by the long exact sequence for cohomology one sees that $\mathcal{I}_C$ is generated by these minors.   □

Remark.   Once again, this illustrates the Hilbert-Burch Theorem.

Problem 4.26. *For reduced irreducible curves C, what constraints are there on the sporadic zeros, i.e. on the numbers $f_C(i_1, i_2)$?*

This problem is the fundamental unsolved problem in studying generic initial ideals of curves in $\mathbf{P}^3$. We will give many examples of general classes of constraints, of which the inequality on the number of sporadic zeros is the first.

Example 4.27. *The generic initial ideal contains more information than the Hilbert function.*

The ideals $I = (x_1^3, x_1^2 x_2, x_1 x_2^2, x_2^3, x_1^2 x_3^2)$ and $J = (x_1^3, x_1^2 x_2, x_1 x_2^2, x_2^4, x_1^2 x_3, x_2^3 x_3)$ are Borel-fixed monomial ideals, hence equal their own generic initial ideals. They have the same Hilbert function, but different generic initial ideals. I am not claiming that they are both generic initial ideals of reduced irreducible curves.

We now want to discuss generic initial ideals for reduced, irreducible, non-degenerate curves of low degree.

Example 4.28. *Curves of degree 3.* The only uniform position candidate for $\mathrm{gin}(I_\Gamma)$ is $(x_1^2, x_1 x_2, x_2^2)$. The genus bound implies that there are no sporadic zeros. Thus $\mathrm{gin}(I_C) = (x_1^2, x_1 x_2, x_2^2)$.

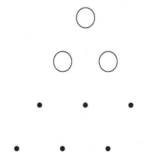

Note that the minimal free resolution of $\mathrm{gin}(I_C)$ is $S^2(-3) \to S^3(-2) \to \mathrm{gin}(I_C) \to 0$. No cancellation is possible, so this is the minimal free resolution of I_C by the Cancellation Theorem. Thus I_C is generated by the 2×2 minors of a 2×3 matrix of linear forms. This is a rational normal curve (i.e. a twisted cubic), which indeed has the generic initial ideal claimed.

Example 4.29. *Curves of degree 4.* The only uniform position candidate for $\text{gin}(I_\Gamma)$ is $(x_1^2, x_1 x_2, x_2^3)$. There is thus at most one sporadic zero. If there are no sporadic zeros, then I_C contains two quadrics which cannot have a common factor, and hence C is a complete intersection of type $(2,2)$. Since $\text{gin}(I_C)$ is Borel-fixed, the only possibilities if there is one sporadic zero are $\text{gin}(I_C) = (x_1^2, x_1 x_2, x_2^4, x_2^3 x_3)$

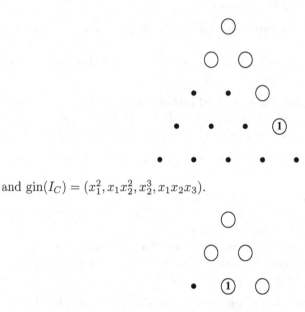

and $\text{gin}(I_C) = (x_1^2, x_1 x_2^2, x_2^3, x_1 x_2 x_3)$.

However, the first one of these has both generators of I_Γ belonging to the image of I_C, and hence the restriction map $I_C \to I_\Gamma$ is surjective, contradicting the existence of a sporadic zero. The second case indeed occurs as the ideal of a general rational quartic, i.e. as a curve of type $(1,3)$ on a smooth quadric.

An essential problem in eliminating false candidates for generic initial ideals of reduced irreducible curves is that, once we have a saturated ideal whose general hyperplane section is a collection of points in uniform position, we might have the ideal of a reduced irreducible curve union a finite set of points. None of our methods thus far have the potential to eliminate such cases. A beautiful idea of Strano allows one to do this.

Let C be a reduced irreducible curve and H a general hyperplane, with $\Gamma = C \cap H$ the general hyperplane section. We will denote $H^k \colon H^1(\mathcal{I}_C(m-k)) \to H^1(\mathcal{I}_C(m))$ the obvious multiplication map.

Definition 4.30. *We will say that an element $\alpha \in H^1(\mathcal{I}_C(m))$ is **primitive** if it does not lie in the image of H.*

Remark. Equivalently, α is primitive if and only if its restriction $\alpha_\Gamma \in H^1(\mathcal{I}_\Gamma(m))$ is non-zero; this is an easy consequence of the restriction sequence.

Proposition 4.31. *(Strano [Str]) Let $\alpha \in H^1(\mathcal{I}_C(m-k))$ be an element of $\ker(H^k)$. Then $\alpha_\Gamma \in (0 : \mathfrak{m}_H^k)$, i.e. α_Γ is annihilated by all polynomials of degree k on H.*

Proof. Since H is a general hyperplane, we may assume that the dimension of $\ker(H^k)$ is constant on a Zariski open set of hyperplanes containing H. If $t_1, \ldots, t_n$ are homogeneous coordinates on the dual projective space of hyperplanes, and we write $H = \sum_{i=1}^n t_i x_i$, then we may extend α to a homogeneously varying element of $H^1(\mathcal{I}_C(m-k))$. If we now differentiate the equation $H^k \alpha \sim 0$ with respect to $\frac{\partial^k}{\partial t_{i_1} \cdots \partial t_{i_k}}$, we obtain $x_{i_1} \cdots x_{i_k} \alpha + H \beta_{i_1 \cdots i_k} \sim 0$ for some $\beta \in H^1(\mathcal{I}_C(m-1))$. Restricting to Γ, we obtain $x_{i_1} \cdots x_{i_n} \alpha_\Gamma \sim 0$. $\qquad\square$

Proposition 4.32. *Let the minimal free resolution of I_Γ be*

$$0 \to \oplus_j S(-b_j) \to \oplus_i S(-a_i) \to I_\Gamma \to 0.$$

Then there exists a non-zero element $\alpha \in H^1(\mathcal{I}_\Gamma(m-k))$ which is annihilated by all polynomials of degree k if and only if there is an element of $\oplus_{b_j \le m+2} H^2(\mathcal{O}_{\mathbf{P}^2}(m-k-b_j))$ that maps to zero in $\oplus_i H^2(\mathcal{O}_{\mathbf{P}^2}(m-k-a_i))$ under the natural map induced by the resolution.

Proof. The sheafification of the resolution of I_Γ is $0 \to \oplus_j \mathcal{O}_{\mathbf{P}^2}(-b_j) \to \oplus_i \mathcal{O}_{\mathbf{P}^2}(-a_i) \to \mathcal{I}_\Gamma \to 0$. From this, we obtain that $H^1(\mathcal{I}_\Gamma(m-k)) \cong \ker(\oplus_j H^2(\mathcal{O}_{\mathbf{P}^2}(m-k-b_j)) \to \oplus_i H^2(\mathcal{O}_{\mathbf{P}^2}(m-k-a_i)))$. Now an element of $H^2(\mathcal{O}_{\mathbf{P}^2}(q))$ is annihilated by all polynomials of degree k if and only if $q \ge -k-2$. Thus our element of the kernel must belong to $\oplus_{b_j \le m+2} H^2(\mathcal{O}_{\mathbf{P}^2}(m-k-b_j))$.

Theorem 4.33. *(Strano) If C is a reduced irreducible curve and has a sporadic zero in degree m, then I_Γ has a syzygy in degree $\le m+2$.*

Proof. If C has a sporadic zero in degree m, then by Proposition 4.23 there is a non-zero $\alpha \in \ker(H : H^1(\mathcal{I}_C(m-1)) \to H^1(\mathcal{I}_C(m)))$. Continuing inductively to see whether α and its preimages belong to $\operatorname{im}(H)$, we obtain a primitive element $\beta \in H^1(\mathcal{I}_C(m-k))$ in the kernel of H^k. Since β is primitive, β_Γ is non-zero, and by Proposition 4.31 it is annihilated by all polynomials of degree k. By Proposition 4.32, $\beta_\Gamma \in \oplus_{b_j \le m+2} H^2(\mathcal{O}_{\mathbf{P}^2}(m-k-b_j))$, and hence $b_j \le m+2$ for some j. $\quad\square$

Corollary 4.35. *(Strano [Str]) If C is a reduced irreducible curve whose hyperplane section has the Hilbert function of a complete intersection of type (m, n), where $m, n > 2$, then C is a complete intersection of type (m, n).*

Proof. Since the Hilbert function determines the generic initial ideal for a set of points in the plane, we know that $\operatorname{gin}(I_\Gamma)$ is the same as that of a complete intersection of type (m, n). By Corollary 4.9, this implies that Γ is a complete intersection of type (m, n). Thus the only syzygy of I_Γ is of degree $m + n$. So C has no sporadic zeros in degrees $\le m + n - 3$. In particular, the restriction map $I_C \to I_\Gamma$ is surjective in degrees m and n (since $m + n - 3 \ge m, n$ by hypothesis), and hence the restriction map $I_C \to I_\Gamma$ is surjective. If F, G are elements of I_C of degrees m, n going to the generators of I_Γ, then $I_C = (F, G)$ and C is a complete intersection of type (m, n). $\square$

Corollary 4.36 (Laudal's Lemma). *(Strano) If C is a reduced irreducible curve of degree d such that $I_{\Gamma,s} \neq 0$ and $I_{C,s} = 0$, then $d \leq s^2 + 1$.*

Proof. We may assume that $I_{\Gamma,s-1} = 0$, as otherwise replace s by $s - 1$. Since $I_{C,s} \to I_{\Gamma,s}$ is not surjective, C has a sporadic zero in degree s, and hence I_Γ has a syzygy in degree $\leq s + 2$. If d_m are the difference sequence for Γ, then $d_s \geq 1$, since $I_{\Gamma,s} \neq 0$. If $d_s = 2$, then since by uniform position the two independent elements of $I_{\Gamma,s}$ cannot have a common factor, so there must be a generator of I_Γ in degree $s + 1$ in order for there to be a syzygy by degree $s + 2$. This forces $d_{s+1} \geq 4$. If $d_s = 1$, then there must be at least 2 generators of I_Γ in degree $s + 1$, as otherwise the syzygy of degree $\leq s + 2$ would give a common factor and violate uniform position. So in this case, $d_{s+1} \geq 4$. If $d_s \geq 3$, then by (1) of Proposition 4.12, $d_{s+1} \geq 5$. So in all cases, we conclude $d_s \geq 1$ and $d_{s+1} \geq 4$. Now by Example 4.16, this implies $d \leq s^2 + 1$. $\square$

In proving Laudal's Lemma, we are far from using the full strength of Strano's method. For example, we have:

Proposition 4.37. *Let C be a reduced irreducible curve with $I_{C,s} = 0$ having difference sequence for Γ given by $(d_s, d_{s+1}, d_{s+2}, \dots) = (1, 4, 6, \dots)$. Then C has no sporadic zeros in degree $s + 1$.*

Proof. By Proposition 4.14, the minimal free resolution of I_Γ must be of the form

$$0 \to S(-(s+2)) \oplus (\oplus_j S(-b_j)) \to S(-s) \oplus S^2(-(s+1)) \oplus (\oplus_i S(-a_i)) \to I_\Gamma \to 0,$$

where $a_i > s + 1$ for all i and $b_j > s + 3$ for all j. Let us introduce the notations $N_k^m = \{\alpha \in H^1(\mathcal{I}_C(k)) \mid H^{m-k}\alpha = 0\}$ and $\bar{N}_k^m = \{\alpha \in H^1(\mathcal{I}_\Gamma(k)) \mid P\alpha = 0 \text{ for all } P \in H^0(\mathcal{O}_H(m-k))\}$. There is an exact sequence

$$0 \to N_{k-1}^k \to N_{k-1}^m \to N_k^m \to \bar{N}_k^m.$$

From the resolution for I_Γ, we see that $\bar{N}_{s-1}^s = \bar{N}_{s-1}^{s+1} \cong \mathbf{C}$ and $\bar{N}_{s-2}^s = \bar{N}_{s-2}^{s+1} \cong \mathbf{C}$, while $\bar{N}_k^s = N_k^{s+1} = 0$ for $k \neq s - 1, s - 2$. Also, $\bar{N}_k^m = 0$ for $m < s$. From the exact sequence above, we conclude that $\dim(N_{s-2}^{s+1}) \leq 1$. If $\alpha \in N_{s-2}^{s+1}$ is non-zero, then because N_{s-2}^{s+1} has dimension 1, the two linear forms which annihilate α_Γ lift to two linear forms l_1, l_2 that annihilate α. If we write $l_i\alpha =_C P_i$, where P_i is a homogeneous polynomial of degree $s - 1$, then $l_1P_2 - l_2P_1 =_C 0$. Since $I_{C,s} = 0$ by hypothesis, we conclude that there exists a polynomial U of degree $s - 2$ such that $P_i = l_iU$ for $i = 1, 2$. Now $l_i(\alpha - U) =_C 0$, and this forces $\alpha - U =_C 0$, and then α is zero in $H^1(\mathcal{I}_C(s-2))$. Thus $N_{s-2}^{s+1} = 0$.

Now by the exact sequence above, $\dim(N_{s-1}^{s+1}) \leq 1$. However, the sporadic zero of degree s gives an element of this space, and hence there is room for no other element. Any sporadic zero of degree $s + 1$ gives a primitive element of some N_k^{s+1}, and hence there cannot be one. $\square$

We now introduce some standard notation. On $\mathbf{P}^n$, if $V = H^0(\mathcal{O}_{\mathbf{P}^n}(1))$, then define a bundle M by $0 \to M \to V \otimes \mathcal{O}_{\mathbf{P}^n} \to \mathcal{O}_{\mathbf{P}^n}(1) \to 0$. One may identify $\wedge^2 V \cong H^0(M(1))$. On $\mathbf{P}^3$, an element of $\wedge^2 V$ of rank 4 gives a surjection $M^* \to \mathcal{O}_{\mathbf{P}^3}(1) \to 0$ whose kernel is a rank 2 vector bundle E. A bundle arising in this way is called a **null-correlation bundle**.

Let $\sigma \in H^0(\mathbf{P}^3, E(s))$, where E is a null-correlation bundle. For $s \geq 2$, a general σ defines a smooth curve C. We now have a resolution $0 \to \wedge^2 E^*(-2s) \to E^*(-s) \to \mathcal{I}_C \to 0$. If H is a general hyperplane and $\Gamma = C \cap H$, we also have $0 \to \wedge^2 E^*|_H(-2s) \to E^*|_H(-s) \to \mathcal{I}_\Gamma \to 0$. From this, we see that $h^0(\mathcal{I}_C(s)) = 0$ and $h^0(\mathcal{I}_\Gamma(s)) = 1$. A Chern class computation shows that $\deg(C) = s^2 + 1$.

Corollary 4.38. *(Gruson-Peskine, Strano) If C is a reduced irreducible curve of degree $s^2 + 1$ such that $I_{C,s} = 0$ and $I_{\Gamma,s} \neq 0$, then if $s \geq 3$, C is a curve arising from a null-correlation bundle.*

Proof. If $i_{\Gamma,s} \geq 2$, then the difference sequence $d_s, d_{s+1}, \ldots$ for Γ is at least $2, 4, 6, \ldots$, and this gives degree $\geq s^2 + 2$. So $i_{\Gamma,s} = 1$, and by Proposition 4.12, $i_{\Gamma,s+1} \geq 4$. Hence the difference sequence is at least $1, 4, 6, \ldots$, and since this gives degree $s^2 + 1$, we must have equality. We use the fact $s \geq 3$ to conclude that $d_{s+2} = 6$, for which we need that $6 \leq s+3$. Now C has one sporadic zero in degree s and none in degree $s+1$. However, since I_Γ is generated in degrees $\leq s+1$ by Proposition 4.12, this implies that C has no sporadic zeros except for the one in degree s, as $I_{C,k} \to I_{\Gamma,k}$ is surjective if $k \geq s + 1$. Thus $H \colon H^1(\mathcal{I}_C(k)) \to H^1(\mathcal{I}_C(k+1))$ is injective for $k \geq s$, and hence $H^1(\mathcal{I}_C(k)) = 0$ for $k \geq s$. Using the fact there is one sporadic zero, we conclude that $H^1(\mathcal{I}_C(s-1))$ has dimension 1, and thus $\alpha \in \ker(H)$ may be taken to be constant. Since $F =_C H\alpha$, we see that F has degree 1 in t, and hence the $x_i \frac{\partial F}{\partial t_j} - x_j \frac{\partial F}{\partial t_i}$ are constant, and so is the linear relation between them. This gives an element of $\wedge^2 V$, and one sees that $\bar{d}F$ is a section of a twist of a null-correlation bundle, and the result follows. $\square$

Comments for students:

This chapter reverses the historical order. The fundamental result was that of Gruson-Peskine, which was then improved to the result of Ellia-Peskine [E-P], [D]. I found this latter result independently, and it is interesting to compare the initial ideal proof with theirs. This section illustrates the strength of the generic initial ideal approach in proving geometric results. Further evidence is found in Michele Cook's powerful generalization [C] of the result of Gruson-Peskine on connectedness of the numerical character [G-P] and the work of Braun and Floystad bounding the degree of smooth surfaces in $\mathbf{P}^4$ which are not of general type.

The topic of curves in $\mathbf{P}^3$ is a very large area with many beautiful results, and I hesitate to attempt to give a list of the principal papers. There have been many fruitful points of view other than the one presented here. I have been most strongly influenced by the work of Gruson and Peskine, and by Strano. Floystad's work on higher gins [F] carries forward the point of view in this chapter; see also the upcoming Berkeley thesis of Rich Liebling [Li].

There is some overlap in this chapter with the approach of M. Amasaki [A].

Algebraists seeking a foothold in this chapter should look at the treatment of the Hilbert-Burch Theorem in Eisenbud's book (pp. 501-502 of [E]). Corollary 4.35 to Strano's Theorem has a beautiful algebraic proof by Huneke and Ulrich [H-U]; perhaps some of the other results in this chapter can be done and perhaps improved by building on their work.

A highly interesting open problem is to generalize the Ellia-Peskine Theorem to points in $\mathbf{P}^3$ and higher.

The diagrams used in this chapter are my invention, although they grew out of Dave Bayer's thesis. I use them much more often than is evident in these notes. The illustrations were created by a Mathematica routine I wrote (which I am happy to share), which produces the diagram if you give it the gin.

5. Gins in the Exterior Algebra

Some of the results discussed in this section have been found independently by Aramova, Herzog, and Hibi [A-H-H]; indeed, it is my impression that they have gone substantially beyond what I present here.

Let V be a vector space spanned by $x_1, \ldots, x_n$ and $E = \wedge^* V = \oplus_k \wedge^k V$ the exterior algebra in V. If $I = \oplus_k I_k$ is a homogeneous ideal in $\wedge^* V$, initial terms, in(I), and gin(I) are defined exactly as before. We do not need to worry about bounding the regularity, since every ideal dies by degree n.

Most of the important theorems for ideals in the symmetric algebra go through for the exterior algebra, but in many cases the results are nicer. For example, Macaulay's bound is different, reflecting the slower growth of the exterior algebra. The bound is realized by the lex segment ideals, and Gotzmann's Persistence Theorem goes through unchanged for this new bound.

Another nice feature of the exterior algebra is that all of the monomial orders, lex and rlex, agree, because all variables appear to at most the first power.

A finitely-generated graded module $M = \oplus_k M_k$ over $\wedge^* V$ has a minimal free resolution of the form

$$\cdots F_3 \to F_2 \to F_1 \to F_0 \to M \to 0,$$

where

$$F_p = \oplus_i E(-q)^{b_{pq}},$$

where for any p only finitely many of the b_{pq} are non-zero. What is different is that this resolution need not be bounded on the left, i.e. F_p may be non-zero for arbitrarily large p. For any M as above, there is an analogue of the Koszul complex $B_k^\bullet$ with $B_k^p = S^p V \otimes M_{k-p}$ and maps given by the composition

$$S^p V \otimes M_{k-p} \to S^{p-1} V \otimes V \otimes M_{k-p} \to S^{p-1} V \otimes M_{k-p+1}.$$

An easy argument shows that

$$b_{pq} = \dim(H^p(B_q^\bullet)).$$

By analogy with the symmetric case, we will call $\max\{q-p \mid b_{pq} \neq 0\}$ the **regularity** of M and denote it $\mathrm{reg}(M)$. For an ideal $I \subseteq E$, we have that $\mathrm{reg}(I) \leq n$.

The basic theory for exterior ideals is essentially the same as that for symmetric ideals, with occasional subtle differences which I will point out as we go along. The same argument proves the existence of $\mathrm{gin}(I)$ and that it is Borel-fixed. The ideal $\mathrm{in}(I)$ is a flat deformation of I as before, and the minimal free resolution of I is obtained from the minimal free resolution of $\mathrm{in}(I)$ by cancelling some twists of E which occur in adjacent slots of the resolution. If we define $(I : H) = \{P \mid H \wedge P \in I\}/(I \cap \{H \wedge Q\})$, then the same argument shows that $\mathrm{gin}(I : H) = (\mathrm{gin}(I) : x_n)$.

The minimal free resolution of a Borel-fixed monomial ideal in E is described by saying that the leading terms of syzygies involving a monomial x^I are precisely $x_i \otimes x^I$ where $i \leq \max\{k \mid k \text{ appears in } I\}$, and that for p'th syzygies the leading terms are exactly $x_{i_p} \otimes x_{i_{p-1}} \otimes \cdot \otimes x_{i_1} \otimes x^I$, where $i_p \leq i_{p-1} \leq \cdots \leq i_1 \leq \max\{k \mid i_k \neq 0\}$. Note the small but important difference from the symmetric case lies in the $\leq$ signs replacing $<$ – this is why resolutions of symmetric modules end after n steps while resolutions of exterior modules just keep growing. Another consequence of this resolution is that we have once again that for any homogeneous ideal I, $\mathrm{reg}(I)$ is equal to the maximal degree of a generator of $\mathrm{gin}(I)$, and the minimal free resolution of I is obtained from the minimal free resolution of the Borel-fixed ideal $\mathrm{gin}(I)$ by cancellation of some adjacent terms with the same twist.

For any integer $c > 0$, we let

$$c = \binom{k_d}{d} + \binom{k_{d-1}}{d-1} + \cdots + \binom{k_\delta}{\delta}$$

be its Macaulay representation. We will denote

$$c^{[d]} = \binom{k_d}{d+1} + \binom{k_{d-1}}{d} + \cdots + \binom{k_\delta}{\delta+1}.$$

Theorem 5.1 (Macaulay's Bound for Exterior Ideals, Kruskal-Katona Theorem). Let $I \subseteq \wedge^*V$ be an ideal with the Hilbert function of $\wedge^*V/I$ denoted by $h(d)$. Then

$$h(d+1) \leq h(d)^{[d]}.$$

Remark. Theorem 5.1 is due to Kruskal-Katona and Schützenberger.

Theorem 5.2 (Hyperplane Restriction Theorem). Let $I \subseteq \wedge^*V$ be an ideal with the Hilbert function of $\wedge^*V/I$ denoted by $h(d)$. Let H be a general hyperplane and I_H denote the restriction of I to H and $h_H(d)$ the Hilbert function of $\wedge^*(V/H)/I_H$. Then

$$h_H(d) \leq h(d)_{<d>}.$$

Proposition 5.3. *Let I be a homogeneous exterior ideal with Hilbert function h. Then for all $d \geq 1$,*

$$h(d) \leq \left(\sum_{j=d-1}^{d} h(j) \right)_{<d>}.$$

Proof. We will prove the foregoing three results at once, doing induction on both d and n. Replacing I by $\mathrm{gin}(I)$ in all cases, we may reduce to the case when I is a Borel-fixed monomial ideal. We let $E, \bar{E}$ denote $\wedge^* V$ and $\wedge^*(V/x_n)$.

Proposition for $d, n-1$ implies Hyperplane Restriction for d, n: We may write $I_d = J_d + x_n J_{d-1}$. Because I is Borel-fixed, $\bar{\mathfrak{m}} J_{d-1} \subseteq J_d$. Let J be the homogeneous ideal in $\bar{E}$ generated by J_{d-1} and J_d. Now $h_{E/I}(d) = h_{\bar{E}/J}(d) + h_{\bar{E}/J}(d-1)$, so by the proposition $h_{\bar{E}/J}(d) \leq h_{E/I}(d)_{<d>}$, but $J_d = I_d|_{x_n}$.

Hyperplane Restriction for d, n and Macaulay's Estimate for $d, n-1$ imply Macaulay's Estimate for d, n: As above, we write $I_d = J_d + x_n J_{d-1}$. Now $I_{d+1} = J_{d+1} + x_n J_d$, and thus $h_{E/I}(d+1) = h_{\bar{E}/J}(d+1) + h_{\bar{E}/J}(d)$. If $h_{E/I}(d)$ has Macaulay representation $k_d, \ldots, k_\delta$, then $h_{\bar{E}/J}(d) \leq \binom{k_d - 1}{d} + \cdots + \binom{k_\delta - 1}{\delta}$. It follows that $h_{\bar{E}/J}(d+1) \leq \binom{k_d - 1}{d+1} + \cdots + \binom{k_\delta - 1}{\delta+1}$, and adding these together gives $h_{E/I}(d+1) \leq \binom{k_d}{d+1} + \cdots + \binom{k_\delta}{\delta+1}$ as desired.

Macaulay's Estimate for $d-1, n$ implies the Proposition for d, n: Let $h(d-1) + h(d)$ have Macaulay representation $u_d, \ldots, u_\mu$. Then $(h(d-1) + h(d))_{<d>} = \binom{u_d - 1}{d} + \cdots + \binom{u_\mu - 1}{\mu}$. If $h(d)$ is strictly larger than this, then, subtracting, $h(d-1) < \binom{u_d - 1}{d-1} + \cdots + \binom{u_\mu - 1}{\mu-1}$. Now using Macaulay's estimate, $h(d) \leq \binom{u_d - 1}{d} + \cdots + \binom{u_\mu - 1}{\mu}$, contradicting our assumption that $h(d)$ was larger than this. $\qquad\square$

Theorem 5.4 (Gotzmann's Persistence Theorem for the Exterior Algebra).
Let I be a homogeneous ideal in the exterior algebra generated in degrees $\leq d+1$. If in Macaulay's estimate,

$$h(d+1) = h(d)^{[d]}, \quad \text{then } \mathrm{reg}(I) \leq d \text{ and } h(k+1) = h(k)^{[k]}$$

for all $k \geq d$.

Proof. By comparing I with $\mathrm{gin}(I)$, we note that the ideal generated by $\mathrm{gin}(I)_d$ must achieve Macaulay's bound in degree d, and thus $\mathrm{gin}(I)$ cannot have a generator in degree $d+1$. By the Crystallization Principle, this means that $\mathrm{gin}(I)$ is generated in degree $\leq d$ and therefore $\mathrm{reg}(I) \leq d$. It further means that the ideal generated by I_d and $\mathrm{gin}(I)$ have the same Hilbert function, and so we may replace I by $\mathrm{gin}(I)$.

We now write $I_d = J_d + x_n J_{d-1}$, where J_d, J_{d-1} involve only $x_1, \ldots, x_{n-1}$. Let J be the homogeneous ideal generated by J_{d-1} and J_d in $\bar{E}$. Since $\bar{\mathfrak{m}} J_{d-1} \subseteq J_d$, we have that $I_{d+1} = J_{d+1} + x_n J_d$, and thus $h_{E/I}(d+1) = h_{\bar{E}/J}(d+1) + h_{\bar{E}/J}(d)$ and $h_{E/I}(d) = h_{\bar{E}/J}(d) + h_{\bar{E}/J}(d-1)$. If $h_{E/I}(d)$ has Macaulay representation $u_d, \ldots, u_\delta$, and $h_{\bar{E}/J}(d-1)$ has Macaulay representation $l_{d-1}, \ldots, l_\nu$ and $h_{\bar{E}/J}(d)$ has Macaulay representation $k_d, \ldots, k_\mu$, then $\binom{u_d}{d+1} + \cdots + \binom{u_\delta}{\delta+1} = \binom{k_d}{d} + \cdots +$

$\binom{k_\mu}{\mu} + \binom{k_d}{d+1} + \cdots + \binom{k_\mu}{\mu+1} = \binom{k_d+1}{d+1} + \cdots + \binom{k_\mu+1}{\mu+1} - \epsilon$, where J fails to achieve Macaulay's bound at degree d by $\epsilon \geq 0$. We know that $\binom{k_d}{d} + \cdots + \binom{k_\mu}{\mu} \leq \binom{u_d-1}{d} + \cdots + \binom{u_\delta-1}{\delta}$ by the Hyperplane Restriction theorem and monotonicity of $^{<d>}$, and thus $\binom{u_d}{d+1} + \cdots + \binom{u_\delta}{\delta+1} \leq \binom{u_d}{d+1} + \cdots + \binom{u_\delta}{\delta+1} - \epsilon$, from which we conclude that $\epsilon = 0$ and $k_i = u_i - 1$ for all i. Thus J achieves Macaulay's bound in degree d, and thus by induction, J achieves Macaulay's bound for all degrees $\geq d$. However, $I_{k+1} = J_{k+1} + x_n J_k$ for all $k \geq d$, and now we can read off that $h_{E/I}(k+1) = \binom{u_d}{k+1} + \cdots + \binom{u_\delta}{k+1+\delta-d}$ for all $k \geq d$, achieving Macaulay's bound for all these degrees.   □

There are several reasons why the exterior algebra is important even to commutative algebraists (Most of my friends who are commutative algebraists have at one time or another expressed the sentiment that the exterior algebra "might as well be commutative"). One reason is that minimal free resolutions of symmetric modules are computed by Koszul cohomology, and the properties of the exterior algebra can be used to deduce purely commutative consequences – an encouraging example is the solution of the Eisenbud-Koh-Stillman conjecture [G4]. Another reason is the following well-known construction, which I learned from Bernd Sturmfels:

Construction 5.5 (Stanley-Reisner Algebra of a Simplicial Complex). There is a 1-1 correspondence between simplicial complexes with n vertices and monomial ideals in $E = \wedge^* V$ where V has dimension n. Given a simplicial complex σ in vertices $e_1, \ldots, e_n$, we include $x^K \in I_\sigma$ if and only if the simplex e^K does not belong to σ. If e^K does not belong to σ, then no simplex containing e^K can belong, so I_σ is an ideal. Conversely, the same procedure, given a monomial ideal I, leads to a simplicial complex σ_I. If $f_k(\sigma)$ is the number of simplices $e^K \in \sigma$ such that $|K| = k$, then $f_k(\sigma) = h_{E/I}(k)$ by construction. This allows one to reinterpret Macaulay's bound on the growth of exterior ideals (=Kruskal-Katona Theorem) as an inequality on the growth of the number of faces of dimension k of a simplicial complex.

Example. *A square.*

Let σ be a square with vertices e_1, e_2, e_3, e_4 and edges $e_1 e_2, e_2 e_3, e_3 e_4, e_4 e_1$. The ideal I_σ is the ideal generated by $x_1 \wedge x_3, x_2 \wedge x_4$.

Definition 5.6. *Let h be a linear form. For any ideal $I \subseteq E$, there is a complex $W^\bullet(E/I, h)$ given by*

$$(E/I)_0 \xrightarrow{\wedge h} (E/I)_1 \xrightarrow{\wedge h} \cdots \xrightarrow{\wedge h} (E/I)_n.$$

Proposition 5.7. *Let $I \subseteq E$ be a monomial ideal. If $h = \sum_i t_i x_i$, $h' = \sum_i s_i x_i$ be two linear forms such that $t_i \neq 0$, $s_i \neq 0$ for all $1 \leq i \leq n$. Then $H^k(W^\bullet(E/I, h)) \cong H^k(W^\bullet(E/I, h'))$ for all $k \geq 0$.*

Proof. The map $x_i \mapsto (s_i/t_i)x_i$ preserves the monomial ideal I and gives an isomorphism of complexes $W^\bullet(E/I, h) \to W^\bullet(E/I, h')$ which induces an isomorphism on the cohomology groups.   □

Definition 5.8. *For a monomial ideal I and $h = \sum_{i=1}^{n} x_i$, we denote by $H^k(E/I)$ the module $H^k(W^\bullet(E/I, h))$.*

Definition 5.9. *Given a simplicial complex σ on the set of vertices $\{1, 2, \ldots, n\}$, let σ_k be the vector space spanned by the k-simplices in σ, $\{e^K \in \sigma \mid |K| = k+1\}$. The complex $\sigma_\bullet$ is defined by*

$$\sigma_n \xrightarrow{\partial} \sigma_{n-1} \xrightarrow{\partial} \cdots \xrightarrow{\partial} \sigma_0,$$

*where $\partial(e^{i_i \cdots i_k}) = \sum_j (-1)^j e^{i_1 \cdots i_{j-1} i_{j+1} \cdots i_k}$. The k'th **simplicial homology** $H_k(\sigma)$ of σ is $H_k(\sigma_\bullet)$. The **reduced simplicial homology** $\bar{H}_k(\sigma)$ is the cohomology of the augmented complex $\bar{\sigma}_\bullet$ obtained by setting $\sigma_{-1} = k$ and defining $\partial(e_i) = 1$ for all i; this differs from $H_k(\sigma)$ only for $k = 0$.*

I am not sure to whom to attribute the following result, which once again I learned from Bernd Sturmfels:

Theorem 5.10. *Let σ be a simplicial complex. Then for all k, the simplicial homology $\bar{H}_k(\sigma) \cong H^{k+1}(E/I_\sigma)^*$.*

Proof. The complexes $\bar{\sigma}_\bullet$ and $W^{\bullet+1}(E/I, h)$ are dual to one another, where $h = \sum_i x_i$. □

Remark. If $\bar{E}$ is the exterior algebra on V/h, then $\oplus_k H^k(E/I_\sigma)$ is an $\bar{E}$-algebra, although the exact structure depends on the choice of h. It is not clear geometrically what this algebra structure corresponds to. It would be nice to have interpretations of the discrete invariants of this $\bar{E}$-module in terms of the geometry and combinatorics of σ.

Example. *Homology of the square.*

Returning to the example of the square, where $I_\sigma = (x_1 \wedge x_3, x_2 \wedge x_4)$, if we take $h = x_1 + x_2 + x_3 + x_4$, then the complex $W^\bullet(E/I, h)$ is $k \to V \to \wedge^2 V/I_{\sigma,2} \to 0 \to 0$. This is exact at the two terms on the left, but fails to be surjective at $\wedge^2 V/I_{\sigma,2}$, where the cokernel is 1-dimensional. By Theorem 5.10, one has $\bar{H}_1(\sigma) = k$ and all the other $\bar{H}_k(\sigma)$ vanish.

Comments for students:

Although differential geometers and algebraic geometers appreciate the exterior algebra because it supplies the correct framework for integration on manifolds via the theory of differential forms, commutative algebraists perhaps undervalue it. A good starting place for seeing how combinatoricists use it is Chapter 2 of Stanley's book [St]. Stanley gives a good set of references to the literature for this topic. I also recommend the paper of Aramova, Herzog, and Hibi [A-H-H].

One way the exterior algebra is useful even in strictly commutative algebra is via the Koszul complex; see my notes [G1] for a survey. I also recommend my

paper [G4] for a hint of ways that more subtle facts about the exterior algebra
can be applied to commutative problems. An especially recommended method of
learning about the exterior algebra is to have some coffee with Bernd Sturmfels.

It would be interesting to see further how the geometry and combinatorics
of a simplicial complex is reflected in its Stanley-Reisner algebra.

A beautiful area where the exterior algebra is used is in the field of exterior
differential systems (A wonderful modern treatment is [B-C-G3].) I suspect that
looking at the initial ideals of the ideals in the exterior algebra arising from exterior
differential systems would prove interesting.

6. Lexicographic Gins and Partial Elimination Ideals

Let $I = \oplus_k I_k$ be a homogeneous ideal in $k[x_1, \dots , x_n]$. In what follows, we will
look at the generic initial ideal of I for the **lexicographic order**. We will also look
at some information that is finer than the gin. The remarkable feature of lex gins
is how much geometric information is encoded in them.

Definition 6.1. If $p \in I_d$ has leading term $\mathrm{in}(p) = x_1^{d_1} \cdots x_n^{d_n}$, we will set $d_1(p) = d_1$,
the **leading power** of x_1 in p. We set

$$\tilde{K}_k(I) = \oplus_d \{p \in I_d \mid d_1(p) \leq k\}.$$

If $p \in \tilde{K}_k(I)$, then we may write uniquely $p = x_1^k \bar{p} + q$ where $d_1(q) < k$. The image
of $\tilde{K}_k(I)$ in $k[x_2, \dots , x_n]$ under the map $p \mapsto \bar{p}$ we will denote $K_k(I)$. We call
$K_k(I)$ the k'th **partial elimination ideal** of I.

Remark. $K_k(I)$ is a homogeneous ideal in $k[x_2, \dots , x_n]$ and $\tilde{K}_k(I)$ is a graded
module over $k[x_2, \dots , x_n]$. The ideal $K_0(I)$ is just the ideal obtained from eliminating the variable x_1; geometrically it corresponds to the procedure of projecting I from the point $p = (1, 0, \dots , 0)$. We note that $f \in \tilde{K}_k(I)_d$ if and only if
$\mathrm{mult}_p(f) \geq d - k$, i.e. iff $f \in (x_2, \dots , x_n)^{d-k} = I_p^{d-k}$, where I_p is the homogeneous
ideal of p.

We now give a geometric interpretation of the $K_k(I)$. Let $Z = \mathrm{Var}(I)$. Let
π denote projection from $(1, 0, \dots , 0)$ mapping $\mathbf{P}^{n-1}$ to $\mathbf{P}^{n-2}$. Set-theoretically,
$K_0(I)$ is the ideal of $\pi(Z)$.

Remark. If we take the initial ideal of I for either the lexicographic order or for
the elimination order $1, n - 1$ (this means to use lexicographic order on x_1, but
then rlex for the remaining variables), then $\mathrm{in}(I) = \sum_k x_1^k \mathrm{in}(K_k(I))$.

Proposition 6.2. Set-theoretically, $K_k(I)$ is the ideal of $\{p \in \pi(Z) \mid \mathrm{mult}_p(\pi(Z)) > k\}$.

Proof. Let $p = (1, 0, \dots , 0)$ and let L be a line through p. Then $f \in \tilde{K}_k(I)$ if and
only if $\mathrm{mult}_p(f) \geq d - k$. If we have the length estimate $l(Z \cdot L) \geq k + 1$, then
$f|_L = 0$ for all $f \in \tilde{K}_k(I)$ and thus L belongs set-theoretically to the zero locus

of $K_k(I)$. Conversely, let P be the ideal of p. For $d >> 0$, the map $P^{d-k} \cap I_d \to H^0(\mathcal{I}_{Z \cdot L + (d-k)p}(d))$ is surjective, and thus there is an element of $\tilde{K}_k(I)$ that does not vanish identically on L if $l(Z \cdot L) \le k$. If $\bar{L}$ is the point corresponding to L under projection from p, then $\text{mult}_{\bar{L}}(\pi(Z)) = l(Z \cdot L)$.   □

Definition 6.3. For a homogeneous ideal I, the Jacobi ideal $J(I)$ is the ideal generated by $\partial f/\partial x_i$ for $f \in I$ and $1 \le i \le n$.

Proposition 6.4. For a general choice of coordinates, $J(K_{k-1}(I)) \subseteq K_k(I)$ for all k.

Proof. We apply Lemma 2.16. Fix a degree d and let $W = k[x_1, \ldots, x_n]_d$. Let W^p be those polynomials spanned by monomials x^I with $i_1 \le k$. Let $V = I_d$ and M the linear map $q \mapsto x_1 \partial q/\partial x_i$, $i \ne 1$. Then $g(t)$ is the one parameter family associated to the substitution $x_i \mapsto x_i + t x_1$. Note that $M(W^p) \subseteq W^{p-1}$ because we increase the power of x_1 by at most one. Further, $V^k(t) = \tilde{K}_k(g(t)(I_d))$. We may identify $V^k(t) + W^{k+1}$ with $K_k(I)_d + W^{k+1}$. If we are using general coordinates, the dimensions are locally constant, and the result now follows.   □

    This elementary result is actually quite powerful. For example:

Proposition 6.5. Let Z be a smooth projective variety. For projection π from a general point of $\mathbf{P}^{n-1}$, if $f \in I_{\pi(Z)}$, then $\partial f/\partial x_i$ vanishes on the singular locus of $\pi(Z)$.

Proof. If $f \in K_0(I_Z)$, then $\partial f/\partial x_i \in K_1(I_Z)$, and one now uses Proposition 6.2 to interpret this geometrically.   □

Remark. In general, for a variety Z of codimension c, it is only the $c \times c$ minors of the matrix of first partials of the generators of the ideal of Z that vanish on the singular locus of Z, so this gives a condition on the singular locus of a general projection (This condition is local, so the conclusion is also true of analytic functions vanishing on Z in a neighborhood of a point.) For example, if Z is two lines in $\mathbf{P}^3$ lying in a plane, then Z does not have this property.

    We recall, if V is a vector space and $M = \oplus_q M_q$ is a finitely-generated graded module over $S(V)$, we denote by $\mathcal{K}_m^\bullet(M, V)$ the Koszul complex

$$\cdots \to \wedge^{p+1} V \otimes M_{m-p-1} \to \wedge^p V \otimes M_{m-p} \to \wedge^{p-1} V \otimes M_{m-p+1} \to \cdots$$

with the indexes set up so that $\mathcal{K}_m^{-p} = \wedge^p V \otimes M_{m-p}$ and denote

$$\mathcal{K}_{p,q}(M, V) = H^{-p}(\mathcal{K}_{p+q}^\bullet(M, V)).$$

Proposition 6.6. Let I be a homogeneous ideal in $k[x_1, \ldots, x_n]$. Let V be the vector space spanned by $x_1, \ldots, x_n$ and $\bar{V}$ the vector space spanned by $x_2, \ldots, x_n$. There is for each m a spectral sequence with $E_1^{p,q} = \mathcal{K}_{-p-q, m+2p+q}(K_{-p}(I), \bar{V})$ abutting to $\mathcal{K}_{-p-q, m+2p+q}(I, \bar{V})$.

Proof. We filter I by $F^p I = \tilde{K}_{-p}(I)$, and thus $\text{Gr}^p I_d = K_{-p}(I)_{d+p}$. It follows (see [G-H], p. 440) that there is a spectral sequence with

$$E_1^{p,q} = H^{p+q}(\mathcal{K}_{m+p}^\bullet(\text{Gr}^p I, \bar{V}))$$

which abuts to $H^{p+q}\mathcal{K}_m^\bullet(I,\bar{V})$. Thus $E_1^{p,q} \cong \mathcal{K}_{-p-q,m+2p+q}(K_{-p}(I),\bar{V})$ and we converge to $\mathcal{K}_{-p-q,m+2p+q}(I,\bar{V})$ as desired.   □

Corollary 6.7. *Let Z be a reduced projective variety in $\mathbf{P}^{n-1}$ such that $p = (1,0,\dots,0)$ is not a component of Z. Then $K_0(I_Z)$ is saturated and $(K_1(I_Z) : \bar{\mathfrak{m}})_d/K_1(I_Z)_d$ injects into $\mathcal{K}_{n-3,d+2}(K_0(I),\bar{V})$.*

Proof. We note that if $f \in (I_Z : \bar{\mathfrak{m}})$, then $x_i f \in I_Z$ for all $i \geq 2$. Since p is not a component of Z, this implies that f vanishes on Z, and therefore $f \in I_Z$. So $(I_Z : \bar{\mathfrak{m}}) = I_Z$. It is now a simple matter to apply the preceding proposition. To give the flavor of the argument, we give a more explicit version of the proof.

If $f \in (K_0(I_Z) : \bar{\mathfrak{m}})$, then $x_i f \in K_0(I_Z)$ for $i \geq 2$. By the preceding paragraph, it follows that $f \in I_Z$ and it is also clear that $f \in k[x_2,\dots,x_n]$, so $f \in K_0(I_Z)$.

If $f \in (K_1(I_Z) : \bar{\mathfrak{m}})_d$, then there exist $g_i \in \tilde{K}_1(I_Z)$ such that $g_i = x_1(x_i f) + h_i$, where $h_i \in k[x_2,\dots,x_n]_{d+2}$. Now $x_i g_j - x_j g_i = x_i h_j - x_j h_i$ gives an element of $\wedge^2 \bar{V}^* \otimes K_0(I)_{d+2} \cong \wedge^{n-3}\bar{V} \otimes K_0(I)_{d+2}$, which represents an element of $\mathcal{K}_{n-3,d+2}(K_0(I),\bar{V})$. If it represents the zero element, then we can write $x_i h_j - x_j h_i = x_i u_j - x_j u_i$ for some $u_i \in K_0(I)$. Since $x_i(h_j - u_j) = x_j(h_i - u_i)$, we conclude that $h_i = u_i + x_i v$ for some polynomial $v \in k[x_2,\dots,x_n]$. Now $g_i - u_i = x_i(x_1 f + v)$, and the left-hand terms belong to I_Z. Thus $x_1 f + v \in (I_Z : \bar{\mathfrak{m}})$ and hence in I_Z. It follows that $f \in K_1(I_Z)$.   □

Remark. The preceding Corollary is just a sample of the implications of Proposition 6.6.

The following proposition and its corollary are joint with David Eisenbud, when we spent a week at Stockholm University thanks to the hospitality of Jan-Erik Roos. We suspect that in some form or other they are probably known.

Given f,g polynomials of the form $f = \sum_{i=0}^{d_1} x_1^i f_{d_1-i}$, $g = \sum_{i=0}^{d_2} x_1^i g_{d_2-i}$, let us consider the f_i, g_i to be new variables. The **Sylvester matrix** $\mathrm{Syl}(f,g)$ is the $(d_1 + d_2) \times (d_1 + d_2)$ matrix

$$\begin{pmatrix} f_0 & f_1 & f_2 & \cdots & 0 & 0 \\ 0 & f_0 & f_1 & \cdots & 0 & 0 \\ \vdots & \vdots & \vdots & \ddots & \vdots & \vdots \\ 0 & 0 & 0 & \cdots & f_{d_1-1} & f_{d_1} \\ g_0 & g_1 & g_2 & \cdots & 0 & 0 \\ 0 & g_0 & g_1 & \cdots & 0 & 0 \\ \vdots & \vdots & \vdots & \ddots & \vdots & \vdots \\ 0 & 0 & 0 & \cdots & g_{d_2-1} & g_{d_2} \end{pmatrix}.$$

Let $\mathrm{Syl}_k(f,g)$ denote the first $(d_1 + d_2 - k)$ columns of $\mathrm{Syl}(f,g)$.

Remark. The determinant of $\mathrm{Syl}(f, g)$ is just the **resultant** of f and g.

Proposition 6.8. *Let $I = (f, g)$, where the coefficients of f and g are independent variables. We take $d_1 \leq d_2$ and $f_0 = 1$.*
(1) The maximal minors of $\mathrm{Syl}_k(f, g)$ define a variety of the expected codimension $k + 1$;
(2) The variety defined by these minors is the set of f, g having a common factor of degree $\geq k + 1$;
(3) $K_k(I)$ is the ideal of maximal minors of $\mathrm{Syl}_k(f, g)$.

Proof. Let $R = k[f_0, \ldots, f_{d_1}, g_0, \ldots, g_{d_2}]$. Then $\mathrm{Syl}_k(f, g)$ is the matrix of the linear map $R[x]_{d_2-1} \oplus R[x]_{d_1-1} \rightarrow R[x]_{d_1+d_2-1}/\mathrm{span}(1, x, \ldots, x^{k-1})$ given by $(A, B) \mapsto Af + Bg$. Thus $\ker(\mathrm{Syl}_{k+1}(f, g)) = \{(A, B) \mid Af + Bg \in \mathrm{span}(1, x, \ldots, x^k)\}$. We want to apply $\mathrm{Syl}_k(f, g)$ to this kernel. $\mathrm{Syl}_k(f, g)$ fails to be surjective for a particular set of values of the f_i, g_i iff the space of solutions of $Af + Bg = 0$ has dimension $\geq k + 1$, i.e. f and g have a common factor of degree $\geq k + 1$. The space of such f, g has the expected codimension in the set of all f, g. It therefore follows that the maximal minors of $\mathrm{Syl}_{k+1}(f, g)$ span the kernel, in the sense that if we pick $d_1 + d_2 - k$ of the rows of $\mathrm{Syl}_{k+1}(f, g)$, taking the coefficients of A, B corresponding to these rows to be the minors and taking the other coefficients of A, B to be zero gives an element of the kernel, and these solutions span the set of all solutions. The image of the solution just given under $\mathrm{Syl}_k(f, g)$ is the $(d_1 + d_2 - k) \times (d_1 + d_2 - k)$ minor of $\mathrm{Syl}_k(f, g)$ determined by these rows. $\qquad\square$

Corollary 6.9. *Let f, g be general polynomials of degrees d_1, d_2 in n variables. Let $I = (f, g)$. For $k \leq n - 3$, $K_k(I)$ is the ideal of maximal minors of $\mathrm{Syl}_k(f, g)$, and this is a determinantal variety having the expected codimension $k + 1$.*

Proof. For a general choice of the f_i and g_i as polynomials in $n - 1$ variables, the matrix Syl_{k+1} drops rank in either the expected codimension $k + 2$ or else is the empty set, which has codimension $n - 1$ in $\mathbf{P}^{n-2}$. The preceding argument is thus OK unless $k + 2 > n - 1$, so the argument works if $k \leq n - 3$. $\qquad\square$

Remark. If $d_1 \leq d_2$, the first d_2 rows of $\mathrm{Syl}_k(f, g)$ have entries of degree 0, and thus for $k \leq d_1$, we get the same ideal of maximal minors if we use only those minors containing all of the first d_2 rows. We may thus express $K_k(I)$ under the conditions of the Proposition or its Corollary as the maximal minors of a $d_1 \times (d_1 - k)$ matrix whose i, j entry has degree $d_2 + 1 - i + j$. We also get that for any n, that $K_{d_1-1}(I)$ is given by the entries of $\mathrm{Syl}_{d_1-1}(f, g)$.

Remark. It is quite easy to find examples where the corollary breaks down if $k > n - 3$, for instance for $K_1(I_Z)$ for Z a set of points in the plane ($k = 1, n = 3$.) The reason that lex gins even of complete intersections f, g are so complicated is that, first of all, if the $K_k(I)$ are given by resultants, then the initial terms of the $K_k(I)$ are the lex gin of a determinantal variety in one fewer variable, and the lex gins of these can be complicated. For k large relative to n, the resultants don't work, and even if they do, as we progress to fewer and fewer variables, any determinantal description we may have will break down as we pass the expected codimension. On the other hand, for f, g, as the number of variables goes up, the $K_k(I)$ stabilize as determinantal varieties. If we use the $1, n-1$ order on the variables (i.e. rlex in the last $n-1$ variables), the gin will stabilize once the lower of the two degrees is $\leq n-2$.

Example 6.10. *A complete intersection curve of type (3,3) in* $\mathbf{P}^3$.

The lex gin of a general complete intersection of two cubics in 4 variables is x_1^3, $x_1^2(x_2, x_3^2, x_3 x_4^2, x_4^4)$, $x_1(x_2^4, x_2^3 x_3^2, x_2^2 x_3 x_4^2, x_2^3 x_4^3, x_2^2 x_3^5, x_2^2 x_3^4 x_4, x_2^2 x_3^3 x_4^2, x_2^2 x_3^2 x_4^4$, $x_2^2 x_3^5, x_2^2 x_4^7, x_2 x_3^8, x_2 x_3^7 x_4^2, x_2 x_3^6 x_4^4, x_2 x_3^5 x_4^6, x_2 x_3^4 x_4^8, x_2 x_3^3 x_4^{10}, x_2 x_3^2 x_4^{12}, x_2 x_3 x_4^{14}$, $x_2 x_4^{16}, x_3^{18})$, x_2^9. The lex gins of $K_2(I)$ and $K_1(I)$ occur in parentheses. The x_2^9 at the end is the lex gin of $K_0(I)$, which is the saturated ideal of the projection $\pi(Z)$ of the curve Z; this is a plane curve of degree 9. By the remark following Corollary 6.9, since $1 \leq 4-3$, $K_1(I_Z)$ is a determinantal ideal of a 3×2 matrix whose columns have degrees 3,2,1 and 4,3,2, and thus whose maximal minors have degrees 4,5,6, and that $K_2(I_Z)$ is given by the entries of the 3×1 matrix with entries of degrees 3,2,1 because $2 = d_1 - 1$. We know that $K_1(I)$ is the saturated ideal of the double point locus of $\pi(Z)$, and can read off that it has degree 18, i.e. there are 18 double points. We could also compute this geometrically, since by the adjunction formula the canonical bundle of Z is $K_Z = \mathcal{O}_Z(3 + 3 - 4) = \mathcal{O}_Z(2)$, and thus $\deg(K_Z) = 2 \cdot 3 \cdot 3 = 18$, so $2g - 2 = 18$, so $g = 10$. The genus of a plane curve of degree d with nodes is $\binom{d-1}{2} - \delta$, where there are δ nodes, and thus $10 = \binom{8}{2} - \delta$, so $\delta = 18$. $K_2(I)$ is not saturated, and it defines the empty ideal, which is expected since there are no triple points for a general projection of a curve in $\mathbf{P}^3$. Nevertheless, $K_2(I)$ contains a line and a conic, and naturally associates to Z and the center of projection p a plane through p and two lines through p in this plane, which project to 2 points in $\mathbf{P}^2$. It is useful to compute the generic initial ideal for the order 1 3, i.e. we use lexicographic order on the first variable, but reverse lex in comparing monomials in the last 3 variables. This gives us the rlex gin of the $K_k(I)$. For this order, the gin of I_Z is: $x_1^3, x_1^2(x_2, x_3^2, x_3 x_4^2, x_4^4), x_1(x_2^4, x_2^3 x_3^2, x_2^2 x_3^4, x_2 x_3^5, x_3^7), x_2^9$. From this, we can read off the numerical character of the set of double points of Z, and see that the 18 double points lie on a quartic and a quintic and an additional sextic – notice that 18 general points do not lie on a quartic, and that 18 general points on a quartic do not lie on a quintic.

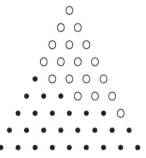

The quartic and the quintic intersect in 20 points, the 18 double points and 2 residual points. The rather remarkable fact, which we leave as an exercise to the reader, is that the 2 residual points are precisely the 2 points determined by the linear and quadratic forms in $K_2(I_Z)$!

Example 6.11. *The general complete intersection curve of type (2,2,2) in* $\mathbf{P}^4$.

We let Z be a general complete intersection curve of 3 quadrics in 5 variables. Geometrically, Z is a canonical curve of genus 5 in $\mathbf{P}^4$. In the 1 4 order, the gin of I_Z is $x_1^2, x_1(x_2, x_3, x_4^2, x_4x_5, x_5^3), x_2^3, x_2^2x_3^2, x_2^2x_3x_4^2, x_2x_3^3, x_3^4$. The last 5 terms are the rlex gin of $\pi(Z)$, which is the saturated ideal of a curve of degree 8 and genus 5 in $\mathbf{P}^3$.

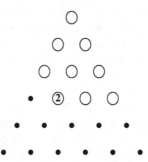

We note that $\pi(Z)$ lies on a cubic V. Now $K_1(I_Z)$ is not saturated and defines the empty variety, reflecting the fact that a general projection of a curve in $\mathbf{P}^4$ into $\mathbf{P}^3$ has no double points. However, the two linear forms define a line L in $\mathbf{P}^3$. The first remarkable fact is that L is contained in V. The two quadratic forms in $K_1(I_Z)$ are the restrictions of the partial derivatives of V to L. The second remarkable fact is that $\pi(Z)$ belongs to the linear series $|2H+2L|$ on V, where H is the hyperplane bundle. We leave to the reader the enjoyable task of proving these two facts. If we continue the story by looking at the ideal of $\pi(Z)$ in the 1 3 order, we find that its gin is $x_2^3, x_2^2(x_3^2, x_3x_4, x_4^2, x_3x_5^2, x_4x_5^2, x_5^4), x_2(x_3^4, x_3^3x_4^2, x_3^2x_4^3, x_3x_4^5, x_4^6), x_3^8$. The last term reflects the fact that the general projection of $\pi(Z)$ into $\mathbf{P}^2$ is a plane curve of degree 8. From the rlex gin of $K_1(I)$ we see that there are 16 double points for this projection, lying on a quartic (and 2 quintics).

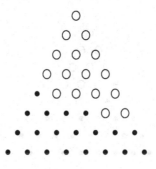

Since Z was a canonical curve, $K_Z = H$, and on the other hand, if D is the double points locus and $\tilde{D}$ is its preimage in $\mathbf{P}^3$, we have that $K_Z \cong 5H - \tilde{D}$, from which we conclude that $\tilde{D} \cong 4H$, and thus that D, counted twice, is the complete intersection of the second projection of Z with a plane quartic. $K_2(I_{\pi(Z)})$ is a regular sequence of 3 conics in $\mathbf{P}^2$.

Remark.   One of the most interesting uses of lex gins, as these examples show, is their fertility in suggesting geometric constructions. Given a variety Z of a certain type and a general linear space, one gets a number of distinguished cones over the linear space. Many of these are not at all obvious geometrically.

We note that there are some interesting further invariants associated to the partial elimination ideals.

Definition 6.12. *For any ideal I and any choice of coordinates, if $f \in K_k(I)_d$, then there is a lifting $\tilde{f} \in \tilde{K}_k(I)_{d+k}$, unique modulo $\tilde{K}_{k-1}(I)$. If we write $\tilde{f} = x_1^k f + x_1^{k-1} g + \cdots$, then the map $f \mapsto g$ gives a well-defined map*

$$\phi_k^1 : K_k(I) \to S^{d+1}\bar{V}/K_{k-1}(I)_{d+1}.$$

If $\phi_k^1(f) = 0$, then we can choose a lifting $\tilde{f}$ of f such that $f = x_1^k f + x_1^{k-2} g + \cdots$, and now $f \mapsto g$ gives a map

$$\phi_k^2 : \ker(\phi_k^1)_d \to S^{d+2}\bar{V}/K_{k-2}(I)_{d+2}.$$

The ϕ_k^ν are defined inductively so that

$$\phi_k^\nu : \ker(\phi_k^{\nu-1})_d \to S^{d+\nu}\bar{V}/K_{k-\nu}(I)_{d+\nu}.$$

There is a hierarchy of further constructions exemplified by, if $f \in K_k(I)_d$, choosing a lifting $\tilde{f} \in \tilde{K}_k(I)_d$ with $\tilde{f} = x_1^k f + x_1^{k-1} g + x_1^{k-2} h + \cdots$. The map $f \mapsto h$ gives a well-defined map

$$\psi_k^1 : K_k(I)_d \to S^{d+2}\bar{V}/(K_{k-2}(I) + \phi_k^1(K_{k-1}(I))).$$

Example 6.13. *Griffiths-Harris Fundamental Forms*

Let $Z \subseteq \mathbf{P}^{n+c}$ be a variety of dimension n and $p \in Z$ a smooth point. Choose coordinates so that $p = (1, 0, 0, \ldots, 0)$. The tangent space to Z at p is cut out by linear forms $l_1, \ldots l_c$ where $l_i \in K_{k_i}(I_Z)$ for some k_i. Thus we may identify $l_1, \ldots, l_c$ with a basis for the dual of the normal space $N_p^*(Z)$, $\bar{V}$ with the cotangent space to projective space at p, and $T_p^* Z$ with $\bar{V}/(l_1, \ldots, l_c)$. The element $\phi_{k_i}^1(l_i) \in S^2\bar{V}/K_{k_i-1}(I)$. However, smoothness at p forces all elements of $K_k(I)$ for all k to be in the ideal generated by $l_1, \ldots, l_c$. Thus $\phi_{k_i}^1(l_i)$ defines a well-defined element of $S^2 T_p^* Z$, and thus a map $II : N_p^* Z \to S^2 T_p^* Z$. This is the **Griffiths-Harris second fundamental form** for Z at p (see [G-H2]). The maps ϕ_k^ν give the $\nu + 1$'st **fundamental form** in $S^{\nu+1} T_p^* Z$ and $\psi_k^1(l_i)$ gives an element of $S^3 T_p^* Z / (T_p^* Z \otimes \mathrm{im}(II))$ which Griffiths and Harris call the **cubic form**. Applying a variant of the proof of Proposition 6.4, one sees that by choosing a general point $p \in Z$, we make the Jacobi ideal of the $\nu + 1$'st fundamental form contained in the ν'th fundamental form – this is a result of Griffiths and Harris. Some highly interesting work involving their construction has been done by J.M. Landsberg ([L], [L2]).

Example 6.14. *A general complete intersection of 2 cubics in* $\mathbf{P}^3$ *projected from a general point.*

If we pick a general point p on a curve Z of type $(3,3)$ in $\mathbf{P}^3$, the gin in the 1 3 order is $x_1^2(x_2, x_3), x_1(x_2^3, x_2^2x_3^2, x_2x_3^4, x_3^6), x_2^8$. Here, the projection $\pi(Z)$ has degree 8 – we lose a degree because the p is on Z. Now $K_1(I_Z)$ is the saturated ideal of the double point locus of $\pi(Z)$ together with the point of $\mathbf{P}^2$ corresponding to the tangent line to Z at p. Computing the genus, we see that there are 11 double points, while the gin of $K_1(I_Z)$ tells us that the 11 double points and the image of p lie on a complete intersection of type $(3,4)$ in $\mathbf{P}^2$. $K_2(I)$ is the saturated ideal of the tangent line to Z at p, i.e. of the image of p under π; in the notation above, l_1, l_2 are the generators of $K_2(I_Z)$. If $II(l_i) = q_i$, then the cubic generator of $K_1(I_Z)$ is $V = l_1q_2 - l_2q_1$. The partials of V in the directions tangent to Z lie in $K_2(I_Z)$, while the partials of V with respect to the other variables, restricted to T_pZ, are $-q_2, q_1$, and hence the second fundamental form. If the cubic forms of l_i is c_i, then the quartic generator of $K_1(I_Z)$ is $W = l_1c_2 - l_2c_1$, and the partials of W, restricted to T_pZ, are the cubic forms $-c_2, c_1$.

Remark. Rob Lazarsfeld points out that if $Z \subseteq \mathbf{P}^{n-1}$, then if we map the blow-up Y of $\mathbf{P}^{n-1}$ at Z to $\phi(Y) \subseteq \mathbf{P}^N$ by the linear system $I_{Z,d}$, and if $p = (1, 0, \ldots, 0)$ is a point off of Z mapping to $\phi(p)$, then the Griffiths-Harris fundamental forms of $\phi(Y)$ at $\phi(p)$ are the $K_k(I)_d$, so that each construction is a special case of the other.

Comments for students:

This chapter had its origins in a conversation I had with Bernd Sturmfels, in which he introduced me to the surprising complexity manifested by the lex gins of even fairly tame ideals. I became curious to "explain" some of this strange behavior. I have not thought about the material here as long as the material in Chapters 1-4, and this chapter would certainly have been larger otherwise. I leave it to the reader to regret or rejoice over this situation.

I have a lot of interesting examples which I have not included here; I encourage students with access to a computer to explore and to try to figure out what is going on geometrically using the lex gin.

References

[A]   M. Amasaki, "Preparatory structure theorem for ideals defining space curves,"
           Publ. RIMS 19 (1983), 493–518.

[A-C-G-H] E. Arbarello, M. Cornalba, P. Griffiths and J. Harris, *Geometry of Algebraic
           Curves*, V. 1, Springer-Verlag, New York, 1985.

[A-H-H]   A. Aramova, J. Herzog and T. Hibi, "Gotzmann theorems for exterior algebras
           and combinatorics," preprint.

[B-C-G3]   R. Bryant, S.S. Chern, R. Gardner, H. Goldschmidt, P. Griffiths, *Exterior
           Differential Systems*, Springer-Verlag, Boston, 1991.

[B-H]   W. Bruns and J. Herzog, *Cohen-Macaulay Rings*, Cambridge University
           Press, Cambridge, 1993.

[B-S]   D. Bayer and M. Stillman, "A criterion for detecting m-regularity," Invent.
           Math. 87 (1987), 1–11.

[C]   M. Cook, "The connectedness of space curve invariants," preprint.

[D]   E. Davis, "0-dimensional subschemes of $\mathbf{P}^2$:lectures on Castelnuovo's func-
           tion," Queen's Papers in pure and appl. math. 76 (1986).

[E]   D. Eisenbud, *Commutative Algebra with a View Toward Algebraic Geometry*,
           Springer-Verlag, New York 1995.

[E-G-H]   D. Eisenbud, M. Green, and J. Harris, "Some conjectures extending Casteln-
           uovo theory," in Astérisque 218 (1993), 187–202.

[E-K]   S. Eliahou and M. Kervaire, "Minimal resolutions of some monomial ideals,"
           J. Alg. 129 (1990), 1–25.

[E-P]   Ph. Ellia and C. Peskine, "Groupes de points de $\mathbf{P}^2$: caractère et position
           uniforme," 111–116 in *Algebraic geometry (L'Aquila, 1988)*, Springer LNM
           1417, Berlin, 1990.

[F]   G. Floystad, "Higher initial images of homogeneous ideals," preprint.

[G1]   M. Green, "Koszul cohomology and geometry," 177–200 in *Lectures on Rie-
           mann Surfaces* (M. Cornalba, X. Gomez-Mont, A. Verjovsky eds.), World
           Scientific, Singapore, 1989.

[G2]   M. Green, "Restrictions of linear series to hyperplanes, and some results of
           Macaulay and Gotzmann," 76-88 in *Algebraic Curves and Projective Geom-
           etry* (E. Ballico and C. Ciliberto, eds.), Springer LNM 1389, Berlin 1989.

[G3]   M. Green, "Infinitesimal methods in Hodge theory," 1–92 in *Algebraic Cycles
           and Hodge Theory* (A. Albano and F. Bardelli, eds.), Springer LNM 1594,
           Berlin, 1994.

[G4]   M. Green, "The Eisenbud-Koh-Stillman conjecture on linear syzygies,"
           preprint.

[G-H]   P. Griffiths, and J. Harris, *Principles of Algebraic Geometry*, Wiley Inter-
           science, New York, 1978.

[G-H2]   P. Griffiths and J. Harris, "Algebraic geometry and local differential geome-
           try," Ann. Sci. École Norm. Sup. 12 (1979), 355–452.

[G-L-P]   L. Gruson, R. Lazarsfeld, and C. Peskine, "On a theorem of Castelnuovo, and
           the equations defining space curves," Inv. Math. 72 (1983), 491–506.

[G-P]   L. Gruson and C. Peskine, "Genre des courbes de l'espace projectif," 31–59 in *Algebraic Geometry: Proc. Symposium University of Tromso, Tromso, 1977,* Springer LNM 687, Berlin, 1978.

[Go]   G. Gotzmann, "Eine Bedingung für die Flachheit und das Hilbertpolynom eines graduierten Ringes," Math. Z. 158 (1978), 61–70.

[H]   R. Hartshorne, *Algebraic Geometry*, Springer-Verlag, New York, 1977.

[H-U]   C. Huneke and B. Ulrich, "General hyperplane sections of algebraic varieties," J. Alg. Geom. 2 (1993), 487–505.

[L]   J.M. Landsberg, "On the second fundamental forms of projective varieties," Inv. Math. 117 (1994), 303–315.

[L2]   J.M. Landsberg, "Differential-geometric characterizations of complete intersections," preprint.

[Li]   R. Liebling, "Classification of space curves using initial ideals," Thesis, University of California at Berkeley, 1996.

[M]   H. Matsumura, *Commutative Ring Theory*, Cambridge University Press, Cambridge, 1986.

[Ri]   C. Rippel, "Generic initial ideal theory for coordinate rings of flag varieties," Thesis (UCLA).

[Ro1]   L. Robbiano, "Term orderings in the polynomial ring," Proc. of Eurocal 1985, LNCS 203 II, Springer-Verlag, Berlin, 1985.

[Ro2]   L. Robbiano, "On the theory of graded structures," J. Symb. comp 2 (1986), 139–170.

[R]   J.-E. Roos, "A description of the homological behavior of families of quadratic forms in four variables," in Iarrobino, Martsinkovsky, and Weyman, *Syzygies and Geometry* (1995), 86–95.

[S]   J.-P. Serre, "Faisceaux algébriques cohérents," Ann. of Math. 61 (1955), 197–278.

[St]   R. Stanley, *Combinatorics and Commutative Algebra*, 2nd edition, Birkhäuser, Boston, 1996.

[Str]   R. Strano, "A characterization of complete intersection curves in $\mathbf{P}^3$," Proc. A.M.S. 104 (1988), no. 3, 711–715.

University of California, Los Angeles
E-mail address: mlg@math.ucla.edu

Tight Closure, Parameter Ideals, and Geometry

Craig Huneke

Foreword

These notes are based on five lectures given at the Summer School on Commutative Algebra held at the CRM in Barcelona during July, 1996. I would like to thank the organizers J. Elias, J. M. Giral, R. M. Miró-Roig, and S. Zarzuela for the excellent job they did. The great success of the Summer School was due mainly to their efforts.

The topic of these notes is tight closure, a notion introduced by Melvin Hochster and myself in late 1986. The notes are intended mainly for students trying to learn something about this subject. Another source for such an introduction is [Hu5], but the material in these notes is largely separate. When overlap does occur, I have tried to give different proofs and new insight. Other useful introductory articles are [Bru] and [Sm8].

1. An Introduction to Tight Closure

As I mentioned in the Foreword, these notes concern the theory of 'tight closure'. Tight closure is an operation upon ideals in rings containing fields, or even upon submodules of a given module. If I is an ideal in a Noetherian equicharacteristic ring [1] R then we denote the tight closure of I in R by I^*. It turns out that I^* is another ideal, containing I, and has the property that it is tightly closed, i.e. $(I^*)^* = I^*$. Thus it is a 'closure' operation.

These notes will concentrate on the tight closure of parameter ideals in positive characteristic. There are several reasons I made this choice of material. Studying the tight closure of parameter ideals connects tight closure to several more classical themes in both algebra and geometry. On the one hand, the existence of Big Cohen-Macaulay algebras is inherent in studying such tight closures, while another piece of information which is hidden in the tight closure of parameter ideals

The author was partially supported by the NSF. I thank Alberto Corso for helping me with both the proofreading and the LATEXing.
[1]'Equicharacteristic' is just another way to say the ring contains a field. Rings such as the integers are not equicharacteristic; the fraction field has characteristic 0, but the residue fields at the maximal ideals have positive characteristic.

is the Kodaira vanishing theorem. Even more: the Briançon-Skoda theorem and information about rational singularities are also all captured by the information provided by the tight closure of parameters. We will discuss all of these topics.

We will also try to give an idea of how elements in the tight closure of ideals arise. We will concentrate on this, along with basic examples and properties. Other sections will discuss the all important idea of test elements, F-rationality and the action of Frobenius on local cohomology, and the relationship of the Kodaira vanishing theorem to tight closure. See [Hu5] for an expanded introduction to the subject, and [HH1]-[HH12] for basic papers on tight closure. A large bibliography is included with much of the work on tight closure and related material.

Most of the results in these notes will be valid in equicharacteristic, i.e. for Noetherian rings containing a field. However, the proofs in these notes will be exclusively in characteristic $p > 0$. The phrase 'characteristic p' always means positive and prime characteristic p. We will use 'q' throughout these notes to denote a variable power of the characteristic p. Tight closure is a method which requires reduction to characteristic p, although the theory has now been developed for any Noetherian ring containing a field.

We set R^o to be the set of all elements of R not in any minimal prime of R. The definition of tight closure for ideals is:

Definition 1.1. Let R be a Noetherian ring of characteristic $p > 0$. Let I be an ideal of R. An element $x \in R$ is said to be in the tight closure of I if there exists an element $c \in R^o$ such that for all large $q = p^e$, $cx^q \in I^{[q]}$, where $I^{[q]}$ is the ideal generated by the qth powers of all elements of I.

It is helpful to see an example of tight closure immediately:

Example 1.2. Let $R = k[X,Y,Z]/(X^3+Y^3+Z^3)$, where k is a field of characteristic $\neq 3$. Let x, y, z denote the image of X, Y, Z in R. R is easily seen to be a two dimensional Gorenstein normal ring. The elements y, z form a homogeneous system of parameters. We claim that $x^2 \in (y, z)^*$. Write $2q = 3k + i$, where i is either 1 or 2. Then $x^{3-i}x^{2q} = x^{3(k+1)} = (-y^3 - z^3)^{k+1}$. A typical monomial in the expansion of the latter expression is $y^{3j}z^{3h}$ with $j+h = k+1$. If both $3j < q$ and $3h < q$, then $3(k+1) = 3(j+h) < 2q \le 3k + 2$. It follows that $x^3x^{2q} \in (y^q, z^q)$ and hence that $x^2 \in (y, z)^*$. Even in this simple example, not everything is known about the tight closure of an arbitrary ideal. See [McD] for a study of this example. McDermott has noted recently that it is not even known whether $xyz \in (x^2, y^2, z^2)^*$ in this example for arbitrary characteristic.

We will later prove that a power of z can be used as a 'test' element for tight closure in this ring. This means that there is a fixed power of z, say z^N, such that for all ideals $I \subseteq R$ and all $a \in I^*$, $z^N a^q \in I^{[q]}$ for all $q = p^e$.

We can use the above remark and Gröbner bases to prove that the tight closure of (y, z) is exactly (y, z, x^2). This argument is due to M. Katzman. We have shown above that $(y, z, x^2) \subseteq (y, z)^*$. The only thing which needs to be proved is that $x \notin (y, z)^*$. Suppose $x \in (y, z)^*$. Choose $q > N$, where N is chosen as in the paragraph above. Let $S = k[X,Y,Z]$ and consider the ideal

$J = (X^3 + Y^3 + Z^3, Y^q, Z^q)$ in S. Fix a monomial ordering by letting $X > Y > Z$ and taking reverse lex order. Then $\mathrm{in}(X^3 + Y^3 + Z^3) = X^3$, and the initial ideal $\mathrm{in}(J) = (X^3, Y^q, Z^q)$. We claim that $Z^N X^q \notin J$. This suffices to reach a contradiction. Write $q = 3k + i$, where i is 1 or 2. Then $X^q = (-1)^k X^i (Y^3 + Z^3)^k$ modulo J, and so it is enough to prove that $\mathrm{in}(Z^N (-1)^k (Y^3 + Z^3)^k) \notin \mathrm{in}(J)$. However, $\mathrm{in}(Z^N (-1)^k X^i (Y^3 + Z^3)^k) = X^i Y^{3k} Z^N \notin \mathrm{in}(J) = (X^3, Y^q, Z^q)$. This contradiction proves that $x \notin (y, z)^*$.

The basic properties of tight closure are summed up in the following theorem:

Theorem 1.3. [HH4] *Let R be a Noetherian ring of characteristic p and let I be an ideal.*

a) $I \subseteq (I^*) = (I^*)^*$. If $I_1 \subseteq I_2 \subseteq R$, then $I_1^* \subseteq I_2^*$.

b) *If R is reduced or if I has positive height, then $x \in R$ is in I^* if and only if there exists $c \in R^o$ such that $cx^q \in I^{[q]}$ for all $q = p^e$.*

c) *An element $x \in R$ is in I^* iff the image of x in R/P is in the tight closure of $(I + P)/P$ for every minimal prime P of R.*

d) $I^* \subseteq \overline{I}$, *the integral closure of I.*

e) *Further let R be a regular local ring. Then $I^* = I$ for every ideal $I \subseteq R$.*

f) *If I is tightly closed, then $I : J$ is tightly closed for every ideal J.*

g) *An intersection of tightly closed ideals is tightly closed.*

h) $I^* J^* \subseteq (IJ)^*$.

Proof. It is clear that $I \subseteq I^*$. Suppose that $x \in (I^*)^*$, and choose $c \in R^o$ such that $cx^q \in (I^*)^{[q]}$ for all large q. Choose generators for I^*, say $(y_1, \ldots, y_n) = I^*$. Choose $d_i \in R^o$ such that $d_i y_i^q \in I^{[q]}$ for all large q. We may write $cx^q = \sum_i s_i y_i^q$ for large q. Multiplying by $d = d_1 \cdots d_n$ then yields that $dcx^q \in I^{[q]}$ for all large q which proves a).

To prove b) suppose that $cx^q \in I^{[q]}$ for all $q \geq q'$. If the height of I is positive simply choose an element $d \in I^{[q']} \cap R^o$. Then $dcx^q \in I^{[q]}$ for all q. We leave the case in which R is reduced as an exercise.

We prove c). One direction is clear: if $x \in I^*$, then this remains true modulo every minimal prime of R since $c \in R^o$. Let $P_1, \ldots, P_n$ be the minimal primes of R. If $c_i' \in R/P_i$ is nonzero we can always lift c_i' to an element $c_i \in R^o$ by using the Prime Avoidance theorem. Suppose that $c_i' \in R/P_i$ is nonzero and such that $c_i' x_i^q \in I_i^{[q]}$ for all large q, where x_i (respectively I_i) represent the images of x (respectively I) in R/P_i. Choose a lifting $c_i \in R^o$ of c_i'. Then $c_i x^q \in I^{[q]} + P_i$ for every i. Choose elements t_i in all the minimal primes except P_i. Set $c = \sum_i c_i t_i$. It is easy to check that $c \in R^o$. Choose $q' \gg 0$ so that $N^{[q']} = 0$, where N is the nilradical of R. Then $cx^q \in I^{[q]} + N$, and so $c^{q'} x^{qq'} \in I^{[qq']}$, which proves that $x \in I^*$.

To prove d), recall a definition of integral closure: The *integral closure* of I in a Noetherian ring is the set of all elements x such that there exists an element $c \in R^o$, the complement of the minimal primes of R, such that $cx^n \in I^n$ for infinitely many n. With this definition, part d) is immediate since $I^{[q]} \subseteq I^q$.

Part e) follows from a theorem of Kunz [Ku1]: a ring R of positive character-istic p is regular iff $R^{1/q}$ is flat over R for all (equivalently for some) $q = p^e, e \geq 1$. Assume this and suppose that $x \in I^*$ for some ideal $I \subseteq R$. Then there exists a $c \in R^o$ such that for all large q, $cx^q \in I^{[q]}$. Taking qth roots yields that $c^{1/q}x \in IR^{1/q}$, and hence that $c^{1/q} \in I :_{R^{1/q}} x = (I :_R x)R^{1/q}$, the last equality following because $R^{1/q}$ is flat over R. It then follows taking qth powers that $c \in (I :_R x)^{[q]}$ for all large q. This contradicts the assumption that $c \in R^o$ unless $(I :_R x) = R$, i.e. $x \in I$. [2]

We prove f). It is easy to prove that a finite intersection of tightly closed ideals is tightly closed. It follows that we may assume that $J = (y)$ is principal. Suppose that $x \in (I : y)^*$, where I is tightly closed. Choose $c \in R^o$ such that $cx^q \in (I : y)^{[q]} \subseteq I^{[q]} : y^q$. Then $c(xy)^q \in I^{[q]}$ implies that $xy \in I^* = I$; hence $x \in I : y$.

Parts g) and h) are easy.   □

The property e) that ideals in regular rings are tightly closed does not char-acterize regular rings, but is important enough to warrant its own name.

Definition 1.4. A Noetherian ring in which every ideal is tightly closed is called *weakly F-regular*. A Noetherian ring R such that R_W is weakly F-regular for every multiplicative system W is called *F-regular*.

It is unfortunate to need the terminology 'weakly' F-regular. If tight closure commutes with localization, then these two definitions are equivalent. We believe this is true, but have been unable to prove it except in special cases. For work on this topic see [Wil, Mac, AHH]. On the other hand, we can do a couple special cases concerning the behavior of tight closure under completion and localization which we will need later in these notes.

The proof of the first part of Proposition 1.5 below requires knowledge of test elements. One problem with the definition of tight closure is that the c in the definition varies with the ideal I and the element x. An element c such that for all I and $x \in I^*$, $cx^q \in I^{[q]}$ for all q is called a *test element*. The existence of such elements is a crucial part of tight closure theory.

One of the main theorems guaranteeing the existence of test elements is (cf. [HH9], (6.2)):

Theorem (Existence of Test Elements). *Let R be a reduced algebra of finite type over an excellent local ring (B, m, K) of characteristic p. Let $c \in R^o$ be such that R_c is F-regular and Gorenstein (e.g. R_c regular). Then c has a power which is a test element for R.*

We return to test elements in Section 3.

---

[2]More generally, from the Buchsbaum-Eisenbud criterion for exactness, it follows that for every Noetherian ring R of characteristic p, $Tor_i^R(M, S) = 0$ for $i > 0$ whenever the projective dimension of M is finite, and where $S = R$ viewed as a module over R via the Frobenius morphism.

Proposition 1.5. (Completion and Localization). *Let (R, m) be an excellent Noetherian local ring of characteristic p.*

a) *Let I be an m-primary ideal. Then $I^* \hat{R} = (I\hat{R})^*$.*
b) *Let I be an ideal generated by a regular sequence, and let W be a multiplicatively closed set. Then $(I^*)_W = (I_W)^*$.*

Proof. First we prove a). Choose an element c such that $(R_{red})_c$ is regular. By replacing c by a power of itself, we can assume that c is a test element for R_{red}, using the Theorem above.

We may choose a power q' of the characteristic p such that the nilradical raised to this power is zero. Lift c from R_{red} back to an element of R which we still call c. This element has the following property: $x \in J^*$ iff for all $q \geq q'$, $cx^q \in J^{[q]}$. We call such an element a q'-weak test element.

Since R is excellent, $(\hat{R}_{red})_c$ will also be regular, and we may assume that c is also a q'-weak test element for $\hat{R}$. Without loss of generality, we can assume that $I = I^*$. If $\hat{I}$ is not tightly closed, there will be an element $z \in (\hat{I})^* \cap R$. Since R and $\hat{R}$ have a common test element, it then follows that $z \in I^* = I$.

Next we prove b). Let $G = \mathrm{gr}_I(R)$ and choose an element $s \in W$ such that $\mathrm{ann}_G s = \mathrm{ann}_G ws$ for all $w \in W$. Suppose that $wu \in I^m$. Assume that $u \in I^r$ but $u \notin I^{r+1}$ where $r \leq m - 1$. We claim that $s^m u \in I^m$. The choice of s shows that $su \in I^{r+1}$. Repeating the argument with su in place of u and continuing gives the claim.

Fix q and an element $u \in I^{[q]} : w$ for some $w \in W$. By induction on h we claim that $d_h = s^{q+h} u \in I^{[q]} + I^{q+h}$. This suffices to prove part b) since when $h = qn$ we get that $s^{(n+1)q} u \in I^{[q]} + I^{q(n+1)} \subseteq I^{[q]}$ as I has n generators.

The case $h = 0$ follows from the above argument. Suppose $d_h \in I^{[q]} + I^{q+h}$ and write $d_h = \sum_i r_i x_i^q + \sum_\nu t_\nu x^\nu$ where ν runs through n-tuples of nonnegative integers all less than q whose sum is $q + h$. Some $w \in W$ multiplies this element into $I^{[q]}$ giving an equation, $w d_h = \sum_i a_i x_i^q = \sum_i w r_i x_i^q + \sum_\nu w t_\nu x^\nu$. Since the x_i form a regular sequence and none of the monomials occurring is formally in the ideal generated by the other monomials, each $w r_\nu \in I$. Then $s r_\nu \in I$ by the choice of s and so $d_{h+1} = s d_h$ is in $I^{[q]} + I^{q+h+1}$.

To finish the proof, consider the equations $w(q) c x^q \in I^{[q]}$, where c is a test element, $q = p^e$ runs through all powers of p, and $w(q) \in W$ depends on q. Choose $s \in W$ as above. Then $s^{(n+1)q} c x^q \in I^{[q]}$, and so $c(s^{n+1}x)^q \in I^{[q]}$ which proves that $s^{n+1} x \in I^*$ and therefore $x \in (I^*)_W$ as needed. □

The main topic we will study is the properties of the tight closure of parameter ideals. We say elements $x_1, \ldots, x_d$ are parameters if they generate an ideal of height d. A *parameter ideal* is an ideal generated by parameters. If (R, m) is local and Noetherian of dimension d, then parameters $x_1, \ldots, x_d$ are said to be a system of parameters, or s.o.p. for short. Equivalently, the radical of the ideal they generate is the maximal ideal m. In Gorenstein rings a system of parameters generates an

irreducible ideal. This fact can be effectively used to compute tight closures in
such rings as part (5) of the next theorem shows.

Theorem 1.6. *Let R be a Noetherian ring of characteristic p.*

(1) *Let I be primary to a maximal ideal m. Then $(IR_m)^* = (I^*)_m$.*

(2) *R is weakly F-regular iff every ideal primary to a maximal ideal is tightly
closed.*

(3) *R is weakly F-regular iff R_m is weakly F-regular for every maximal ideal m
of R.*

(4) *Assume that R has a test element c. Then the tight closure of an arbitrary
ideal I is the intersection of the tight closure of ideals primary to maximal
ideals.*

(5) *Suppose that R is local Gorenstein with maximal ideal m. Then R is weakly
F-regular (in fact even F-regular) iff an ideal generated by a system of parameters is tightly closed.*

(6) *Let S be a weakly F-regular ring and suppose that $R \subseteq S$ is a direct summand
of S (as an R-module). Then R is weakly F-regular.*

Proof. The only point to (1) is that if $c \in R^\circ$ and $z \in R$ such that $cz^q \in I^{[q]}R_m$
then $cz^q \in I^{[q]}$ since I is m-primary. Since every ideal in a Noetherian ring is an
intersection of ideals primary to maximal ideals Part (2) follows from (1), while
(3) is immediate from (1) and (2).

To prove (4), write I as the intersection of ideals of the form $I + m^n$ where m
is maximal. Clearly $I^* \subseteq \cap_{n,m}(I + m^n)^*$ where the intersection is over all maximal
m and all integers n. The reverse inclusion also holds. For if $x \notin I^*$ there exists a
fixed power of p, say q, such that $cx^q \notin I^{[q]}$. Choose a maximal ideal m such that
the same equation holds after localizing at m. Then lift back to R to obtain that
$cx^q \notin I^{[q]} + m^n$ for all $n \gg 0$. Hence $cx^q \notin (I + m^n)^{[q]}$. Thus $x \notin (I + m^n)^*$.

We prove (5). By part (2), it suffices to prove every m-primary ideal I is
tightly closed. Let $(x_1, \ldots, x_d)$ be an ideal generated by a system of parameters
which we know to be tightly closed. Using the fact that the x_i form a regular
sequence one can prove that $(x_1^t, \ldots, x_d^t)$ is tightly closed for all t. Choose $t \gg 0$
such that $J = (x_1^t, \ldots, x_d^t) \subseteq I$. By part (f) in Theorem1.3, $I = J : (J : I)$ is
tightly closed (the equality following since R is Gorenstein). This proves that R is
weakly F-regular. We will prove R is even F-regular in (6.2).

Let $I \subseteq R$ and suppose that $x \in I^*$. It follows that $x \in (IS)^* = IS$ and so
$x \in IS \cap R = I$. This proves (6). $\square$

To paraphrase, every regular ring is F-regular. A local Gorenstein ring is
F-regular iff the ideal generated by a single system of parameters is tightly closed.

Example 1.7. By Theorem 1.6, in a Gorenstein local ring if the ideal generated by
a single system of parameters is tightly closed, then so is every ideal. In particular,
if there exists an element u which generates the socle of an ideal I generated by a
s.o.p. such that $u \notin I^*$, then every ideal is tightly closed.

For example, let $R = k[X, Y, Z]/(X^2+Y^3+Z^5)$, where k is a field of positive characteristic $p > 5$. Let x, y, z denote the image of X, Y, Z in R. R is easily seen to be a two dimensional Gorenstein normal ring. The elements y, z form a homogeneous system of parameters. In this case the ideal generated by y and z is tightly closed. One only needs to check that $x \notin (y, z)^*$. We will later see that x is a test element in R. Knowing this, it suffices to see that $x(x^p) \notin (y^p, z^p)$ to prove that x is not in the tight closure of (y, z). Set $k = \frac{p+1}{2}$. Then $x^{p+1} = x^{2k} = (-y^3 - z^5)^k$. A typical term in the expansion of $(-y^3 - z^5)^k$ is $\binom{k}{j} y^{3j} z^{5(k-j)}$. If $x^{p+1} \in (y^p, z^p)$, then either $\binom{k}{j} = 0$, $3j \geq p$ or $5(k - j) \geq p$. Since $k < p$, the binomial coefficient is not zero. Notice that $15k = 5(3j) + 3(5(k - j)) < 5p + 3p = 8p$, since $k = \frac{p+1}{2}$. It follows we may choose a j such that both $3j < p$ and $5(k - j) < p$. For such a choice of j, we see that $y^{3j} z^{5(k-j)} \notin (y^p, z^p)$, which proves that x is not in the tight closure of (y, z).

Example 1.8. Let $R = k[X, Y, Z]/(X^2 + Y^3 + Z^7)$, where k is a field of characteristic ≥ 5. Let x, y, z denote the image of X, Y, Z in R. R is easily seen to be a two dimensional Gorenstein normal ring. The elements y, z form a homogeneous system of parameters. Although the equation has changed only slightly from that of Example 1.7, it is now the case that $x \in (y, z)^*$. To see this directly, let $k = \frac{q+1}{2}$. Then $xx^q = x^{2k} = (-1)^k (y^3 + z^7)^k$. A monomial in the expansion of $(y^3 + z^7)^k$ has the form $y^{3j} z^{7(k-j)}$. If both $3j < p$ and $7(k - j) < p$ then $21j + 21(k - j) < 7p + 3p$, which implies that $21(\frac{p+1}{2}) < 10p$, a contradiction. It follows that $x^{q+1} \in (y^q, z^q)$ for all q and $x \in (y, z)^*$.

2. How Does Tight Closure Arise?

One of the basic questions which we will partially answer in this lecture is, 'where does tight closure come from'? The examples in Section 1 do little to answer this question. We know of essentially four ways in which tight closure arises: through contractions from finite extensions, through the integral closures of powers of ideals, through the failure of the ring to be Cohen-Macaulay (briefly, 'colon-capturing'), and through the persistence of tight closure. On top of these four, there are other less obvious consequences of putting all four together. Probably the most fundamental way in which tight closure arises is through the extension and contraction of ideals in finite extensions. See [HH10, (5.22)] and [Hu5, (1.7)].

Theorem 2.1. (Tight closure from contractions). *Let $S \subseteq T$ be a module-finite extension of Noetherian domains of positive characteristic p. Let $I \subseteq S$ be an ideal. Then $(IT)^* \cap S \subseteq I^*$.*

Proof. Let $w \in (IT)^* \cap S$. We can choose an S-linear map $\phi : T \to S$ such that $\phi(1) = d \in S - \{0\}$ and we can choose $c \in T - \{0\}$ such that $cw^q \in (IT)^{[q]}$ for all q.[3] We can choose a nonzero multiple of c in S, and so we may assume that $c \in S - \{0\}$. It follows that for all q, cw^q is a T-linear combination of elements of $I^{[q]}$. Applying the map ϕ, we find that $dcw^q \in I^{[q]}$ for all q, which shows that w is in the tight closure of I. $\qquad\square$

Example 2.2. Consider the hypersurface R defined by $x^3 + y^3 + z^3 = 0$. In (1.2) we saw that the tight closure of (y, z) is (y, z, x^2). The fact that x^2 is in the tight closure of (y, z) is no 'accident'; in fact we shall see that it can be explained by every single way in which we know tight closure arises. We first apply Theorem 2.1 as follows.

First suppose that $p \equiv 2 \mod(3)$. Write $p = 3k + 2$, so that $x^{2p} = x^{6k+4} = x \cdot (x^3)^{2k+1} = x(-1)^{2k+1}(y^3 + z^3)^{2k+1}$. We claim that $x^{2p} \in (y^p, z^p)$. To see this it suffices to prove that each monomial in the expansion of $(y^3 + z^3)^{2k+1}$ has an exponent at least p. But if $i + j = 2k + 1$ and $3i \leq p - 1$ and $3j \leq p - 1$ then $3(2k + 1) = 3(i + j) \leq 2p - 2$, a contradiction. Hence $x^2 \in (y, z)R^{1/p} \cap R$. By Theorem 2.1, this proves that $x^2 \in (y, z)^*$.

The case in which $p \equiv 1 \mod(3)$ is not so clear. In that case R is F-pure, which implies that for every ideal $I \subseteq R$, $IR^{1/p} \cap R = I$. However, one can show that there is a finite extension of R, say S, such that $x^2 \in (y, z)S \cap R$. The fraction field of S is an Artin-Schreier extension of the fraction field of R. It is a good exercise to try to find S.

It follows from Theorem 2.1 that if S is an arbitrary integral extension of a Noetherian domain R of positive characteristic, and $I \subseteq R$, then $IS \cap R \subseteq I^*$. For every element of $IS \cap R$ is in $IT \cap R$ for some finite extension T of R, and then we may apply Theorem 2.1. For a domain R by R^+, the *absolute integral closure* of R, we mean the integral closure of R in an algebraic closure of its fraction field. The discussion above proves that for any ideal, $IR^+ \cap R \subseteq I^*$. It is one of the important open questions of tight closure whether equality always holds. The best that is presently known is a result due to Karen Smith [Sm4]: if I is generated by parameters in an excellent local domain of characteristic p, then $IR^+ \cap R = I^*$. We will sketch the proof of this theorem in Section 7. The ring R^+ has amazing properties, e.g., the sum of any two prime ideals is either prime or the whole ring. If R is local, excellent, and of positive characteristic, R^+ is even a big Cohen-Macaulay algebra [HH7]. The study of such absolute integral closures is closely tied to the theory of tight closure. See [HH7, Ho8].

A second way in which tight closure arises is through what can be thought of as 'colon-capturing'. Elements in colon ideals which would be in the base ideal if the ring were Cohen-Macaulay, are at least in the tight closure of the base ideal. Specifically:

---

[3]Hochster has developed the idea of this proof considerably in his paper [Ho8] where he introduces the concept of 'solid closure'.

Theorem 2.3. (Tight closure via colon-capturing). *Let (R, m) be a local equidimensional ring of prime characteristic which is a homomorphic image of a Cohen-Macaulay local ring S. Let $x_1, \ldots, x_t$ be parameters in R. Then*

    a) $(x_1, \ldots, x_{t-1}) :_R x_t \subseteq (x_1, \ldots, x_{t-1})^*$.

    b) $(x_1^n, \ldots, x_t^n) :_R (x_1 \cdots x_t)^{n-1} \subseteq (x_1, \ldots, x_t)^*$.

Proof. We need a lemma which can easily be proved using prime avoidance:

Lemma 2.4. *Let S be a catenary Noetherian ring and let Q be a proper ideal of S of height m. Set $R = S/Q$, and assume that R is equidimensional. Let $x_1, \ldots, x_k$ be parameters in R. Then there exist elements $z_1, \ldots, z_m \in Q$ and lifting y_i of x_i such that any j element subset of the y_i together with all the z_i has height $m + j$. Furthermore the z_i may be chosen so that there exists an element $c \notin Q$ with $cQ^n \subseteq (z_1, \ldots, z_m)$ for some n.*

Write $R = S/Q$, and choose z_i and y_i as in Lemma 2.4. Pick an element $c \notin Q$ such that $cQ^{q'} \subseteq (z_1, \ldots, z_m)$, where $m = \text{height}(Q)$ and q' is a power of p.

We first prove a). Suppose that $r = r_t \in (x_1, \ldots, x_{t-1}) :_R x_t$. Then there is a relation $\sum_{1 \le i \le t} r_i x_i = 0$. Lift the r_i to $s_i \in S$. The relation $\sum_{1 \le i \le t} r_i^q x_i^q = 0$ becomes in S the relation $\sum_{1 \le i \le t} s_i^q y_i^q \in Q$. Raising this equation to the q'th power and multiplying by c gives the relation, $\sum_{1 \le i \le t} c s_i^{q'q} y_i^{q'q} = \sum_i t_i z_i$ for some $t_i \in S$. Now the y_i together with the z_i form a regular sequence, so we must have that $c s_t^{q'q} \in (z_1, \ldots, z_n, y_1^{q'q}, \ldots, y_{t-1}^{q'q})$. Let d be the image of c in R. Then $d r_t^q \in (x_1^q, \ldots, x_{t-1}^q)$ for all $q \ge q'$, which finishes the proof.

The proof of b) is similar. Let $r \in (x_1^n, \ldots, x_t^n) :_R (x_1 \cdots x_t)^{n-1}$. Lift r to $s \in S$. There will be a relation in S,

$$s(y_1 \cdots y_t)^{n-1} = \sum_{1 \le i \le t} s_i(y_i)^n + w$$

where $w \in Q$. Taking a $q'q$th power and multiplying by c yields an equation,

$$c s^{q'q}(y_1 \cdots y_t)^{q'q(n-1)} = \sum_{1 \le i \le t} (s_i)^{qq'} (y_i)^{qq'n} + \sum_{1 \le i \le m} t_i z_i.$$

As the y_i together with the z_i form a regular sequence, it follows that

$$c s^{q'q} \in ((y_1)^{q'q}, \ldots, (y_t)^{q'q}, z_1, \ldots, z_m).$$

Going modulo Q then yields that $r \in (x_1, \ldots, x_t)^*$. $\square$

Example 2.5. We return to the example R, the hypersurface $x^3 + y^3 + z^3 = 0$. We know that $x^2 \in (y, z)^*$. It is not clear why Theorem 2.3 should relate to this fact since R is Cohen-Macaulay, and Theorem 2.3 gives no information in the case in which the ring is Cohen-Macaulay. However let $S = R[mt]$, the Rees algebras of the homogeneous maximal ideal m of R. In S, $y, zt, z + yt$ are parameters, and $(z + yt)x^2t = x^2(zt) + y(x^2t^2)$, so that $x^2t \in (y, zt) : (z + yt)$. By Theorem 2.3, it follows that $x^2t \in (y, zt)^*$. There is a homomorphism ϕ from S back to R by first embedding S in $R[t]$, and then setting $t = 1$. By persistence, Theorem 2.13, it follows that $x^2 = \phi(x^2 t) \in \phi((y, zt))^* = (y, z)^*$.

The idea of colon-capturing has been developed a great deal in the notion of phantom homology. Homology is *phantom* if it is in the tight closure of 0 in the homology module. To study (in fact even define) phantom homology requires a discussion of the tight closure of a submodule of a given module. See [Ab1, HH8, HH11, Hu5] for developments in this line.

A third way in which tight closure arises is through the integral closure of powers of ideals. We first recall the definition of integral closures.

Definition 2.6. Given an ideal I, an element x is in the *integral closure* of I, $\overline{I}$, if x satisfies an equation of the form $x^k + a_1 x^{k-1} + \cdots + a_k = 0$ where $a_i \in I^i$.

An equivalent definition for the integral closure in a Noetherian ring R brings it closer to the definition of tight closure. An element x is in $\overline{I}$ iff there exists an element c, not in any minimal prime of R, such that for infinitely many N (equivalently for all large N), $cx^N \in I^N$.

Assume now that R is a Noetherian local ring of characteristic p. The next theorem gives a tight closure version of the theorem of Briançon and Skoda. It appears in [HH4, Theorem 5.4].[4]

Theorem 2.7. (Tight Closure Briançon-Skoda). *Let R be a ring of characteristic p. Let I be any ideal generated by n elements. For all $w \geq 0$,*

$$\overline{I^{n+w}} \subseteq (I^{w+1})^* .$$

Proof. Write $I = (a_1, \ldots, a_n)$. Let $z \in \overline{I^{n+w}}$. There exists an element $c \in R^\circ$ such that $cz^N \in I^{(n+w)N}$ for all large N.

We claim for all $h \geq 0$, $I^{nh+wh} \subseteq (a_1^h, \ldots, a_n^h)^{w+1} I^{h(n-1)}$. Consider a monomial v in the ideal I^{nh+wh} in which the exponent of a_i is b_i. Let c_i be the integer part of b_i/h for each i. Then $c_i + 1 > b_i/h$, and so $\sum_i (c_i + 1) = \sum_i c_i + n >$

---

[4]The theorem of Briançon and Skoda was proved in response to a question of Mather: let $\mathbf{O}_n = \mathbb{C}\{z_1, \ldots, z_n\}$ be the ring of convergent power series in n variables. Let $f \in \mathbf{O}_n$ be a non-unit (i.e., f vanishes at the origin). The Jacobian ideal of f is $j(f) = (\partial f/\partial z_1, \ldots, \partial f/\partial z_n)\mathbf{O}_n$. Since $f \in \overline{j(f)}$ there is an integer k such that $f^k \in j(f)$. Mather asked if there a bound for k which works for all non-units f. Briançon and Skoda answered this question affirmatively with the following stronger result ([BrS]): Let $I \subseteq \mathbf{O}_n$ be an ideal which can be generated by d elements. Then for every $w \geq 0$ $\overline{I^{d+w}} \subseteq I^{w+1}$. Since $j(f)$ has at most n generators, applying the theorem with $I = j(f)$ and $w = 0$ gives $f^n \in j(f)$, answering Mather's question. The ring $\mathbf{O}_n$ is a regular local ring so one may ask if the entirely algebraic statement of the theorem remains true in any regular ring. Lipman and Sathaye succeeded in proving the same theorem for arbitrary regular local rings [LS]. Lipman and Teissier partly extended this theorem to rings having rational singularities, cf. [LT]. Since then, there has been considerable activity in proving more precise theorems of this type, e.g., see [HH10, AH1, AH2, AHT, RS, Sw1, Sw2, Sw3, L2].

$\sum_i (b_i/h) = n + w$, so that $\sum_i c_i > w$, i.e. $\sum_i c_i \geq w + 1$ whence

$$\prod_i (a_i^h)^{c_i} \in (a_1^h, \ldots, a_n^h)^{w+1},$$

and divides the given monomial generator v of I^{nh+wh}.

It follows that $cz^N \in I^{(n+w)N} \subseteq (a_1^N, \ldots, a_n^N)^{w+1} I^{N(n-1)}$. Put $N = q = p^e$. We obtain that $cz^q \in (I^{w+1})^{[q]}$ by ignoring the term $I^{N(n-1)}$ in the containment of the above line. Hence $z \in (I^{w+1})^*$ as claimed. $\qquad\square$

Remark 2.8. There is a useful form of the tight closure Briançon-Skoda theorem for graded rings and homogeneous parameter ideals which we shall later need. If R is a graded ring, by $R_{\geq n}$ we denote the ideal generated by all forms of R of degree at least n. This theorem appears in [HH10, Theorem 7.7] and in [Sm7, Proposition 3.3]. In the latter paper it is generalized to the case of an arbitrary homogeneous m-primary ideal.

Theorem 2.9. (Graded Briançon-Skoda). *Let R be a Noetherian nonnegatively graded ring of positive Krull dimension d over a field $k = R_0$ of positive characteristic p. Let $f_1, \ldots, f_d$ be a homogeneous system of parameters of degrees $n_1, \ldots, n_d$. Set $N = n_1 + \cdots + n_d$. Then $R_{\geq N} \subseteq I^*$, where $I = (f_1, \ldots, f_d)$.*

Proof. We first do the case in which the degree of f_i is n for all $1 \leq i \leq d$. In this case $N = dn$, and we claim that $R_{\geq N} \subseteq \overline{I^d} \subseteq I^*$, the last containment following from Theorem 2.7.

To see the first containment, $R_{\geq N} \subseteq \overline{I^d}$, observe that R is module-finite over $S = k[f_1, \ldots, f_d]$. A form g of degree $\delta \geq dn$ satisfies an integral equation,

$$g^m + b_1 g^{m-1} + \cdots + b_m = 0 \tag{1}$$

where $b_i \in S$. We may assume that the b_i are homogeneous and $\deg(b_i) = i\delta$. This means that $b_i \in R_{i\delta} \cap S$, and hence $b_i \in (f_1, \ldots, f_d)^{i\delta/n} \subseteq (f_1, \ldots, f_d)^{id}$. In particular, (1) then proves that $g \in \overline{I^d}$ as required.

Now suppose that the f_i have possibly unequal degrees. Choose nonnegative integers m_i such that $F_1 = f_1^{m_1+1}, \ldots, F_d = f_d^{m_d+1}$ have equal degrees, say T. Let $g \in R_{\geq N}$ where $N = n_1 + \cdots + n_d$. Then $(g f_1^{m_1} \cdots f_d^{m_d})^q$ has degree at least qdT and hence by the first case is in the tight closure of $(F_1^q, \ldots, F_d^q)$. There is then a $c \in R^\circ$ such that

$$c(g f_1^{m_1} \cdots f_d^{m_d})^q \in (f_1^{q(m_1+1)}, \ldots, f_d^{q(m_d+1)})$$

and so

$$cg^q \in (f_1^{q(m_1+1)}, \ldots, f_d^{q(m_d+1)}) :_R (f_1^{m_1} \cdots f_d^{m_d})^q$$

which by Theorem 2.3 b) is contained in $(f_1^q, \ldots, f_d^q)^*$. Multiplying by a test element d, we obtain that $cdg^q \in (f_1^q, \ldots, f_d^q)$, and finally that $g \in (f_1, \ldots, f_d)^*$. $\qquad\square$

Example 2.10. In the hypersurface $x^3 + y^3 + z^3 = 0$ of (1.2) we saw that $x^2 \in (y, z)^*$. This follows at once from the graded Briançon-Skoda Theorem 2.9 since the degree of x^2 is $2 = \deg(y) + \deg(z)$. Likewise, in Example 1.8, we saw that $x \in (y, z)^*$ in the hypersurface $x^2 + y^3 + z^7 = 0$. This hypersurface is weighted homogeneous with the degree of $x = 21 > \deg(y) + \deg(z) = 14 + 6$.

A final way in which tight closure arises is through the property of *persistence*. This means that elements in the tight closure of an ideal stay in the tight closure of the image of the ideal under any homomorphism. This property of tight closure is not obvious from the definition: the element c in the definition of tight closure might well go to zero under the given homomorphism. The proof of persistence needs the following theorem on the existence of test elements, whose proof will be deferred until the next section.

Definition 2.11. A Noetherian ring R of characteristic p is said to be *F-finite* if R is a finite module over R^p.

The next theorem appears in [HH9, Theorem 6.20], while the persistence theorem is essentially [HH9, Theorem 6.24].

Theorem 2.12. (Existence of Test Elements). *Let R be either a reduced algebra of finite type over an excellent local ring (B, m, K) of characteristic p, or a reduced ring of characteristic p which is F-finite. Let $c \in R^\circ$ be such that R_c is F-regular and Gorenstein (e.g. if R_c is regular). Then c has a power which is a test element for R.*

We can now prove,

Theorem 2.13. (Persistence of Tight Closure). *Let $\phi : R \to S$ be a homomorphism of Noetherian rings of characteristic p. Let I be an ideal of R and let $w \in R$ be an element in I^*. Assume either that R is essentially of finite type over an excellent local ring, or that R_{red} is F-finite. Then $\phi(w)$ is in the tight closure of IS.*

Proof. If there is a counterexample to the theorem, there is one in which S is a domain, for if $\phi(w)$ is not in the tight closure of IS, this will remain true when S is replaced by S/P for a suitable minimal prime P. Thus, we may assume that S is a domain. Let $Q = \text{Ker}\,(R \to S)$.

Then we may replace S by R/Q as well. For if tight closure is preserved when we pass to R/Q, it will also be preserved when we pass to S since R/Q embeds in S. (The definition of tight closure immediately proves there is no problem with Theorem 2.13 as long as a test element does not go into a minimal prime of S under ϕ, which is certainly the case if the map is an injective map of domains). Thus, there is no loss of generality in supposing that $S = R/Q$ for a suitable prime Q of R. Let $Q = Q_h \supseteq Q_{h-1} \supseteq \cdots \supseteq Q_0$ be a saturated chain of prime ideals of R descending from Q such that Q_0 is a minimal prime of R. We shall prove by induction on i that tight closure is preserved when we pass from R to R/Q_i, $0 \le i \le h$. For $i = 0$ this is clear, for tight closure is always preserved when one kills a minimal prime. To carry through the inductive step, we may replace R by R/Q_{i-1}.

To complete the proof, it suffices to show that if R is a domain and Q is a height one prime ideal of R, then tight closure is preserved when we pass to $S = R/Q$. To see this, let R' be the integral closure of R in its fraction field (which is module-finite over R, since R is excellent), and let Q' be a prime ideal of R' which lies over Q, so that $R/Q \to R'/Q'$ is injective and module-finite. Now, tight closure is obviously preserved when we pass from R to $R' \supseteq R$. Moreover, since R' is excellent normal and Q' is height one (so that $R'_{Q'}$ is regular), there is an element $c \in R' - Q'$ such that R'_c is regular. After replacing c by a power we see that we may assume that c is a test element for R' not in Q'. It follows that tight closure will be preserved when we pass from R' to $T = R'/Q'$. Thus, the image of w in T is in the tight closure of IT. To finish the proof, we must show that this implies that the image of w in S is in the tight closure of IS. (Here, $S \subseteq T$ is a module-finite extension of domains.) This follows from Theorem 2.1. $\square$

We end this section by giving a few of the consequences of how tight closure arises.

Theorem 2.14. (Briançon-Skoda). *Let (R, m) be a regular local ring containing a field. Let I be any ideal of R which is generated by l elements. Then for any $w \geq 0$*

$$\overline{I^{l+w}} \subseteq I^{w+1}.$$

Proof. In positive characteristic this follows immediately from Theorem 2.7 and Theorem 1.3 e). The equicharacteristic zero case follows using reduction to characteristic p. $\square$

We should note, however, that this theorem is known in mixed characteristic. See [LT, LS].

Theorem 2.15. (Weakly F-regular is Cohen-Macaulay). *Let R be a local ring which is the homomorphic image of a Cohen-Macaulay ring and of positive characteristic. If R is weakly F-regular, or even if ideals generated by parameters are tightly closed[5] then R is Cohen-Macaulay.*

Proof. The proof is immediate from part a) of Theorem 2.3. $\square$

Theorem 2.16. *Let S be a regular ring of characteristic p, and suppose that R is a direct summand of S. Then R is Cohen-Macaulay.*

Proof. It is easy to reduce to the case in which R is complete local and S is a regular domain. Since every ideal of R is contracted from S, and since every ideal of S is tightly closed, it follows that R is weakly F-regular, i.e. every ideal is tightly closed (see Theorem 1.6(6)). It follows immediately that R is Cohen-Macaulay by Theorem 2.15. $\square$

---

[5]Such rings are said to be F-rational. We will study these in the eighth section.

It is also worth pointing out that the 'Monomial Conjecture' (see [Ho2, 6])
follows at once from Theorem 2.3.

Theorem 2.17. (The Monomial Conjecture). *Let (R, m) be a local ring containing a field and let $x_1, \ldots, x_d$ be a system of parameters. Then for all $t \geq 1$,
$(x_1 \cdots x_d)^{t-1} \notin (x_1^t, \ldots, x_d^t)$.*

Proof. Suppose that R has characteristic p. We may reduce at once to the case
in which R is complete. If $(x_1 \cdots x_d)^{t-1} \in (x_1^t, \ldots, x_d^t)$, it follows from Theorem 2.3 b) that $1 \in (x_1, \ldots, x_d)^*$. But then there is an element $c \in R^o$ such that
$c1^q = c \in (x_1^q, \ldots, x_d^q)$ for all large q. This is a contradiction as intersecting the
ideals $(x_1^q, \ldots, x_d^q)$ over all q yields 0. The general case follows from reduction to
characteristic p.   □

It is an important question to find a definition of a 'tight' closure in mixed
characteristic, or even to find a definition in equicharacteristic 0 which does not
refer back to characteristic p. It is easy to list the properties which a good closure
operation should have to give it the same force as tight closure. The closure should
be persistent. Every ideal in a regular local ring should be closed under the operation. The operation should capture the colon in the sense of Theorem 2.3. The
closure of an ideal should contain the integral closure of the dth power of itself,
where d is the dimension of the ring. The expansion and contraction of an ideal in
a module-finite extension should be in the closure of the ideal. A final property a
good closure operation should have is a theory of test elements: if $c \in R$ is such
that every ideal in R_c is closed under the operation, then there should be a fixed
power of c which multiplies the closure of every ideal $I \subseteq R$ back into I. We do not
quite know this for tight closure, but we do know many cases where it is true (see
Theorem 2.12 and the next section). A closure with all these properties gives a
very powerful tool. Unfortunately, it is far from clear how to define such a closure
except through reduction to characteristic p.

3. The Test Ideal I

The existence of test elements is one of the most important theorems in the theory
of tight closure. Recall their definition:

Definition 3.1. Let R be a Noetherian ring of characteristic p. An element $c \in R^o$
is said to be a *test element* if for all ideals I and all $x \in I^*$, $cx^q \in I^{[q]}$ for all
$q = p^e$. An element $c \in R^o$ is said to be a test element for parameter ideals, or a
parameter test element for short, if for all ideals I generated by parameters, and
all $x \in I^*$, $cx^q \in I^{[q]}$ for all $q = p^e$.

For (parameter) test elements it is easy to see that the definition could just
read that $cx \in I$ for all (parameter) ideals I. Moreover, both are properly defined
to include the case of tight closures of modules, but we have elected not to do
so in these notes. If the ring is approximately Gorenstein, the notions agree. Any

excellent normal domain is approximately Gorenstein [Ho4]. The ring R needs to be reduced to have test elements.

The best result to date about the existence of test elements was given as Theorem 2.12. We repeat it here (cf. [HH9], (6.2)):

Theorem (Existence of Test Elements). *Let R be a reduced algebra of finite type over an excellent local ring (B, m, K) of characteristic p. Let $c \in R^o$ be such that R_c is F-regular and Gorenstein (e.g. R_c regular). Then c has a power which is a test element for R.*

Of course, the best theorem one wants is one which says that if R is reduced (and excellent, perhaps) and if R_c is weakly F-regular, then c has a power which is a test element. This is an open question.

The proof of the above theorem is beyond the scope of these notes. However, one of the main cases is when the Frobenius map is a finite map, and in this case complete details can be given. The discussion below is taken from [Hu5]. Kunz [Ku1] has shown that if R is an F-finite Noetherian ring of characteristic p, then R is excellent. Moreover the property of the Frobenius map being finite passes to finitely generated R-algebras.

Let (A, m) be a regular local ring of characteristic p. Assume that $A^{1/p}$ is finite over A as a module. Since A is regular, the Frobenius map is flat, and hence $A^{1/p}$ is a flat A-module. As it is finitely presented and A is local, it is actually free. It follows that for all $q = p^e$, $A^{1/q}$ is also free over A. Let $d \in A$ be nonzero. For sufficiently large $q = p^e$, $m^{[q]}$ does not contain d. Taking qth roots yields that $d^{1/q} \notin mA^{1/q}$. Since $A^{1/q}$ is free over A it follows that one may use $d^{1/q}$ as part of a free basis of $A^{1/q}$. In particular, there is an A-linear homomorphism $\phi : A^{1/q} \to A$ which sends $d^{1/q}$ to 1.

Next suppose that A is a regular domain but not necessarily local. Taking pth roots commutes with localization, so that $A^{1/q}$ will be projective over A for all $q = p^e$. Fix a maximal ideal m of A and a nonzero element $d \in A$. There will be a power of p, say $q = q(m)$, depending upon m, such that $d \notin m^{[q]}$, and so $d^{1/q} \notin mA^{1/q}$. One wants a uniform q working for all maximal ideals (see Exercise 11.4 for a more general result). By the paragraph above, there is a homomorphism from $A_m^{1/q}$ to A_m sending $d^{1/q}$ to 1. Clearing denominators one sees that there is an element $r_m \notin m$ such that there is an A-linear map $\phi_m : A_{r_m}^{1/q(m)} \to A_{r_m}$ sending $d^{1/q}$ to 1. The ideal generated by all such r_m is not contained in any maximal ideal so that there are finitely many of them, say $r_{m_1}, \dots, r_{m_k}$, which generate the unit ideal. Set $q = \max\{q(m_i)\}$. Let m be an arbitrary maximal ideal of A. Some r_{m_i} is not contained in m, say $r = r_{m_1}$. As there is an A_r-linear map from $A_r^{1/q(m_1)} \to A_r$ sending $d^{1/q(m_1)}$ to 1, *a fortiori* there is such a map from $A_m^{1/q(m_1)} \to A_m$. In particular, $d \notin m^{[q]}$, which proves the existence of a uniform q.

The existence of such a q proves that for each maximal ideal m of A, there is an element $r = r_m$ and an A_r-linear map from $A_r^{1/q} \to A_r$ sending $d^{1/q}$ to 1. For each such r, there is a power r^{N_r}, such that there is an A linear map ϕ_r from

$A^{1/q} \to A$ sending $d^{1/q}$ to r^{N_r}. There exists a finite number of such r generating the unit ideal, so we may express $1 = \sum s_i r_i^N$, where N is taken larger than all N_{r_i}. Taking $\phi = \sum s_i \phi_{r_i}$ gives an A-linear map taking $d^{1/q}$ to 1.

Theorem 3.2. [HH3, Theorem 3.4] *Let R be an F-finite reduced ring of characteristic p. Let c be any nonzero element of R such that R_c is regular. Then c has a power which is a test element.*

Proof. Since R_c is regular, the discussion preceding this theorem proves that for every nonzero element $d \in R$ there is a sufficiently high power of p, say Q, such that there exists an R_c-linear map from $R_c^{1/Q}$ to R_c sending $d^{1/Q}$ to 1. Lifting back to R, one obtains an R-linear map from $R^{1/Q}$ to R sending $d^{1/Q}$ to a power of c. Taking $d = 1$ yields an R-linear map from $R^{1/Q}$ to R sending 1 to c^N for some N. The embedding of $R^{1/p}$ into $R^{1/Q}$ composed with this R linear map yields an R-linear map ϕ from $R^{1/p}$ to R sending 1 to c^N. Relabel this power of c as c. Then there is an R-linear map ϕ from $R^{1/p}$ to R sending 1 to c. We claim that for any such c, c^2 is a test element, except in characteristic 2, where c^3 will be a test element.

Let I be an arbitrary ideal of R, and let $z \in I^*$. There is an element $d \in R$, not in any minimal prime of R such that for all q, $dz^q \in I^{[q]}$. Using the results of the paragraphs above, there is a power of p, say q', and an R-linear map α from $R^{1/q'} \to R$ sending $d^{1/q'}$ to c^N for some N. In this case, $c^N z^q \in I^{[q]}$ for all q. Simply take q'th roots of the equation $dz^{qq'} \in I^{[qq']}$ to obtain that $d^{1/q'} z^q \in I^{[q]} R^{1/q'}$. Applying α yields that $c^N z^q \in I^{[q]}$ for all q. The problem is we must prove that this power N can be chosen independently of the element z and the ideal I.

Choose N least with the property that $c^N z^q \in I^{[q]}$ for all q. Write $N = p(\lfloor N/p \rfloor) + i$. Taking pth roots yields that $c^{\lfloor N/p \rfloor + i/p} z^q \in I^{[q]} R^{1/p}$ for all q. Hence $c^{\lfloor N/p \rfloor + 1} z^q \in I^{[q]} R^{1/p}$ for all q. Applying ϕ we obtain that $c^{\lfloor N/p \rfloor + 2} z^q \in I^{[q]}$ for all q. As N was chosen least, we must have that $\lfloor N/p \rfloor + 2 \geq N$. It easily follows that in odd characteristics, $N \leq 2$ and if $p = 2$, $N \leq 3$. $\square$

A ring R is F-finite if R is essentially of finite type over a perfect field K, or if R is complete with perfect residue field K. In fact, all one needs in both cases is that $K^{1/p}$ be finite over K. Thus, this is not a very restrictive hypothesis. The more general theorem of the existence of test elements for excellent local rings follows from Theorem 2.1 by passing to the complete case and expanding the residue field to make it finite over its pth powers. However, this is difficult since one must be able to control the fibers of this base change well enough to control what happens to tight closures.

As I noted at the beginning of this section, every reduced ring of characteristic p which is essentially of finite type over an excellent local ring has abundant test elements in the sense that if R_d is regular, then d has a power which is a test element. An open question is:

Question 3.3. (Existence of Test Elements). Let R be a reduced excellent ring of finite Krull dimension and of characteristic p. Does R have a test element?

Definition 3.4. The *test ideal* of a reduced ring R, denoted $\tau(R)$, is the ideal generated by all test elements.

It follows that any element of $\tau(R)$ not in R^o is a test element. If R is a reduced graded ring over a field such that R has an isolated singularity at its irrelevant ideal, then it follows from Theorem 3.2 that the test ideal is primary to the irrelevant ideal. A particularly important question is what ideal it is.

Example 3.5. Let (R, m) be a one-dimensional complete local domain, and let C be the conductor ideal. In [Cow] it is noted that C is the test ideal. Let I be an arbitrary ideal of R. The tight closure of I is simply the integral closure by the tight closure Briançon-Skoda Theorem. But the integral closure of I is $IS \cap R$, where S is the integral closure of R. Then $C(IS \cap R) \subseteq CSI = CI$. Hence C is contained in the test ideal. Conversely, if $x(IS \cap R) \subseteq I$ for every ideal I of R it easily follows that $x \in C$.

Example 3.6. (F-purity and the test ideal). A ring of characteristic p is *F-pure* if the Frobenius map is pure, i.e. if whenever we tensor the Frobenius map $F : R \to R$ with an R-module M, the ensuing map is injective. If R is F-finite, then this is equivalent to the condition that $u^p \in I^{[p]}$ implies $u \in I$.

Suppose that R is F-pure and has an isolated singularity at m. Then the test ideal is either R or m. To prove this first observe that the test ideal τ will be m-primary, by Theorem 2.12. Choose a power of p, say q', such that $m^{q'} \subseteq \tau$. If I is an ideal and $u \in I^*$, then $\tau u^q \subseteq I^{[q]}$ for large q. It follows that $m^{[q']}I^{[qq']} \subseteq I^{[qq']}$ and hence that $mu^q \subseteq (I^{[q]})^F$ for all large q. Since R is F-pure, it follows that $mu^q \subseteq I^{[q]}$ for all large q and the test ideal contains m. If R is not weakly F-regular, then the test ideal must be exactly m.

A criterion of Fedder gives a nice condition for a quotient of a regular ring to be F-pure:

Theorem 3.7. [Fe3] *Let R be a regular local ring of characteristic p which is F-finite. The quotient ring R/I is F-pure iff $I^{[p]} : I \not\subseteq m^{[p]}$.*

Janet Cowden has given a new proof of this theorem [Cow], and used her method to show that if R is F-pure and F-finite with test ideal τ, then R/τ is also F-pure. If $I = (f)$ is a hypersurface, Fedder's criterion becomes particularly easy to apply. In this case $I^{[p]} : I = (f^{p-1})$ and the criterion just reads that $f^{p-1} \notin m^{[p]}$ iff R is F-pure.

The condition that the test ideal be the whole ring simply says that the ring is weakly F-regular. When R is Gorenstein, this condition is closely related to the condition that R have rational singularities. We shall discuss this connection at length in Sections 6 and 7. The next interesting case is when the test ideal is the maximal ideal. In dimension 1, with algebraically closed residue field of characteristic at least 5, Cowden has classified complete domains whose test ideal is the maximal ideal. The only one up to isomorphism is $k[[t^2, t^3]]$. Cowden [Cow] has also classified 2-dimensional normal Gorenstein complete local domains with algebraically closed field of sufficiently large characteristic whose test ideal is the maximal ideal. These rings turn out to be exactly the class of minimal elliptic singularities.

An important source of test elements can be deduced (see [HH10, (8.22)]) from a theorem of Lipman and Sathaye. This theorem states:

Theorem 3.8. *Let R be a regular Noetherian domain with quotient field K. Let L be a finite separable field extension of K, and let S be a finitely generated R-subalgebra of L. Set $J_{S/R} = J = 0^{th}$ Fitting ideal of the S-module of Kähler R-differentials $\Omega_{S/R}$. Let T be the integral closure of S. Then $JT \subseteq S$.*

We can use this to prove the following:

Theorem 3.9. *Let R be a Noetherian domain of characteristic p which is module-finite over a regular subring A and whose fraction field is separable over the fraction field of A. Then every nonzero element of $J_{R/A}$ is a test element.*

The proof of this needs several steps. We first prove the following lemma:

Lemma 3.10. *Let R be a Noetherian domain which is a module-finite and generically smooth extension of a regular domain A. Then $A^{1/q}[R] \cong A^{1/q} \otimes_A R$ is flat over R.*

Proof. The condition that R is generically smooth over A implies that there is an element $d \in A^\circ$ such that R_d is smooth over A_d. Set $S = R_d$ and $B = A_d$. We first claim that $S \otimes_B B^{1/q} \cong S[B^{1/q}]$. This is a local question on B so without loss of generality we may assume that B is local. Both $S \otimes_B B^{1/q}$ and $S[B^{1/q}]$ are free over B, and by Nakayama's lemma it suffices to see we have an isomorphism after killing the maximal ideal n of B. But then S/nS is a finite separable extension of B/n and this is well-known.

The natural map of $A^{1/q} \otimes_A R$ into $R^{1/q}$ has image $A^{1/q}[R]$. This map becomes an isomorphism after inverting d by the argument above. Since both of these rings are torsion-free over A, the result follows.   □

Lemma 3.11. *Let the notation be as in Theorem 3.9. Let $c \in J_{R/A}$. Then for all $q = p^e$, $cR^{1/q} \subseteq A^{1/q}[R]$.*

Proof. The point is that the relative Jacobian ideal $J_{R/A}$ is equal to the relative Jacobian ideal $J_{A^{1/q}[R]/A^{1/q}}$ since $A^{1/q}$ is flat over A, and Lemma 3.10 identifies $A^{1/q}[R]$ with $A^{1/q} \otimes_A R$. Hence any element in $J_{R/A}$ multiplies the integral closure of $A^{1/q}[R]$ back into $A^{1/q}[R]$. Since $R^{1/q}$ is contained in the integral closure of $A^{1/q}[R]$, clearly any nonzero element in $J_{R/A}$ satisfies the conditions of the lemma.   □

Proof of Theorem 3.9. Let $c \in J_{R/A}$. Then for all $q = p^e$, $cR^{1/q} \subseteq A^{1/q}[R]$, by Lemma 3.11. Let $x \in I^*$. It suffices to prove that $cx \in I$, since if this holds for all ideals I and all $x \in I^*$ then c will be a test element. Assume that $R \ne I : cx$. There exists some nonzero element $d \in R$ such that $dx^q \in I^{[q]}$ for all q. By taking a multiple of d we may assume that $d \in A$. Then $d^{1/q}x \in IR^{1/q}$, and by multiplying by c we obtain that $cd^{1/q}x \in IA^{1/q}[R]$. Since $A^{1/q}$ is flat over A, by base change $A^{1/q}[R]$ is flat over R (see Lemma 3.10). Therefore $d^{1/q} \in (I :_R cx)A^{1/q}[R]$. Raising to the qth power, one obtains that $d \in m^{[q]}$ which is a contradiction. $\qquad\square$

Example 3.12. Let $R = k[X_1, \ldots, X_n]/(F)$ where $F = X_1^{a_1} + \cdots + X_n^{a_n}$. We can choose a regular subring A_i inside R where $A_i = k[X_1, \ldots, X_{i-1}, X_{i+1}, \ldots, X_d]$. and obtain that provided the characteristic p of k does not divide the product $a_1 \cdots a_n$, that R is generically smooth over all A_i. The relative Jacobian ideal J_{R/A_i} is simply the partial of F with respect to X_i. It follows that the test ideal of R contains $(X_1^{a_1-1}, \ldots, X_n^{a_n-1})$.

4. The Test Ideal II: the Gorenstein Case

In this section we will discuss several results which concern the test ideal of a Gorenstein ring. In this case, the duality between ideals in R and submodules of the injective hull of the residue field of R plays an important role and gives us extra punch. We are aiming for two results. The first result is a result of Smith which gives that the test ideal localizes in the Gorenstein case. This will be an important ingredient in Section 7 of the proof that the tight closure and plus closure of parameters agree. The second says that when the test ideal is m-primary, we can identity R/τ with the Matlis dual of the tight closure of a system of parameters of test elements modulo the ideal generated by the parameters. The proofs in this section rest on the identification of the injective hull of the residue field of a Gorenstein local ring with the highest local cohomology of the ring. We begin with a general discussion of local cohomology.

Discussion: Local Cohomology. Fix any set of elements $x_1, \ldots, x_d$ for a ring R. Let I denote the ideal they generate and let x denote their product. The local cohomology module $H^i_I(R)$ can be computed as the cohomology at the ith spot of the complex

$$0 \longrightarrow R \longrightarrow \bigoplus R_{x_i} \longrightarrow \bigoplus R_{x_{i_1} x_{i_2}} \longrightarrow \cdots \longrightarrow R_{x_1 x_2 \cdots x_d} \longrightarrow 0. \qquad (2)$$

We will use the notation $\eta = [\frac{z}{x^t}]$ to denote the image of the fraction $\frac{z}{x^t} \in R_{x_1 x_2 \cdots x_d}$ in the highest local cohomology $H^d_I(R)$. If R is graded and the x_i are homogeneous, then the individual modules appearing in the sequence (2) are also graded, and consequently so are the local cohomology modules. A typical homogeneous element in the last local cohomology, $\eta = [\frac{z}{x^t}]$ will have degree equal to $\deg(z) - t(\sum_{1 \le i \le d} \deg(x_i))$. If η is such an element, $\eta = 0$ iff there exists an integer

n such that $zx^n \in (x_1^{t+n}, x_2^{t+n}, \ldots, x_d^{t+n})$. If R is Cohen-Macaulay, and $x_1, \ldots, x_d$ form a system of parameters, then $\eta = 0$ iff $z \in (x_1^t, x_2^t, \ldots, x_d^t)$.

If R has characteristic p, then the Frobenius acts on R and all its localizations, and is compatible with the maps in the complex above, and thus Frobenius acts on the local cohomology. Specifically, applying the Frobenius to an element $\eta = [\frac{z}{x^t}] \in H_m^d(R)$ gives the element $F(\eta) = [\frac{z^p}{x^{tp}}]$. If η is homogeneous of degree N, then evidently the degree of $F(\eta)$ is pN. We denote $F(\eta)$ by η^p. Since the Frobenius acts on the local cohomology, we can easily define the tight closure of submodules of the local cohomology module. In fact, there is a definition for submodules of arbitrary modules in general. In these notes of particular importance will be the tight closure of 0 in the highest local cohomology module $H_m^d(R)$. We say that $\eta \in H_m^d(R)$ is in the *tight closure of* 0 if there exists an element $c \in R^o$ such that $c\eta^q = 0$ for all $q = p^e \gg 0$. Translating this back to R yields that if $\eta = [\frac{z}{x^t}]$, then η is in the tight closure of 0, denoted $0^*_{H_m^d(R)}$, iff $z \in (x_1^t, \ldots, x_d^t)^*$, using Theorem 2.3.

If (R, m) is Gorenstein, then it is well-known that an injective hull of the residue field of R can be identified with $H_m^d(R)$, where $d = \dim R$. A crucial point in our discussion of the test ideal is the following proposition from [HH4, (8.23)]:

Proposition 4.1. *Let (R, m) be a Gorenstein local ring of positive characteristic. Let τ be the test ideal of R. Then $\tau = \operatorname{Ann}_R(0^*_{H_m^d(R)})$.*

Proof. We proved above that $\tau \subseteq \operatorname{Ann}_R(0^*_{H_m^d(R)})$. Conversely suppose that $c \in \operatorname{Ann}_R(0^*_{H_m^d(R)}) \cap R^o$. If $c \notin \tau$, there is an ideal $I \subseteq R$ and an element $x \in I^*$ such that $cx \notin I$. Replace I by an ideal J containing I and maximal with respect to not containing cx. Then $x \in J^*$, $cx \notin J$, and J is m-primary and irreducible. Then R/J embeds in $H_m^d(R)$ as this module is an injective hull of the residue field of R. But then the image of z in $H_m^d(R)$ under this embedding is in $0^*_{H_m^d(R)}$, and by assumption τ must kill it. $\square$

This Proposition gives us a very powerful corollary:

Corollary 4.2. *Let (R, m) be a Gorenstein local ring of positive characteristic with test ideal τ. Let $x_1, \ldots, x_d$ be an arbitrary system of parameters generating an ideal I, and set $J = (x_1, \ldots, x_i)$ for some $1 \le i \le d - 1$. Then*

(1) $I : I^* = I + \tau$, and $I : \tau = I^*$,

(2) $J : \tau = J^*$,

(3) $\bigcap_K (K : K^*) = \tau$, *where the intersection runs over all ideals K generated by a full system of parameters,*

(4) *if J is an ideal such that for all ideals K generated by a system of parameters $K : J = K^*$, then $J = \tau$.*

Proof. To prove (1), it suffices to prove that $I : \tau = I^*$, since then duality forces $I : I^* = I : (I : \tau) = I + \tau$. Suppose that $z\tau \subseteq I$. We need to prove that $z \in I^*$, and it suffices to prove that $\eta = [\frac{z}{x}] \in 0^*_{H_m^d(R)}$ by the discussion above. By

Proposition 4.1, $\tau = \mathrm{Ann}_R(0^*_{H^d_m(R)})$, and hence by duality, $0^*_{H^d_m(R)} = \mathrm{Ann}_{H^d_m(R)}\tau$. Since $\tau\eta = 0$, the corollary follows.

Write $I_t = J + (x^t_{i+1}, \ldots, x^t_d)$. Suppose that $u \in J : \tau$. Then $u \in I_t : \tau$, and by part (1) it then follows that $u \in I^*_t$. The proof of (2) will be finished by proving that $\cap_t I^*_t = J^*$. Suppose that $z \in \cap_t I^*_t$, and let c be a test element. Then for all t and all q, $cz^q \in J^{[q]} + (x^{tq}_{i+1}, \ldots, x^{tq}_d)$. Intersecting over t yields that for all q, $cz^q \in J^{[q]}$, which proves the claim.

To prove (3), we first observe that $\tau \subseteq \cap_K(K : K^*)$ by definition of the test ideal. To prove that converse, suppose that $u \in \cap_K(K : K^*)$. We can prove u is a test element by proving that $u \in \mathrm{Ann}_R(0^*_{H^d_m(R)})$. But if η is in $0^*_{H^d_m(R)}$, we can represent $\eta = [\frac{z}{x}]$ for some system of parameters $x_1, \ldots, x_d$, and $\eta \in 0^*_{H^d_m(R)}$ is equivalent to $z \in K^*$, where $K = (x_1, \ldots, x_d)$. Our assumption on u then shows that $uz \in K$, and hence $u\eta = 0$.

By (3) it suffices to prove that $J = \cap_K(K : K^*)$ where the intersection runs over all ideals K generated by a full system of parameters. Since $K : J = K^*$, duality yields that $K : K^* = J + K$. Hence $\cap_K(K : K^*) = \cap_K(J + K) = J$ as J is separated. It follows that $J = \tau$.   □

Corollary 4.3. *Let (R, m) be either a complete local Gorenstein ring of characteristic p, or a nonnegatively graded Gorenstein ring over a field of characteristic p. Suppose that $x_1, \ldots, x_d$ is a system of parameters of R, homogeneous in the second case, which are test elements. Then $\tau = (x_1, \ldots, x_d) :_R (x_1, \ldots, x_d)^*$, where τ is the test ideal of R.*

Proof. Clearly $\tau \subseteq (x_1, \ldots, x_d) :_R (x_1, \ldots, x_d)^*$. To prove the reverse inclusion, choose an element $c \in (x_1, \ldots, x_d) :_R (x_1, \ldots, x_d)^*$. Let J be an arbitrary ideal of R and let $x \in J^*$. Suppose that $cx \notin J$. This continues to hold after localizing at some maximal ideal M. If $M \neq m$, then since x_i are test elements it follows that $x \in I_M$. Hence $M = m$, and we may now assume that R is local with maximal ideal m.

The proof now follows at once from Corollary 4.2(1), using the assumption that the x_i are test elements and hence in τ.   □

The next theorem is a crucial ingredient of the proof of Theorem 7.1. One of the important unsolved problems of tight closure theory is whether tight closure commutes with localization. There are a few cases in which this is known. If the ideal I is generated by a regular sequence, then $(I^*)_W = (I_W)^*$ for an arbitrary multiplicatively closed subset W of R. Other cases are in [AHH]. Theorem 4.4 gives us the next best thing in the Gorenstein case: the test ideal commutes with localization.

Theorem 4.4. [Sm4, (4.1)] *Let (R, m) be a Gorenstein local ring of characteristic p with test ideal τ. Let P be an arbitrary prime ideal of R. Then the test ideal of R_P is τ_P.*

Proof. Choose arbitrary parameters $x_1, \ldots, x_i$ inside P which form a system of parameters in R_P. Let I be the ideal they generate in R. By Corollary 4.2 (4) applied to the ring R_P, it suffices to prove that $I_P : \tau_P = (I_P)^* = (I^*)_P$. The last equality is true because tight closure commutes with localization for ideals generated by regular sequences by Proposition 1.5 b). To prove this equality it is enough to prove that $I : \tau = I^*$, and this holds by Corollary 4.2 (2).   □

There are many special properties enjoyed by the tight closure of ideals generated by test elements which are parameters. We need the following lemma:

Lemma 4.5. *Let $x_1, \ldots, x_d$ be a regular sequence of elements which are test elements for parameter ideals. Then*

$$(x_1^t, \ldots, x_d^t)^* = (x_1^t, \ldots, x_d^t) + (x_1 \cdots x_d)^{t-1}(x_1, \ldots, x_d)^*.$$

Proof of Lemma 4.5. Let $u \in (x_1^t, \ldots, x_d^t)^*$. Since the x_i are test elements for parameters, $u \in (x_1^t, \ldots, x_d^t) : (x_1, \ldots, x_d) = (x_1^t, \ldots, x_d^t) + (x_1 \cdots x_d)^{t-1}R$. Writing u as $s(x_1 \cdots x_d)^{t-1}$ modulo $(x_1^t, \ldots, x_d^t)$, one sees that $s(x_1 \cdots x_d)^{t-1} \in (x_1^t, \ldots, x_d^t)^*$. By Theorem 2.3 b), one obtains that $s \in (x_1, \ldots, x_d)^*$.   □

Theorem 4.6. *Let (R, m) be either a complete local Gorenstein ring of characteristic p, or a nonnegatively graded Gorenstein ring over a field of characteristic p. Suppose that $x_1, \ldots, x_d$ is a system of parameters of R, homogeneous in the second case, which are test elements. Set $I = (x_1, \ldots, x_d)$. Then R/τ and I^*/I are Matlis dual, where τ is the test ideal of R.*

Proof. We give the proof in the local case. Set $I = (x_1, \ldots, x_d)$. We need to prove that $\text{Hom}_R(R/\tau, E) \cong I^*/I$, where E is an injective hull of the residue field of R. Note that $E = H_m^d(R)$ as R is Gorenstein. We may represent an element in this local cohomology by $\eta = [\frac{z}{x_1^t \cdots x_d^t}]$, for some $t \geq 1$. Define a homomorphism $f : I^*/I \to E$ by sending the coset of an element $u \in I^*$ to $f(u) = [\frac{u}{x_1 \cdots x_d}] \in E$. Observe that f is injective, for if $f(u) = 0$, then for some n, $(x_1 \cdots x_d)^n u \in (x_1^{n+1}, \ldots, x_d^{n+1})$, which forces $u \in I$ as R is Cohen-Macaulay.

Let $\eta = [\frac{z}{x_1^t \cdots x_d^t}] \in \text{Hom}_R(R/\tau, E)$, which we identify with the annihilator of τ in E. Then $\tau\eta = 0$ or equivalently $\tau z \subseteq (x_1^t, \ldots, x_d^t)$. By Corollary 4.2, we then know that $z \in (x_1^t, \ldots, x_d^t)^* = (x_1^t, \ldots, x_d^t) + (x_1 \cdots x_d)^{t-1}I^*$, by Lemma 4.5. Write $z = u + (x_1 \cdots x_d)^{t-1}v$ for some $u \in (x_1^t, \ldots, x_d^t)$ and $v \in I^*$. Then $\eta = [\frac{z}{x_1^t \cdots x_d^t}] = [\frac{u + (x_1 \cdots x_d)^{t-1}v}{x_1^t \cdots x_d^t}] = [\frac{v}{x_1 \cdots x_d}]$. This proves the map f defined above is also surjective and finishes the proof of the Theorem.   □

Example 4.7. Let R be the hypersurface $x^3 + y^3 + z^3 = 0$ over a perfect field of characteristic p. Then R is F-pure iff $p \equiv 1 \mod(3)$. By Fedder's theorem (3.7) above, we need to prove that $(X^3 + Y^3 + Z^3)^{p-1} \notin (X, Y, Z)^{[p]}$. Since $(X, Y, Z)^{3p-3} \subseteq (X^p, Y^p, Z^p, (XYZ)^{p-1})$, whether $(X^3 + Y^3 + Z^3)^{p-1} \notin (X^p, Y^p, Z^p)$ is simply a question of whether there exist integers $i + j + k = p - 1$ such that $X^{3i}Y^{3j}Z^{3k} = (XYZ)^{p-1}$, and such integers can be chosen iff $p \equiv 1 \mod(3)$.

Example 4.8. Again let R be the hypersurface $x^3 + y^3 + z^3 = 0$. Assume that the characteristic of k is at least 5, and further that k is algebraically closed. If $p \equiv 1 \mod(3)$, then R is F-pure by (4.7) and consequently Example 3.6 proves that the test ideal is exactly m, since we know that R is not weakly F-regular. In fact, even when $p \equiv 2 \mod(3)$, the test ideal is exactly $m = (x, y, z)$.

One can compute by hand many examples of the tight closure of parameter ideals in hypersurfaces. An important feature of these calculations is the degree of the socle element in the highest local cohomology. This degree is the *a-invariant*. Precisely, the a-invariant of a graded ring is the largest integer n such that the nth graded piece of the highest local cohomology $H_M^d(R)$ is nonzero ($d = \dim(R)$). It is particularly easy to compute the a-invariant in the case $R = k[X_1, \ldots, X_n]/(f_1, \ldots, f_g)$ is a graded complete intersection. The a-invariant of R can be calculated to be

$$a(R) = \sum_{1 \leq i \leq g} \deg(f_i) - \sum_{1 \leq i \leq n} \deg(X_i).$$

This calculation can be done from the fact that if S is a polynomial ring in n-variables, and if $R = S/I$ is a graded quotient of S, then the a-invariant of R is the maximum of the absolute values of the last twists in a minimal graded S-free resolution of R plus the a-invariant of S, which is $-\sum_{1 \leq i \leq n} \deg(X_i)$.

The following chart gives some data from the calculations of tight closure:

Equation	degrees of x, y, z	a-inv	Test Ideal	$(y, z)^*$
$x^3 + y^3 + z^3 = 0$	$1, 1, 1$	0	m	(y, z, x^2)
$x^2 + y^3 + yz^4 + z^6 = 0$	$3, 2, 1$	0	m	(y, z, x)
$x^4 + y^4 + z^4 = 0$	$1, 1, 1$	1	m^2	(y, z, x^2)
$x^5 + y^5 + z^5 = 0$	$1, 1, 1$	2	m^3	(y, z, x^2)
$x^2 + y^3 + z^7 = 0$	$21, 14, 6$	1	m	(y, z, x)
$x^2 + y^3 + yz^8 = 0$	$6, 4, 1$	1	(x, y, z^2)	(y, z, x)
$x^2 + y^3 + z^5 = 0$	$15, 10, 6$	-1	R	(y, z)

After computing many examples of the test ideal in graded hypersurface rings with isolated singularities, one reaches a rather amazing conclusion: the test ideal is always a power of the maximal ideal when the ring is generated by 1-forms. More generally, the test ideal always seems to consist of all forms of degree greater than some fixed integer N. Closer inspection reveals that the integer N seems to be the a-invariant of R.

Summarizing empirical evidence, we reach the following question:

Question 4.9. Let R be a Noetherian nonnegatively graded Gorenstein ring over a field of characteristic 0 or $p \gg 0$. Assume that R has an isolated singularity at the irrelevant ideal. Is the test ideal $R_{\geq a+1}$, where a is the a-invariant of R?

We will give a proof of this question in Section 5 for hypersurfaces with isolated singularity. This 'conjecture' turns out to be closely related to a conjecture concerning what the tight closure of parameter ideals looks like, which in turn is related to the Kodaira vanishing theorem. N. Hara [Ha5] has recently proved results which in conjunction with the results of [HS] imply an affirmative answer to (4.9).

5. The Tight Closure of Parameter Ideals

In this section we will begin our concentration upon the tight closure of parameter ideals. Throughout, (R, m) will denote either a Noetherian local ring of characteristic p, or a nonnegatively Noetherian graded ring over a field $R_0 = k$ of characteristic p, with m the irrelevant ideal generated by all forms of positive degree. We fix the dimension d of R, and let $x_1, \ldots, x_d$ be a system of parameters, homogeneous in the case in which R is graded. Our basic questions are:

What is the tight closure of $(x_1, \ldots, x_d)$? Where does it come from?

There are several ways we can list from Section 2:

Remark 5.1. Suppose that R is a domain, and let R^+ be the algebraic closure of R in an algebraic closure of its fraction field. Using Theorem 2.1 it follows that $(x_1, \ldots, x_d)R^+ \cap R \subseteq (x_1, \ldots, x_d)^*$. A subideal of this 'plus' closure is often easier to study, namely the Frobenius closure, $(x_1, \ldots, x_d)^F$. The Frobenius closure I^F of an ideal I is the set of elements u such that there exists a $q = p^e$ with $u^q \in I^{[q]}$.

A remarkable theorem of Smith [Sm4] says that in fact

$$(x_1, \ldots, x_i)R^+ \cap R = (x_1, \ldots, x_i)^*$$

for all i provided R is an excellent local (or graded) domain. We will prove this result in Section 7. In some sense, we don't need to look any further! However, it is extremely difficult to understand what lies in this plus closure. It is possible that for every ideal J of R, $JR^+ \cap R = J^*$. We know of no counterexample. However, we do not know this even for 2-dimensional normal local rings, even for the ring $k[x, y, z]/(x^3 + y^3 + z^3)$ with char $k \neq 3$.

Remark 5.2. The main theorem of [HH7] states that if (R, m) is an excellent local domain of positive characteristic, then R^+ is a big Cohen-Macaulay algebra for R. It is worth noting that this follows at once from colon-capturing and Smith's theorem[6]. For let $x_1, \ldots, x_d$ be a full system of parameters. To see that R^+ is Cohen-Macaulay it suffices to prove that $(x_1, \ldots, x_i) :_{R^+} x_{i+1} = (x_1, \ldots, x_i)R^+$ for $0 \leq i \leq d - 1$. Let $u \in (x_1, \ldots, x_i) :_{R^+} x_{i+1}$. After extending R by a finite integral extension we may assume that $u \in R$. By Theorem 2.3, $(x_1, \ldots, x_i) :_R x_{i+1} \subset (x_1, \ldots, x_i)^* = (x_1, \ldots, x_i)R^+ \cap R$, the latter equality coming from Smith's

---

[6]Smith uses that R^+ is Cohen-Macaulay to prove her result, so her result does not give a new proof that R^+ is Cohen-Macaulay, but rather is a good way of understanding the fact that R^+ is Cohen-Macaulay and how this fact is related to the theory of tight closure.

theorem. It follows that $u \in (x_1, \ldots, x_i)R^+$, which proves that R^+ is Cohen-Macaulay.

Instead of looking at the entire plus closure, which is difficult to work with, one can instead focus upon the Frobenius closure, I^F. Recall this is the set of elements u such that there exists a $q = p^e$ such that $u^q \in I^{[q]}$. Ultimately studying this piece of the tight closure of parameters ideals relates to the Kodaira vanishing theorem.

To obtain another piece of the tight closure we may use the colon capturing of Theorem 2.3. We need a definition to help us:

Definition 5.3. Let S be a ring and let $x_1, \ldots, x_l$ be parameters in S generating an ideal I. The notation I^{lim} indicates the ideal of all elements $z \in S$ for which there exists an integer s with $x^{s-1}z \in (x_1^s, x_2^s, \ldots, x_l^s)$, where $x = x_1 \cdots x_l$ is the product of all the x_i.

We call I^{lim} the *limit closure* of I. It is an ideal containing I. This definition actually arises from a consideration of local cohomology. We will discuss this connection later in this section.

Remark 5.4. Using Theorem 2.3 b) it follows that if $I = (x_1, \ldots, x_d)$ is a system of parameters, then $I^{lim} \subseteq I^*$. Notice that if R is Cohen-Macaulay, $I^{lim} = I$.

Remark 5.5. Let $I = (x_1, \ldots, x_d)$ be a system of parameters. Using Theorem 2.7 we know that $\overline{I^d} \subseteq I^*$. In the graded case this Briançon-Skoda theorem takes the particularly nice form of Theorem 2.9: setting $D = \sum_i \deg(x_i)$, we have that $R_{\geq D} \subseteq (x_1, \ldots, x_d)^*$.

Remark 5.6. If $J \subseteq I = (x_1, \ldots, x_d)$, then $J^* \subseteq I^*$. Of particular interest is specializing this concept to subsequences of the parameters. In particular, define

$$I^{germ} = \sum_i (x_1, \ldots, x_{i-1}, x_{i+1}, \ldots, x_d)^*.$$

Then $I^{germ} \subseteq I^*$. Although it is not obvious, both the germ and the limit closure of parameters do not depend upon the choice of parameters, only on the ideal they generate.

Summarizing, we have the following containment for the tight closure of a parameter ideal I:

$$\overline{I^d} + I^{germ} + I^F + I^{lim} \subseteq I^*.$$

In the graded case we can make a slightly stronger statement:

$$R_{\geq D} + I^{germ} + I^F + I^{lim} \subseteq I^*, \tag{3}$$

where $D = \sum_i \deg(x_i)$. When do we get equality? What are the relationships between the various terms on the right hand side of these containments? If R is Cohen-Macaulay $I^{lim} = I$. This is clear since any system of parameters $x_1, \ldots, x_d$ form a regular sequence.

The germ of I is closely related to the unmixed parts of parameters when there are enough test elements.

Definition 5.7. The *equidimensional hull* of an ideal I in a Noetherian ring R is by definition the intersection of the minimal primary components of maximal dimension. We denote this intersection by I^{unm}.

The sense of this definition is that if R is biequidimensional and catenary, and if I is an ideal generated by parameters, then I^{unm} is exactly the intersection of all minimal primary components of I, its unmixed part.

Proposition 5.8. *Let (R, m) be an equidimensional graded or local ring essentially of finite type over a field. Assume that R has an m-primary ideal of test elements for parameter ideals. If I is a parameter ideal of height less than the dimension of R, then $I^{unm} = I^*$.*

Proof. Let $\{x_1, x_2, \ldots, x_l\}$ be the parameters generating I, where $l < \dim(R)$. By assumption every element of m has some power that is a test element for I. Thus, we can find an element c such that $\{x_1, x_2, \ldots, x_l, c\}$ also form parameters, and such that c is a test element for parameter ideals. Let $P_1, P_2, \ldots, P_s$ be the minimal primary components of $I = (x_1, x_2, \ldots, x_l)R$, and let $Q_1, \ldots, Q_r$ be the embedded primary components. Because I has height l, every minimal primary component must have height l by the Krull principal ideal theorem so the equidimensional hull of I is simply the intersection of the P_i's.

The minimal primes of I are the radicals $\sqrt{P_i}$ of the P_i. Choose an element $z \in (x_1, x_2, \ldots, x_l)^*$. Then $cz \in (x_1, x_2, \ldots, x_l) \subset \bigcap_{i=1}^{s} P_i$. But since each P_i is primary and c is not in any $\sqrt{P_i}$, we must have that $z \in \bigcap_{i=1}^{s} P_i$, the equidimensional hull of I.

For the converse, let $z \in \bigcap P_i$. Let c be contained in each Q_i but no P_i. Then $cz \in \bigcap P_i \cap \bigcap Q_i = (x_1, \ldots, x_l)R$. By the fundamental 'colon capturing' property of tight closure given in Theorem 2.3, we conclude that $z \in (x_1, \ldots, x_l)^*$. Thus $(x_1, \ldots, x_l)^* = \bigcap_{i=1}^{s} P_i$, as claimed.   □

It follows that if R has an m-primary ideal of test elements for parameters and is Cohen-Macaulay then $I^{germ} = I$. For partial systems of parameters are always unmixed in a Cohen-Macaulay ring. Applying Proposition 5.8 then gives that $I^{germ} = I$.

Let R be graded and Cohen-Macaulay with an m-primary ideal of test elements. The containments of (3) then simply reduce to saying that $I^F + R_{\geq D} \subseteq I^*$. In fact, equality occurs in this case:

Proposition 5.9. *Let R be a graded ring over a field $R_0 = k$ of characteristic p, and suppose that $x_1, \ldots, x_d$ are a homogeneous system of parameters which are test elements. If R is Cohen-Macaulay then*

$$(x_1, \ldots, x_d)^* = (x_1, \ldots, x_d)^F + R_{\geq D},$$

where $D = \sum_i deg(x_i)$.

Proof. Theorem 2.9 proves that $(x_1, \ldots, x_d)^F + R_{\geq D} \subseteq (x_1, \ldots, x_d)^*$ in general. Let $u \in (x_1, \ldots, x_d)^*$. By Lemma 4.5, $u^q \in (x_1^q, \ldots, x_d^q)^* \subseteq (x_1^q, \ldots, x_d^q, y^{q-1})$, where $y = x_1 \cdots x_d$. Suppose that $\deg(u) = n < D$. Then for large q, $nq < D(q-1)$, and hence $u^q \in (x_1^q, \ldots, x_d^q)$. $\qquad\square$

Can we do even better? For instance, is it possible that elements in the Frobenius closure but not in the ideal must have high degree? Before we take up this question in more detail, let us backtrack a bit to the non Cohen-Macaulay case. If R has an m-primary ideal of test elements for parameters, then it turns out that while one cannot remove either of the terms I^{lim} or I^{germ} from (3), they collapse to give the same answer, at least for parameters which are test elements for parameters. This follows from work of Goto and Yamagishi [GY] on unconditioned strong d-sequences, d^+-sequences for short.

Definition 5.10. [Hu6], [GY] Let R be a commutative ring. A sequence of elements $x_1, \ldots, x_n$ is said to be a *d-sequence* if for every $0 \leq i \leq n-1$ and $k > i$,

$$(x_1, \ldots, x_i) : x_{i+1}x_k = (x_1, \ldots, x_i) : x_k.$$

A sequence $x_1, \ldots, x_n$ is said to be a *strong d-sequence* if $x_1^{m_1}, \ldots, x_n^{m_n}$ is a d-sequence for every $m_i \geq 1$. Finally a sequence is said to be a *d^+-sequence* if every permutation of it is a strong d-sequence.

Remark 5.11. Goto and Yamagishi prove a great many properties of such sequences. Such sequences are particularly convenient to use in tight closure theory because of the following observation: if $x_1, \ldots, x_n$ are parameters which are test elements for parameter tight closure, then they are a d^+-sequence. Since every power of these elements and every rearrangement of them are still parameters which are parameter test elements, it suffices to prove they are a d-sequence. If $u \in (x_1, \ldots, x_i) : x_{i+1}x_k$, then as $x_{i+1}x_k$ is a parameter modulo $x_1, \ldots, x_i$, we may apply Theorem 2.3 to see that $u \in (x_1, \ldots, x_i)^*$. The assumption that x_k is a test element for such ideals then implies that $u \in (x_1, \ldots, x_i) : x_k$.

Theorem 5.12. (Germ = Limit Closure). *Let (R, m) be as above, and assume that $x_1, \ldots, x_d$ are a system of parameters which are test elements for parameter tight closure. Let I be the ideal they generate. Then $I^{germ} = I^{lim}$.*

Proof. We first prove that $I^{germ} \subseteq I^{lim}$. Let $u \in (x_1, \ldots, x_{i-1}, x_{i+1}, \ldots, x_d)^*$ for some i. Since x_i is a test element, $x_i u \in (x_1, \ldots, x_{i-1}, x_{i+1}, \ldots, x_d)$. But then $x_1 \cdots x_d u \in (x_1^2, \ldots, x_d^2)$ which means that $u \in I^{lim}$.

Conversely we use the following 'Monomial property' of d^+-sequences proved by Goto and Yamagishi [GY, Theorem 2.3]:

Theorem 5.13. *Let $n_1, \ldots, n_s, m_1, \ldots, m_s$ be positive integers and suppose that $x_1, \ldots, x_s$ are a d^+-sequence. The ideal $(a_1^{n_1 + m_1}, \ldots, a_s^{n_s + m_s}) : (\prod_i a_i^{m_i})$ is equal to*

$$\sum_i (a_1^{n_1}, \ldots, a_{i-1}^{n_{i-1}}, a_{i+1}^{n_{i+1}}, \ldots, a_s^{n_s}) : a_i + (a_1^{n_1}, \ldots, a_s^{n_s}).$$

Suppose that $u \in I^{lim}$, and choose t such that $ux^t \in (x_1^{t+1}, \ldots, x_d^{t+1})$, where $x = x_1 \cdots x_d$. By Remark 5.11, $x_1, \ldots, x_d$ form a d^+-sequence. Applying (5.13) with $s = d$, $m_1 = \cdots = m_d = t$, and $n_1 = \cdots = n_d = 1$ gives that $u \in \sum_i (a_1, \ldots, a_{i-1}, a_{i+1}, \ldots, a_s) : a_i + I = \sum_i (a_1, \ldots, a_{i-1}, a_{i+1}, \ldots, a_s) : a_i$. It follows that $I^{germ} = I^{lim}$. $\square$

The various parts of the tight closure of parameter ideals can be understood best by going to the limit, i.e. looking at the highest local cohomology of R. This point of view has been used very effectively by both Smith and Hara. In addition, the local cohomology modules of an ideal I in R play an important role in the study of F-rational rings.

Remark 5.14. The term $R_{\geq D}$ in (3) can be explained in terms of the local cohomology. Let R be graded. If $x_1, \ldots, x_d$ are homogeneous parameters of degrees d_i with $D = \sum_i d_i$, then $z \in R_{\geq D}$ iff the degree of $\eta = [\frac{z}{x}] \in H_m^d(R)$ is greater than or equal to $D - \sum_i d_i = 0$. It follows that $0^*_{H_m^d(R)}$ contains all elements of nonnegative degree, and this is the meaning of the Briançon-Skoda part of the tight closure of parameter ideals. The most interesting part of the tight closure of parameters is what occurs in the negative part of the local cohomology.

An important observation is that if $x_1, \ldots, x_d$ is a homogeneous system of parameter for the graded ring R, then an element $u \in (x_1, \ldots, x_d)^*$ iff $\eta = [\frac{u}{x}] \in 0^*_{H_m^d(R)}$. (Here, as above, x represents the product of the elements x_i.) It follows that there is a correspondence between studying the tight closure of parameter ideals and studying the 'fundamental' submodule $0^*_{H_m^d(R)}$ of the highest local cohomology. By restricting our attention to $0^*_{H_m^d(R)}$ we in some sense are studying the tight closure of all parameters ideals at the same time. Under this correspondence we have seen that the limit closure is simply the set of elements going to zero.

Theorem 5.15. (Strong Vanishing Theorem). *Let R be a nonnegatively graded ring over a field $R_0 = K$ of characteristic 0 or characteristic $p \gg 0$.[7] Assume that R has an m-primary ideal of parameter test elements. Let $x_1, \ldots, x_d$ be a homogeneous system of parameters of degrees $\delta_1, \ldots, \delta_d$. Set $\delta = \sum_i \delta_i$. Then*

$$(x_1, \ldots, x_d)^* = (x_1, \ldots, x_d)^{germ} + R_{\geq \delta}.$$

This was originally a conjecture which arose through a reinterpretation of the Kodaira vanishing theorem in [Sm6] and [HS]. The last section will explain this connection. This 'conjecture' has now been proved by Nobuo Hara [Ha5]. The next section will give applications of this theorem, and give a direct proof in the case in which R is a hypersurface.

---

[7] At this point we will not be precise about what this means.

6. The Strong Vanishing Theorem

In this section we will give some of the consequences of the Strong Vanishing Theorem, discuss equivalent formulations, and prove this theorem for hypersurfaces with isolated singularity. This theorem was recently proved by N. Hara using techniques of Deligne and Illusie [DI].

We begin by restricting to the Cohen-Macaulay case.

Theorem 6.1. *Let R be a nonnegatively graded Cohen-Macaulay ring over a field $R_0 = K$ of characteristic 0 or characteristic $p \gg 0$ Assume that R has an m-primary ideal of parameter test elements. Let $x_1, \ldots, x_d$ be a homogeneous system of parameters of degrees $\delta_1, \ldots, \delta_d$. Set $\delta = \sum_i \delta_i$. Then*

$$(x_1, \ldots, x_d)^* = (x_1, \ldots, x_d) + R_{\geq \delta}.$$

Proof. This is nothing more than a restatement of the Vanishing Theorem 5.15, using the fact that $I^{germ} = I$ if I is Cohen-Macaulay.

The next theorem gives several equivalent forms of the Strong Vanishing Theorem. Probably the most appealing is the fourth equivalence.

Theorem 6.2. *Let (R, m) be a d-dimensional $\mathbb{N}$-graded Noetherian domain over a field S_0 of characteristic $p > 0$. Assume that R has an m-primary ideal of test elements for parameter ideals. Then the following are equivalent:*

(1) $(x_1, x_2, \ldots, x_d)^* = (x_1, x_2, \ldots, x_d)^{lim} + R_{\geq \delta}$ *for all homogeneous systems of parameters $x_1, x_2, \ldots, x_d$ for R where δ is the sum of the degrees of the x_j's.*

(2) $(x_1, x_2, \ldots, x_d)^* = (x_1, x_2, \ldots, x_d)^{germ} + R_{\geq \delta}$ *for all homogeneous systems of parameters $x_1, x_2, \ldots, x_d$ for R where δ is the sum of the degrees of the x_j's.*

(3) *The tight closure of zero in $H_m^d(R)$ has no non-zero elements of negative degrees. In particular, the tight closure of zero in $H_m^d(R)$ is precisely the submodule of elements of non-negative degrees.*

(4) *The Frobenius acts injectively on $H_m^d(S)$ in negative degrees.*

Further assume that R is Cohen-Macaulay. Then (1)-(4) above are equivalent to:

(5) $(x_1, x_2, \ldots, x_d)^* = (x_1, x_2, \ldots, x_d) + R_{\geq \delta}$ *for a fixed homogeneous system of parameters $x_1, x_2, \ldots, x_d$ for R which are test elements for parameters, where δ is the sum of the degrees of the x_j's.*

Proof. The equivalence of (1) and (2) follows from Theorem 5.12. The equivalence of (2) and (3) follows from the discussion in Remark 5.14.

That (3) implies (4) is obvious: if any negative degree element of $H_m^d(S)$ is in the kernel of the Frobenius map, then it is in the tight closure of zero in $H_m^d(S)$.

To see that (4) implies (3), suppose that Frobenius acts injectively on the negative degree pieces and η is an element of negative degree in the tight closure of zero in $H_m^d(S)$. Each of the non-zero elements η^q has a non-zero multiple in the socle of $H_m^d(S)$. Because the socle is of fixed degree, the degrees of these multipliers must be getting larger and larger as q goes to infinity (because the degree of η^q is going to $-\infty$ as q gets larger). But all elements of large degree are test elements, so they must actually kill η^q.

Clearly (1) implies (5). Assume (5). To prove (2), notice that any element of the highest local cohomology can be represented in the form $[\frac{z}{x_1^t \cdots x_d^t}]$ for some choice of t. If we prove the system of elements $x_1^t, \ldots, x_d^t$ satisfies the conclusion of (5), then (2) follows as in the proof of (2) implies (3).

Suppose that $z \in (x_1^t, \ldots, x_d^t)^*$ and the degree of z is strictly smaller than $t\delta$. Writing z in $(x_1^t, \ldots, x_d^t) + (x_1 \cdots x_d)^{t-1}(x_1, \ldots, x_d)^*$ by Lemma 4.5, we see that if there is a nonzero contribution from the last term, say $s(x_1 \cdots x_d)^{t-1}$, then the degree of s will be strictly smaller than δ, and the lemma shows that $s \in (x_1, \ldots, x_d)^*$. This contradicts assumption (5) and finishes the proof. $\qquad \square$

The Strong Vanishing Theorem turns out to answer Question 4.9 concerning the test ideal. The following theorem is a slightly special case of [HS, Theorem 5.4]:

Theorem 6.3. (The Test Ideal and Vanishing Theorem). *Let (R, m) be an $\mathbb{N}$-graded Noetherian domain over a field $k = R_0$ of characteristic $p > 0$ and of dimension d. Assume that R is Gorenstein and has an isolated singularity at m. Further assume that the Strong Vanishing Theorem holds for R.[8] Then test ideal for R is $R_{\geq a+1}$, where a is the a-invariant of R.*

Proof. Since R has an isolated singularity, we know that there exists an m-primary ideal of test elements, and hence we can choose homogeneous parameters $x_1, \ldots, x_d$ which are test elements. By Corollary 4.3, the test ideal is exactly

$$\tau = (x_1, \ldots, x_d) :_R (x_1, \ldots, x_d)^*.$$

As R is Gorenstein, so are the 0-dimensional rings $R/(x_1, \ldots, x_d)$, and these rings are graded. The socle sits in degree $a + D$, where $D = \sum_{1 \leq i \leq d} \deg(x_i)$ and a is the a-invariant of R.

By the Strong Vanishing Theorem, $(x_1, \ldots, x_d)^* = (x_1, \ldots, x_d)^{germ} + R_{\geq D} = (x_1, \ldots, x_d) + R_{\geq D}$, since R is Cohen-Macaulay. If $u \in (x_1, \ldots, x_d) :_R (x_1, \ldots, x_d)^*$, then $uR_{\geq D} \subseteq (x_1, \ldots, x_d)$. As u has a nonzero multiple in the socle, it follows that $\deg(u) \geq a+1$. Conversely, if $\deg(u) \geq a+1$ then $u(x_1, \ldots, x_d)^* \subseteq u((x_1, \ldots, x_d) + R_{\geq D}) \subseteq (x_1, \ldots, x_d) + R_{\geq a+D+1} \subseteq (x_1, \ldots, x_d)$. Hence $\tau = R_{\geq a+1}$. $\qquad \square$

It is worth remarking that in the Gorenstein case, the statement of Theorem 6.3 is in fact equivalent to the Strong Vanishing Theorem. For suppose that $\tau = R_{\geq a+1}$. Let $u \in (x_1, \ldots, x_d)$ with $\deg(u) < D$, where D is the sum of the degrees of the homogeneous parameters $x_1, \ldots, x_d$. Since $\tau u \subseteq (x_1, \ldots, x_d)$, we obtain that $uR_{\geq a+1} \subseteq (x_1, \ldots, x_d)$. But if $u \notin (x_1, \ldots, x_d)$, then u must have a multiple uv in the socle, necessarily of degree $D + a$. The degree of v is at least $a+1$ since $\deg(u) < D$. Hence $u \in (x_1, \ldots, x_d)$, which proves the Strong Vanishing Theorem.

---

[8]The Strong Vanishing Theorem holds in characteristic 0 or for 'large' characteristic p.

It is possible to give a direct proof of the Strong Vanishing Theorem for the case of hypersurfaces with isolated singularity, which does not involve any more machinery. This was first proved essentially by Fedder, and the proof here, while new, uses Fedder's ideas.

Theorem 6.4. (Strong Vanishing for Hypersurfaces). *Let R be the ring $k[X_0, \dots, X_d]/(f)$. Assume that R is an isolated singularity which is quasihomogeneous, where k is a field of characteristic p. Assume that the partial derivatives $f_i = \frac{\partial f}{\partial X_i}$ form a system of parameters in R where $1 \le i \le d$. Further assume that $p > (d-1)(\deg(f)) - \sum_{1 \le i \le n} \deg(X_i)$. Let $y_1, \dots, y_d$ be a homogeneous system of parameters of degrees $a_1, \dots, a_d$. Set $A = a_1 + \cdots + a_d$. Then*

$$(y_1, \dots, y_d)^* = (y_1, \dots, y_d) + R_{\ge A}.$$

Proof. To prove Theorem 6.4 it suffices to prove this for a single system of parameters which are test elements, by Theorem 6.2.5. As R has an isolated singularity, without loss of generality we may assume that the partial derivatives $f_i = \frac{\partial f}{\partial X_i}$ for $1 \le i \le d$ form a system of parameters, which are clearly homogeneous, and which are test elements by Theorem 3.9.

Choose an element $u \in (f_1, \dots, f_d)^*$ of minimal degree such that $u \notin (f_1, \dots, f_d)$. Set $\delta = \sum_{1 \le i \le d} \deg(f_i)$. If $u \in R_{\ge \delta}$ we are done, so we may assume that $\deg(u) < \delta$. By Lemma 4.5,

$$u^p \in (f_1^p, \dots, f_d^p)^* = (f_1^p, \dots, f_d^p) + (f_1 \cdots f_d)^{p-1}(f_1, \dots, f_d)^*.$$

Write

$$u^p = \sum_{1 \le i \le d} r_i f_i^p + (f_1 \cdots f_d)^{p-1} s, \tag{4}$$

where $s \in (f_1, \dots, f_d)^*$. In particular, $\deg(s) \ge \deg(u)$. We claim that $u^p \in (f_1^p, \dots, f_d^p)$. If not, the term $(f_1 \cdots f_d)^{p-1} s$ in (4) must be nonzero, and so $p(\deg(u)) \ge (p-1)\delta + \deg(u)$ which contradicts our assumption that $\deg(u) < \delta$. We may rewrite (4) as

$$u^p = \sum_{1 \le i \le d} r_i f_i^p.$$

We now lift the latter equation back to $S = k[X_0, \dots, X_d]$ and write

$$u^p = \sum_{1 \le i \le d} r_i f_i^p + r f^i, \tag{5}$$

where $1 \le i$. Choose i maximal such that $u^p \in (f_1^p, \dots, f_d^p, f^i)$. If $i \ge p$, we obtain that $u^p \in (f_1, \dots, f_d, f)^{[p]}$, and as S is regular we then find that $u \in (f_1, \dots, f_d)R$ as needed.

Assume that (5) holds with $i < p$, and we'll prove $u^p \in (f_1^p, \dots, f_d^p, f^{i+1})$. Write D_j for the differential $\frac{\partial}{\partial X_j}$. Recall that $D_j(f) = f_j$. Applying D_j

to (5) yields an equation which shows that $if^{i-1}f_jr \in ((f_1, \ldots, f_d, f)^{[p]}, f^i)S$. Since $i < p$ we may invert i. Moreover this holds for every $1 \le j \le d$. We can write

$$r \in ((f_1, \ldots, f_d, f)^{[p]}, f^i)S :_S ((f_1, \ldots, f_d, f^{i-1})).$$

Since $f, f_1, \ldots, f_d$ form a regular sequence, $r \in ((f_1, \ldots, f_d, f)^{[p]}, F^{p-1}, f)S$, where $F = (f_1 \cdots f_d)$. It suffices to prove that the term involving F^{p-1} is 0 since in that case $r \in ((f_1, \ldots, f_d, f)^{[p]}, f)S$ which forces $u^p \in (f_1^p, \ldots, f_d^p, f^{i+1})S$ using (5). The degree of the term involving F^{p-1} is at least degree $(p-1)\delta$. The degree of r is at most $p\deg(u) - i\deg(f) \le p\deg(u) - \deg(f)$. It follows that if there is a nonzero contribution from the term F^{p-1} to r then

$$p(\delta - \deg(u)) \le \delta - \deg(f).$$

Since $(\delta - \deg(u)) > 0$ this does not hold for $p > \delta - \deg(f)$. $\qquad\square$

Remark 6.5. A similar proof gives the Strong Vanishing Theorem for complete intersections which are quasi-homogeneous with isolated singularity. This was essentially done by Fedder with a different proof. See also [Ha4].

Example 6.6. Let $S = k[X, Y, Z]$ and $f = X^2 + Y^3 + Z^5$. In terms of the proof of Theorem 6.4 we may choose the partials $\frac{\partial f}{\partial X}$ and $\frac{\partial f}{\partial Y}$ to be the system of parameters for $R = S/Sf$. We must then assume that the characteristic p of k exceeds $(d-1)(\deg(f)) - \deg(X) - \deg(Y) = 30 - 10 - 15 = 5$.

Write small x, y, z for the images of X, Y, Z in R. Apply Theorem 6.4 to the parameters y, z of R. We get that the tight closure of (y, z) is contained in $(y, z) + R_{\ge 16}$. As $R_{\ge 16} \subseteq (y, z)R$, it follows that the test ideal for parameters is the whole ring if the characteristic of k is at least 7.

Another example is helpful:

Example 6.7. Consider the three hypersurfaces: $R_1 = k[X, Y, Z]/(X^2 + Y^3 + YZ^4)$, $R_2 = k[X, Y, Z]/(X^2 + Y^3 + Z^7)$, and $R_3 = k[X, Y, Z, U]/(X^5 + Y^5 + Z^5 + U^5)$. We claim that in each case, the test ideal is the maximal ideal, at least for large enough characteristic.

R_1 is graded with the weights, $\deg(x) = 3$, $\deg(y) = 2$, and $\deg(z) = 1$. The a-invariant is then $0 = 6 - 1 - 2 - 3$. It follows that the test ideal is the maximal ideal, provided the characteristic is at least $6 - 2 - 3 = 1$. Thus the Strong Vanishing Theorem holds in R_1 for all characteristics.

R_2 is graded with the weights, $\deg(x) = 21$, $\deg(y) = 14$, and $\deg(z) = 6$. The a-invariant is then $1 = 42 - 21 - 14 - 6$. It follows that the test ideal is the maximal ideal, provided the characteristic is greater than $42 - 21 - 14 = 7$. Hence the Strong Vanishing Theorem holds in R_2 for all characteristics at least 11.

R_3 is graded with all the weights 1, and the a-invariant is then $1 = 5 - 1 - 1 - 1 - 1$. It follows that the test ideal is the maximal ideal, provided the characteristic is greater than $2(5) - 1 - 1 - 1 = 7$. The Strong Vanishing Theorem holds in R_3 for all characteristics at least 11.

7. Plus Closure

In this section we will sketch the proof of the the main theorem from [Sm4]. Recall that if R is a domain, then by R^+ we denote the integral closure of R in an algebraic closure of its fraction field. We will use the fact that R^+ is Cohen-Macaulay if R is an excellent local domain of characteristic $p > 0$. This fact is the main result of [HH4]. Theorem 5.1 of [Sm4] states:

Theorem 7.1. *Let R be a locally excellent Noetherian domain of characteristic $p > 0$. If I is any parameter ideal of R, then $I^* = IR^+ \cap R$.*

The proof of this Theorem rests on a thorough understanding of the role played by the tight closure of zero in the highest local cohomology of a ring R.

Before beginning the proof we need to introduce another submodule of the highest local cohomology of a d-dimensional Noetherian local ring (R, m).

Definition 7.2. The plus closure of 0 in $H_m^d(R)$, denoted $0^+_{H_m^d(R)}$, is the kernel of the natural map

$$\phi : H_m^d(R) \to H_m^d(R) \otimes_R R^+ \cong H_m^d(R^+).$$

We have already seen that an element $z \in (x_1, \ldots, x_d)^*$ determines an element $[\frac{z}{x_1 \cdots x_d}] \in 0^*_{H_m^d(R)}$. The next Proposition delineates some of the basic properties we will use.

Proposition 7.3. *Let (R, m) be an excellent local domain of dimension d and characteristic $p > 0$, and let R^+ be as above.*

(1) *Let $x_1, \ldots, x_d$ be a system of parameters generating an ideal I. Then*

$$[\frac{z}{x_1 \cdots x_d}] \in 0^+_{H_m^d(R)} \text{ iff } z \in I^+.$$

(2) $0^+_{H_m^d(R)} \subseteq 0^*_{H_m^d(R)}$.

(3) $0^+_{H_m^d(R)} = 0^*_{H_m^d(R)}$ *iff $I^+ = I^*$ for all ideals I generated by systems of parameters $x_1, \ldots, x_d$.*

Proof. First suppose that $z \in I^+$. Then $z \in IR^+$ and so $[\frac{z}{x_1 \cdots x_d}] = 0$ in $H_m^d(R^+)$. Conversely, suppose that $[\frac{z}{x_1 \cdots x_d}] = 0$ in $H_m^d(R^+)$. Since R^+ is Cohen-Macaulay by [HH4], it then follows that $z \in IR^+ \cap R = I^+$. This proves 1).

To prove (2), let $\eta \in 0^+_{H_m^d(R)}$ and write $\eta = [\frac{z}{x_1 \cdots x_d}]$ for some system of parameters $x_1, \ldots, x_d$. By (1), we know that $z \in I^+ \subseteq I^*$. It then follows that $\eta \in 0^*_{H_m^d(R)}$.

If $0^+_{H_m^d(R)} = 0^*_{H_m^d(R)}$, and $z \in I^*$, then $\eta = [\frac{z}{x_1 \cdots x_d}] \in 0^*_{H_m^d(R)} = 0^+_{H_m^d(R)}$ implies by 1) that $z \in I^+$. Hence $I^* = I^+$. The converse is even easier. $\square$

A crucial step in the proof of the main theorem of this section is the fact that if $0^*_{H^d_m(R)}/0^+_{H^d_m(R)}$ has finite length then it must be zero! (See [Sm4, Theorem 5.1].)

A slightly more general result explains this phenomena. Recall that if $\eta = [\frac{z}{x_1 \cdots x_d}] \in H^d_m(R)$, then $\eta^q = [\frac{z^q}{x_1^q \cdots x_d^q}]$.

Lemma 7.4. *Let (R, m) be an excellent local domain of positive characteristic p and dimension $d > 1$. Set $K = 0^+_{H^d_m(R)}$. Suppose that $\eta \in H^d_m(R)$, and set N_e equal to the submodule of $H^d_m(R)$ spanned by K together with $\eta, \eta^p, \ldots, \eta^{p^e}$. Suppose that $N_{e-1} = N_e$ for some e. Then $\eta \in 0^+_{H^d_m(R)}$.*

Proof. Suppose that $N_e = N_{e-1}$, and set $q = p^e$. We then have an equation,

$$\eta^q = r_1 \eta^{q_1} + \cdots + r_j \eta^{q_j} + \theta, \tag{6}$$

where $q_j < \cdots < q_1 < q$ are powers of p, and $\theta \in 0^+_{H^d_m(R)}$. We now pass to R^+. The element θ becomes 0 by definition of $0^+_{H^d_m(R)}$, and we may rewrite (6) as

$$[\frac{z^q}{x_1^q \cdots x_d^q}] = [\frac{r_1 z^{q_1} x^{q-q_1}}{x_1^q \cdots x_d^q}] + \cdots + [\frac{r_j z^{q_j} x^{q-q_j}}{x_1^q \cdots x_d^q}],$$

where $x = x_1 \cdots x_d$ and the equation holds in $H^d_m(R^+)$. The fact that R^+ is Cohen-Macaulay then gives us that

$$z^q = r_1 z^{q_1} x^{q-q_1} + \cdots + r_j z^{q_j} x^{q-q_j} + w, \tag{7}$$

where $w \in (x_1^q, \ldots, x_d^q)R^+$. The beautiful fact we now need to use is that equation (7) forces $z \in (x_1, \ldots, x_d)R^+$. This is the statement of the modified 'Equational Lemma': see [HH4, 2.2] and [Sm4, (5.3)]. However $z \in (x_1, \ldots, x_d)R^+$ means that $\eta = [\frac{z}{x_1 \cdots x_d}]$ is 0 in $H^d_m(R^+)$, and hence $\eta \in 0^+_{H^d_m(R)}$. $\qquad \square$

Proposition 7.5. *Let (R, m) be an excellent local domain of positive characteristic p and dimension $d > 1$. If the R-module $0^*_{H^d_m(R)}/0^+_{H^d_m(R)}$ has finite length then it vanishes.*

Proof. Let $\eta = [\frac{z}{x_1 \cdots x_d}] \in 0^*_{H^d_m(R)}$. Define N_e as in Lemma 7.4. The fact that $\eta \in 0^*_{H^d_m(R)}$ implies that for all q, $\eta^q \in 0^*_{H^d_m(R)}$. In particular, $0^+_{H^d_m(R)} \subseteq N_1 \subseteq N_2 \subseteq \cdots \subseteq 0^*_{H^d_m(R)}$. The condition that $0^*_{H^d_m(R)}/0^+_{H^d_m(R)}$ has finite length forces the chain of submodules N_e to stabilize, and then Lemma 7.4 gives that $\eta \in 0^+_{H^d_m(R)}$. $\qquad \square$

Proof of Theorem 7.1. The proof of (7.1) proceeds by induction on the dimension of R. Smith makes several technical reductions of this problem which are somewhat laborious. However, one eventually reaches the situation in which a minimal dimensional counterexample (R, m) is complete, normal and local. We next need to reduce to the Gorenstein case. Accordingly, let (R, m) be complete local and normal, and suppose that $z \in I^* \notin IR^+ \cap R$ for an ideal $I = (x_1, \ldots, x_d)$ generated by a system of parameters. Choose a coefficient field K and let $A = K[[x_1, \ldots, x_d]] \subseteq R$ be the complete subring generated by the x_i over K. Of course, A is a regular local ring. Set $B = A[z]$. The inclusions $A \subseteq B \subseteq R$ prove that B is a complete local ring of the same dimension d as both A and R. Moreover, B is Gorenstein, in fact is isomorphic with $A[Z]/(f)$ for some nonzero element $f \in A[Z]$. Theorem 2.1 shows that $z \in ((x_1, \ldots, x_d)R)^* \cap B \subseteq ((x_1, \ldots, x_d)B)^*$. On the other hand, $z \notin IB^+ \cap B$, since $IB^+ = IR^+$, and $z \notin IR^+$. Thus we have reached the situation in which a minimal dimensional counterexample is Gorenstein.

By Proposition 7.3 (3) it suffices to prove that $0^+_{H^d_m(R)} = 0^*_{H^d_m(R)}$, where d is the dimension of R. By Proposition 7.5 it then suffices to prove that $0^*_{H^d_m(R)}/0^+_{H^d_m(R)}$ has finite length, and we do this by considering their Matlis duals.

By Theorem 4.6, the Matlis dual of $0^*_{H^d_m(R)}$ is τ, the test ideal of R. Denote the Matlis dual of $0^+_{H^d_m(R)}$ by J. We have that $\tau \subseteq J$, and we wish to prove that J/τ has finite length. Let P be any prime ideal not equal to m.

Choose any element $c/1 \in J_P$. Without loss of generality we may assume that $c \in J$. It suffices to prove that c kills the tight closure of 0 in $H^n_{PR_P}(R_P)$, where $n = \dim(R_P)$, since in that case $c/1$ is in the test ideal of R_P which by Theorem 4.4 is exactly τ_P. In other words, if $x_1, \ldots, x_n$ are parameters in R which form a system of parameters in R_P, we need to prove that $c((x_1, \ldots, x_n)_P)^* = c((x_1, \ldots, x_n)^*)_P \subseteq (x_1, \ldots, x_n)_P$. However, the induction implies that the plus closure is equal to the tight closure for parameter ideals in R_P. If $z \in (x_1, \ldots, x_n)^*$, then $z/1 \in ((x_1, \ldots, x_n)_P)^+ = ((x_1, \ldots, x_n)^+)_P$ (it is elementary to see that plus closure commutes with localization) and so there exists an element $u \notin P$ such that $uz \in (x_1, \ldots, x_n)R^+$. Extend $x_1, \ldots, x_n$ to a full system of parameters $x_1, \ldots, x_d$. Then $uz \in (x_1, \ldots, x_d)R^+$ and so $\eta = [\frac{uz}{x_1 \cdots x_d}] \in 0^+_{H^d_m(R)}$. By choice of c, we then obtain that $c\eta = 0$ in $H^d_m(R)$, and since R is Cohen-Macaulay it follows that $cuz \in (x_1, \ldots, x_d)$. Since we may vary the parameters $x_{n+1}, \ldots, x_d$ (in particular we may raise them to arbitrary powers), we obtain that $cuz \in (x_1, \ldots, x_n)$. Hence $cz \in (x_1, \ldots, x_n)_P$. It follows that $c \in \tau_P$, and this finishes the proof of Theorem 7.1. □

Remark 7.6. Although R^+ is not a Noetherian ring, and so the definition of tight closure does not pertain to this ring, it is nonetheless interesting to see what Theorem 7.1 means in the context of this ring. Essentially it says that ideals generated by parameters in R^+ are tightly closed. For suppose that (R, m) is a complete local domain of characteristic p and let $x_1, \ldots, x_d$ be a system of parameters in R^+. Suppose that $z \in ((x_1, \ldots, x_d)R^+)^*$, where we take this to mean as usual that there exists a nonzero element c such that $cz^q \in (x_1^q, \ldots, x_d^q)R^+$ for all large q. Choose a local Noetherian complete ring (S, n), finite over R containing $x_1, \ldots, x_d, c, z$. Then $cz^q \in S \cap (x_1^q, \ldots, x_d^q)R^+ \subseteq ((x_1^q, \ldots, x_d^q)S)^*$. Choosing a test element d for S then gives that for all large q, $dcz^q \in (x_1^q, \ldots, x_d^q)S$. Hence $z \in ((x_1, \ldots, x_d)S)^*$ which by the main theorem of this section forces $z \in (x_1, \ldots, x_d)S^+ = (x_1, \ldots, x_d)R^+$. (Observe that the elements $x_1, \ldots, x_d$ must form parameters in S.) It then follows that we may think of the ideal $(x_1, \ldots, x_d)R^+$ as being tightly closed. It turns out that Noetherian rings whose parameter ideals are tightly closed are essentially those with rational singularities. The next section discusses these rings.

8. F-Rational Rings

In this section we will discuss rings in which every ideal generated by parameters is tightly closed. Such rings are said to be *F-rational*. It turns out that F-rational rings are closely related to rational singularities. Indeed, if R is a Noetherian ring over a field of characteristic 0, then R is 'F-rational type'[9] implies that R has rational singularities [Sm2]. Recently, using methods from the Deligne-Illusie proof of Kodaira vanishing, N. Hara has proved that the converse is also true for such rings [Ha5].

Lemma 8.1. *Let (R, m) be an equidimensional local Noetherian ring of characteristic p which is a homomorphic image of a Cohen-Macaulay ring and let $x_1, \ldots, x_d \in m$ be part of system of parameters. If the ideal $(x_1, \ldots, x_d)R$ is tightly closed then so is the ideal $(x_1, \ldots, x_i)R$ for $0 \leq i \leq d$.*

Proof. The proof of the claim reduces, by reverse induction on i, to the case where $d \geq 1$ and $i = d - 1$. Let r be any element in the tight closure of $J = (x_1, \ldots, x_{d-1})R$. Then $r \in (J + x_d R)^* = J + x_d R$ by hypothesis, say $r = j + x_d u$. It follows that $r - j \in J^* + J = J^*$ and so $u \in J^* :_R x_d R$. Then $x_d u \in J^*$ and so there exists $c_0 \in R^o$ such that $c_0(x_d u)^q \in J^{[q]}$ for all large q. Thus, for all large q, we have that $c_0 u^q \in J^{[q]} :_R x_d^q R = (x_1^q, \ldots, x_{d-1}^q)R :_R x_d^q R$. Theorem 2.3 gives then that $c_0 u^q \in (x_1^q, \ldots, x_{d-1}^q)^*$, which implies that $u \in J^*$. Thus, $J^* = J + x_d J^*$, and the fact that $J^* = J$ now follows from Nakayama's lemma. $\square$

We summarize the basic properties of F-rational rings. See [Hu5, Section 4]:

Theorem 8.2. [FeW], [HH10] *For Noetherian rings of characteristic p the following hold. In parts d)–g) assume either that R is locally excellent or is a homomorphic image of a Cohen-Macaulay ring.*

   a) *A weakly F-regular ring is F-rational.*
   b) *An F-rational ring is normal.*
   c) *If R is local and excellent, then R is F-rational iff $\hat{R}$ is F-rational.*
   d) *If R is F-rational, then R is Cohen-Macaulay.*
   e) *A local ring (R, m) is F-rational if and only if it is equidimensional and the ideal generated by one system of parameters is tightly closed.*

---

[9] *F*-rational type means that after expressing relevant data over a finitely generated $\mathbb{Z}$-algebra (instead of a field) and reducing modulo the maximal ideals in a dense open set, the corresponding algebra (now in positive characteristic) is *F*-rational.

f) R is F-rational if and only if its localization at every maximal ideal is F-rational.

g) A localization of an F-rational ring R is F-rational. In particular, a localization of a weakly F-regular ring which is locally excellent is F-rational.

Proof of Theorem 8.2. (See [Hu5, Section 4]) a) is immediate from the definitions.

To prove part b) we show the stronger statement that if R is a ring such that no minimal prime is maximal and every height one principal ideal is tightly closed, then R is normal. Let N denote the nilradical of R. We first prove that $N = 0$ as follows. The tight closure of every ideal contains N. For any fixed minimal prime P choose an element $x \in R^o$ such that x is not invertible modulo P. The assumption concerning the spectrum of R guarantees that such a choice is possible. The product of all such elements, say y, is in R^o and is not invertible modulo every minimal prime. Then $\cap_n(y^n) = \cap_n(y^n)^*$ contains N, and is annihilated by an element of the form $z = 1 - ry$. But then $z \in R^o$ and by the same argument $N(1 - sz) = 0$ for some $s \in R$. Hence $N = 0$. Thus R is reduced. Let r/s be in the total quotient field of R. If r/s is integral over R, then $r \in \overline{(s)}$, and the integral closure is the same as the tight closure for principal ideals of height at least one (see the definition of integral closure given in this text, or use Theorem 2.7). Hence $r \in (s)$, and $r/s \in R$.

To prove c) suppose that R is F-rational. Any system of parameters in $\hat{R}$ comes from one in R. Let I be the ideal they generate. Using Lemma 1.5 we obtain that $I^*\hat{R} = (I\hat{R})^*$. It follows that $\hat{R}$ is F-rational. Conversely, if $\hat{R}$ is F-rational the faithful flatness of the map from R to $\hat{R}$ proves that R is F-rational.

To prove d) we may assume that R is a homomorphic image of a Cohen-Macaulay local ring since in the excellent case we can complete R by using c). Choose a system of parameters $x_1, \ldots, x_d$. We need to prove they form a regular sequence. If not, there is an i such that $(x_1, \ldots, x_i) : x_{i+1} \neq (x_1, \ldots, x_i)$. But by Theorem 2.1 $(x_1, \ldots, x_i) : x_{i+1} \subseteq (x_1, \ldots, x_i)^* = (x_1, \ldots, x_i)$, the last equality coming from the assumption that R is F-rational.

To prove e) we can again assume that R is a homomorphic image of a Cohen-Macaulay ring since in the case R is excellent we can complete R by c) for both directions. The second condition in e) is obviously necessary. Assume that R is equidimensional and the ideal generated by a single system of parameters $x_1, \ldots, x_d$ is tightly closed. Let $y_1, \ldots, y_n \in R$ be a system of parameters for R. To show that $(y_1, \ldots, y_i)R$ is tightly closed, it will suffice to show that $(y_1, \ldots, y_n)R$ is tightly closed. But $H_m^n(R) \cong \varinjlim R/(x_1^t, \ldots, x_n^t) \cong \varinjlim R/(y_1^t, \ldots, y_n^t)$ where the map from the term indexed by t to that indexed by $t+1$ is induced by multiplication by $x_1 \cdots x_n$ (respectively, by $y_1 \cdots y_n$) and is injective. Thus, $R/(y_1, \ldots, y_n)$ injects into $R/(x_1^t, \ldots, x_n^t)$ for any sufficiently large t. To show that $(y_1, \ldots, y_n)R$ is tightly closed in R, it suffices to show that 0 is tightly closed in $R/(y_1, \ldots, y_n)R$, and, hence, to show that 0 is tightly closed in $R/(x_1^t, \ldots, x_n^t)R$. Thus, we have reduced to the case where $y_i = x_i^t$. The argument will be finished by showing that if $(x_1, \ldots, x_n)R$ is tightly closed then so is $(x_1^t, \ldots, x_n^t)R$ for every t.

Let $z \in (x_1^n, \ldots, x_t^n)^*$. If there is an i such that $zx_i \notin (x_1^n, \ldots, x_t^n)$ one can re-place z by this new element. Eventually we may assume that $z(x_1, \ldots, x_t) \subseteq (x_1^n, \ldots, x_t^n)$. Since the x_i form a regular sequence, $(x_1^n, \ldots, x_t^n) : (x_1, \ldots, x_t) = (x_1^n, \ldots, x_t^n, y^{n-1})$, where $y = x_1 \cdots x_t$. Without loss of generality we may assume that $z = uy^{n-1}$. Choose $c \in R^o$ such that $cz^q \in (x_1^{nq}, \ldots, x_t^{nq})$ for all large q. Then $cu^q \in (x_1^{nq}, \ldots, x_t^{nq}) : y^{(n-1)q} = (x_1^q, \ldots, x_t^q)$, the last equality following be-cause the x_i form a regular sequence. Hence $u \in (x_1, \ldots, x_t)^* = (x_1, \ldots, x_t)$. This implies that $z \in (x_1^n, \ldots, x_t^n)$.

To prove f) first suppose that R_m is F-rational for every maximal ideal m. Suppose that $(x_1, \ldots, x_n)R$ has height n in R and that y is in its tight closure but not in the ideal. Then all this can be preserved while localizing at a suitable maximal ideal, giving a contradiction. Thus, R is F-rational. The converse follows immediately from part g).

To prove g), suppose that R is F-rational, and let P be any prime ideal of R. Choose $x_1, \ldots, x_n \in P$ to be an R-sequence, where n is the height of P (which is the same as the depth of R on P). The images $x_1/1, \ldots, x_n/1$ in R_P will be a system of parameters. If we can show that the ideal $(x_1/1, \ldots, x_n/1)R_P$ is tightly closed in R_P, it will follow from part e) that R_P is F-rational. This follows from Proposition 1.5.

It now follows that R is F-rational iff all its localizations at primes are, and this implies that every localization of R is F-rational. $\qquad\square$

A criterion for F-rationality in terms of the highest local cohomology module is given in [Sm2]. The criterion states:

Theorem 8.3. (F-rationality and Local Cohomology). *Let (R, m) be a d-dimensional excellent local Cohen-Macaulay ring of characteristic p. R is F-rational iff $H_m^d(R)$ has no proper nontrivial submodules stable under the action of Frobenius.*

Proof. Assume first that R is not F-rational. Choose a system of parameters, $x_1, \ldots, x_d$ for R and an element $z \in (x_1, \ldots, x_d)^*$ but not in the ideal I generated by these parameters. Consider the element $\eta = [\frac{z}{x}] \in H_m^d(R)$. Since R is Cohen-Macaulay, this element is nonzero in $H_m^d(R)$. Applying Frobenius repeatedly to η gives elements $[\frac{z^q}{x^q}]$. Let N be the submodule of $H_m^d(R)$ spanned by these elements. Clearly N is stable under Frobenius. Since $z \in (x_1, \ldots, x_d)^*$ there exists an element $c \in R^o$ such that $cz^q \in I^{[q]}$. This then implies that $cN = 0$. The highest local cohomology module of R is always faithful, and so N must be a proper submodule of $H_m^d(R)$.

Conversely, assume that R is F-rational. Since R is excellent, by Theorem 8.2 c) we can complete R and assume R is complete. This does not change the highest local cohomology of R. Suppose that there is a nonzero proper submodule $N \subseteq H_m^d(R)$ stable under Frobenius. The Matlis dual of $H_m^d(R)$ is the canonical module of R (see [HH13] and [HeK]), and since R is reduced this module has constant rank 1. The canonical module maps onto the Matlis dual of N, and the kernel of this surjection must be a rank 1 nonzero submodule of the canonical module of R.

In particular the dual of N is the cokernel of an injective map of rank 1 torsion-free modules over R and is therefore torsion. Hence there is an element $c \in R^o$ which annihilates the Matlis dual of N. Then $cN = 0$ also. Now chose any nonzero element $\eta = [\frac{z}{x}] \in N$, where $x_1, \ldots, x_d$ are parameters in R and x represents their product. All Frobenius powers of η are killed by c. Hence $[\frac{cz^q}{x^q}] = 0$ in $H_m^d(R)$. This implies that for large t, $cz^q x^t \in (x_1^{t+q}, x_2^{t+q}, \ldots, x_d^{t+q})$. As R is Cohen-Macaulay we then obtain that $cz^q \in (x_1^q, \ldots, x_d^q)$ which gives that $z \in (x_1, \ldots, x_d)^*$. Since R is F-rational, $z \in (x_1, \ldots, x_d)$. Then $\eta = 0$, a contradiction.   □

9. Rational Singularities

In this section we discuss the relationship between rational singularities and F-rationality. We begin with a discussion of rational singularities.

Definition 9.1. Let R be a normal local ring which is essentially of finite type over a field k of characteristic 0. Let $f : Z \to X = \mathrm{Spec}(R)$ be a resolution of singularities of X (such resolutions are known to exist in this case by Hironaka). R is said to be (or have) a *rational singularity* if $R^j f_*(\mathcal{O}_Z) = 0$ for all $j > 0$.

This definition is independent of the resolution of singularities Z. The higher direct images of the structure sheaf of Z are sometimes hard to understand. If R has an isolated singularity (i.e., R_P is regular for all nonmaximal primes P), then $R^j f_*(\mathcal{O}_Z) \cong H_m^{j+1}(R)$ for $1 \leq j \leq \dim(R) - 2$. Since $\mathrm{Spec}(R)$ is affine, $R^j f_*(\mathcal{O}_Z) = H^j(Z, \mathcal{O}_Z)$, the usual sheaf cohomology of Z. Moreover, if R is a rational singularity, then R must be Cohen-Macaulay and normal.

There are many examples of rational singularities. Any regular local ring has rational singularities. In the homogeneous case, the a-invariant plays an important role:

Theorem 9.2. [Fl, W5] *Let R be a nonnegatively graded ring over a field $k = R_0$ of characteristic 0. Then R has rational singularities (i.e., R_P is a rational singularity for every prime P of R) iff the following conditions hold:*

(1) *R is Cohen-Macaulay and normal.*
(2) *R_P has a rational singularity for all primes $P \neq M$, where M is the unique homogeneous maximal ideal.*
(3) *The a-invariant, $a(R)$, is negative.*

Example 9.3.

9.3.1 Set $R = k[X, Y, Z]/(X^2 + Y^3 + Z^5)$, where k is a field of finite characteristic not equal to 2, 3, or 5. R is Cohen-Macaulay, normal (use the Jacobian criterion) and graded. Set $\deg(X) = 15$, $\deg(Y) = 10$, and $\deg(Z) = 6$. The degree of $X^2 + Y^3 + Z^5$ is 30, while the sum of the degrees of the variables is 31. The a-invariant is $-1 = 30 - 31$. R is a rational singularity.

9.3.2 Let $R = k[X, Y, Z]/(X^3 + Y^3 + Z^3)$, where k is a field of characteristic $\neq 3$. This has the usual grading, and the degree of $X^3 + Y^3 + Z^3$ is 3. The a-invariant is $0 = 3 - 3$. Hence R does not have a rational singularity.

9.3.3 Let $R = k[X, Y, Z]/(X^2 + Y^3 + Z^7)$, where k is a field of characteristic $\neq 2, 3, 5$. We can grade R by setting $\deg(X) = 21$, $\deg(Y) = 14$, and $\deg(Z) = 6$. Then the sum of the degrees of the variables is 41, while the degree of $X^2 + Y^3 + Z^7$ is 42. The a-invariant is $1 = 42 - 41$. R is not a rational singularity.

9.3.4 If R is a two-dimensional regular local ring and I is an integrally closed ideal, then the blow-up of I has rational singularities (see [L3]). This is false in dimension three.

9.3.5 Perhaps the largest class of rational singularities comes from invariant theory. By a theorem of Boutot [Bou] if R has rational singularities and is finite type over the complex numbers and S is a direct summand of R, then S also has rational singularities. In particular if a reductive groups acts linearly on a polynomial ring over the complex numbers, then the ring of invariants has rational singularities.

9.3.6 A recent theorem of Lipman [L1] connects rational singularities to the Cohen-Macaulay property of Rees algebras. Suppose that (R, m) is a local Cohen-Macaulay ring which is essentially of finite type over the complex numbers. Choose an ideal I such that the blowup of I is smooth, i.e. such that $X = \mathrm{Proj}(R[It])$ is a desingularization of $\mathrm{Spec}(R)$. Then R has rational singularities iff there exists an integer N such that $R[I^N t]$ is Cohen-Macaulay.

9.3.7 It is possible that a partial converse to (9.3.6) is true. Let R be a ring with rational singularities (in equicharacteristic zero), and suppose that the multi-Rees ring $R[I_1 t_1, \ldots, I_n t_n]$ is both Cohen-Macaulay and normal. Then does $R[I_1 t_1, \ldots, I_n t_n]$ have only rational singularities? For $n = 1$ if we in addition assume that the blowup of I has rational singularities, then Lipman's work gives a positive answer. If this is true, then the Rees ring $R[(I_1 \cdots I_n)t]$ must also be Cohen-Macaulay, because of Boutot's theorem (9.3.5) and the fact that this latter ring is a direct summand of $R[I_1 t_1, \ldots, I_n t_n]$.

Let us ignore for the moment the problems which exist between the translation from fields of characteristic 0 to positive characteristic. In any case, a theory of tight closure exists for rings containing fields of any characteristic, and the theorems in these notes carry over from characteristic p to equicharacteristic 0. See [HH12, Ho9] for proofs. We then can obtain easily:

Theorem 9.4. *Let (R, m) be a nonnegatively graded ring over a field R_0 of characteristic 0 with an isolated singularity at m. R is F-rational iff R has rational singularities.*

Proof. If R is F-rational, then R is Cohen-Macaulay and normal by Theorem 8.2. To prove that R has rational singularities, we then need to prove only that the a-invariant a of R is negative, by Theorem 9.2. If $x_1, \ldots, x_d$ is a homogeneous system of parameters with $D = \sum_i \deg(x_i)$, then since $R_{\geq D} \subseteq (x_1, \ldots, x_d)^* =$

$(x_1, \ldots, x_d)$, we see that the highest local cohomology lives only in negative degree. This means that $a < 0$, and hence R has rational singularities.

Consider the converse. If R has rational singularities then R is Cohen-Macaulay and normal and $a < 0$, where a is the a-invariant of R. Let $x_1, \ldots, x_d$ be a homogeneous system of parameters. The Strong Vanishing Theorem (in the Cohen-Macaulay and characteristic 0 case) says that

$$(x_1, \ldots, x_d)^* = (x_1, \ldots, x_d) + R_{\geq D},$$

where $D = \sum_i \deg(x_i)$. However, we have seen that $a < 0$ exactly means that $R_{\geq D} \subseteq (x_1, \ldots, x_d)$. It follows that R is F-rational. $\qquad\square$

In fact, combining work of Smith [Sm2] and Hara [Ha5] gives the following remarkable result:

Theorem 9.5. *Let X be a scheme of finite type over a field of characteristic 0. X has F-rational type iff X has rational singularities.*

Example 9.6. Let R be the hypersurface $x^2 + y^3 + z^5 = 0$. If the characteristic of the base field K is at least 7, then R satisfies the Strong Vanishing Theorem by Theorem 6.4. In particular, R is F-rational, since the tight closure of (y, z) is exactly (y, z) together with everything of degree at least the sum of the degrees of y and z. In this case, the degree of y is 10, and the degree of z is 6. Hence $(y, z)^* = (y, z) + R_{\geq 16} = (y, z)$. (Notice the degree of x is 15.) R is known to be a rational singularity if the characteristic is at least 7–see Example 6.6. However, in characteristic 2, for example, we see that $x^2 \in (y^2, z^2)$. The element $[\frac{x}{yz}] \in H_m^2(R)$ has degree -1, and applying the Frobenius gives the element $[\frac{x^2}{y^2 z^2}] \in H_m^2(R)$ which is zero since $x^2 \in (y^2, z^2)$. Hence the Frobenius does not act injectively on the negative degree piece of the top local cohomology. This shows that the Strong Vanishing Theorem is false if one fixes the characteristic. But the Theorem is still valid: it does hold for this example whenever the characteristic exceeds 7.

Although we have concentrated on hypersurfaces of dimension two, the theorems hold in arbitrary dimension. For example, if $R = k[X_0, \ldots, X_d]/(X_0^N + \cdots + X_d^N)$ then R has an isolated singularity provided the characteristic p of k does not divide N. The theorem on hypersurfaces guarantees than R satisfies the strong vanishing conjecture if $p > (d-1)N - d$. Since the a-invariant is $a = N - d - 1$ we obtain for such large p that the test ideal is exactly m^{N-d+2}, where m is the irrelevant ideal.

By the theorem of Boutot, direct summands of rational singularities also have rational singularities, at least for schemes of finite type over the complex numbers. It was hoped that direct summands of F-rational rings are F-rational. However, K. Watanabe recently gave a counterexample. This example is closely tied to the failure of the Kodaira vanishing theorem in finite characteristics, as the next section will relate. This counterexample does not contradict the theorems of Hara or Smith. These theorems are ultimately about characteristic 0 or rings of characteristic $p \gg 0$.

Example 9.7. This is a sketch of Watanabe's example. Let A be the hypersurface $x^2 + y^3 + z^5 = 0$ in characteristic 2. Let $B = k[s,t]$ and let C be the Segre product of A and B, i.e. the subring generated by all elements in $A[s,t]$ of the form af where $a \in A$, $f \in k[s,t]$ and $\deg(a) = \deg(f)$. A is a direct summand of C. We have seen that A is not F-rational, so it is enough to see that C is F-rational. C is the invariant ring of an appropriate torus action on $D = A[s,t]$. In particular C is normal. The local cohomology $H_M^i(C)$, where M is the irrelevant ideal can be computed as the invariants of $H_N^i(D)$ of the same torus, where N is the irrelevant ideal of D. Using this it is not difficult to prove that the C is Cohen-Macaulay with a-invariant -2. Since the Frobenius acts injectively on the local cohomology of C, it follows that C is F-rational.

10. The Kodaira Vanishing Theorem

In this section we discuss the relationship between the Kodaira Vanishing theorem and the Strong Vanishing Theorem of Section 5. All of this material is taken from the paper [HS], which in turn was motivated by [Sm6]. The starting point was the realization that the Kodaira vanishing theorem actually is equivalent to a statement concerning the tight closure of partial systems of parameters, which is discussed in [Sm6].

Recall that the Kodaira Vanishing Theorem states that if X is a smooth projective variety over a field k of characteristic 0 and if $\mathcal{L}$ is an ample line bundle on X, then $H^i(X, \mathcal{L}^{-1}) = 0$ for i less than the dimension of X.

The following definition comes from [Sm6].

Definition 10.1. For an ample invertible sheaf $\mathcal{L}$ on a normal irreducible projective variety X, define the *section ring for X with respect to $\mathcal{L}$* to be the ring

$$S_{\mathcal{L}} = \bigoplus_{n \in \mathbb{N}} H^0(X, \mathcal{L}^n).$$

The first point in the translation to Commutative Algebra is to change the sheaf cohomology to the local cohomology of the section ring:

Theorem 10.2. *Let $S_{\mathcal{L}}$ be a section ring for a pair $(X, \mathcal{L})$, where $\mathcal{L}$ is an ample invertible sheaf on a projective variety X over a field k. If X is smooth and k has characteristic zero, then for all $i < \dim S_{\mathcal{L}}$, the graded local cohomology modules*

$$H_m^i(S_{\mathcal{L}})$$

have no non-zero graded components of negative degree.

To interpret the vanishing of graded pieces of local cohomology in terms of the tight closure of parameter ideals, we need to study the unmixed part of parameter ideals. The following proposition from [HS] is crucial to the interpretation we seek. The statement and proof are taken from [HS, (2.7)].

Proposition 10.3. *Let (S, m) be an equidimensional graded ring. Assume that for every $P \in \operatorname{Spec} S - \{m\}$, the local ring S_P is Cohen-Macaulay. The following are equivalent:*

(1) *$[H_m^i(S)]_n = 0$ for all $n < 0$ and all i less than the dimension of S.*
(2) *For every (homogeneous) parameter ideal I of height i and for all $n < 0$ and $j < i$, $[H_I^j(S)]_n = 0$. Further if $\eta \in [H_I^i(S)]_n$ with $n < 0$ and $\eta \neq 0$, then $\operatorname{Ann}(\eta)$ has the same height as I.*
(3) *For every parameter ideal I of height i, the following holds: If $\eta \in [H_I^i(S)]_n$ with $n < 0$ and $\eta \neq 0$, then $\operatorname{Ann}(\eta)$ has the same height as I.*
(4) *For every parameter ideal I, the following holds: $I^{unm} \subseteq I^{lim} + S_{\geq \delta}$, where δ is the sum of the degrees of a set of parameters generating I.*

Proof. Throughout the proof, $x_1, \ldots, x_i$ will denote a fixed set of parameters generating the height i parameter ideal I. Let δ be the sum of the degrees of these parameters and let d be the dimension of S. We will prove that $(1) \Rightarrow (2) \Rightarrow (3) \Rightarrow (4) \Rightarrow (3) \Rightarrow (2) \Rightarrow (1)$.

Assuming (1), we prove (2) by descending induction on i, the height of I. If $i = d$, then the second condition is immediate as in any case, the element η is annihilated by a power of I, and hence $\operatorname{Ann}(\eta)$ has height at least d. Because $\eta \neq 0$, the height cannot be more than d. The first statement is the same as that in (1), as the radical of I is m.

Assume we have proved (2) for parameter ideals of height greater than i. Fix $\eta \in [H_I^i(S)]_n$ with $n < 0$. If the annihilator of η has height at least $i+1$, then since the annihilator contains a power of I, there exists an element x_{i+1} annihilating η such that $x_1, \ldots, x_{i+1}$ form parameters. If the annihilator of η has height i, choose any x_{i+1} such that $x_1, \ldots, x_{i+1}$ form parameters. In either case, let J be the ideal generated by $x_1, \ldots, x_{i+1}$. Recall that there is a long exact sequence of graded modules and degree preserving

$$\cdots \to H_J^j(S) \to H_I^j(S) \to (H_I^j(S))_{x_{i+1}} \to \cdots$$

If $j \leq i - 1$, then $(H_I^j(S))_{x_{i+1}} = 0$, since $S_{x_{i+1}}$ is Cohen-Macaulay and I is a parameter ideal of height i. The induction applied to J then gives that $H_J^j(S)$ is zero in negative degree, and consequently the same is true for $H_I^j(S)$. For the $j = i$ case, assume on the contrary that the annihilator of η is height at least $i + 1$. The image of η under the localization map in the long exact sequence is 0 by choice of x_{i+1}. Hence η is in the image of $H_J^i(S)$. Since the degree of η is negative and $i \leq (i + 1) - 1$, the induction then proves that $\eta = 0$.

Clearly (2) implies (3). Assume (3). Write $I = I^{unm} \cap K$, where the height of K is at least $i + 1$. Choose an element $x_{i+1} \in K$ such that $x_1, \ldots, x_{i+1}$ are parameters. It easily follows that $I^{unm} = I :_R x_{i+1}$. Let $z \in I^{unm}$ and assume that $\deg(z) < \delta$. Write $z x_{i+1} = \sum_{1 \leq j \leq i} z_j x_j$. Consider the element $\eta = [\frac{z}{(x_1 \cdots x_i)}] \in H_I^i(S)$. The formula in the above line shows that η is killed by x_{i+1} so that the height of its annihilator is at least $i + 1$. The degree of η is $\deg(z) - \delta < 0$, so that by (2), $\eta = 0$. Therefore there exists an integer $t \geq 0$ such that $z(x_1 \cdots x_i)^t \in (x_1^{t+1}, \ldots, x_i^{t+1})$, which implies that $z \in I^{lim}$, and proves (4).

Assume (4), and let $\eta = [\frac{z}{(x_1 \cdots x_i)}] \in H_I^i(S)$ be such that its annihilator has height at least $i + 1$, and such that $\deg(\eta) < 0$. After replacing z by $zx_1^t \cdots x_i^t$ and x_j by x_j^{t+1}, we may assume that there exists an ideal J of height at least $i + 1$ such that $Jz \subseteq I$. It follows from primary decomposition that $z \in I^{unm}$. The assumption that the degree of η is negative implies that the degree of z is strictly less than δ, the sum of the degrees of the x_j. (4) then says that $z \in I^{lim}$, which forces $\eta = 0$.

Now assume (3). By ascending induction on i we will prove that $(H_I^j(S))$ is zero in negative degree if $j < i$. If $i = 1$, we need to prove that $H^0_{(x_1)}(S)$ is zero in negative degrees, which is true since this local cohomology module sits in S. Suppose we have shown this to be true for $x_1, \ldots, x_{i-1}$. Let J be the ideal they generate. If $j < i-1$, then $H_I^j(S)$ is isomorphic with $H_J^j(S)$ ($I = (x_1, \ldots, x_i)$) since $H_J^k(S)_{x_i} = 0$ for $k = j, j - 1$ as S_{x_i} is Cohen-Macaulay. The induction assumption gives the conclusion for these values. If $j = i - 1$, then $H_I^{i-1}(S)$ may be identified with the kernel of the localization map from $H_J^{i-1}(S) \rightarrow (H_J^{i-1}(S))_{x_i}$. If η is in this kernel, then η is annihilated by a power of J, together with a power of x_i, and therefore has annihilator of height at least $i > \mathrm{ht}(J)$. If further the degree of η is negative, it follows that $\eta = 0$, since by (3) the annihilator of any non-zero element of $H_I^{i-1}(S)$ of negative degree should have height exactly $i - 1$. This proves that $H_J^{i-1}(S)$ is zero in negative degrees, as required.

Finally, assume (2). Condition (1) follows immediately from the case $i = d$. $\square$

The following theorem is from [HS, (3.7)].

Theorem 10.4. (Kodaira Vanishing and Tight Closure). *Let $S_{\mathcal{L}}$ be the section ring for an irreducible projective variety X over a field k with respect to an ample invertible sheaf $\mathcal{L}$. Let $x_1, x_2, \ldots, x_i$ be parameters in $S_{\mathcal{L}}$ with $i < d = dim(S_{\mathcal{L}})$. If X is non-singular and k has characteristic zero, then*

$$(x_1, x_2, \ldots, x_i)^* \subseteq (x_1, x_2, \ldots, x_i)^{lim} + (S_{\mathcal{L}})_{\geq \delta},$$

where δ is the sum of the degrees of the $x_j, 1 \leq j \leq i$.

Proof. Let I be the ideal generated by $x_1, \ldots, x_i$. The assertion of this Theorem is exactly the statement (4) in Proposition 10.3, using from Proposition 5.8 that $(x_1, x_2, \ldots, x_i)^* = (x_1, x_2, \ldots, x_i)^{unm}$. The equivalent statement (1) of Proposition 10.3 is exactly the Kodaira Vanishing Theorem as in (10.2) above. The hypothesis of Proposition 10.3 is satisfied because if X is Cohen-Macaulay, then $S_{\mathcal{L}}$ has (at worst) an isolated non-Cohen-Macaulay point at the irrelevant ideal m. $\square$

The relationship between Kodaira vanishing and the Strong Vanishing Theorem should now be clear: all the Strong Vanishing Theorem does is push the tight closure interpretation of Kodaira to a *full* system of parameters. Kodaira vanishing says that the negative part of the local cohomology $H^i_m(S)$ is zero if $i < \dim(S)$. This is wildly false for the top local cohomology. Most of it lives in negative degree. But once a tight closure formulation is given, it becomes clear what piece of the highest local cohomology must vanish.

Another perspective can be obtained from the equivalent formulation of the Strong Vanishing Theorem given in Theorem 6.2. It is equivalent to saying that the Frobenius acts injectively on the negative part of the top local cohomology. What about the lower local cohomology? Since S is assumed to be Cohen-Macaulay on the punctured spectrum and is equidimensional, it is well-known that the local cohomology modules $H^i_m(S)$ have finite length. In particular, there is a positive integer N such that $H^i_m(S)_n = 0$ for $n < -N$. Suppose that we knew the Frobenius acted injectively in negative degree. If $\eta \in H^i_m(S)$ has negative degree j, then applying Frobenius e times gives us an element of degree jp^e, and for large e, $jp^e < -N$. It follows that $\eta = 0$. In other words, the statement that the Frobenius acts injectively in negative degrees on the lower local cohomology forces it to be zero.

In [HS, (3.14)] it is shown that the Strong Vanishing Theorem implies the Kodaira vanishing theorem. The basic idea of this proof is to use a Bertini theorem to concentrate 'bad' homology in the top local cohomology.

References

[Ab1]   Aberbach, I., *Finite phantom projective dimension*, Amer. J. Math. **116** (1994), 447–477.

[Ab2]   Aberbach, I., *Test elements in excellent rings with an application to the uniform Artin-Rees property*, Proc. Amer. Math. Soc. **118** (1993), 355–363.

[Ab3]   Aberbach, I., *Tight closure in F-rational rings*, Nagoya Math. J. **135** (1994), 43–54.

[Ab4]   Aberbach, I., *Arithmetic Macaulayfications using ideals of dimension one*, Illinois J. Math., **40** (1996), 518–526.

[AHH]   Aberbach, I., Hochster, M., and Huneke, C., *Localization of tight closure and modules of finite phantom projective dimension*, J. Reine Angew. Math. (Crelle's Journal) **434** (1993), 67–114.

[AH1]   Aberbach, I. and Huneke, C., *An improved Briançon-Skoda theorem with applications to the Cohen-Macaulayness of Rees rings*, Math. Ann. **297** (1993), 343–369.

[AH2]   Aberbach, I. and Huneke, C., *A theorem of Briançon-Skoda type for regular local rings containing a field*, Proc. Amer. Math. Soc. **124** (1996), 707–713.

[AHS]   Aberbach, I., Huneke, C. and Smith, K., *Arithmetic Macaulayfication and tight closure* Illinois J. Math. **40** (1996), 310–329.

[AHT]   Aberbach, I., Huneke, C., and Trung, N.V., *Reduction numbers, Briançon-Skoda theorems and the depth of Rees rings*, Compositio Math. **97** (1995), 403–434.

[And]   André, M., *Cinq exposés sur la désingularization*, preprint.

[Ar1]   Artin, M., *Algebraic approximation of structures over complete local rings*, Publ. Math. I.H.E.S. (Paris) **36** (1969), 23–56.

[Ar2]   Artin, M., *On the joins of Hensel rings*, Advances in Math. **7** (1971), 282–296.

[ArR]   Artin, M. and Rotthaus, C., *A structure theorem for power series rings*, in Algebraic Geometry and Commutative Algebra: in honor of Masayoshi Nagata, Vol. I, Kinokuniya, Tokyo, 1988, 35–44.

[Ba]   Barger, S.F., *A theory of grade for commutative rings*, Proc. Amer. Math. Soc. **36** (1972), 365–368.

[Bou]   Boutot, J.-F., *Singularités rationelles et quotients par les groupes réductifs*, Invent. Math. **88**, (1987), 65–68.

[BrS]   Briançon, J. and Skoda, H., *Sur la clôture intégrale d'un idéal de germes de fonctions holomorphes en un point de C^n*, C. R. Acad. Sci. Paris Sér. A **278** (1974), 949–951.

[Br1]   Brodmann, M., *A macaulayfication of unmixed domains*, J. Algebra **44** (1977), 221–234.

[Br2]   Brodmann, M., *Local cohomology of certain Rees and Form rings II*, J. Algebra **86** (1984), 457–493.

[Br3]   Brodmann, M., *Asymptotic stability of $Ass(M/I^n M)$*, Proc. Amer. Math. Soc. **74** (1979), 16–18.

[Br4]   Brodmann, M., *A few remarks on 'Macaulayfication' of sheaves*, preprint, 1995.

[Bru]   Bruns, W., *Tight closure*, Bull. Amer. Math. Soc. **33** (1996), 447–458.

[BH]   Bruns, W. and Herzog, J., *Cohen-Macaulay Rings*, vol. 39, Cambridge studies in advanced mathematics, 1993.

[BV]   Bruns, W. and Vetter, U., *Determinantal Rings*, vol. 1327, Springer-Verlag Lecture Notes in Math., 1988.

[BuE]   Buchsbaum, D. and Eisenbud, D., *What makes a complex exact*, J. Algebra **25** (1973), 259–268.

[BP]   Buchweitz, R. and Pardue, K., *Hilbert-Kunz functions*, in preparation.

[Ch]   Chang, S.-t., *The asymptotic behavior of Hilbert-Kunz functions and their generalizations*, Thesis, University of Michigan, 1993.

[Co1]   Conca, A., *Hilbert-Kunz function of monomial ideals and binomial hypersurfaces*, Manu. Math. **90** (1996), 287–300.

[Co2]   Conca, A., *The a-invariant of determinantal rings*, Math. J. Toyama University **18**, (1995), 47–63.

[CHe]   Conca, A. and Herzog, J., *Ladder determinantal ideals have rational singularities*, preprint.

[Con]   Contessa, M., *On the Hilbert-Kunz function and Koszul homology*, J. Algebra **177** (1995), 757–766.

[Cow]   Cowden, J., developing thesis.

[DI]   Deligne, P. and Illusie, L., *Relèvements modulo p^2 et decomposition du complexe de de Rham*, Invent. Math. **89** (1987), 247–270.

[DO1]   Duncan, A. J. and O'Carroll, L., *A full uniform Artin-Rees theorem*, J. reine angew. Math. **394** (1989), 203–207.

[DO2]   Duncan, A. J. and O'Carroll, L., *On Zariski regularity, the vanishing of Tor, and a uniform Artin-Rees theorem*, Topics in Algebra, Banach Center Publications, Warsaw **26** (1990), 49–55.

[Du1]   Dutta, S. P., *On the canonical element conjecture*, Trans. Amer. Math. Soc. **299** (1987), 803–811.

[Du2]   Dutta, S. P., *Frobenius and multiplicities*, J. Algebra **85**, (1983), 424–448.

[EH]   Eisenbud, D. and Hochster, M., *A Nullstellensatz with nilpotents and Zariski's main lemma on holomorphic functions*, J. Algebra **58** (1979), 157–161.

[EvG]   Evans, E.G. and Griffith, P., *The syzygy problem*, Annals of Math. **114** (1981), 323–333.

[Fa1]   Faltings, G., *A contribution to the theory of formal meromorphic functions*, Nagoya Math. J. **77** (1980), 99–106.

[Fa2]   Faltings, G., *Über die Annulatoren lokaler Kohomologiegruppen*, Arch. Math. **30** (1978), 473–476.

[Fa3]   Faltings, G., *Über Macaulayfizierung*, Math. Ann. **238** (1978), 175–192.

[Fe1]   Fedder, R., *A Frobenius characterization of rational singularity in 2-dimensional graded rings*, Trans. Amer. Math. Soc. **340** (1993), 655–668.

[Fe2]   Fedder, R., *F-purity and rational singularity in graded complete intersection rings*, Trans. Amer. Math. Soc. **301** (1987), 47–61.

[Fe3]   Fedder, R., *F-purity and rational singularity*, Trans. Amer. Math. Soc. **278** (1983), 461–480.

[FeW]   Fedder, R. and Watanabe, K.-i., *A characterization of F-regularity in terms of F-purity*, in Commutative Algebra, Math. Sci. Research Inst. Publ. **15**, Springer-Verlag, New York · Berlin · Heidelberg, 1989, pp. 227–245.

[Fl]   Flenner, H., *Rationale quasihomogene Singularitäten*, Arch. Math. **36** (1981), 35–44.

[Gi]   Gibson, G.J., *Seminormality and F-purity in local rings*, Osaka J. Math. **26** (1989), 245–251.

[GS]   Gillet, H. and Soulé, C., *K-théorie et nullité des multiplicités d'intersection*, C. R. Acad. Sci., Paris, Series I **300 (no. 3)** (1985), 71–74.

[Gla1]   Glassbrenner, D. J., *Invariant rings of group actions, determinantal rings, and tight closure*, Thesis, University of Michigan, 1992.

[Gla2]   Glassbrenner, D. J., *Strong F-regularity in images of regular rings*, Proc. Amer. Math. Soc. (to appear).

[GlaS]   Glassbrenner, D. J and Smith, K., *Singularities of ladder determinantal varieties*, J. Pure Appl. Alg. **101** (1995), 59–75.

[Go]   Goto, S., *A problem on Noetherian local rings of characteristic p*, Proc. Amer. Math. Soc. **64** (1977), 199–205.

[GW]   Goto, S. and Watanabe, K.-i., *The structure of 1-dimensional F-pure rings*, J. Algebra **49** (1977), 415–421.

[GY]   Goto, S. and Yamagishi, K., *The theory of unconditioned strong d-sequences and modules of finite local cohomology*, preprint.

[GraR]   Grauert, H. and Riemenschneider, O., *Verschwindungsätze für analytische kohomologiegruppen auf komplexen Räuman*, Invent. Math. **11** (1970), 263–290.

[Ha]   Han, C., *The Hilbert-Kunz function of a diagonal hypersurface*, Thesis, Brandeis University, 1992.

[HaMo]   Han, C. and Monsky, P., *Some surprising Hilbert-Kunz functions*, Math. Z. **214** (1993), 119–135.

[Ha1]   Hara, N., *Classification of two dimensional F-regular and F-pure singularities*, preprint.

[Ha2]   Hara, N., *F-regularity and F-purity of graded rings*, J. Algebra **172** (1995), 804–818.

[Ha3]   Hara, N., *A characterization of graded rational singularities in terms of injectivety of Frobenius maps*, preprint, 1996.

[Ha4]   Hara, N., *F-injectivity in negative degree and tight closure in graded complete intersection rings*, C.R. Math. Acad. Sci. Canada **17(6)** (1995), 247–252.

[Ha5]   Hara, N., *A characterization of rational singularities in terms of the injectivity of Frobenius maps*, preprint, 1996.

[HW]   Hara, N. and Watanabe, K-i., *The injectivity of Frobenius acting on cohomology and local cohomology modules*, Manu. Math. **90** (1996), 301–316.

[He1]   Herzog, J., *Ringe der Characteristik p und Frobeniusfunktoren*, Math. Z. **140** (1974), 67–78.

[HeK]   Herzog, J. and Kunz, E., *Der kanonische Modul eines Cohen-Macaulay Rings*, Lecture Notes in Math., Springer-Verlag **238** (1971).

[Ho1]   Hochster, M., *Contracted ideals from integral extensions of regular rings*, Nagoya Math. J. **51** (1973), 25–43.

[Ho2]   Hochster, M., *Topics in the homological theory of modules over commutative rings*, C.B.M.S. Regional Conf. Ser. in Math. No. **24**, A.M.S., Providence, R.I., 1975.

[Ho3]   Hochster, M., *Big Cohen-Macaulay modules and algebras and embeddability in rings of Witt vectors*, in Proceedings of the Queen's University Commutative Algebra Conference Queen's Papers in Pure and Applied Math. **42**, 1975, pp. 106–195.

[Ho4]   Hochster, M., *Cyclic purity versus purity in excellent Noetherian rings*, Trans. Amer. Math. Soc. **231** (1977), 463–488.

[Ho5]   Hochster, M., *Some applications of the Frobenius in characteristic 0*, Bull. Amer. Math. Soc. **84** (1978), 886–912.

[Ho6]   Hochster, M., *Cohen-Macaulay rings and modules*, Proc. of the International Congress of Mathematicians, Helsinki, Finland, Vol. I, Academia Scientarium Fennica, 1980, pp. 291–298.

[Ho7]   Hochster, M., *Canonical elements in local cohomology modules and the direct summand conjecture*, J. Algebra **84** (1983), 503–553.

[Ho8]   Hochster, M., *Solid closure*, Contemp. Math. **159** (1994), 103–172.

[Ho9]   Hochster, M., *Tight closure in equicharacteristic, big Cohen-Macaulay algebras and solid closure*, Contemp. Math. **159** (1994), 173–196.

[Ho10]   Hochster, M., *The notion of tight closure in equal characteristic zero, an Appendix to Tight Closure and Its Applications, by C. Huneke*, CBMS Regional Conference Series, AMS **88** (1996), 94–106.

[HoE]   Hochster, M., and Eagon, J.A., *Cohen-Macaulay rings, invariant theory, and the generic perfection of determinantal loci*, Am. J. Math. **93** (1971), 1020–1058.

[HH1]   Hochster, M. and Huneke, C., *Tightly closed ideals*, Bull. Amer. Math. Soc. **18** (1988), 45–48.

[HH2]   Hochster, M. and Huneke, C., *Tight closure*, in Commutative Algebra, Math. Sci. Research Inst. Publ. **15**, Springer-Verlag, New York · Berlin · Heidelberg, 1989, pp. 305–324.

[HH3]   Hochster, M. and Huneke, C., *Tight closure and strong F-regularity*, Mémoires de la Société Mathématique de France, numéro **38** (1989), 119–133.

[HH4]   Hochster, M. and Huneke, C., *Tight closure, invariant theory, and the Briançon-Skoda theorem*, J. Amer. Math. Soc. **3** (1990), 31–116.

[HH5]   Hochster, M. and Huneke, C., *Absolute integral closures are big Cohen-Macaulay algebras in characteristic p*, Bull. Amer. Math. Soc. (New Series) **24** (1991), 137–143.

[HH6]   Hochster, M. and Huneke, C., *Tight closure and elements of small order in integral extensions*, J. of Pure and Appl. Algebra **71** (1991), 233–247.

[HH7]   Hochster, M. and Huneke, C., *Infinite integral extensions and big Cohen-Macaulay algebras*, Annals of Math. **135** (1992), 53–89.

[HH8]   Hochster, M. and Huneke, C., *Phantom homology*, Memoirs Amer. Math. Soc. Vol. **103**, No. **490** (1993), 1–91.

[HH9]   Hochster, M. and Huneke, C., *F-regularity, test elements, and smooth base change*, Trans. Amer. Math. Soc. **346** (1994), 1–62.

[HH10]   Hochster, M. and Huneke, C., *Tight closures of parameter ideals and splitting in module-finite extensions*, J. Alg. Geom. **3** (1994), 599–670.

[HH11]   Hochster, M. and Huneke, C., *Applications of the existence of big Cohen-Macaulay algebras*, Advances Math. **113** (1995), 45–117.

[HH12]   Hochster, M. and Huneke, C., *Tight closure in equal characteristic zero*, in preparation.

[HH13]   Hochster, M. and Huneke, C., *Indecomposable canonical modules and connectedness*, Contemp. Math. **159** (1994), 197–208.

[HR1]   Hochster, M. and Roberts, J.L., *Rings of invariants of reductive groups acting on regular rings are Cohen-Macaulay*, Advances Math. **13** (1974), 115–175.

[HR2]   Hochster, M. and Roberts, J.L., *The purity of the Frobenius and local cohomology*, Advances in Math. **21** (1976), 117–172.

[Hu1]   Huneke, C., *An algebraist commuting in Berkeley*, Mathematical Intelligencer **11** (1989), 40–52.

[Hu2]   Huneke, C., *Uniform bounds in Noetherian rings*, Invent. Math. **107** (1992), 203–223.

[Hu3]   Huneke, C., *Absolute integral closures and big Cohen-Macaulay algebras*, in Proc. of the 1990 Intern. Congress of Mathematicians, Kyoto 1990, Vol. I, Math. Soc. of Japan, Springer-Verlag, New York · Berlin · Heidelberg, 1991, pp. 339–349.

[Hu4]   Huneke, C., *Hilbert functions and symbolic powers*, Michigan Math. J. **34** (1987), 293–318.

[Hu5]   Huneke, C., *Tight Closure and its Applications*, CBMS Lecture Notes in Mathematics, vol. 88, American Math. Soc., Providence, 1996.

[Hu6]   Huneke, C., *The theory of d-sequences and powers of ideals*, Advances Math. **46** (1982), 249–279.

[HS]   Huneke, C. and Smith, K., *Kodaira vanishing and tight closure*, to appear, Crelle's Journal.

[HSw]   Huneke, C. and Swanson, I., *Cores of ideals in 2-dimensional regular local rings*, Michigan Math. J. **42** (1995), 193–208.

[It]   Itoh, S., *Integral closures of ideals generated by regular sequences*, J. Algebra **117** (1988), 390–401.

[Kat]   Katz, D., *Complexes acyclic up to integral closure*, preprint.

[Ka1]   Katzman, M., *Finiteness of $\cup_e AssF^e(M)$ and its connections to tight closure*, Illinois J. Math. **40** (1996), 330–337.

[Ka2]   Katzman, M., *Finite criteria for weak F-regularity*, Illinois J. Math. **40** (1996), 454–463.

[Ku1]   Kunz, E., *On Noetherian rings of characteristic p*, Am. J. Math. **98** (1976), 999–1013.

[Ku2]   Kunz, E., *Characterizations of regular local rings of characteristic p*, Amer. J. Math. **41** (1969), 772–784.

[Kur]   Kurano, K., *Macaulayfication using tight closure*, preprint, 1995.

[La]   Lai, Y., *On the relation type of systems of parameters*, J. Algebra **175** (1995), 339–358.

[L1]   Lipman, J., *Cohen-Macaulayness in graded algebras*, Math. Res. Letters **1** (1994), 149–157.

[L2]   Lipman, J., *Adjoints of ideals in regular local rings*, Math. Res. Letters **1** (1994), 1–17.

[L3]   Lipman, J., *Rational singularities ...* , Publ. Math. I.H.E.S. **36** (1969), 195–279.

[LS]   Lipman J. and Sathaye, A., *Jacobian ideals and a theorem of Briançon-Skoda*, Michigan Math. J. **28** (1981), 199–222.

[LT]   Lipman, J. and Teissier, B., *Pseudo-rational local rings and a theorem of Briançon-Skoda about integral closures of ideals*, Michigan Math. J. **28** (1981), 97–116.

[Ma]   Ma, F., *Splitting in integral extensions, Cohen-Macaulay modules and algebras*, J. Algebra **116** (1988), 176–195.

[Mac]   MacCrimmon, B., Thesis, Univ. of Michigan (1996).

[Mat]   Matsumura, H., *Commutative Algebra*, Benjamin, 1970.

[McD]   McDermott, M., *Tight Closure, Plus Closure, and Frobenius Closure in Cubical Cones*, Thesis, Univ. of Michigan, (1996).

[MR]   Mehta, V.B. and Ramanathan, A., *Frobenius splitting and cohomology vanishing for Schubert varieties*, Annals Math. **122** (1985), 27–40.

[MS]   Mehta, V.B. and Srinivas, V., *Normal F-pure surface singularities*, J. Algebra **143** (1991), 130–143.

[Mo]   Monsky, P., *The Hilbert-Kunz function*, Math. Ann. **263** (1983), 43-49.

[Mu]   Murthy, M.P., *A note on factorial rings*, Arch. Math. **15** (1964), 418–420.

[Nak]   Nakamura, Y., *On numerical invariants of Noetherian local rings of character-istic p*, J. Math. Kyoto Univ. **34** (1994), 1–13.

[NR]   Northcott, D.G. and Rees, D., *Reductions of ideals in local rings*, Proc. Camb. Phil. Soc. **50** (1954), 145–158.

[O1]   O'Carroll, L., *A uniform Artin-Rees theorem and Zariski's main lemma on holomorphic functions*, Invent. Math. **90** (1987), 674–682.

[O2]   O'Carroll, L., *A note on Artin-Rees numbers*, to appear, Bull. London Math. Soc.

[Og]   Ogoma, T., *General Néron desingularization following an idea of Popescu*, J. Algebra, to appear.

[PS1]   Peskine, C. and Szpiro, L., *Dimension projective finie et cohomologie locale*, I.H.E.S. Publ. Math. (Paris) **42** (1973), 323–395.

[PS2]   Peskine, C. and Szpiro, L., *Syzygies et multiplicités*, C. R. Acad. Sci. Paris Sér. A **278** (1974), 1421–1424.

[PS3]   Peskine, C. and Szpiro, L., *Sur la topologie des sous-schémas fermés d'un schéma localement noethérien, définis comme support d'un faisceau cohérent loclaement de dimension projective finie*, C. R. Acad. Sci. Paris Sér. A **269** (1969), 49–51.

[Po1]   Popescu, D., *General Néron desingularization*, Nagoya Math. J. **100** (1985), 97–126.

[Po2]   Popescu, D., *General Néron desingularization and approximation*, Nagoya Math. J. **104** (1986), 85–115.

[Ra]   Ratliff, L.J. Jr., *On prime divisors of I^n, n large*, Michigan Math. J. **23** (1976), 337–352.

[R1]   Rees, D., *Reduction of modules*, Math. Proc. Camb. Phil. Soc. **101** (1987), 431–449.

[R2]   Rees, D., *Two classical theorems of ideal theory*, Math. Proc. Camb. Phil. Soc. **52** (1956), 155–157.

[R3]   Rees, D., *A note on analytically unramified local rings*, J. London Math. Soc. **36** (1961), 24–28.

[RS]   Rees, D. and Sally, J. *General elements and joint reductions*, Michigan Math. J. **35** (1988), 241–254.

[Ro1]   Roberts, P., *Two applications of dualizing complexes over local rings*, Ann. Sci. Ec. Norm. Sup. **9** (1976), 103–106.

[Ro2]   Roberts, P., *Cohen-Macaulay complexes and an analytic proof of the new in-tersection conjecture*, J. Algebra **66** (1980), 225–230.

[Ro3]   Roberts, P., *The vanishing of intersection multiplicities of perfect complexes*, Bull. Amer. Math. Soc. **13** (1985), 127–130.

[Ro4]   Roberts, P., *Le théorème d'intersection*, C. R. Acad. Sc. Paris Sér. I **304** (1987), 177–180.

[Ro5] Roberts, P., *A computation of local cohomology*, Cont. Math., **159** (1994), 351–356.

[Ro6] Roberts, P., *Intersection theorems*, in Commutative Algebra, Math. Sci. Research Inst. Publ. **15**, Springer-Verlag, New York · Berlin · Heidelberg, 1989, pp. 417–436.

[Rot] Rotthaus, C., *On the approximation property of excellent rings*, Invent. Math. **88** (1987), 39–63.

[SS] Sancho de Salas, J. B., *Blowing-up morphisms with Cohen-Macaulay associated graded rings*, Géomètrie Algèbrique et applications I, Géomètrie et calcul algèbrique, Deuxième conférence internationale de La Rabida, Travaux en Cours no 22, Hermann, Paris, 1987, 201–209.

[Sch] Schenzel, P., *Cohomological annihilators*, Math. Proc. Camb. Philos. Soc. **91** (1982), 345–350.

[Se] Seibert, G., *Complexes with homology of finite length and Frobenius functors*, J. Algebra **125**, (1989), 278–287.

[Sh] Sharp, R. Y., *Cohen-Macaulay properties for balanced big Cohen-Macaulay modules*, Math. Proc. Camb. Philos. Soc. **90** (1981), 229–238.

[Sk] Skoda, H., *Applications des techniques L^2 a la théorie des idéaux d'une algèbre de fonctions holomorphes avec poids*, Ann. Scient. Ec. Norm. Sup. 4ème série t. **5** (1972), 545–579.

[Sm1] Smith, K. E., *Tight closure of parameter ideals and F-rationality*, Thesis, University of Michigan, 1993.

[Sm2] Smith, K. E., *F-rational rings have rational singularities*, to appear, Amer. J. Math.

[Sm3] Smith, K. E., *Test ideals in local rings*, Trans. Amer. Math. Soc. **347** (1995), 3453–3472.

[Sm4] Smith, K. E., *Tight closure of parameter ideals*, Invent. Math. **115** (1994), 41–60.

[Sm5] Smith, K. E., *Tight closure and graded integral extensions*, J. Algebra **175**, (1995), 568–574.

[Sm6] Smith, K. E., *Fujita's freeness conjecture in terms of local cohomology*, preprint.

[Sm7] Smith, K. E., *Tight closure in graded rings*, preprint.

[Sm8] Smith, K. E., *Vanishing, singularities and effective bounds via prime characteristic local algebra*, preprint.

[Sm9] Smith, K. E., *The D-module structure of F-split rings* Math. Res. Letters **2** (1995), 377–386.

[Sp] Spivakovsky, M., *Smoothing of ring homomorphisms, approximation theorems, and the Bass-Quillen conjecture*, preprint.

[Str] Strooker, J., *Homological Questions in Local Algebra* London Math. Soc. Lecture Note Series **145** (1990).

[Sw1] Swanson, I., *Joint reductions, tight closure, and the Briançon-Skoda theorem* J. Algebra **147** (1992), 128–136.

[Sw2] Swanson, I., *Tight closure, joint reductions, and mixed multiplicities*, Thesis, Purdue University, 1992.

[Sw3]   Swanson, I., *Joint reductions, tight closure, and the Briançon-Skoda theorem II* J. Algebra **170** (1994), 567–583.

[TW]   Tomari, M. and Watanabe, K., *Filtered rings, filtered blowing-ups and normal two-dimensional singularities with 'star-shaped' resolution*, Publ. Res. Inst. Math. Sci. Kyoto Univ. **25** (1989), 681–740.

[VV]   Valabrega, P. and Valla, G., *Form rings and regular sequences*, Nagoya Math. J. **72** (1978), 93–101.

[Vel1]   Velez, J., *Openness of the F-rational locus, smooth base change, and Koh's conjecture*, Thesis, University of Michigan, 1993.

[Vel2]   Velez, J., *Openness of the F-rational locus and smooth base change*, to appear, J. Algebra.

[Wa]   Wang, H., *On the Relation-Type conjecture*, preprint.

[W1]   Watanabe, K.-I., *Study of F-purity in dimension two*, in Algebraic Geometry and Commutative Algebra in honor of Masayoshi Nagata, Vol. II, Kinokuniya, Tokyo, 1988, pp. 791–800.

[W2]   Watanabe, K.-I., *F-regular and F-pure normal graded rings*, J. of Pure and Applied Algebra **71** (1991), 341–350.

[W3]   Watanabe, K.-I., *F-regular and F-pure rings vs. log-terminal and log-canonical singularities*, preprint.

[W4]   Watanabe, K.-I., *Infinite cyclic covers of strongly F-regular rings*, Contemp. Math. **159** (1994), 423–432.

[W5]   Watanabe, K.-I., *Rational singularities with K^*-action*, Lecture Notes in Pure and Applied Math., Dekker **84** (1983), 339–351.

[Wi]   Wickham, C.G., *Annihilation of homology of certain finite free complexes*, Thesis, University of Utah, 1991.

[Wil]   Williams, L., *Uniform stability of kernels of Koszul cohomology indexed by the Frobenius endomorphism*, J. Algebra **172** (1995), 721–743.

[Y]   Yoshino, Y., *Skew-polynomial rings of Frobenius type and the theory of tight closure*, Comm. Alg. **22**, (1994), 2473–2502.

Department of Mathematics
Purdue University
West Lafayette, Indiana 47907, U.S.A.
E-mail address: huneke@math.purdue.edu

On the Use of Local Cohomology in Algebra and Geometry

Peter Schenzel

Introduction

Local cohomology is a useful tool in several branches of commutative algebra and algebraic geometry. The main aim of this series of lectures is to illustrate a few of these techniques. The material presented in the sequel needs some basic knowledge about commutative resp. homological algebra. The basic chapters of the textbooks [9], [28], and [48] are a recommended reading for the preparation. The author's intention was to present applications of local cohomology in addition to the examples in these textbooks as well as those of [7].

Several times the author applies spectral sequence techniques for the proofs. Often people claim that it is possible to avoid spectral sequence arguments in the proofs for certain results. The present author believes that these techniques are quite natural. They will give deep insights in the underlying structure. So he forced these kinds of arguments even in cases where he knows more 'elementary' proofs. He has the hope to interest more researchers working in commutative algebra for such a powerful technique. As an introduction to spectral sequences he suggests the study of the corresponding chapters in the textbooks [9] and [48].

In the first section there is an introduction to local duality and dualizing complexes. There is a consequent use of the Čech complexes. In the main result, see 1.6, there is a family of dualities, including Matlis duality and duality for a dualizing complex of a complete local ring. This approach does not use 'sophisticated' prerequisites like derived categories. It is based on a few results about complexes and flat resp. injective modules. As applications there are a proof of the local duality theorem and vanishing theorems of the local cohomology of the canonical module. In particular it follows that a factorial domain is a Cohen-Macaulay ring provided it is a 'half way' Cohen-Macaulay ring. The first section concludes with a discussion of the cohomological annihilators $\operatorname{Ann} H_{\mathfrak{a}}^n(M)$ of a finitely generated A-module M and an ideal $\mathfrak{a}$. The consideration of these annihilators provides more subtle information than vanishing results.

Section 2 is concerned with the structure of the local cohomology modules in 'small' resp. 'large' homological dimensions. The 'small' homological dimension has to do with ideal transforms. To this end there is a generalization of Chevalley's theorem about the equivalence of ideal topologies. This is applied in order to prove

Grothendieck's finiteness result for ideal transforms. The structure of particular cases of ideal transforms of certain Rees rings is a main technical tool for the study of asymptotic prime divisors. On the other side of the range, i.e. the 'large' homological dimensions, there is a proof of the Lichtenbaum-Hartshorne vanishing theorem for local cohomology. In fact the non-vanishing of the d-dimensional local cohomology of a d-dimensional local ring is the obstruction for the equivalence of a certain topology to the adic topology. The Lichtenbaum-Hartshorne vanishing theorem is a helpful tool for the proof of a connectedness result invented by G. Faltings. We do not relate our considerations to a more detailed study of the cohomological dimension of an ideal. For results on cohomological dimensions see R. Hartshorne's article [16]. For more recent developments compare C. Huneke's and G. Lyubeznik's work in [22].

The third Section is devoted to the study of finite free resolutions of an A-module M in terms of its local cohomology modules. There are length estimates for $\operatorname{Ext}_A^n(M, N)$ and $\operatorname{Tor}_n^A(M, N)$ for two finitely generated A-modules M, N such that $M \otimes_A N$ is of finite length. This leads to an equality of the Auslander-Buchsbaum type, first studied by M. Auslander in [1], and a Cohen-Macaulay criterion. Moreover there are estimates of the Betti numbers of M in terms of the Betti numbers of the modules of deficiency of M. More subtle considerations are included in the case of graded modules over graded rings. This leads to the study of the Castelnuovo-Mumford regularity and a generalization of M. Green's duality result for certain Betti numbers of M and its canonical module K_M.

The author's aim is to present several pictures about the powerful tool of local cohomology in different fields of commutative algebra and algebraic geometry. Of course the collection of known applications is not exhausted. The reader may feel a challenge to continue with the study of local cohomology in his own field. In most of the cases the author tried to present basic ideas of an application. It was not his goal to present the most sophisticated generalization. The author expects further applications of local cohomology in the forthcoming textbook [6].

The present contribution has grown out of the author's series of lectures held at the Summer School on Commutative Algebra at CRM in Bellaterra, July 16 - 26, 1996. The author thanks the organizers of the Summer School at Centre de Recerca Matemàtica for bringing together all of the participants at this exciting meeting. For the author it was a great pleasure to present a series of lectures in the nice and stimulating atmosphere of this Summer School. During the meeting there were a lot of opportunities for discussions with several people; this made this School so exciting for the author. Among them the author wants to thank Luchezar Avramov, Hans-Bjørn Foxby, José-Maria Giral, Craig Huneke, David A. Jorgensen, Ruth Kantorovitz, Leif Melkersson, Claudia Miller, who drew the author's attention to several improvements of his original text. The author wants to thank also the staff members of the Centre de Recerca Matemàtica for their effort to make the stay in Bellaterra so pleasant. Finally he wants to thank R. Y. Sharp for a careful reading of the manuscript and several suggestions for an improvement of the text.

1. A Guide to Duality

1.1. Local Duality. Let A denote a commutative Noetherian ring. Let C denote a complex of A-modules. For an integer $k \in \mathbb{Z}$ let $C[k]$ denote the complex C shifted k places to the left and the sign of differentials changed to $(-1)^k$, i.e.

$$(C[k])^n = C^{k+n} \quad \text{and} \quad d_{C[k]} = (-1)^k d_C.$$

Moreover note that $H^n(C[k]) = H^{n+k}(C)$.

For a homomorphism $f : C \to D$ of two complexes of A-modules let us consider the mapping cone $M(f)$. This is the complex $C \oplus D[-1]$ with the boundary map $d_{M(f)}$ given by the following matrix

$$\begin{pmatrix} d_C & 0 \\ -f & -d_D \end{pmatrix}$$

where d_C resp. d_D denote the boundary maps of C and D resp. Note that $(M(f), d_{M(f)})$ forms indeed a complex.

There is a natural short exact sequence of complexes

$$0 \to D[-1] \xrightarrow{i} M(f) \xrightarrow{p} C \to 0,$$

where $i(b) = (0, -b)$ and $p(a, b) = a$. Clearly these homomorphisms make i and p into homomorphisms of complexes. Because $H^{n+1}(D[-1]) = H^n(D)$ the connecting homomorphism δ provides a map $\delta^{\cdot} : H^{\cdot}(C) \to H^{\cdot}(D)$. By an obvious observation it follows that $\delta^{\cdot} = H^{\cdot}(f)$. Note that $f : C \to D$ induces an isomorphism on cohomology if and only if $M(f)$ is an exact complex.

Let M, N be two A-modules considered as complexes concentrated in homological degree zero. Let $f : M \to N$ be a homomorphism. Then the mapping cone of f is

$$M(f) : \quad \cdots \to 0 \to M \xrightarrow{-f} N \to 0 \to \cdots$$

with the cohomology modules given by

$$H^n(M(f)) \simeq \begin{cases} \ker f & i = 0, \\ \operatorname{coker} f & i = 1, \\ 0 & \text{otherwise.} \end{cases}$$

This basic observation yields the following result:

Lemma 1.1. Let $f : C \to D$ be a homomorphism of complexes. Then there is a short exact sequence

$$0 \to H^1(M(H^{n-1}(f))) \to H^n(M(f)) \to H^0(M(H^n(f))) \to 0$$

for all $n \in \mathbb{Z}$.

Proof. This is an immediate consequence of the long exact cohomology sequence. Recall that the connecting homomorphism is $H^{\cdot}(f)$. □

For a complex C and $x \in A$ let $C \xrightarrow{x} C$ denote the multiplication map induced by x, i.e. the map on C^n is given by multiplication with x. Furthermore let $C \to C \otimes_A A_x$ denote the natural map induced by the localization, i.e. the map on C^n is given by $C^n \xrightarrow{i} C^n \otimes_A A_x$, where for an A-module M the map i is the natural map $i : M \to M \otimes_A A_x$.

In the following let us use the previous consideration in order to construct the Koszul and Čech complexes with respect to a system of elements $\underline{x} = x_1, \dots, x_r$ of A. To this end we consider the ring A as a complex concentrated in degree zero. Then define

$$K^{\cdot}(x; A) = M(A \xrightarrow{x} A) \text{ and } K^{\cdot}_x(A) = M(A \to A_x).$$

Note that both of these complexes are bounded in degree 0 and 1. Inductively put

$$\begin{aligned} K^{\cdot}(\underline{x}; A) &= M(K^{\cdot}(\underline{y}; A) \xrightarrow{x} K^{\cdot}(\underline{y}; A)) \quad \text{and} \\ K^{\cdot}_{\underline{x}}(A) &= M(K^{\cdot}_{\underline{y}}(A) \to K^{\cdot}_{\underline{y}}(A) \otimes_A A_x), \end{aligned}$$

where $\underline{y} = x_1, \dots, x_{r-1}$ and $x = x_r$. For an A-module M finally define

$$K^{\cdot}(\underline{x}; M) = K^{\cdot}(\underline{x}; A) \otimes_A M \text{ and } K^{\cdot}_{\underline{x}}(M) = K^{\cdot}_{\underline{x}}(A) \otimes_A M.$$

Call them (co-) Koszul complex resp. Čech complex of $\underline{x}$ with respect to M. Obviously the Čech complex is bounded. It has the following structure

$$0 \to M \to \oplus_i M_{x_i} \to \oplus_{i<j} M_{x_i x_j} \to \dots \to M_{x_1 \cdots x_r} \to 0$$

with the corresponding boundary maps.

It is well known that there is an isomorphism of complexes

$$K^{\cdot}_{\underline{x}}(M) \simeq \varinjlim K^{\cdot}(\underline{x}^{(n)}; M),$$

where $\underline{x}^{(n)} = x_1^n, \dots, x_r^n$, see [14]. The direct maps in the direct system are induced in a natural way by the inductive construction of the complex. The proof follows by induction on the number of elements.

The importance of the Čech complex is its close relation to the local cohomology. For an ideal $\mathfrak{a}$ of A let $\Gamma_{\mathfrak{a}}$ denote the section functor with respect to $\mathfrak{a}$. That is, $\Gamma_{\mathfrak{a}}$ is the subfunctor of the identity functor given by

$$\Gamma_{\mathfrak{a}}(M) = \{m \in M \mid \operatorname{Supp} Am \subseteq V(\mathfrak{a})\}.$$

The right derived functors of $\Gamma_{\mathfrak{a}}$ are denoted by $H^i_{\mathfrak{a}}$, $i \in \mathbb{N}$. They are called the local cohomology functors with respect to $\mathfrak{a}$.

Lemma 1.2. *Let $\mathfrak{a}$ resp. S be an ideal resp. a multiplicatively closed set of A. Let E denote an injective A-module. Then*

a) *$\Gamma_{\mathfrak{a}}(E)$ is an injective A-module,*
b) *the natural map $E \to E_S, e \mapsto \frac{e}{1}$, is surjective, and*
c) *the localization E_S is an injective A-module.*

Proof. The first statement is an easy consequence of Matlis' Structure Theorem about injective modules. In order to prove b) let $\frac{e}{s} \in E_S$ be an arbitrary element. In the case s is A-regular it follows easily that there is an $f \in E$ such that $e = fs$ and $\frac{f}{1} = \frac{e}{s}$, which proves the claim. In the general case choose $n \in \mathbb{N}$ such that $0 :_A s^n = 0 :_A s^{n+1}$. Recall that A is a Noetherian ring. Then consider the injective $(A/0 :_A s^n)$-module $\operatorname{Hom}_A(A/0 :_A s^n, E)$. Moreover

$$\operatorname{Hom}_A(A/0 :_A s^n, E) \simeq E/0 :_E s^n$$

because $A/0 :_A s^n \simeq s^n A$. Therefore it turns out that $E/0 :_E s^n$ has the structure of an injective $A/0 :_A s^n$-module. But now s acts on $A/0 :_A s^n$ as a regular element. Therefore for $e + 0 :_E s^n$ there exists an $f + 0 :_E s^n$, $f \in E$, such that $e - fs \in 0 :_E s^n$. But this proves $\frac{e}{s} = \frac{f}{1}$, i.e. the surjectivity of the considered map.

By Matlis' Structure Theorem and because the localization commutes with direct sums it is enough to prove statement c) for $E = E_A(A/\mathfrak{p})$, the injective hull of $A/\mathfrak{p}$, $\mathfrak{p} \in \operatorname{Spec} A$. In the case $S \cap \mathfrak{p} \neq \emptyset$ it follows that $E_S = 0$. So let $S \cap \mathfrak{p} = \emptyset$. Then any $s \in S$ acts regularly on E. Therefore the map in b) is an isomorphism.   □

For simplicity put $H_{\underline{x}}^n(M) = H^n(K_{\underline{x}}(M))$. This could give a misunderstanding to $H_{\mathfrak{a}}^n(M)$. But in fact both are isomorphic as shown in the sequel.

Theorem 1.3. *Let $\underline{x} = x_1, \ldots, x_r$ denote a system of elements of A with $\mathfrak{a} = \underline{x}A$. Then there are functorial isomorphisms $H_{\underline{x}}^n(M) \simeq H_{\mathfrak{a}}^n(M)$ for any A-module M and any $n \in \mathbb{Z}$.*

Proof. First note that $H_{\underline{x}}^0(M) \simeq \Gamma_{\mathfrak{a}}(M)$ as is easily seen by the structure of $K_{\underline{x}}^{\cdot}(M)$. Furthermore let $0 \to M' \to M \to M'' \to 0$ be a short exact sequence of A-modules. Because $K_{\underline{x}}^{\cdot}(A)$ consists of flat A-modules the induced sequence of complexes

$$0 \to K_{\underline{x}}^{\cdot}(M') \to K_{\underline{x}}^{\cdot}(M) \to K_{\underline{x}}^{\cdot}(M'') \to 0$$

is exact. That is, $H_{\underline{x}}^n(\cdot)$ forms a connected sequence of functors. Therefore, in order to prove the claim it will be enough to prove that $H_{\underline{x}}^n(E) = 0$ for $n > 0$ and an injective A-module E. This will be proved by induction on r. For $r = 1$ it is a particular case of 1.2. Let $r > 1$. Put $\underline{y} = x_1, \ldots, x_{r-1}$ and $x = x_r$. Then 1.1 provides a short exact sequence

$$0 \to H_x^1(H_{\underline{y}}^{n-1}(E)) \to H_{\underline{x}}^n(E) \to H_x^0(H_{\underline{y}}^n(E)) \to 0.$$

For the case $n > 2$ the claim follows by the induction hypothesis. In the remaining case $n = 1$ note that $H_{\underline{y}}^0(E) = \Gamma_{\underline{y}A}(E)$ is an injective A-module, see 1.2. So the induction hypothesis applies again.   □

Together with 1.1 the previous result provides a short exact sequence describing the behaviour of local cohomology under enlarging the number of generators of an ideal.

Corollary 1.4. *Let $\mathfrak{a}$ resp. x denote an ideal resp. an element of A. Then for $n \in \mathbb{N}$ there is a functorial short exact sequence*

$$0 \to H^1_{xA}(H^{n-1}_{\mathfrak{a}}(M)) \to H^n_{(\mathfrak{a},xA)}(M) \to H^0_{xA}(H^n_{\mathfrak{a}}(M)) \to 0$$

for any A-module M.

The previous theorem provides a structural result about local cohomology functors with support in $\mathfrak{m}$ in the case of a local ring $(A, \mathfrak{m})$. To this end let $E = E_A(A/\mathfrak{m})$ denote the injective hull of the residue field. Furthermore it provides also a change of ring theorem.

Corollary 1.5. a) *Let $(A, \mathfrak{m})$ denote a local ring. Then $H^n_{\mathfrak{m}}(M)$, $n \in \mathbb{N}$, is an Artinian A-module for any finitely generated A-module M.*
b) *Let $A \to B$ denote a homomorphism of Noetherian rings. Let $\mathfrak{a}$ be an ideal of A. For a B-module M there are A-isomorphisms $H^n_{\mathfrak{a}}(M) \simeq H^n_{\mathfrak{a}B}(M)$ for all $n \in \mathbb{N}$. Here in the first local cohomology module M is considered as an A-module.*

Proof. a) Let $E^{\cdot}(M)$ denote the minimal injective resolution of M. Then by Matlis' Structure Theorem on injective A-modules it follows that $\Gamma_{\mathfrak{m}}(E^{\cdot}(M))$ is a complex consisting of finitely many copies of E in each homological degree. Therefore $H^n_{\mathfrak{m}}(M)$ is – as a subquotient of an Artinian A-module – an Artinian module.
b) Let $\underline{x} = x_1, \ldots, x_r$ denote a generating set of $\mathfrak{a}$. Let $\underline{y} = y_1, \ldots, y_r$ denote the images in B. Then there is the following isomorphism $\bar{K}^{\cdot}_{\underline{x}}(M) \simeq K^{\cdot}_{\underline{y}}(B) \otimes_B M$, where both sides are considered as complexes of A-modules. This proves the claim. $\square$

In the following let $T(\cdot) = \mathrm{Hom}_A(\cdot, E)$ denote the Matlis duality functor for a local ring $(A, \mathfrak{m})$. An exceptional rôle is played by the complex $D^{\cdot}_{\underline{x}} = T(K^{\cdot}_{\underline{x}})$ with $K^{\cdot}_{\underline{x}} = K^{\cdot}_{\underline{x}}(A)$ as follows by the following theorem. In some sense it extends the Matlis duality.

For two complexes C, D consider the single complex $\mathrm{Hom}_A(C, D)$ associated to the corresponding double complex. To be more precise let

$$\mathrm{Hom}_A(C, D)^n = \prod_{i \in \mathbb{Z}} \mathrm{Hom}_A(C^i, D^{i+n}).$$

The n-th boundary map restricted to $\mathrm{Hom}_A(C^i, D^{i+n})$ is given by

$$\mathrm{Hom}_A(d^{i-1}_C, D^{i+n}) + (-1)^{n+1} \mathrm{Hom}_A(C^i, d^{i+n}_D).$$

Note that this induces a boundary map on $\mathrm{Hom}_A(C, D)$. Moreover, it is easy to see that $H^0(\mathrm{Hom}_A(C, D))$ is isomorphic to the homotopy equivalence classes of homomorphisms of the complexes C and D.

Theorem 1.6. *There is a functorial map*

$$M \otimes_A \hat{A} \to \mathrm{Hom}_A(\mathrm{Hom}_A(M, D^{\cdot}_{\underline{x}}), D^{\cdot}_{\underline{x}}),$$

which induces an isomorphism in cohomology for any finitely generated A-module M.

Proof. Let M be an arbitrary A-module. First note that for two A-modules X and Y there is a functorial map

$$M \otimes_A \operatorname{Hom}_A(X, Y) \quad \to \quad \operatorname{Hom}_A(\operatorname{Hom}_A(M, X), Y),$$
$$m \otimes f \quad \mapsto \quad (m \otimes f)(g) = f(g(m))$$

for $m \in M, f \in \operatorname{Hom}_A(X, Y)$ and $g \in \operatorname{Hom}_A(M, X)$. It induces an isomorphism for a finitely generated A-module M provided Y is an injective A-module. Because $D_{\underline{x}}^{\cdot}$ is a bounded complex of injective A-modules it induces a functorial isomorphism of complexes

$$M \otimes_A \operatorname{Hom}_A(D_{\underline{x}}^{\cdot}, D_{\underline{x}}^{\cdot}) \xrightarrow{\sim} \operatorname{Hom}_A(\operatorname{Hom}_A(M, D_{\underline{x}}^{\cdot}), D_{\underline{x}}^{\cdot})$$

for any finitely generated A-module M.

Now consider the complex $\operatorname{Hom}_A(D_{\underline{x}}^{\cdot}, D_{\underline{x}}^{\cdot})$. It is isomorphic to $T(K_{\underline{x}}^{\cdot} \otimes_A T(K_{\underline{x}}^{\cdot}))$. Continue with the investigation of the natural map

$$f : K_{\underline{x}}^{\cdot} \otimes_A T(K_{\underline{x}}^{\cdot}) \to E$$

defined in homological degree zero. In the following we abbreviate

$$C := K_{\underline{x}}^{\cdot} \otimes_A T(K_{\underline{x}}^{\cdot}).$$

We claim that f induces an isomorphism in cohomology. To this end consider the spectral sequence

$$E_1^{ij} = H^i(K_{\underline{x}}^{\cdot} \otimes_A T(K_{\underline{x}}^{-j})) \Rightarrow E^{i+j} = H^{i+j}(C).$$

Because $T(K_{\underline{x}}^{-j})$ is an injective A-module it follows that

$$E_1^{ij} = H_{\mathfrak{a}}^i(T(K_{\underline{x}}^{-j})) = 0$$

for all $i \neq 0$ as shown in 1.3. Here $\mathfrak{a}$ denotes the ideal generated by $\underline{x}$. Let $i = 0$. Then we have $E_1^{0j} = \Gamma_{\mathfrak{a}}(T(K_{\underline{x}}^{-j}))$. This implies that

$$E_1^{0j} = \varinjlim \operatorname{Hom}_A(A/\mathfrak{a}^n \otimes K_{\underline{x}}^{-j}, E).$$

Therefore $E_1^{0j} = 0$ for $j \neq 0$ because of $(A/\mathfrak{a}^n) \otimes_A A_x = 0$ for an element $x \in \mathfrak{a}$. Finally

$$E_1^{00} = \varinjlim \operatorname{Hom}_A(A/\mathfrak{a}^n, E) = E,$$

because E is an Artinian A-module. This proves the claim as is easily seen.

Moreover C is a complex of injective A-modules as follows by view of 1.2. Therefore the mapping cone $M(f)$ is an exact complex of injective A-modules. Furthermore let $g = T(f)$ denote the natural map

$$g : \hat{A} \to \operatorname{Hom}_A(D_{\underline{x}}^{\cdot}, D_{\underline{x}}^{\cdot}).$$

This induces an isomorphism in cohomology because the mapping cone $M(g) = T(M(f))$ is an exact complex. But now $\operatorname{Hom}_A(D_{\underline{x}}^{\cdot}, D_{\underline{x}}^{\cdot})$ is a complex of flat A-modules. So there is a sequence of functorial maps

$$M \otimes_A \hat{A} \to M \otimes_A \operatorname{Hom}_A(D_{\underline{x}}^{\cdot}, D_{\underline{x}}^{\cdot}) \xrightarrow{\sim} \operatorname{Hom}_A(\operatorname{Hom}_A(M, D_{\underline{x}}^{\cdot}), D_{\underline{x}}^{\cdot}).$$

In order to prove the statement it is enough to show that the first map induces an isomorphism in cohomology. To this end note that the mapping cone $M(g)$ is a bounded exact complex of flat A-modules. But now $M \otimes_A M(g) \simeq M(1_M \otimes g)$ is an exact complex. Therefore the map

$$1_M \otimes g : M \otimes_A \hat{A} \to M \otimes_A \operatorname{Hom}_A(D_{\underline{x}}^{\cdot}, D_{\underline{x}}^{\cdot})$$

induces an isomorphism in cohomology. □

In the case of a complete local ring $(A, \mathfrak{m})$ and a system $\underline{x} = x_1, \ldots, x_r$ of elements such that $\mathfrak{m} = \operatorname{Rad} \underline{x}A$ it follows that $D_{\underline{x}}^{\cdot}$ is a bounded complex of injective A-modules with finitely generated cohomology such that the natural map

$$M \to \operatorname{Hom}_A(\operatorname{Hom}_A(M, D_{\underline{x}}^{\cdot}), D_{\underline{x}}^{\cdot})$$

induces an isomorphism. Such a complex is called a dualizing complex. By virtue of this observation call $D_{\underline{x}}^{\cdot}$ a quasi-dualizing complex with support in $V(\mathfrak{a})$, where $\mathfrak{a} = \operatorname{Rad} \underline{x}A$. While the dualizing complex does not exist always, there are no restrictions about the existence of quasi-dualizing complexes with support in $V(\mathfrak{a})$. The isomorphisms in 1.6 are a common generalization of the Matlis duality obtained for $r = 0$ and the duality for a dualizing complex.

The most important feature of the dualizing complex is the local duality theorem first proved by A. Grothendieck. As an application of our considerations let us derive another proof.

Theorem 1.7. (Local Duality) *Let $(A, \mathfrak{m})$ denote a local ring. Let $\underline{x} = x_1, \ldots, x_r$ be a system of elements such that $\mathfrak{m} = \operatorname{Rad} \underline{x}A$. Then there are functorial isomorphisms*

$$H_{\mathfrak{m}}^n(M) \simeq \operatorname{Hom}_A(H^{-n}(\operatorname{Hom}_A(M, D_{\underline{x}}^{\cdot})), E), \ n \in \mathbb{Z},$$

for a finitely generated A-module M.

Proof. First note that $H_{\mathfrak{m}}^n(M) \simeq H^n(K_{\underline{x}}^{\cdot} \otimes_A M)$ by 1.3. Now $D_{\underline{x}}^{\cdot}$ is a bounded complex of injective A-modules. As in the proof of 1.6 there is a functorial map

$$M \otimes_A \operatorname{Hom}_A(D_{\underline{x}}^{\cdot}, E) \to \operatorname{Hom}_A(\operatorname{Hom}_A(M, D_{\underline{x}}^{\cdot}), E),$$

which is an isomorphism of complexes for any finitely generated A-module M. Now consider the functorial map

$$K_{\underline{x}}^{\cdot} \to \operatorname{Hom}_A(D_{\underline{x}}^{\cdot}, E) = T^2(K_{\underline{x}}^{\cdot}).$$

By 1.5 and the Matlis duality it induces an isomorphism in cohomology. Because $K_{\underline{x}}^{\cdot}$ and $\operatorname{Hom}_A(D_{\underline{x}}^{\cdot}, E)$ are complexes of flat A-modules the natural map

$$M \otimes_A K_{\underline{x}}^{\cdot} \to M \otimes_A \operatorname{Hom}_A(D_{\underline{x}}^{\cdot}, E)$$

induces an isomorphism in cohomology by the same argument as in 1.6. The proof follows now by putting together both of the maps. □

In this form of local duality the complex $D_{\underline{x}}^{\cdot}$ plays the rôle of the dualizing complex. In the next section there are a few more statements about dualizing complexes.

1.2. Dualizing Complexes and Some Vanishing Theorems. Let $(A, \mathfrak{m})$ denote a local ring. For a non-zero finitely generated A-module M there are the well-known vanishing results $\operatorname{depth} M = \min\{n \in \mathbb{Z} \mid H_{\mathfrak{m}}^n(M) \neq 0\}$ and $\dim M = \max\{n \in \mathbb{Z} \mid H_{\mathfrak{m}}^n(M) \neq 0\}$ shown by A. Grothendieck. In the following we recall two more subtle vanishing results on $H_{\mathfrak{m}}^n(M)$. To this end let us first investigate a few consequences of local duality.

Theorem 1.8. *Suppose that the local ring $(A, \mathfrak{m})$ is the factor ring of a Gorenstein ring $(B, \mathfrak{n})$ with $r = \dim B$. Then there are functorial isomorphisms*

$$H_{\mathfrak{m}}^n(M) \simeq \operatorname{Hom}_A(\operatorname{Ext}_B^{r-n}(M, B), E), \quad n \in \mathbb{Z},$$

for any finitely generated A-module M, where E denotes the injective hull of the residue field.

Proof. By 1.5 one may assume without loss of generality that A itself is a Gorenstein ring. Let $\underline{x} = x_1, \ldots, x_r$ denote a system of parameters of A. Under this additional Gorenstein assumption it follows that $K_{\underline{x}}^{\cdot}$ is a flat resolution of $H_{\mathfrak{m}}^r(A)[-r] \simeq E[-r]$, where E denotes the injective hull of the residue field. Therefore $D_{\underline{x}}^{\cdot}$ is an injective resolution of $\hat{A}[r]$. By definition it turns out that

$$H^{-n}(\operatorname{Hom}(M, D_{\underline{x}}^{\cdot})) \simeq \operatorname{Ext}_A^{r-n}(M, \hat{A})$$

for all $n \in \mathbb{Z}$. Because of $T(\operatorname{Ext}_A^{r-n}(M, \hat{A})) \simeq T(\operatorname{Ext}_A^{r-n}(M, A))$ this proves the claim. $\qquad\square$

In the situation of 1.8 introduce a few abbreviations. For $n \in \mathbb{Z}$ put

$$K_M^n = \operatorname{Ext}_B^{r-n}(M, B).$$

Moreover for $n = \dim M$ we often write K_M instead of $K_M^{\dim M}$. The module K_M is called the canonical module of M. In the case of $M = A$ it coincides with the classical definition of the canonical module of A. By the Matlis duality and by 1.8 the modules K_M^n do not depend – up to isomorphisms – on the presentation of the Gorenstein ring B. Clearly $K_M^n = 0$ for all $n > \dim M$ and $n < 0$. Moreover we have the isomorphism

$$K_M^n \otimes_A \hat{A} \simeq H^{-n}(\operatorname{Hom}_A(M, D_{\underline{x}}^{\cdot})), \ n \in \mathbb{Z},$$

as follows by view of 1.7. The advantage of K_M^n lies in the fact that it is – in contrast to $H^{-n}(\operatorname{Hom}_A(M, D_{\underline{x}}^{\cdot}))$ – a finitely generated A-module.

For a finitely generated A-module M say it satisfies *Serre's condition S_k*, $k \in \mathbb{N}$, provided

$$\operatorname{depth} M_{\mathfrak{p}} \geq \min\{k, \dim M_{\mathfrak{p}}\} \text{ for all } \mathfrak{p} \in \operatorname{Supp} M.$$

Note that M satisfies S_1 if and only if it is unmixed. M is a Cohen-Macaulay module if and only if it satisfies S_k for all $k \in \mathbb{N}$.

Lemma 1.9. *Let M denote a finitely generated A-module. The finitely generated A-modules K_M^n satisfy the following properties:*

a) $(K_M^n)_\mathfrak{p} \simeq K_{M_\mathfrak{p}}^{n-\dim A/\mathfrak{p}}$ for any $\mathfrak{p} \in \operatorname{Supp} M$, i.e. $\dim K_M^n \le n$ for all $n \in \mathbb{Z}$.

b) If $\dim M_\mathfrak{p} + \dim A/\mathfrak{p} = \dim M$ for some $\mathfrak{p} \in \operatorname{Supp} M$, then $(K_M)_\mathfrak{p} \simeq K_{M_\mathfrak{p}}$.

c) $\operatorname{Ass} K_M = \{\mathfrak{p} \in \operatorname{Ass} M \mid \dim A/\mathfrak{p} = \dim M\}$, i.e. $\dim M = \dim K_M$.

d) Suppose that M is equidimensional. Then M satisfies condition S_k if and only if $\dim K_M^n \le n - k$ for all $0 \le n < \dim M$.

e) K_M satisfies S_2.

Proof. Let $\mathfrak{p} \in \operatorname{Supp} M$ denote a prime ideal. Then $(K_M^n)_\mathfrak{p} \simeq K_{M_\mathfrak{p}}^{n-\dim A/\mathfrak{p}}$, as is easily seen by the presentation as an Ext-module. Therefore $\dim K_M^n \le n$.

Let $E^{\cdot}(B)$ denote the minimal injective resolution of B as a B-module. Then $(\operatorname{Hom}_B(M, E^{\cdot}(B)))^n = 0$ for all $n < r - \dim M$ and

$$\operatorname{Ass}_A H^{r-\dim M}(\operatorname{Hom}_B(M, E^{\cdot}(B))) = \{\mathfrak{p} \in \operatorname{Ass} M \mid \dim A/\mathfrak{p} = \dim M\}.$$

Putting this together the proofs of a), b), and c) follow immediately.

In order to prove d) first note that $\dim M_\mathfrak{p} + \dim A/\mathfrak{p} = \dim M$ for all $\mathfrak{p} \in \operatorname{Supp} M$ since M is equidimensional. Suppose there is an integer n with $0 \le n < \dim M$ and $\mathfrak{p} \in \operatorname{Supp} K_M^n$ such that $\dim A/\mathfrak{p} > n - k \ge 0$. This implies $H_{\mathfrak{p} A_\mathfrak{p}}^{n-\dim A/\mathfrak{p}}(M_\mathfrak{p}) \ne 0$, see 1.8. Therefore

$$\begin{aligned}
\operatorname{depth} M_\mathfrak{p} &\le n - \dim A/\mathfrak{p} < \dim M - \dim A/\mathfrak{p} = \dim M_\mathfrak{p} \text{ and} \\
\operatorname{depth} M_\mathfrak{p} &\le n - \dim A/\mathfrak{p} < k,
\end{aligned}$$

in contradiction to S_k. Conversely suppose there is a $\mathfrak{p} \in \operatorname{Supp} M$ such that $\operatorname{depth} M_\mathfrak{p} < \min\{k, \dim M_\mathfrak{p}\}$. Then $(K_M^n)_\mathfrak{p} \ne 0$ for $n = \dim A/\mathfrak{p} + \operatorname{depth} M_\mathfrak{p}$ and $\dim A/\mathfrak{p} = n - \operatorname{depth} M_\mathfrak{p} > n - k$, a contradiction. This finishes the proof of the statement in d).

In order to prove e) it is enough to show that $\operatorname{depth} K_M \ge \min\{2, \dim K_M\}$. Note that $(K_M)_\mathfrak{p} \simeq K_{M_\mathfrak{p}}$ for all $\mathfrak{p} \in \operatorname{Supp} K_M$. This follows because K_M is unmixed by c) and because $\operatorname{Supp} K_M$ is catenarian, i. e. $\dim M_\mathfrak{p} + \dim A/\mathfrak{p} = \dim M$ for all $\mathfrak{p} \in \operatorname{Supp} K_M$.

Without loss of generality we may assume that there is an M-regular element $x \in \mathfrak{m}$. Then the short exact sequence $0 \to M \xrightarrow{x} M \to M/xM \to 0$ induces an injection $0 \to K_M/xK_M \to K_{M/xM}$, which proves the claim. $\square$

Another reading of a) in 1.9 is that $H_\mathfrak{m}^{i+\dim A/\mathfrak{p}}(M) \ne 0$ provided $H_{\mathfrak{p} A_\mathfrak{p}}^i(M_\mathfrak{p}) \ne 0$. This is true for an arbitrary local ring as shown by R. Y. Sharp, see [46, Theorem (4.8)].

Proposition 1.10. Let $\mathfrak{p}$ be a t-dimensional prime ideal in a local ring $(A, \mathfrak{m})$. Let M denote a finitely generated A-module such that $H_{\mathfrak{p} A_\mathfrak{p}}^i(M_\mathfrak{p}) \ne 0$ for a certain $i \in \mathbb{N}$. Then $H_\mathfrak{m}^{i+t}(M) \ne 0$.

Proof. Choose $P \in V(\mathfrak{p}\widehat{A})$ a prime ideal such that $\dim \widehat{A}/P = \dim \widehat{A}/\mathfrak{p}\widehat{A} = t$. In particular this implies $P \cap A = \mathfrak{p}$ and that $\mathfrak{p}\widehat{A}_P$ is a $P\widehat{A}_P$-primary ideal. These data induce a faithful flat ring homomorphism $A_\mathfrak{p} \to \widehat{A}_P$. It yields that

$$0 \ne H_{\mathfrak{p} A_\mathfrak{p}}^i(M_\mathfrak{p}) \otimes_{A_\mathfrak{p}} \widehat{A}_P \simeq H_{\mathfrak{p}\widehat{A}_P}^i(M_\mathfrak{p} \otimes_{A_\mathfrak{p}} \widehat{A}_P).$$

But now there is the following canonical isomorphism $M_{\mathfrak{p}} \otimes_{A_{\mathfrak{p}}} \widehat{A}_P \simeq (M \otimes_A \widehat{A})_P$. Because $\mathfrak{p}\widehat{A}_P$ is a $P\widehat{A}_P$-primary ideal it follows that $0 \neq H^i_{P\widehat{A}_P}((M \otimes_A \widehat{A})_P)$. By Cohen's Structure Theorem $\widehat{A}$ is a homomorphic image of a Gorenstein ring. By the faithful flatness of $A \to \widehat{A}$ and by the corresponding result for a homomorphic image of a Gorenstein ring, see 1.9, it turns out that

$$H^{i+t}_{\mathfrak{m}}(M) \otimes \widehat{A} \simeq H^{i+t}_{\mathfrak{m}\widehat{A}}(M \otimes_A \widehat{A}) \neq 0,$$

which finally proves the claim.   □

Note that the previous result has the following consequence. Let $\mathfrak{p} \in \operatorname{Supp} M$ denote a prime ideal. Then

$$\operatorname{depth} M \leq \dim A/\mathfrak{p} + \operatorname{depth} M_{\mathfrak{p}}$$

for a finitely generated A-module M. This follows by the non-vanishing of the local cohomology for the depth of a module.

As above let $(A, \mathfrak{m})$ denote a local ring which is the factor ring of a Gorenstein ring $(B, \mathfrak{n})$. Let $E^{\cdot}(B)$ denote the minimal injective resolution of the Gorenstein ring B as a B-module. The complex $D_A = \operatorname{Hom}_B(A, E^{\cdot}(B))$ is a bounded complex of injective A-modules and finitely generated cohomology modules $H^n(D_A) \simeq \operatorname{Ext}^n_B(A, B)$.

Theorem 1.11. *The complex D_A is a dualizing complex of A. That is, there is a functorial map*

$$M \to \operatorname{Hom}_A(\operatorname{Hom}_A(M, D_A), D_A)$$

that induces an isomorphism in cohomology for any finitely generated A-module M.

Proof. Because D_A is a bounded complex of injective A-modules there is an isomorphism of complexes

$$M \otimes_A \operatorname{Hom}_A(D_A, D_A) \xrightarrow{\sim} \operatorname{Hom}_A(\operatorname{Hom}_A(M, D_A), D_A)$$

for any finitely generated A-module M as shown above. By similar arguments as before there is a natural map $A \to \operatorname{Hom}_A(D_A, D_A)$. Because both of the complexes involved – A as well as $\operatorname{Hom}_A(D_A, D_A)$ – are complexes of flat A-modules it will be enough to show that this map induces an isomorphism in cohomology in order to prove the statement.

Next consider the isomorphism of complexes

$$\operatorname{Hom}_A(D_A, D_A) \xrightarrow{\sim} \operatorname{Hom}_B(\operatorname{Hom}_B(A, E^{\cdot}(B)), E^{\cdot}(B)).$$

Therefore, in order to show the claim it will be enough to prove the statement for the Gorenstein ring B. First it will be shown that the natural map

$$j_B : B \to \operatorname{Hom}_B(E^{\cdot}(B), E^{\cdot}(B))$$

induces an isomorphism in cohomology. To this end consider the natural map $i_B : B \to E^{\cdot}(B)$. It induces an isomorphism of complexes. That is, the mapping cone $M(i_B)$ is exact. Therefore

$$\mathrm{Hom}_B(M(i_B), E^{\cdot}(B)) = M(\mathrm{Hom}_B(i_B, E^{\cdot}(B)))$$

is also an exact complex. Hence

$$\mathrm{Hom}_B(i_B, E^{\cdot}(B)) : \mathrm{Hom}_B(E^{\cdot}(B), E^{\cdot}(B)) \to E^{\cdot}(B)$$

induces an isomorphism in cohomology. Now it is easy to check that the composition of the homomorphisms

$$B \to \mathrm{Hom}_B(E^{\cdot}(B), E^{\cdot}(B)) \to E^{\cdot}(B)$$

is just i_B. Therefore j_B induces an isomorphism in cohomology. Moreover

$$\mathrm{Hom}_B(E^{\cdot}(B), E^{\cdot}(B))$$

is a complex of flat B-modules. Therefore the natural map j_B induces a homomorphism of complexes $M \otimes_B B \to M \otimes \mathrm{Hom}_B(E^{\cdot}(B), E^{\cdot}(B))$ which induces – by the same argument as above – an isomorphism in cohomology.   □

By his recent result on Macaulayfications, see [25], T. Kawasaki proved the converse of 1.11, namely, that A is the quotient of a Gorenstein ring provided A possesses a dualizing complex.

In the following there is a characterization when a certain complex is a dualizing complex. To this end recall the following induction procedure well-suited to homological arguments.

Proposition 1.12. *Let* $\mathfrak{P}$ *denote a property of finitely generated A-modules, where* $(A, \mathfrak{m}, k)$ *denotes a local ring with residue field* k. *Suppose that* $\mathfrak{P}$ *satisfies the following properties:*

a) *The residue field* k *has* $\mathfrak{P}$.
b) *If* $0 \to M' \to M \to M'' \to 0$ *denotes a short exact sequence of finitely generated A-modules such that* M' *and* M'' *have* $\mathfrak{P}$, *then so does* M.
c) *If* x *is an M-regular element such that* M/xM *has* $\mathfrak{P}$, *then so does* M.

Then any finitely generated A-module M *has* $\mathfrak{P}$.

The proof is easy. We leave it as an exercise to the interested reader. In the following we will use this arguments in order to sketch the characterization of dualizing complexes.

Theorem 1.13. *Let* D *denote a bounded complex of injective A-modules. Assume that* D *has finitely generated cohomology modules. Then* D *is a dualizing complex if and only if*

$$H^n(\mathrm{Hom}_A(k, D)) \simeq \begin{cases} 0 \text{ for } & n \neq t \\ k \text{ for } & n = t \end{cases}$$

for a certain integer $t \in \mathbb{Z}$.

Proof. Suppose that D is a dualizing complex. Then – by definition – the natural homomorphism

$$k \to \mathrm{Hom}_A(\mathrm{Hom}_A(k, D), D))$$

induces an isomorphism in cohomology. Furthermore $\mathrm{Hom}_A(k, D)$ is a complex consisting of k-vector spaces and whose cohomology modules are finite dimensional k-vector spaces. For $i \in \mathbb{Z}$ let $H^i = H^i(\mathrm{Hom}_A(k, D))$ and $h_i = \dim_k H_i$. Then it is easy to see that that there is an isomorphism of complexes $H^\cdot \xrightarrow{\sim} \mathrm{Hom}_A(k, D)$, where $H^\cdot$ denotes the complex consisting of H^i and the zero homomorphisms as boundary maps. Then for $n \in \mathbb{Z}$ it follows that

$$\dim_k H^n(\mathrm{Hom}_A(\mathrm{Hom}_A(k, D), D)) = \sum_{i \in \mathbb{Z}} h_i h_{i+n}.$$

As easily seen this implies the existence of an integer $t \in \mathbb{Z}$ such that $h_t = 1$ and $h_i = 0$ for any $i \neq t$.

In order to prove the converse one has to show that the natural homomorphism

$$M \to \mathrm{Hom}_A(\mathrm{Hom}_A(M, D), D)$$

induces an isomorphism in cohomology for any finitely generated A-module M. To this end we proceed by 1.12. By the assumption it follows immediately that a) is true. In order to prove b) recall that $\mathrm{Hom}_A(\mathrm{Hom}_A(\cdot, D), D)$ transforms short exact sequences into short exact sequences of complexes. This holds because D is a bounded complex of injective A-modules. Finally c) is true because the cohomology modules of $\mathrm{Hom}_A(\mathrm{Hom}_A(M, D), D)$ are finitely generated A-modules. Then one might apply Nakayama's Lemma. □

In the case the integer t in 1.13 is equal to zero call D a normalized dualizing complex.

It is noteworthy to say that 1.12 does not apply for the proof of 1.6. In general it will be not true that the complex $\mathrm{Hom}(M, D_{\underline{x}})$ has finitely generated cohomology. So the Nakayama Lemma does not apply in proving condition c) in 1.12.

Under the previous assumptions on A and B with $r = \dim B$ let $D(M)$ denote the complex $\mathrm{Hom}_A(M, D_A)$, where M denotes a finitely generated A-module M. Then there is an isomorphism $D(M) \xrightarrow{\sim} \mathrm{Hom}_B(M, E^\cdot(M))$. Therefore

$$H^n(D(M)) \simeq \mathrm{Ext}_B^n(M, B) \text{ for all } n \in \mathbb{Z}.$$

This implies $H^{r-d}(D(M)) \simeq K_M$ and $H^{r-n}(D(M)) \simeq K_M^n$ for all $n \neq d = \dim M$. Because of $D(M)^n = 0$ for all $n < r - d$ there is a natural homomorphism of complexes

$$i_M : K_M[d - r] \to D(M),$$

where K_M is considered as a complex concentrated in homological degree zero. So the mapping cone $M(i_M)$ provides a short exact sequence of complexes

$$0 \to D(M)[-1] \to M(i_M) \to K_M[d - r] \to 0.$$

Therefore we see that $H^{r-n+1}(M(i_M)) \simeq K_M^n$ for all $0 \leq n < \dim M$ and $H^{r-n+1}(M(i_M)) = 0$ for all $n < 0$ and all $n \geq \dim M$. By applying the functor $D(\cdot) := \operatorname{Hom}_A(\cdot, D_A)$ it induces a short exact sequence of complexes

$$0 \to D(K_M)[r-d] \to D(M(i_M)) \to D^2(M)[1] \to 0.$$

Recall that D_A is a complex consisting of injective A-modules. By 1.11 and the definition of K_{K_M} this yields an exact sequence

$$0 \to H^{-1}(D(M(i_M))) \to M \xrightarrow{\tau_M} K_{K_M} \to H^0(D(M(i_M))) \to 0$$

and isomorphisms $H^n(D(M(i_M))) \simeq K_{K_M}^{d-n}$ for all $n \geq 1$.

Note that in the particular case of $M = A$ the homomorphism τ_A coincides with the natural map

$$A \to \operatorname{Hom}_A(K_A, K_A), \ a \mapsto f_a,$$

of the ring into the endomorphism ring of its canonical module. Here f_a denotes the multiplication map by a.

Theorem 1.14. *Let M denote a finitely generated, equidimensional A-module with $d = \dim M$, where A is a factor ring of a Gorenstein ring. Then for an integer $k \geq 1$ the following statements are equivalent:*

(i) *M satisfies condition S_k.*

(ii) *The natural map $\tau_M : M \to K_{K_M}$ is bijective (resp. injective for $k = 1$) and $H_{\mathfrak{m}}^n(K_M) = 0$ for all $d - k + 2 \leq n < d$.*

Proof. First recall that $H^n(D(M(i_M))) \simeq K_{K_M}^{d-n}$ for all $n \geq 1$. By the local duality it follows that $T(K_{K_M}^n) \simeq H_{\mathfrak{m}}^n(K_M)$ for all $n \in \mathbb{Z}$. By virtue of the short exact sequence and the above isomorphisms the statement in (ii) is equivalent to

(iii) $H^n(D(M(i_M))) = 0$ *for all* $-1 \leq n < k - 1$.

Next show that (i) $\Rightarrow$ (iii). First recall that $D(M(i_M)) \xrightarrow{\sim} \operatorname{Hom}_B(M(i_M), E^\cdot(B))$. Then there is the following spectral sequence

$$E_2^{i,j} = \operatorname{Ext}_B^i(H^{-j}(M(i_M)), B) \Rightarrow E^n = H^n(D(M(i_M)))$$

in order to compute $H^n(D(M(i_M)))$. Moreover it follows that

$$H^{-j}(M(i_M)) \simeq \begin{cases} K_M^{r+j+1} & \text{for } 0 \leq r+j+1 < \dim M \text{ and} \\ 0 & \text{otherwise.} \end{cases}$$

By the assumptions and 1.9 it implies $\dim K_M^{r+j+1} \leq r+j+1-k$ for all $j \in \mathbb{Z}$. As a consequence of 1.8 it turns out that $E_2^{i,j} = 0$ for all $i, j \in \mathbb{Z}$ satisfying $i+j < k-1$. That is the spectral sequence proves the condition (iii).

In order to prove the reverse conclusion first note that $\dim M_{\mathfrak{p}} + \dim A/\mathfrak{p} = \dim M$ for all prime ideals $\mathfrak{p} \in \operatorname{Supp} M$ since M is equi-dimensional. Then by 1.9 it follows easily

$$D(M(i_M)) \otimes_A A_{\mathfrak{p}} \xrightarrow{\sim} D(M(i_{M_{\mathfrak{p}}})).$$

This means that the claim is a local question. So by induction we have to show that depth $M \geq k$. By induction hypothesis we know that $\dim K_M^j \leq j - k$ for all $j > k$ and $\dim K_M^j \leq 0$ for all $0 \leq j \leq k$. Then the above spectral sequences degenerates partially to the isomorphisms $\text{Ext}_B^r(K_M^{j+1}, B) \simeq H^j(D(M(i_M)))$ for all $j < k - 1$. Recall that $\text{Ext}_B^r(N, B)$ is the onliest possible non-vanishing Ext module for N a B-module of finite length. By the local duality this implies $H_{\mathfrak{m}}^{j+1}(M) = 0$ for all $j < k - 1$ and $\text{depth}_A M \geq k$, as required. $\qquad \square$

It turns out that for a module M satisfying S_2 the natural map $\tau_M : M \to K_{K_M}$ is an isomorphism. In the case of the canonical module K_A this means that the endomorphism ring of K_A is isomorphic to A if and only if A satisfies S_2. The previous result has a dual statement which characterizes the vanishing of the local cohomology modules below the dimension of the module.

Corollary 1.15. *With the notation of 1.14 suppose that the A-module M satisfies the condition S_2. For an integer $k \geq 2$ the following conditions are equivalent:*

(i) K_M *satisfies condition S_k.*
(ii) $H_{\mathfrak{m}}^n(M) = 0$ *for all $d - k + 2 \leq n < d$.*

Proof. This is just a consequence of 1.14 and the remark that $\tau_M : M \to K_{K_M}$ is an isomorphism. $\qquad \square$

There are several further applications of these vanishing results in the case A is a quasi-Gorenstein ring or in liaison. For the details compare [41]. We conclude with one of them, a Cohen-Macaulay characterization for a quasi-Gorenstein ring. To this end let us call a local ring $(A, \mathfrak{m})$ that is a quotient of a Gorenstein ring B a quasi-Gorenstein ring, provided $K_A \simeq A$. Note that a Cohen-Macaulay quasi-Gorenstein ring is a Gorenstein ring. A local factorial ring that is a quotient of a Gorenstein ring is a quasi-Gorenstein ring, see [32].

Theorem 1.16. *Let $(A, \mathfrak{m})$ denote a quasi-Gorenstein ring such that*

$$\text{depth } A_{\mathfrak{p}} \geq \min\{\dim A, \frac{1}{2} \dim A_{\mathfrak{p}} + 1\} \text{ for all } \mathfrak{p} \in \text{Spec } A.$$

Then A is a Gorenstein ring.

Proof. Let $d = \dim A$. It is enough to show that A is a Cohen-Macaulay ring. This is true for $d \leq 3$ by the assumption. By induction $A_{\mathfrak{p}}$ is a Cohen-Macaulay ring for all prime ideals $\mathfrak{p} \neq \mathfrak{m}$. Therefore by the assumption the local ring $(A, \mathfrak{m})$ satisfies the condition S_k for $k \geq \frac{1}{2} \dim A + 1$. Because of $K_A \simeq A$ the result 1.15 implies that $H_{\mathfrak{m}}^n(A) = 0$ for all $n < \dim A$, which proves the Cohen-Macaulayness of A. $\qquad \square$

The previous result says something about the difficulty to construct non-Cohen-Macaulay factorial domains. If such a ring is 'half way' Cohen-Macaulay it is a Cohen-Macaulay ring. Originally this result was proved by R. Hartshorne and A. Ogus, see [17].

1.3. Cohomological Annihilators.

Vanishing results on local cohomology modules provide strong information. More subtle information comes from consideration of their annihilators. This will be sampled in this subsection.

To this end we have to use a certain generalization of the notion of a regular sequence. First let us summarize basic facts about filter regular sequences. Let M denote a finitely generated A-module over $(A, \mathfrak{m})$, a local Noetherian ring.

A system of elements $\underline{x} = x_1, \ldots, x_r \subseteq \mathfrak{m}$ is called a filter regular sequence of M (or M-filter regular sequence), if

$$x_i \notin \mathfrak{p} \quad \text{for all } \mathfrak{p} \in (\operatorname{Ass} M/(x_1, \ldots, x_{i-1})M) \setminus \{\mathfrak{m}\}$$

for all $i = 1, \ldots, r$. This is equivalent to saying that the A-modules

$$(x_1, \ldots, x_{i-1})M : x_i/(x_1, \ldots, x_{i-1})M, \quad i = 1, \ldots, r,$$

are of finite length. Moreover $\underline{x}$ is an M-filter regular sequence if and only if $\{\frac{x_1}{1}, \ldots, \frac{x_i}{1}\} \in A_{\mathfrak{p}}$ is an $M_{\mathfrak{p}}$-regular sequence for all $\mathfrak{p} \in (V(x_1, \ldots, x_i) \cap \operatorname{Supp} M) \setminus \{\mathfrak{m}\}$ and $i = 1, \ldots, r$.

Lemma 1.17. *Let M denote a finitely generated A-module. Suppose that $\underline{x} = x_1, \ldots, x_r$ denotes an M-filter regular sequence.*

a) *$H^i(\underline{x}; M)$ is an A-module of finite length for all $i < r$.*
b) *$H_i(\underline{x}; M)$ is an A-module of finite length for all $i > 0$.*
c) *$\operatorname{Supp} H_{\mathfrak{c}}^i(M) \subseteq V(\mathfrak{m})$ for all $0 \leq i < r$, where $\mathfrak{c} = (x_1, \ldots, x_r)A$.*

Proof. Because of the self duality of the Koszul complexes it will be enough to prove one of the first two statements. Now note that

$$\operatorname{Supp} H^i(\underline{x}; M) \subseteq V(\underline{x}) \cap \operatorname{Supp} M, \quad i \in \mathbb{Z}.$$

On the other hand $\underline{x}$ is an M-regular sequence if and only if $H^i(\underline{x}; M) = 0$ for all $i < r$. Then the result a) follows by a localization argument of the Koszul complexes. In order to prove c) note that $\operatorname{Supp} H_{\mathfrak{c}}^i(M) \subseteq V(\underline{x}) \cap \operatorname{Supp} M$. Let $\mathfrak{p}$ be a non-maximal prime ideal in $V(\underline{x}) \cap \operatorname{Supp} M$. By a localization argument it follows that

$$H_{\mathfrak{c}}^i(M) \otimes_A A_{\mathfrak{p}} \simeq H_{\mathfrak{c} A_{\mathfrak{p}}}^i(M_{\mathfrak{p}}) = 0 \quad \text{for } i < r,$$

since $\{\frac{x_1}{1}, \ldots, \frac{x_r}{1}\}$ is an $M_{\mathfrak{p}}$-regular sequence. $\square$

Let M denote a finitely generated A-module. Let $\mathfrak{a}$ denote an ideal of $(A, \mathfrak{m})$. The vanishing resp. non-vanishing of the local cohomology modules $H_{\mathfrak{a}}^n(M)$ provides useful local information on M. For a more subtle consideration the annihilators of $H_{\mathfrak{a}}^n(M)$ are of some interest. For a finitely generated A-module M let

$$\mathfrak{a}_n(M) := \operatorname{Ann}_A H_{\mathfrak{a}}^n(M), \, n \in \mathbb{Z},$$

denote the n-th cohomological annihilator of M with respect to $\mathfrak{a}$.

Now we relate the cohomological annihilators of M to those of M modulo a bunch of generic hyperplane sections.

Theorem 1.18. *Let $\underline{x} = x_1, \ldots, x_r$ denote an M-filter regular sequence. Then*

$$\mathfrak{a}_n(M) \cdot \ldots \cdot \mathfrak{a}_{n+r}(M) \subseteq \mathfrak{a}_n(M/\underline{x}M) \quad \text{for all integers } n.$$

Proof. Let $K^{\cdot} := K_{\underline{y}}^{\cdot}$ denote the Čech complex of A with respect to a system of generators $\underline{y} = y_1, \ldots, y_s$ of the ideal $\mathfrak{a}$. Let $K^{\cdot}(\underline{x}; A) \otimes_A M$ be the Koszul-co-complex of M with respect to $\underline{x}$. Put

$$C^{\cdot} := (K^{\cdot} \otimes_A M) \otimes_A K^{\cdot}(\underline{x}; A) \simeq K^{\cdot} \otimes_A K^{\cdot}(\underline{x}; M).$$

There are two spectral sequences for computing the cohomology of $C^{\cdot}$. First consider

$$E_2^{ij} = H^i(K^{\cdot} \otimes_A H^j(\underline{x}; M)) \Rightarrow E^{i+j} = H^{i+j}(C^{\cdot}).$$

Note that $H^i(K^{\cdot} \otimes_A N) \simeq H^i_{\mathfrak{a}}(N)$, $i \in \mathbb{Z}$, for a finitely generated A-module N, see 1.3. Therefore

$$E_2^{ij} \simeq H^i_{\mathfrak{a}}(H^j(\underline{x}; M)) \quad \text{for all } i, j \in \mathbb{Z}.$$

By 1.17 the A-modules $H^i(\underline{x}; M)$, $i < r$, are of finite length. So there are the following isomorphisms

$$E_2^{ij} \simeq \begin{cases} 0 & \text{for } i \neq 0 \quad \text{and } j \neq r, \\ H^j(\underline{x}; M) & \text{for } i = 0 \quad \text{and } j \neq r, \\ H^i_{\mathfrak{a}}(M/\underline{x}M) & \text{for } j = r. \end{cases}$$

To this end recall that $H^r(\underline{x}; M) \simeq M/\underline{x}M$. By virtue of the spectral sequence it turns out that

$$E_\infty^{ij} = 0 \quad \text{for all } i \neq 0, \quad j \neq r.$$

Because of the subsequent stages of the spectral sequence

$$E_k^{i-k,r+k-1} \to E_k^{ir} \to E_k^{i+k,r-k+1}$$

and $E_k^{i-k,r+k-1} = E_k^{i+k,r-k+1} = 0$ for all $k \geq 2$ it yields that $E_\infty^{ir} \simeq H^i_{\mathfrak{a}}(M/\underline{x}M)$. By a similar consideration we obtain that $E_\infty^{0j} \simeq H^j(\underline{x}; M)$ for all $j \neq r$. Therefore there are the following isomorphisms

$$H^i(C^{\cdot}) \simeq \begin{cases} H^i(\underline{x}; M) & \text{for } 0 \leq i < r, \\ H^{i-r}_{\mathfrak{a}}(M/\underline{x}M) & \text{for } r \leq i \leq d, \\ 0 & \text{otherwise}, \end{cases}$$

where $d = \dim M$. On the other hand there is the spectral sequence

$${}'E_2^{ij} = H^j(K^{\cdot}(\underline{x}; A) \otimes_A H^i_{\mathfrak{a}}(M)) \Rightarrow {}'E^{i+j} = H^{i+j}(C^{\cdot}).$$

Because of ${}'E_2^{ij} = H^j(\underline{x}; H^i_{\mathfrak{a}}(M))$ it follows that ${}'E_2^{ij} = 0$ for all $j < 0$ and $j > r$. By the construction of the Koszul complex ${}'E_2^{ij}$ is a subquotient of the direct sum of copies of $H^i_{\mathfrak{a}}(M)$. Therefore $\mathfrak{a}_i(M)({}'E_2^{ij}) = 0$ for all $i, j \in \mathbb{Z}$. Whence it implies that $\mathfrak{a}_i(M)({}'E_\infty^{ij}) = 0$ for all $i, j \in \mathbb{Z}$. By view of the filtration of $H^{i+j}(C^{\cdot})$ defined by ${}'E_\infty^{ij}$ it follows that

$$\begin{aligned} \mathfrak{a}_0(M) \cdot \ldots \cdot \mathfrak{a}_i(M) H^i(C^{\cdot}) &= 0 \quad \text{for } 0 \leq i < r \text{ and} \\ \mathfrak{a}_{i-r}(M) \cdot \ldots \cdot \mathfrak{a}_i(M) H^i(C^{\cdot}) &= 0 \quad \text{for } r \leq i \leq d. \end{aligned}$$

Hence, the above computation of $H^i(C^{\cdot})$ proves the claim. $\qquad\square$

For a filter regular sequence $\underline{x} = x_1, \ldots, x_r$ the proof of Theorem 3.3 provides that $\mathfrak{a}_0(M) \cdot \ldots \cdot \mathfrak{a}_i(M) H^i(x_1, \ldots, x_r; M) = 0$ for all $i < r$. Because of the finite length of $H^i(\underline{x}; M)$ for all $i < r$ this is a particular case of the results shown in [41]. Moreover the notion of M-filter regular sequences provides an interesting expression of the local cohomology modules of M.

Lemma 1.19. *Let $\underline{x} = x_1, \ldots, x_r$ be an M-filter regular sequence contained in $\mathfrak{a}$. Put $\mathfrak{c} = (x_1, \ldots, x_r)A$. Then there are the following isomorphisms*

$$H_{\mathfrak{a}}^i(M) \simeq \begin{cases} H_{\mathfrak{c}}^i(M) & \text{for } 0 \leq i < r, \\ H_{\mathfrak{a}}^{i-r}(H_{\mathfrak{c}}^r(M)) & \text{for } r \leq i \leq d, \end{cases}$$

where $d = \dim_A M$.

Proof. Consider the spectral sequence

$$E_2^{ij} = H_{\mathfrak{a}}^i(H_{\mathfrak{c}}^j(M)) \Rightarrow E^{i+j} = H_{\mathfrak{a}}^{i+j}(M).$$

By 1.17 we have that $\operatorname{Supp} H_{\mathfrak{c}}^j(M) \subseteq V(\mathfrak{m})$ for all $j < r$. Whence $E_2^{ij} = 0$ for all $i \neq 0$ and $j \neq r$. Furthermore, $E_2^{0j} = H_{\mathfrak{c}}^j(M)$ for $j \neq r$ and $E_2^{ir} = H_{\mathfrak{a}}^i(H_{\mathfrak{c}}^r(M))$. An argument similar to that of the proof given in 1.18 yields that

$$E_\infty^{0j} \simeq H_{\mathfrak{c}}^j(M) \quad \text{and} \quad E_\infty^{ir} \simeq H_{\mathfrak{a}}^i(H_{\mathfrak{c}}^r(M)).$$

Because of $E_\infty^{0j} = 0$ for $j > r$ the spectral sequence proves the claim. □

Let $\underline{x} = x_1, \ldots, x_r$ be a system of elements of A. For the following results put $\underline{x}^{(k)} = x_1^k, \ldots, x_r^k$ for an integer $k \in \mathbb{N}$.

Corollary 1.20. *Let $\underline{x} = x_1, \ldots, x_r$ be an M-filter regular sequence contained in $\mathfrak{a}$. The multiplication by $x_1 \cdots x_r$ induces a direct system $\{H_{\mathfrak{a}}^i(M/\underline{x}^{(k)}M)\}_{k \in \mathbb{N}}$, such that*

$$H_{\mathfrak{a}}^{i+r}(M) \simeq \varinjlim H_{\mathfrak{a}}^i(M/\underline{x}^{(k)}M)$$

for all $i \geq 0$.

Proof. There is a direct system $\{M/\underline{x}^{(k)}M\}_{k \in \mathbb{N}}$ with homomorphisms induced by the multiplication by $x_1 \cdots x_r$. By [13] there is an isomorphism

$$H_{\mathfrak{c}}^r(M) \simeq \varinjlim M/\underline{x}^{(k)}M.$$

Then the claim follows by 1.19 since the local cohomology commutes with direct limits. □

In order to produce an 'upper' approximation of $\mathfrak{a}_i(M/\underline{x}M)$, $\underline{x} = x_1, \ldots, x_r$, an M-filter regular sequence, a few preliminaries are necessary. For a given i and $j = 0, 1, \ldots, r$ set

$$\mathfrak{a}_{ij}(\underline{x}; M) = \bigcap_{k_1, \ldots, k_j \geq 1} \mathfrak{a}_i(M/(x_1^{k_1}, \ldots, x_j^{k_j})M).$$

Furthermore define $\mathfrak{a}_i(\underline{x}; M) = \bigcap_{j=0}^r \mathfrak{a}_{ij}(\underline{x}; M)$. The next result relates the cohomological annihilators of M to those of $M/\underline{x}M$.

Corollary 1.21. *Let $\underline{x} = x_1, \ldots, x_r$ be an M-filter regular sequence contained in $\mathfrak{a}$. Then*

$$\mathfrak{a}_i(M) \cdot \ldots \cdot \mathfrak{a}_{i+r}(M) \subseteq \mathfrak{a}_i(\underline{x}; M) \subseteq \mathfrak{a}_i(M) \cap \ldots \cap \mathfrak{a}_{i+r}(M)$$

for all $0 \leq i \leq d - r$. In particular, $\mathfrak{a}_i(\underline{x}; M)$ and $\mathfrak{a}_i(M) \cap \ldots \cap \mathfrak{a}_{i+r}(M)$ have the same radical.

Proof. By 1.18 it follows that $\mathfrak{a}_i(M) \cdot \ldots \cdot \mathfrak{a}_{i+j}(M) \subseteq \mathfrak{a}_{ij}(\underline{x}; M)$ for $j = 0, 1, \ldots, r$. Recall that $x_1^{k_1}, \ldots, x_j^{k_j}$ forms an M-filter regular sequence, provided that $\underline{x} = x_1, \ldots, x_r$ is an M-filter regular sequence. Whence the first inclusion is true. Moreover, by 1.20 it yields that

$$\mathfrak{a}_i(\underline{x}; M) \subseteq \mathfrak{a}_{ij}(\underline{x}; M) \subseteq \mathfrak{a}_{i+j}(M)$$

for all $j = 0, 1, \ldots, r$. This proves the second containment relation. □

The results of this section generalize those for the cohomological annihilators $\mathfrak{m}_n(M)$ of $H^n_{\mathfrak{m}}(M)$, $n \in \mathbb{Z}$, investigated in [41].

2. A Few Applications of Local Cohomology

2.1. On Ideal Topologies. Let S denote a multiplicatively closed set of a Noetherian ring A. For an ideal $\mathfrak{a}$ of A put $\mathfrak{a}_S = \mathfrak{a}A_S \cap A$. For an integer $n \in \mathbb{N}$ let $\mathfrak{a}_S^{(n)} = \mathfrak{a}^n A_S \cap A$ denote the n-th symbolic power of $\mathfrak{a}$ with respect to S. Note that this generalizes the notion of the n-th symbolic power $\mathfrak{p}^{(n)} = \mathfrak{p}^n A_{\mathfrak{p}} \cap A$ of a prime ideal $\mathfrak{p}$ of A. The ideal $\mathfrak{a}_S$ is the so-called S-component of $\mathfrak{a}$, i.e.

$$\mathfrak{a}_S = \{r \in A \mid rs \in \mathfrak{a} \text{ for some } s \in S\}.$$

So the primary decomposition of $\mathfrak{a}_S$ consists of the intersection of all primary components of $\mathfrak{a}$ that do not meet S. In other words

$$\mathrm{Ass}_A A/\mathfrak{a}_S = \{\mathfrak{p} \in \mathrm{Ass}_A A/\mathfrak{a} \mid \mathfrak{p} \cap S = \emptyset\}.$$

Moreover it is easily seen that

$$\mathrm{Ass}_A \mathfrak{a}_S/\mathfrak{a} = \{\mathfrak{p} \in \mathrm{Ass}_A A/\mathfrak{a} \mid \mathfrak{p} \cap S \neq \emptyset\}.$$

However $\mathrm{Supp}_A \mathfrak{a}_S/\mathfrak{a} \subseteq V(\mathfrak{b})$, where $\mathfrak{b} = \prod_{\mathfrak{p} \in \mathrm{Ass}(\mathfrak{a}_S/\mathfrak{a})} \mathfrak{p}$. Whence it turns out that $\mathfrak{a}_S = \mathfrak{a} :_A \langle \mathfrak{b} \rangle$, where the last colon ideal denotes the stable value of the ascending chain of ideals

$$\mathfrak{a} \subseteq \mathfrak{a} :_A \mathfrak{b} \subseteq \mathfrak{a} :_A \mathfrak{b}^2 \subseteq \ldots.$$

Obviously $\mathfrak{a} :_A \langle \mathfrak{b} \rangle = \mathfrak{a} :_A \langle \mathfrak{b}' \rangle$ for two ideals $\mathfrak{b}, \mathfrak{b}'$ with the same radical. On the other hand for two ideals $\mathfrak{a}, \mathfrak{b}$ it follows that

$$\mathfrak{a} :_A \langle \mathfrak{b} \rangle = \mathfrak{a}_S, \quad \text{where } S = \cap_{\mathfrak{p} \in \mathrm{Ass} A/\mathfrak{a} \backslash V(\mathfrak{b})} A \backslash \mathfrak{p}.$$

There is a deep interest in comparing the topology defined by $\{\mathfrak{a}_S^{(n)}\}_{n \in \mathbb{N}}$ with the $\mathfrak{a}$-adic topology. To this end we shall use the following variation of Chevalley's theorem, see [35].

Theorem 2.1. *Let* $\mathfrak{a}$ *denote an ideal of a local ring* $(A, \mathfrak{m})$. *Let* $\{\mathfrak{b}_n\}_{n \in \mathbb{N}}$ *denote a descending sequence of ideals. Suppose that the following conditions are satisfied:*

a) *A is $\mathfrak{a}$-adically complete,*
b) *$\bigcap_{n \in \mathbb{N}} \mathfrak{b}_n = (0)$, i.e. the filtration is separated, and*
c) *for all $m \in \mathbb{N}$ the family of ideals $\{\mathfrak{b}_n A/\mathfrak{a}^m\}_{n \in \mathbb{N}}$ satisfies the descending chain condition.*

Then for any $m \in \mathbb{N}$ there exists an integer $n = n(m)$ such $\mathfrak{b}_n \subset \mathfrak{a}^m$.

Proof. The assumption in c) guarantees that for any given $m \in \mathbb{N}$ there is an integer $n = n(m)$ such that

$$\mathfrak{b}_n + \mathfrak{a}^m = \mathfrak{b}_{n+k} + \mathfrak{a}^m \quad \text{for all } k \geq 1.$$

Call the ideal at the stable value $\mathfrak{c}_m$. Now suppose the conclusion is not true, i.e. $\mathfrak{b}_n \not\subset \mathfrak{a}^m$ for all $n \in \mathbb{N}$ and a fixed $m \in \mathbb{N}$. Therefore $\mathfrak{c}_m \neq \mathfrak{a}^m$. Moreover it is easily seen that $\mathfrak{c}_{m+1} + \mathfrak{a}^m = \mathfrak{c}_m$. Now construct inductively a series $(x_m)_{m \in \mathbb{N}}$ satisfying the following properties

$$x_m \in \mathfrak{c}_m \setminus \mathfrak{a}^m \quad \text{and} \quad x_{m+1} \equiv x_m \mod \mathfrak{a}^m.$$

Therefore $(x_m)_{m \in \mathbb{N}}$ is a convergent series with a limit $0 \neq x \in A$. This follows since A is $\mathfrak{a}$-adically complete by a). By definition that means for any $m \in \mathbb{N}$ there exists an $l \in \mathbb{N}$ such that $x - x_n \in \mathfrak{a}^m$ for all $n \geq l = l(m)$. Because of $x_n \in \mathfrak{c}_n$ this provides that $x \in \bigcap_{m \in \mathbb{N}} \bigcap_{n \in \mathbb{N}} (\mathfrak{c}_n + \mathfrak{a}^m)$. By Krull's Intersection Theorem and assumption b) it follows $x = 0$, a contradiction. $\qquad\square$

As a first application compare the $\mathfrak{a}$-adic topology with the topology derived by cutting the $\mathfrak{m}$-torsion of the powers of $\mathfrak{a}$.

Corollary 2.2. *For an ideal $\mathfrak{a}$ of a local ring $(A, \mathfrak{m})$ the following conditions are equivalent:*

(i) *$\{\mathfrak{a}^n :_A \langle \mathfrak{m} \rangle\}_{n \in \mathbb{N}}$ is equivalent to the $\mathfrak{a}$-adic topology.*
(ii) *$\bigcap_{n \in \mathbb{N}} (\mathfrak{a}^n \widehat{A} : \langle \mathfrak{m} \widehat{A} \rangle) = 0$, where $\widehat{A}$ denotes the $\mathfrak{m}$-adic completion of A.*
(iii) *$\text{height}(\mathfrak{a} \widehat{A} + \mathfrak{p}/\mathfrak{p}) < \dim \widehat{A}/\mathfrak{p}$ for all $\mathfrak{p} \in \text{Ass} \widehat{A}$.*

Proof. Without loss of generality we may assume that A is a complete local ring. The conclusion (i) $\Rightarrow$ (ii) is obviously true by Krull's Intersection Theorem. Let us prove (ii) $\Rightarrow$ (iii). Suppose there is a $\mathfrak{p} \in \text{Ass} A$ such that $\mathfrak{a} + \mathfrak{p}$ is $\mathfrak{m}$-primary. Then

$$0 \neq 0 : \langle \mathfrak{p} \rangle \subseteq \bigcap_{n \in \mathbb{N}} (\mathfrak{a}^n : \langle \mathfrak{p} \rangle) = \bigcap_{n \in \mathbb{N}} (\mathfrak{a}^n : \langle \mathfrak{p} + \mathfrak{a} \rangle) = \bigcap_{n \in \mathbb{N}} (\mathfrak{a}^n : \langle \mathfrak{m} \rangle),$$

a contradiction. Finally we prove the implication (iii) $\Rightarrow$ (i).

First note that because A is complete it is also $\mathfrak{a}$-adically complete. Moreover for a given $m \in \mathbb{N}$ the sequence $\{(\mathfrak{a}^n : \langle \mathfrak{m} \rangle) A/\mathfrak{a}^m\}_{n \in \mathbb{N}}$ satisfies the descending chain condition. Note that $((\mathfrak{a}^n : \langle \mathfrak{m} \rangle) + \mathfrak{a}^m)/\mathfrak{a}^m$ is a module of finite length for all large $n \in \mathbb{N}$. Suppose that (i) is not true. By virtue of 2.1 this means that $0 \neq \bigcap_{n \in \mathbb{N}} (\mathfrak{a}^n : \langle \mathfrak{m} \rangle)$, since the conditions a) and c) are satisfied.

Now choose
$$\mathfrak{p} \in \mathrm{Ass}_A(\cap_{n\in\mathbb{N}}(\mathfrak{a}^n : \langle\mathfrak{m}\rangle))$$
an associated prime ideal. Then $\mathfrak{p} = 0 :_A x$ for some $0 \neq x \in \cap_{n\in\mathbb{N}}(\mathfrak{a}^n : \langle\mathfrak{m}\rangle)$. Therefore $\mathfrak{p} \in \mathrm{Ass}\, A$.

By the Artin-Rees Lemma there exists a $k \in \mathbb{N}$ such that $\mathfrak{a}^k \cap x A \subseteq x\mathfrak{a}$. By the choice of x there is an integer $l \in \mathbb{N}$ such that $\mathfrak{m}^l x \subseteq \mathfrak{a}^k$. Therefore
$$\mathfrak{m}^l x \subseteq \mathfrak{a}^k \cap x A \subseteq x\mathfrak{a},$$
which implies $\mathfrak{m}^l \subseteq \mathfrak{a} + \mathfrak{p}$, in contradiction to assumption (iii).   □

A remarkable improvement of 2.2 was shown by I. Swanson, see [47]. Under the equivalent conditions of 2.2 she proved the existence of a $k \in \mathbb{N}$ such that $\mathfrak{a}^{nk} :_A \langle\mathfrak{m}\rangle \subseteq \mathfrak{a}^n$ for all $n \in \mathbb{N}$.

In the following let us describe the obstruction for the equivalence of both of the topologies considered in 2.2. To this end let $u(\mathfrak{a})$ denote the intersection of those primary components $\mathfrak{q}$ of 0 in A such that the associated prime ideal $\mathfrak{p}$ satisfies $\dim A/(\mathfrak{a} + \mathfrak{p}) > 0$.

Proposition 2.3. *Let $\mathfrak{a}$ denote an ideal of a local ring $(A, \mathfrak{m})$. Then it follows that* $u(\mathfrak{a}) = \cap_{n\in\mathbb{N}}(\mathfrak{a}^n : \langle\mathfrak{m}\rangle)$.

Proof. Let $x \in \cap_{n\in\mathbb{N}}(\mathfrak{a}^n : \langle\mathfrak{m}\rangle)$ be an arbitrary element. Then it is easily seen that $\frac{x}{1} \in \cap_{n\in\mathbb{N}}\mathfrak{a}^n A_\mathfrak{p} = 0$ for every prime ideal $\mathfrak{p} \in V(\mathfrak{a}) \setminus \{\mathfrak{m}\}$. That is $x \in 0_\mathfrak{p}$ for every $\mathfrak{p} \in V(\mathfrak{a}) \setminus \{\mathfrak{m}\}$. By taking the intersection over all those prime ideals it follows $x \in u(\mathfrak{a})$. But this means $\cap_{n\in\mathbb{N}}(\mathfrak{a}^n : \langle\mathfrak{m}\rangle) \subseteq u(\mathfrak{a})$.

In order to prove the converse containment relation let $\mathfrak{c} = \prod \mathfrak{p}$, where the product is taken over all prime ideals $\mathfrak{p} \in \mathrm{Ass}\, A$ such that $\dim A/(\mathfrak{a}+\mathfrak{p}) = 0$. Then $u(\mathfrak{a}) = 0 :_A \langle\mathfrak{c}\rangle$ and
$$0 :_A \langle\mathfrak{c}\rangle \subseteq \cap_{n\in\mathbb{N}}(\mathfrak{a}^n : \langle\mathfrak{c}\rangle) = \cap_{n\in\mathbb{N}}(\mathfrak{a}^n : \langle\mathfrak{c} + \mathfrak{a}\rangle) = \cap_{n\in\mathbb{N}}(\mathfrak{a}^n : \langle\mathfrak{m}\rangle)$$
because of $\mathrm{Rad}(\mathfrak{c} + \mathfrak{a}) = \mathfrak{m}$, as is easily seen.   □

Now consider the case of a principal ideal, important for the applications in the following.

Corollary 2.4. *Let $\mathfrak{a}$ denote an ideal of a commutative Noetherian ring A. For a regular element $x \in \mathfrak{a}$ the following conditions are equivalent:*

(i) $\{x^n A :_A \langle\mathfrak{a}\rangle\}_{n\in\mathbb{N}}$ *is equivalent to the xA-adic topology.*

(ii) $\dim \widehat{A}_P/\mathfrak{p} > 1$ *for all $P \in \mathrm{Ass}\, A/xA \cap V(\mathfrak{a})$ and all $\mathfrak{p} \in \mathrm{Ass}\, \widehat{A}_P$.*

Proof. First prove the implication (i) $\Rightarrow$ (ii). Suppose that there are prime ideals $P \in \mathrm{Ass}\, A/xA \cap V(\mathfrak{a})$ and $\mathfrak{p} \in \mathrm{Ass}\, \widehat{A}_P$ such that $\dim \widehat{A}_P/\mathfrak{p} = 1$. Because x is an $\widehat{A}_P$-regular element this means that $\dim \widehat{A}_P/(x\widehat{A}_P + \mathfrak{p}) = 0$. Note that $\frac{x}{1}$ is not a unit in $\widehat{A}_P$. Now replace $\widehat{A}_P$ by A. Then by 2.2 there is an $n \in \mathbb{N}$ such that $x^m A :_A \langle\mathfrak{m}\rangle \not\subseteq x^n A$ for all $m \geq n$. Therefore $x^m A :_A \langle\mathfrak{a}\rangle \not\subseteq x^n A$ for all $m \geq n$ since $x^m A :_A \langle\mathfrak{m}\rangle \subseteq x^m A :_A \langle\mathfrak{a}\rangle$. This contradicts the assumption in (i).

In order to prove that (ii) $\Rightarrow$(i) consider the ideals

$$E_{m,n} = (x^m A :_A \langle \mathfrak{a} \rangle + x^n A)/x^n A \subseteq A/x^n A$$

for a given n and all $m \geq n$. Obviously $\mathrm{Ass}_A E_{m,n} \subseteq \mathrm{Ass}_A A/xA \cap V(\mathfrak{a})$. Moreover $E_{m+1,n} \subseteq E_{m,n}$. That means, for a fixed $n \in \mathbb{N}$ the set $\mathrm{Ass}_A E_{m,n}$ becomes an eventually stable set of prime ideals, say X_n. The claim will follow provided $X_n = \emptyset$. Suppose that $X_n \neq \emptyset$. By a localization argument and changing notation one might assume that $X_n = \{\mathfrak{m}\}$, the maximal ideal of a local ring $(A, \mathfrak{m})$. Therefore $\mathrm{Supp}\, E_{m,n} = V(\mathfrak{m})$ for any fixed $n \in \mathbb{N}$ and all large m. Whence

$$x^m A :_A \langle \mathfrak{a} \rangle \subseteq x^n A :_A \langle \mathfrak{m} \rangle$$

for a given n and all large m. By assumption (ii) and 2.2 it follows that for a given k there is an integer n such that $x^n A :_A \langle \mathfrak{m} \rangle \subseteq x^k A$. Therefore $X_n = \emptyset$, a contradiction to the choice of X_n. $\qquad\square$

While condition (ii) looks rather technical one should try to simplify it under reasonable conditions on A. Say a local ring $(A, \mathfrak{m})$ satisfies condition (C) provided

$$\dim A/P = \dim \hat{A}/\mathfrak{p} \text{ for all } P \in \mathrm{Ass}\, A \text{ and all } \mathfrak{p} \in \mathrm{Ass}\, \hat{A}/P\hat{A}.$$

This is equivalent to saying that A/P is an unmixed local ring for any prime ideal $P \in \mathrm{Ass}\, A$.

Say that a commutative Noetherian A satisfies locally condition (C) provided any localization $A_{\mathfrak{p}}, \mathfrak{p} \in \mathrm{Spec}\, A$, satisfies condition (C). Let A be an unmixed ring resp. a factor ring of a Cohen-Macaulay ring. Then it follows that A satisfies locally (C), see [35]. In particular, by Cohen's Structure Theorem it turns out that a complete local ring satisfies locally condition (C).

Proposition 2.5. *Let* $(A, \mathfrak{m})$ *denote a local ring satisfying condition* (C). *Then it satisfies also locally* (C).

Proof. By definition condition (C) implies that, for $P \in \mathrm{Ass}\, A, A/P$ is unmixed, i.e., $\dim \hat{A}/P = \dim \hat{A}/\mathfrak{p}\hat{A}$ for all $\mathfrak{p} \in \mathrm{Ass}\, \hat{A}/P\hat{A}$. Now unmixedness localizes, i.e. for any $Q \in V(P)$ the local ring A_Q/PA_Q is again unmixed, see [33]. Therefore A_Q satisfies condition (C) for any $Q \in \mathrm{Spec}\, A$. Recall that a for prime ideal $Q \in V(P)$ we have $\mathfrak{p} \in \mathrm{Ass}\, A$ and $\mathfrak{p} \subseteq Q$ if and only if $\mathfrak{p}A_Q \in \mathrm{Ass}\, A_Q$. $\qquad\square$

Now use the results about condition (C) in order to simplify the result in 2.4.

Corollary 2.6. *Let* A *denote a commutative Noetherian ring satisfying locally the condition* (C). *Let* $x \in \mathfrak{a}$ *be a regular element. Then* $\{x^n A :_A \langle \mathfrak{a} \rangle\}_{n \in \mathbb{N}}$ *is equivalent to the* xA-*adic topology if and only if* $\mathrm{height}(\mathfrak{a} + \mathfrak{p}/\mathfrak{p}) > 1$ *for all* $\mathfrak{p} \in \mathrm{Ass}\, A$. *In particular, in the case of a local ring* $(A, \mathfrak{m})$ *this holds if and only if* $\mathrm{height}(\mathfrak{a}\hat{A} + \mathfrak{p}/\mathfrak{p}) > 1$ *for all* $\mathfrak{p} \in \mathrm{Ass}\, \hat{A}$.

Proof. Let $\mathrm{height}(\mathfrak{a} + \mathfrak{p}/\mathfrak{p}) > 1$ for all primes $\mathfrak{p} \in \mathrm{Ass}\, A$. Then $\dim A_P/\mathfrak{p} > 1$ for all prime ideals $P \in \mathrm{Ass}\, A/xA \cap V(\mathfrak{a})$ and $\mathfrak{p} \in \mathrm{Ass}\, A_P$. Because of condition (C) for A_P it implies that $\dim \widehat{A_P}/\mathfrak{q} > 1$ for all $\mathfrak{q} \in \operatorname{Ass} \widehat{A_P}/\mathfrak{p}\widehat{A_P}$. But now

$$\operatorname{Ass} \widehat{A_P} = \cup_{\mathfrak{p} \in \operatorname{Ass} A_P} \operatorname{Ass} \widehat{A_P}/\mathfrak{p}\widehat{A_P},$$

see [28, Theorem 23.2], which proves the first part of the result in view of 2.4.

Conversely suppose the equivalence of the ideal topologies and let $\operatorname{height}(\mathfrak{a} + \mathfrak{p}/\mathfrak{p}) = 1$ for some $\mathfrak{p} \in \operatorname{Ass} A$. Then there is a prime ideal $P \in V(\mathfrak{a})$ such that $\dim A_P/\mathfrak{p}A_P = 1$. Because $\frac{x}{1}$ is an A_P-regular element it follows that $P \in \operatorname{Ass} A/xA \cap V(\mathfrak{a})$. So condition (C) provides a contradiction by 2.4. □

2.2. On Ideal Transforms. In this subsection let us discuss the behaviour of certain intermediate rings lying between a commutative Noetherian ring and its full ring of quotients. To this end let $x \in A$ be a non-zero divisor and $A \subseteq B \subseteq A_x$ an intermediate ring.

Lemma 2.7. *For an intermediate ring $A \subseteq B \subseteq A_x$ the following conditions are equivalent:*

(i) *B is a finitely generated A-module.*
(ii) *There is a $k \in \mathbb{N}$ such that $x^k B \subseteq A$.*
(iii) *There is a $k \in \mathbb{N}$ such that $x^{k+1}B \cap A \subseteq xA$.*
(iv) *$\{x^n B \cap A\}_{n \in \mathbb{N}}$ is equivalent to the xA-adic topology.*
(v) *There is a $k \in \mathbb{N}$ such that $x^{n+k}B \cap A = x^n(x^k B \cap A)$ for all $n \geq 1$.*

Proof. The implication (i) $\Rightarrow$ (v) is a consequence of the Artin-Rees lemma. The implications (v) $\Rightarrow$ (iv) $\Rightarrow$ (iii) $\Rightarrow$ (ii) are easy to see. Finally we have (ii) $\Rightarrow$ (i) since B is as an A-submodule of the finitely generated A-module $\frac{1}{x^k} A$ finitely generated. □

In the following there is some need for the ideal transform. To this end let $\mathfrak{a}$ denote a regular ideal of a commutative Noetherian ring A. Define

$$T_{\mathfrak{a}}(A) = \{q \in Q(A) \mid \operatorname{Supp}(Aq + A/A) \subseteq V(\mathfrak{a})\},$$

where $Q(A)$ denotes the full ring of quotients of A. It follows that

$$T_{\mathfrak{a}}(A) = \{q \in Q(A) \mid \mathfrak{a}^n q \subseteq A \text{ for some } n \in \mathbb{N}\},$$

Note that $A \subseteq T_{\mathfrak{a}}(A) \subseteq A_x$, where $x \in \mathfrak{a}$ is a non-zero divisor. Moreover, suppose that $\mathfrak{a} = (a_1, \ldots, a_s)A$, where each of the a_i's is a non-zero divisor. Then $T_{\mathfrak{a}}(A) = \cap_{i=1}^{s} A_{a_i}$ as is easily seen. Moreover one might define $T_{\mathfrak{a}}(B)$ for an arbitrary intermediate ring $A \subseteq B \subseteq Q(A)$ in a corresponding way.

Ideal transforms were first studied by M. Nagata in connection with Hilbert's 14th problem, see [34]. It is of some interest to describe when $T_{\mathfrak{a}}(A)$ is an A-algebra of finite type. As a first step towards this direction consider when it is a finitely generated A-module.

Lemma 2.8. *Let $\mathfrak{a}$ denote a regular ideal of a commutative ring A. Let $x \in \mathfrak{a}$ be a non-zero divisor. Then the following conditions are satisfied:*

a) $T_\mathfrak{a}(T_\mathfrak{a}(A)) = T_\mathfrak{a}(A)$,
b) $\mathrm{Ass}_A\, T_\mathfrak{a}(A)/A = \mathrm{Ass}\, A/xA \cap V(\mathfrak{a})$, and
c) $rT_\mathfrak{a}(A) \cap A = rA : \langle \mathfrak{a} \rangle$ for any regular element $r \in A$.

Proof. The first claim is obvious by definition. In order to prove b) let $\mathfrak{p} = A : q$ for some $q = \frac{s}{x^n} \in T_\mathfrak{a}(A) \setminus A$. Then $\mathfrak{p} = x^n A :_A s$ and $s \notin x^n A$. That is $\mathfrak{p} \in \mathrm{Ass}\, A/xA$. Furthermore $\mathfrak{a} \subseteq \mathfrak{p}$ since $\mathfrak{a}^k q \subseteq A$ and $\mathfrak{a}^k \subseteq A : q = \mathfrak{p}$, for some $k \in \mathbb{N}$. The reverse conclusion follows by similar arguments.

In order to prove c) first note that $rT_\mathfrak{a}(A) \cap A \subseteq rA :_A \langle \mathfrak{a} \rangle$ as easily seen. For the reverse inclusion note that $rT_\mathfrak{a}(A) : \langle \mathfrak{a} \rangle = rT_\mathfrak{a}(A)$. □

As a consequence of 2.8 it follows that $\mathrm{Ass}_A\, T_\mathfrak{a}(A)/A = \mathrm{Ass}\, \mathrm{Ext}^1_A(A/\mathfrak{a}, A)$. Therefore $T_\mathfrak{a}(A) = A$ if and only if grade $\mathfrak{a} > 1$. There is a relation of ideal transforms to a more functorial construction. First note that $\mathrm{Hom}_A(\mathfrak{a}, A) \simeq A :_{Q(A)} \mathfrak{a}$ for a regular ideal $\mathfrak{a}$ in A. Therefore $\varinjlim \mathrm{Hom}_A(\mathfrak{a}^n, A) \simeq \cup_{n \in \mathbb{N}} (A :_{Q(A)} \mathfrak{a}^n) = T_\mathfrak{a}(A)$. This yields a short exact sequence

$$0 \to A \to T_\mathfrak{a}(A) \to H^1_\mathfrak{a}(A) \to 0,$$

where the monomorphism is just the inclusion map. Therefore $T_\mathfrak{a}(A)/A \simeq H^1_\mathfrak{a}(A)$. So the ideal transform enables another approach to $H^1_\mathfrak{a}(A)$.

It is of a particular interest when $T_\mathfrak{a}(A)$ or - equivalently $H^1_\mathfrak{a}(A)$ - is a finitely generated A-module. In the following there is a generalization of A. Grothendieck's finiteness result, see [13].

Theorem 2.9. (Grothendieck's Finiteness Result) *Let $\mathfrak{a}$ denote a regular ideal of a commutative Noetherian ring A. Then the following conditions are equivalent:*

(i) $T_\mathfrak{a}(A)$ *is a finitely generated A-module.*
(ii) $\dim \widehat{A_P}/\mathfrak{p} > 1$ *for all* $P \in \mathrm{Ass}_A\, \mathrm{Ext}^1_A(A/\mathfrak{a}, A)$ *and all* $\mathfrak{p} \in \mathrm{Ass}\, \widehat{A_P}$.

Proof. By 2.7 and 2.8 the statement in condition (i) is equivalent to the fact that $\{x^n A :_A \langle \mathfrak{a} \rangle\}_{n \in \mathbb{N}}$ is equivalent to the xA-adic topology for a non-zero divisor $x \in \mathfrak{a}$. Note that $T_\mathfrak{a}(A) \subseteq A_x$. By 2.4 this proves the statement because of

$$\mathrm{Ass}\, A/xA \cap V(\mathfrak{a}) = \mathrm{Ass}\, \mathrm{Ext}^1_A(A/\mathfrak{a}, A)$$

as mentioned above. □

Under the additional assumption of condition (C) on A there is a further simplification of the finiteness result.

Corollary 2.10. a) *Suppose that A is a factor ring of a Cohen-Macaulay ring. Then $T_\mathfrak{a}(A)$ is a finitely generated A-module if and only if* height$(\mathfrak{a} + \mathfrak{p}/\mathfrak{p}) > 1$ *for all $\mathfrak{p} \in \mathrm{Ass}\, A$. In particular $T_\mathfrak{a}(A)$ is a finitely generated A-module if and only if $T_{\mathfrak{a}+\mathfrak{p}/\mathfrak{p}}(A/\mathfrak{p})$ is a finitely generated $A/\mathfrak{p}$-module for all $\mathfrak{p} \in \mathrm{Ass}\, A$.*
b) *Suppose that $(A, \mathfrak{m})$ is a local ring. Then $T_\mathfrak{a}(A)$ is a finitely generated A-module if and only if* height$(\mathfrak{a}\widehat{A} + \mathfrak{p}/\mathfrak{p}) > 1$ *for all $\mathfrak{p} \in \mathrm{Ass}\, \widehat{A}$.*

Proof. It is a consequence of 2.9 with the aid of 2.6. □

In the case of a local ring $(A, \mathfrak{m})$ which is a factor ring of a Cohen-Macaulay ring the finiteness of $H^1_{\mathfrak{m}}(A)$ is therefore equivalent to $\dim A/\mathfrak{p} > 1$ for all prime ideals $\mathfrak{p} \in \mathrm{Ass}\, A$.

A more difficult problem is a characterization of when the ideal transform $T_{\mathfrak{a}}(A)$ is an A-algebra of finite type. This does not hold even in the case of a polynomial ring over a field as shown by M. Nagata, see [34], in the context of Hilbert's 14th problem.

2.3. Asymptotic Prime Divisors. In the following we apply some of the previous considerations to the study of asymptotic prime ideals. To this end there is a short excursion about graded algebras.

For a commutative Noetherian ring A let $F = \{\mathfrak{a}_n\}_{n \in \mathbb{Z}}$ denote a filtration of ideals, i.e. a family of ideals satisfying the following conditions:

a) $\mathfrak{a}_n = A$ for all $n \leq 0$,
b) $\mathfrak{a}_{n+1} \subseteq \mathfrak{a}_n$ for all $n \in \mathbb{Z}$, and
c) $\mathfrak{a}_n \mathfrak{a}_m \subseteq \mathfrak{a}_{n+m}$ for all $n, m \in \mathbb{Z}$.

Then one may form $R(F)$, the Rees ring associated to F, i.e. $R(F) = \oplus_{n \in \mathbb{Z}} \mathfrak{a}_n t^n \subseteq A[t, t^{-1}]$, where t denotes an indeterminate. Let $\mathfrak{a} = (a_1, \ldots, a_s)A$ denote an ideal of A. Then F is called an $\mathfrak{a}$-admissible filtration, whenever $\mathfrak{a}^n \subseteq \mathfrak{a}_n$ for all $n \in \mathbb{Z}$. For an $\mathfrak{a}$-admissible filtration it is easily seen that $R(F)$ is an $R(\mathfrak{a})$-module, where

$$R(\mathfrak{a}) = \oplus_{n \in \mathbb{Z}} \mathfrak{a}^n t^n = A[a_1 t, \ldots, a_s t, t^{-1}]$$

denotes the (extended) Rees ring of A with respect to $\mathfrak{a}$. Note that $R(\mathfrak{a})$ is the Rees ring associated to the $\mathfrak{a}$-adic filtration $F = \{\mathfrak{a}^n\}_{n \in \mathbb{Z}}$.

There are several possibilities to associate an $\mathfrak{a}$-admissible filtration F to a given ideal $\mathfrak{a}$. One of these is defined for a multiplicatively closed subset S of A. Let $\mathfrak{a}_S^{(n)}, n \in \mathbb{N}$, denote the n-th symbolic power of $\mathfrak{a}$ with respect to S. Then $F = \{\mathfrak{a}_S^{(n)}\}_{n \in \mathbb{Z}}$ forms an $\mathfrak{a}$-admissible filtration. The corresponding Rees ring $R_S(\mathfrak{a}) := R(F)$ is called the symbolic Rees ring of $\mathfrak{a}$ with respect to S. In the case of $S = A \setminus \mathfrak{p}$ for a prime ideal $\mathfrak{p}$ of A write $R_S(\mathfrak{p})$ instead of $R_{A \setminus \mathfrak{p}}(\mathfrak{p})$.

Let F denote an $\mathfrak{a}$-admissible filtration. It follows that $R(F)$ is a finitely generated $R(\mathfrak{a})$-module if and only if there is an integer $k \in \mathbb{N}$ such that $\mathfrak{a}_{n+k} = \mathfrak{a}^n \mathfrak{a}_k$ for all $n \in \mathbb{N}$. Equivalently this holds if and only if $\mathfrak{a}_{n+k} \subseteq \mathfrak{a}^n$ for all $n \in \mathbb{N}$ and a certain integer $k \geq 0$. This behaviour sometimes is called linear equivalence of F to the $\mathfrak{a}$-adic topology.

For an integer $k \in \mathbb{N}$ let $F_k = \{\mathfrak{a}_{nk}\}_{n \in \mathbb{Z}}$. Then $R(F_k) \simeq R^{(k)}(F)$, where $R^{(k)}(F) = \oplus_{n \in \mathbb{Z}} \mathfrak{a}_{nk} t^{nk}$ denotes the k-th Veronesean subring of $R(F)$. Before we continue with the study of ideal transforms there is a characterization of when $R(F)$ is an A-algebra of finite type.

Proposition 2.11. *Let $F = \{\mathfrak{a}_n\}_{n \in \mathbb{Z}}$ denote a filtration of ideals. Then the following conditions are equivalent:*

(i) *$R(F)$ is an A-algebra of finite type.*
(ii) *There is a $k \in \mathbb{N}$ such that $R(F_k)$ is an A-algebra of finite type.*

(iii) *There is a $k \in \mathbb{N}$ such that $R(F_k)$ is a finitely generated $R(\mathfrak{a}_k)$-module.*
(iv) *There is a $k \in \mathbb{N}$ such that $\mathfrak{a}_{nk} = (\mathfrak{a}_k)^n$ for all $n \geq k$.*
(v) *There is a $k \in \mathbb{N}$ such that $\mathfrak{a}_{n+k} = \mathfrak{a}_n \mathfrak{a}_k$ for all $n \geq k$.*

Proof. First show (i) $\Rightarrow$ (v). By the assumption there is an $r \in \mathbb{N}$ such that $R(F) = A[\mathfrak{a}_1 t, \ldots, \mathfrak{a}_r t^r]$. Put $l = r!$ and $k = rl$. Then it follows that $\mathfrak{a}_n = \sum \mathfrak{a}_1^{n_1} \cdots \mathfrak{a}_r^{n_r}$, where the sum is taken over all $n_1, \ldots, n_r$ such that $\sum_{i=1}^r i n_i \geq n$. For $n \geq k$ it is easy to see that there is an integer $1 \leq i \leq r$ such that $n_i \geq \frac{l}{i}$. Whence $\mathfrak{a}_n \subseteq \mathfrak{a}_{n-l}\mathfrak{a}_l$ for any $n \geq k$. That means $\mathfrak{a}_{n+k} = \mathfrak{a}_n \mathfrak{a}_k$ for any $n \geq k$ as is easily seen.

While the implication (v) $\Rightarrow$ (iv) holds trivially the implication (iv) $\Rightarrow$ (iii) is a consequence of the Artin-Rees lemma. In order to show (iii) $\Rightarrow$ (ii) note that $R(\mathfrak{a}_k)$ is an A-algebra of finite type.

Finally show (ii) $\Rightarrow$ (i). For $0 \leq i < k$ it follows that $\mathfrak{A}_i = \oplus_{n \in \mathbb{Z}} \mathfrak{a}_{nk+i} t^{nk}$ is an ideal of $R^{(k)}(F)$, and $R^{(k)}(F)$ is isomorphic to $R(F_k)$. So $\mathfrak{A}_i, 0 \leq i < r$, is a finitely generated $R^{(k)}(F)$-module. Because of $R(F) = \oplus_{i=0}^{k-1} \mathfrak{A}_i t^i$ it turns out that $R(F)$ is a finitely generated $R^{(k)}(F)$-module. This proves that $R(F)$ is an A-algebra of finite type. $\square$

The implication (i) $\Rightarrow$ (v) was shown by D. Rees, see [37]. In the case of a local ring (ii) $\Rightarrow$ (i) was proved by a different argument in [43].

Before we shall continue with the study of certain ideal transforms consider two applications of the Artin-Rees Lemma. They will be useful in the study of the Ratliff-Rush closure of an ideal.

Proposition 2.12. *Let $\mathfrak{a}, \mathfrak{b}, \mathfrak{b}_1, \ldots, \mathfrak{b}_t$, $t \in \mathbb{N}$, denote ideals of a commutative Noetherian ring A.*

a) *There is a $k \in \mathbb{N}$ such that*

$$\cap_{i=1}^t (\mathfrak{a}^{n+k} + \mathfrak{b}_i) = \mathfrak{a}^n (\cap_{i=1}^t (\mathfrak{a}^k + \mathfrak{b}_i)) + \cap_{i=1}^t \mathfrak{b}_i \text{ for all } n \geq 1.$$

b) *There is a $k \in \mathbb{N}$ such that*

$$\mathfrak{a}^{n+k} :_A \mathfrak{b} = \mathfrak{a}^n (\mathfrak{a}^k :_A \mathfrak{b}) + 0 :_A \mathfrak{b} \text{ for all } n \geq 1.$$

Proof. In order to prove a) consider the natural injective homomorphism of finitely generated A-modules

$$A/ \cap_{i=1}^t \mathfrak{b}_i \to \oplus_{i=1}^t A/\mathfrak{b}_i, \quad a + \cap_{i=1}^t \mathfrak{b}_i \mapsto (a + \mathfrak{b}_1, \ldots, a + \mathfrak{b}_t).$$

Then the Artin-Rees Lemma provides the existence of an integer $k \in \mathbb{N}$ such that

$$\mathfrak{a}^{n+k}(\oplus_{i=1}^t A/\mathfrak{b}_i) \cap (A/ \cap_{i=1}^t \mathfrak{b}_i) = \mathfrak{a}^n (\mathfrak{a}^k (\oplus_{i=1}^t A/\mathfrak{b}_i) \cap (A/ \cap_{i=1}^t \mathfrak{b}_i))$$

for all $n \in \mathbb{N}$. In fact this proves the statement a).

For the proof of b) let $\mathfrak{b} = (b_1, \ldots, b_s)A$. Then by the Artin-Rees Lemma there is a $c \in \mathbb{N}$ such that

$$\mathfrak{a}^{n+c} :_A b_i \subseteq \mathfrak{a}^n + (0 :_A b_i)$$

for all $n \in \mathbb{N}$ and $i = 1, \ldots, s$. Because of the statement in a) there exists a $d \in \mathbb{N}$ such that

$$\cap_{i=1}^{s}(\mathfrak{a}^{n+d} + 0 :_A b_i) \subseteq \mathfrak{a}^n + \cap_{i=1}^{s}(0 :_A b_i) = \mathfrak{a}^n + (0 :_A \mathfrak{b}).$$

But now we have that $\cap_{i=1}^{s}(\mathfrak{a}^n :_A b_i) = \mathfrak{a}^n :_A \mathfrak{b}$ for all $n \in \mathbb{N}$. So finally there exists a $k \in \mathbb{N}$ such that $\mathfrak{a}^{n+k} :_A \mathfrak{b} \subseteq \mathfrak{a}^n + (0 :_A \mathfrak{b})$ for all $n \in \mathbb{N}$. By passing to $A/0 :_A \mathfrak{b}$ the Artin-Rees Lemma proves the claim in b). □

As a first sample of ideal transforms consider $T_{(\mathfrak{a}t,t^{-1})}(R(\mathfrak{a}))$. But now we have that $T_{(\mathfrak{a}t,t^{-1})}(R(\mathfrak{a})) \subseteq A[t,t^{-1}]$. So it is an easy exercise to prove that the n-th graded piece of the ideal transform is given by

$$T_{(\mathfrak{a}t,t^{-1})}(R(\mathfrak{a}))_n = \begin{cases} A & \text{for } n \leq 0, \\ (\mathfrak{a}^n)^* & \text{for } n > 0, \end{cases}$$

where $(\mathfrak{a}^n)^* = \cup_{m \in \mathbb{N}}(\mathfrak{a}^{n+m} : \mathfrak{a}^m)$ denotes the Ratliff-Rush closure of $\mathfrak{a}^n$. In the following put $R^*(\mathfrak{a}) = \oplus_{n \in \mathbb{Z}}(\mathfrak{a}^n)^* t^n$. A few basic results of the Ratliff-Rush closure are listed in the following result.

Lemma 2.13. *Let $\mathfrak{a}$ be an ideal of a commutative Noetherian ring A.*
  a) *There is an integer $k \in \mathbb{N}$ such that $(\mathfrak{a}^n)^* = \mathfrak{a}^n + 0 :_A \langle \mathfrak{a} \rangle$ for all $n \geq k$. In particular $(\mathfrak{a}^n)^* = \mathfrak{a}^n$ for all $n \geq k$ provided $\mathfrak{a}$ is a regular ideal.*
  b) *$(\mathfrak{a}^{n+1})^* :_A \mathfrak{a} = (\mathfrak{a}^n)^*$ for all $n \in \mathbb{N}$.*
  c) *$T_{(\mathfrak{a}t,t^{-1})}(R(\mathfrak{a}))$ is a finitely generated $R(\mathfrak{a})$-module if and only if $\mathfrak{a}$ is a regular ideal.*

Proof. Fix an integer $n \in \mathbb{N}$. Then for a sufficiently large integer m it follows that $0 :_A \mathfrak{a}^m = 0 :_A \langle \mathfrak{a} \rangle$ and $(\mathfrak{a}^n)^* = (\mathfrak{a}^{n+m} + 0 :_A \langle \mathfrak{a} \rangle) :_A \mathfrak{a}^m$. Therefore, by passing to $A/0 :_A \langle \mathfrak{a} \rangle$ we may assume that $\mathfrak{a}$ is a regular ideal in order to prove a). Then by 2.12 it follows that

$$\oplus_{n \in \mathbb{Z}}(\mathfrak{a}^{n+1} :_A \mathfrak{a})t^n$$

is a finitely generated $R(\mathfrak{a})$-module. Therefore the Artin-Rees Lemma provides the existence of an integer $k \in \mathbb{N}$ such that $\mathfrak{a}^{n+k+1} :_A \mathfrak{a} = \mathfrak{a}^n(\mathfrak{a}^{k+1} :_A \mathfrak{a})$ for all $n \geq 1$. Therefore $\mathfrak{a}^{n+k+1} :_A \mathfrak{a} = \mathfrak{a}^{n+k}$ for all $n \geq 1$. This proves the claim in a).

The statement in b) follows easily by the definitions. Finally c) is a consequence of a) and the Artin-Rees Lemma. □

Next let $(A, \mathfrak{m})$ denote a local Noetherian ring. For an ideal $\mathfrak{a}$ of A consider the ideal transform $T_{(\mathfrak{m},t^{-1})}(R(\mathfrak{a}))$. It is easily seen that its n-th graded component has the following form

$$T_{(\mathfrak{m},t^{-1})}(R(\mathfrak{a}))_n = \begin{cases} A & \text{for } n \leq 0, \\ \mathfrak{a}^n :_A \langle \mathfrak{m} \rangle & \text{for } n > 0. \end{cases}$$

Therefore the finiteness of $T_{(\mathfrak{m},t^{-1})}(R(\mathfrak{a}))$ yields some information about the existence of an integer $k \in \mathbb{N}$ such that $\mathfrak{a}^{n+k} : \langle \mathfrak{m} \rangle \subseteq \mathfrak{a}^n$ for all $n \geq 1$ as it is clear by the Artin-Rees Lemma. This is a sharpening of the problem on the equivalence of the topologies investigated at the beginning.

In the following let $l(\mathfrak{a})$ denote the analytic spread of $\mathfrak{a}$, i.e.

$$l(\mathfrak{a}) = \dim R(\mathfrak{a})/(\mathfrak{m}, t^{-1})R(\mathfrak{a}),$$

see D. G. Northcott and D. Rees [36] for basic results. Recall that

$$\text{height } \mathfrak{a} \le l(\mathfrak{a}) \le \dim A.$$

Moreover, $l(\mathfrak{a}) = \dim gr_A(\mathfrak{a})/\mathfrak{m} gr_A(\mathfrak{a})$, where $gr_A(\mathfrak{a}) = \oplus_{n \in \mathbb{N}} \mathfrak{a}^n/\mathfrak{a}^{n+1}$ denotes the form ring with respect to $\mathfrak{a}$.

Theorem 2.14. *Let $\mathfrak{a}$ denote an ideal of a local ring $(A, \mathfrak{m})$.*

a) *The ideal transform $T_{(\mathfrak{m},t^{-1})}(R(\mathfrak{a}))$ is a finitely generated $R(\mathfrak{a})$-module if and only if*

$$l(\mathfrak{a}\widehat{A} + \mathfrak{p}/\mathfrak{p}) < \dim \widehat{A}/\mathfrak{p} \text{ for all } \mathfrak{p} \in \text{Ass } \widehat{A}.$$

b) *$T_{(\mathfrak{m},t^{-1})}(R(\mathfrak{a}))$ is an A-algebra of finite type if and only if there is a $k \in \mathbb{N}$ such that*

$$l(\mathfrak{a}^k \widehat{A} : \langle \mathfrak{m}\widehat{A} \rangle + \mathfrak{p}/\mathfrak{p}) < \dim \widehat{A}/\mathfrak{p} \text{ for all } \mathfrak{p} \in \text{Ass } \widehat{A}.$$

Proof. At first prove a). As a consequence of the Artin-Rees Lemma the ideal transform $T_{(\mathfrak{m},t^{-1})}(R(\mathfrak{a}))$ is finitely generated over $R(\mathfrak{a})$ if and only if the corresponding result holds for $\mathfrak{a}\widehat{A}$ in $(\widehat{A}, \widehat{\mathfrak{m}})$. Therefore, without loss of generality we may assume that A is a complete local ring.

So we may assume that $R(\mathfrak{a})$ is the quotient of a Cohen-Macaulay ring. Furthermore there is a 1-to-1 correspondence between the associated prime ideals $\mathfrak{P}$ of $R(\mathfrak{a})$ and the associated prime ideals $\mathfrak{p}$ of A given by

$$\mathfrak{P} \mapsto \mathfrak{p} = \mathfrak{P} \cap A \text{ resp. } \mathfrak{p} \mapsto \oplus_{n \in \mathbb{Z}} (\mathfrak{a}^n \cap \mathfrak{p}) t^n.$$

By virtue of 2.10 $T_{(\mathfrak{m},t^{-1})}(R(\mathfrak{a}))$ is a finitely generated $R(\mathfrak{a})$-module if and only if $T_{(\mathfrak{m}/\mathfrak{p},t^{-1})}(R(\mathfrak{a} + \mathfrak{p}/\mathfrak{p}))$ is a finitely generated $R(\mathfrak{a} + \mathfrak{p}/\mathfrak{p})$-module for all $\mathfrak{p} \in \text{Ass } A$. That is, without loss of generality we may assume $(A, \mathfrak{m})$ a complete local domain after changing the notation. But under this assumption $T_{(\mathfrak{m},t^{-1})}(R(\mathfrak{a}))$ is a finitely generated $R(\mathfrak{a})$-module if and only if $\text{height}(\mathfrak{m}, t^{-1})R(\mathfrak{a}) > 1$. Finally A is a universally catenarian domain. Therefore it holds

$$\text{height}(\mathfrak{m}, t^{-1})R(\mathfrak{a}) = \dim R(\mathfrak{a}) - \dim R(\mathfrak{a})/(\mathfrak{m}, t^{-1})R(\mathfrak{a}).$$

Because of $\dim R(\mathfrak{a}) = \dim A + 1$ and $\dim R(\mathfrak{a})/(\mathfrak{m}, t^{-1})R(\mathfrak{a}) = l(\mathfrak{a})$ this completes the proof.

With the aid of statement a) the conclusion in b) follows by virtue of 2.11 □

As above let $\mathfrak{a}$ denote an ideal of a commutative Noetherian ring A. Let $\text{As}(\mathfrak{a})$ resp. $\text{Bs}(\mathfrak{a})$ denote the ultimately constant values of $\text{Ass } A/\mathfrak{a}^n$ resp. $\text{Ass } \mathfrak{a}^n/\mathfrak{a}^{n+1}$ for all large $n \in \mathbb{N}$, as shown by M. Brodmann in [4], see also [42].

As it will be shown in the following the previous result 2.14 has to do with the property $\mathfrak{m} \in \text{As}(\mathfrak{a})$ for an ideal $\mathfrak{a}$ of a local ring $(A, \mathfrak{m})$. To this end we modify a result originally shown by L. Burch, see [8]. Further results in this direction were shown by C. Huneke, see [21].

Theorem 2.15. *Suppose that $(A, \mathfrak{m})$ denotes a local Noetherian ring. Then the following results are true:*

a) *If $\mathfrak{m} \notin \mathrm{Bs}(\mathfrak{a})$, then $l(\mathfrak{a}) < \dim A$.*
b) *The converse is true provided A is a universally catenarian domain and $gr_A(\mathfrak{a})$ is unmixed.*

Proof. In order to show a) first note that the natural epimorphism

$$\phi_n : \mathfrak{a}^n/\mathfrak{a}^{n+1} \to \mathfrak{a}^n \bar{A}/\mathfrak{a}^{n+1}\bar{A} \ \text{ with } \ \bar{A} = A/0 :_A \langle \mathfrak{a} \rangle$$

is an isomorphism for all large $n \in \mathbb{N}$. This follows easily by the Artin-Rees Lemma. By passing to $\bar{A}$ one might assume that $\mathfrak{a}$ is a regular ideal. Then $\mathfrak{m} \notin \mathrm{As}(\mathfrak{a})$ because of $\mathrm{As}(\mathfrak{a}) = \mathrm{Bs}(\mathfrak{a})$ for the regular ideal $\mathfrak{a}$, see [30].

Next investigate the Noetherian ring $R^*(\mathfrak{a})$. Now we claim that

$$\mathrm{Ass}\, A/(\mathfrak{a}^n)^* \subseteq \mathrm{Ass}\, A/(\mathfrak{a}^{n+1})^* \ \text{ for all } \ n \in \mathbb{N}.$$

To this end note that $(\mathfrak{a}^{n+1})^* : \mathfrak{a} = (\mathfrak{a}^n)^*$ for all $n \in \mathbb{N}$, see 2.13. Let $\mathfrak{a} = (a_1, \ldots, a_s)A$. Then the natural homomorphism

$$A/(\mathfrak{a}^n)^* \to \oplus_{i=1}^s A/(\mathfrak{a}^{n+1})^*, \quad r + (\mathfrak{a}^n)^* \mapsto (ra_i + (\mathfrak{a}^{n+1})^*)$$

is injective for all $n \in \mathbb{N}$. Therefore $\mathrm{Ass}\, A/(\mathfrak{a}^n)^* \subseteq \mathrm{Ass}\, A/(\mathfrak{a}^{n+1})^*$, as required. Because of $(\mathfrak{a}^n)^* = \mathfrak{a}^n$ for all large n it turns out that $\mathfrak{m} \notin \mathrm{Ass}\, A/(\mathfrak{a}^n)^*$ for all $n \in \mathbb{N}$. Because of $T_{(\mathfrak{m},t^{-1})}(R^*(\mathfrak{a})) = \oplus_{n \in \mathbb{Z}}((\mathfrak{a}^n)^* : \langle \mathfrak{m} \rangle)t^n$ it follows that $T_{(\mathfrak{m},t^{-1})}(R^*(\mathfrak{a})) = R^*(\mathfrak{a})$. By 2.8 this means that $\mathrm{grade}(\mathfrak{m}, t^{-1})R^*(\mathfrak{a}) > 1$. But now

$$1 < \mathrm{height}(\mathfrak{m}, t^{-1})R^*(\mathfrak{a}) \le \dim R^*(\mathfrak{a}) - \dim R^*(\mathfrak{a})/(\mathfrak{m}, t^{-1}).$$

Because $R^*(\mathfrak{a})$ is a finitely generated $R(\mathfrak{a})$-module it implies that

$$\dim R^*(\mathfrak{a}) = \dim A + 1 \ \text{ and } \ \dim R^*(\mathfrak{a})/(\mathfrak{m}, t^{-1})R^*(\mathfrak{a}) = l(\mathfrak{a}),$$

which finally proves the claim a).

In order to prove b) first note that $\mathrm{height}(\mathfrak{m}, t^{-1})R^*(\mathfrak{a}) = \dim R^* - l(\mathfrak{a})$ since A is universally catenarian and $gr_A(\mathfrak{a})$ is unmixed. Since $(\mathfrak{a}^{n+1})^* :_A \mathfrak{a} = (\mathfrak{a}^n)^*$ for all $n \in \mathbb{N}$ there is no prime ideal $\mathfrak{P}$ of $R(\mathfrak{a})$ associated to $R^*(\mathfrak{a})/(t^{-1})R^*(\mathfrak{a})$ that contains $(\mathfrak{a}t, t^{-1})R(\mathfrak{a})$. Because of $(\mathfrak{a}^n)^* = \mathfrak{a}^n + 0 :_A \langle \mathfrak{a} \rangle$ for all sufficiently large n, see 2.13, it is easy to see that the kernel and the cokernel of the natural graded homomorphism

$$R(\mathfrak{a})/(t^{-1})R(\mathfrak{a}) \to R(\mathfrak{a})^*/(t^{-1})R(\mathfrak{a})^*$$

are finitely generated $R(\mathfrak{a})$-modules whose support is contained in $V((\mathfrak{a}t, t^{-1})R(\mathfrak{a}))$. This implies

$$\mathrm{Ass}\, R^*(\mathfrak{a})/(t^{-1})R^*(\mathfrak{a}) = \{\mathfrak{P} \in \mathrm{Ass}\, R(\mathfrak{a})/(t^{-1})R(\mathfrak{a}) \mid \mathfrak{P} \not\supseteq (\mathfrak{a}t, t^{-1})R(\mathfrak{a})\}$$

as is easily seen. By the assumption it follows that $R^*(\mathfrak{a})/(t^{-1})R^*(\mathfrak{a})$ is unmixed. Therefore

$$\mathrm{grade}(\mathfrak{m}, t^{-1})R^*(\mathfrak{a}) > 1 \ \text{ and } \ T_{(\mathfrak{m},t^{-1})}(R^*(\mathfrak{a})) = R^*(\mathfrak{a}),$$

see 2.8. By definition this means $(\mathfrak{a}^n)^* : \langle\mathfrak{m}\rangle = (\mathfrak{a}^n)^*$ for all $n \in \mathbb{N}$. Because of $(\mathfrak{a}^n)^* = \mathfrak{a}^n$ for all large n this proves the statement.   □

A corresponding result is true for the integral closures $\overline{\mathfrak{a}^n}$ of $\mathfrak{a}^n$. To this end let $\overline{R}(\mathfrak{a})$ denote the integral closure of $R(\mathfrak{a})$ in $A[t, t^{-1}]$. Then

$$\overline{R}(\mathfrak{a})_n = \begin{cases} \overline{\mathfrak{a}^n} & \text{for } n > 0, \\ A & \text{for } n \leq 0, \end{cases}$$

where $\overline{\mathfrak{a}^n}$ denotes the integral closure of $\mathfrak{a}^n$, i.e. the ideal of all elements $x \in A$ satisfying an equation $x^m + a_1 x^{m-1} + \ldots + a_m = 0$, where $a_i \in (\mathfrak{a}^n)^i, i = 1, \ldots, m$. Note that $\mathrm{Ass}\, A/\overline{\mathfrak{a}^n}$ is an increasing sequence that becomes eventually stable for large n, as shown by L. J. Ratliff, see [40]. Call $\overline{\mathrm{As}}(\mathfrak{a})$ the stable value.

Theorem 2.16. *Let $\mathfrak{a}$ denote an ideal of a local ring $(A, \mathfrak{m})$. Then the following conditions are true:*

  a) *If $\mathfrak{m} \notin \overline{\mathrm{As}}(\mathfrak{a})$, then $l(\mathfrak{a}) < \dim A$.*
  b) *The converse is true, provided A is a universally catenarian domain.*

Proof. First note that $\overline{\mathfrak{a}^n} : \langle\mathfrak{m}\rangle = \overline{\mathfrak{a}^n}$ for all $n \in \mathbb{N}$ since $\mathrm{Ass}\, A/\overline{\mathfrak{a}^n}, n \in \mathbb{N}$, forms an increasing sequence. Hence it follows that $T_{(\mathfrak{m}, t^{-1})}(\overline{R}(\mathfrak{a})) = \overline{R}(\mathfrak{a})$. By 2.8 it implies $\mathrm{height}(\mathfrak{m}, t^{-1})\overline{R}(\mathfrak{a}) > 1$. Therefore

$$1 < \mathrm{height}(\mathfrak{m}, t^{-1})R(\mathfrak{a}) \leq \dim R(\mathfrak{a}) - l(\mathfrak{a}),$$

which proves the claim.

In order to prove the converse first note that $\mathrm{height}(\mathfrak{m}, t^{-1})R(\mathfrak{a}) = \dim R(\mathfrak{a}) - l(\mathfrak{a})$, since A is a universally catenarian domain. Therefore the assumption implies that

$$1 < \mathrm{height}(\mathfrak{m}, t^{-1})R(\mathfrak{a}) = \mathrm{height}(\mathfrak{m}, t^{-1})\overline{R}(\mathfrak{a}).$$

But now $\overline{R}(\mathfrak{a})$ is a Krull domain. Hence any associated prime ideal of the principal ideal $(t^{-1})\overline{R}(\mathfrak{a})$ is of height 1. Whence by 2.8 it follows that $T_{(\mathfrak{m}, t^{-1})}(\overline{R}(\mathfrak{a})) = \overline{R}(\mathfrak{a})$. That is, $\overline{\mathfrak{a}^n} : \langle\mathfrak{m}\rangle = \overline{\mathfrak{a}^n}$ for all $n \in \mathbb{N}$, as required.   □

The statements of 2.16 were shown by J. Lipman, see [27]. It extends in a straightforward way to an ideal $\mathfrak{a}$ of an arbitrary local ring $(A, \mathfrak{m})$. This was done by S. McAdam in [29], see also [42] for a different approach. In order to describe this result let $\mathrm{mAss}\, A$ denote the set of minimal prime ideals of $\mathrm{Ass}\, A$.

Corollary 2.17. *Let $\mathfrak{a}$ denote an ideal of a local ring $(A, \mathfrak{a})$. Then $\mathfrak{m} \in \overline{\mathrm{As}}(\mathfrak{a})$ if and only if $l(\mathfrak{a}\widehat{A} + \mathfrak{p}/\mathfrak{p}) < \dim \widehat{A}/\mathfrak{p}$ for all $\mathfrak{p} \in \mathrm{mAss}\, \widehat{A}$.*

Proof. First note that $\mathfrak{m} \in \mathrm{Ass}\, A/\overline{\mathfrak{a}}$ if and only if $\mathfrak{m}\widehat{A} \in \mathrm{Ass}\, \widehat{A}/\overline{\mathfrak{a}\widehat{A}}$, see [40]. Furthermore $\mathfrak{m} \in \mathrm{Ass}\, A/\overline{\mathfrak{a}}$ if and only if there is a minimal prime ideal $\mathfrak{p} \in \mathrm{mAss}\, A$ such that $\mathfrak{m}/\mathfrak{p} \in \mathrm{Ass}(A/\mathfrak{p})/\overline{(\mathfrak{a}A/\mathfrak{p})}$, see e.g. [42]. So the claim follows by 2.16 since the ring $\widehat{A}/\mathfrak{p}$ is – as a complete local domain – a universally catenarian domain.   □

Some of the previous ideas will be applied to the comparison of the ordinary powers of an ideal $\mathfrak{a}$ to the S-symbolic powers $\{\mathfrak{a}_S^{(n)}\}_{n\in\mathbb{N}}$ for a multiplicatively closed subset S of the ring A. To this end use also the symbolic Rees ring $R_S(\mathfrak{a}) = \oplus_{n\in\mathbb{Z}}\mathfrak{a}_S^{(n)}t^n$ of $\mathfrak{a}$ with respect to S.

Corollary 2.18. *Let S denote a multiplicatively closed subset of A. Let $\mathfrak{a}$ denote a regular ideal of A. Suppose that the following conditions are satisfied:*

a) *A is a universally catenarian domain,*
b) *$\operatorname{depth} gr_A(\mathfrak{a})_{\mathfrak{P}} \geq \min\{1, \dim gr_A(\mathfrak{a})_{\mathfrak{P}}\}$ for all $\mathfrak{P} \not\supseteq gr_A(\mathfrak{a})_+$, and*
c) *$l(\mathfrak{a}A_{\mathfrak{p}}) < \dim A_{\mathfrak{p}}$ for all $\mathfrak{p} \in As(\mathfrak{a})$ with $\mathfrak{p} \cap S \neq \emptyset$.*

Then $\mathfrak{a}^n = \mathfrak{a}_S^{(n)}$ for all sufficiently large $n \in \mathbb{N}$.

Proof. As shown in the proof of Theorem 2.15 the assumption b) implies that $R^*(\mathfrak{a})/(t^{-1})R^*(\mathfrak{a})$ is unmixed. Furthermore recall that

$$\operatorname{Ass} \mathfrak{a}_S^{(n)}/\mathfrak{a}^n = \{\mathfrak{p} \in \operatorname{Ass} A/\mathfrak{a}^n \mid \mathfrak{p} \cap S \neq \emptyset\}.$$

Therefore, for large n the set $\operatorname{Ass} \mathfrak{a}_S^{(n)}/\mathfrak{a}^n$ will stabilize to a finite set, say $T(\mathfrak{a})$. The claim says that $T(\mathfrak{a}) = \emptyset$. Suppose that $T(\mathfrak{a}) \neq \emptyset$. Now recall that the claim is a local question. Hence without loss of generality we may assume that $(A, \mathfrak{m})$ is a local ring and $T(\mathfrak{a}) = \{\mathfrak{m}\}$. Whence $\mathfrak{a}_S^{(n)} = \mathfrak{a}^n : \langle \mathfrak{m} \rangle$ for all large $n \in \mathbb{N}$.

But now investigate $R^*(\mathfrak{a})$ and $T_{(\mathfrak{m},t^{-1})}(R^*(\mathfrak{a}))$. Since A is universally catenarian and $R^*(\mathfrak{a})/(t^{-1})R^*(\mathfrak{a})$ is unmixed it follows by c) that

$$1 < \dim R^*(\mathfrak{a}) - l(\mathfrak{a}) = \operatorname{height}(\mathfrak{m}, t^{-1})R^*(\mathfrak{a}).$$

Therefore $T_{(\mathfrak{m},t^{-1})}(R^*(\mathfrak{a})) = R^*(\mathfrak{a})$ and $(\mathfrak{a}^n)^* : \langle \mathfrak{m} \rangle = (\mathfrak{a}^n)^*$ for all $n \in \mathbb{N}$. Moreover $(\mathfrak{a}^n)^* = \mathfrak{a}^n$ for all large n. Putting together all of these equalities it follows that $T(\mathfrak{a}) = \emptyset$, contracting the choice of $\mathfrak{m}$. $\square$

Suppose that condition b) in 2.18 holds for any homogeneous prime ideal. That means that $gr_A(\mathfrak{a})$ is unmixed with respect to the height. Then the conclusion of 2.18 holds for all $n \in \mathbb{N}$. This follows by a slight modification of the proof of 2.18. To this end one has to replace $R^*(\mathfrak{a})$ by $R(\mathfrak{a})$.

In order to conclude with this section let us relate the finiteness conditions of the symbolic Rees ring $R_S(\mathfrak{a})$ to the existence of an ideal $\mathfrak{b}$ whose n-th symbolic power with respect to S coincides with its ordinary power for all large $n \in \mathbb{N}$.

Theorem 2.19. *Let $\mathfrak{a}$ resp. S denote an ideal resp. a multiplicatively closed subset of a commutative Noetherian ring A.*

a) *$R_S(\mathfrak{a})$ is a finitely generated $R(\mathfrak{a})$-module if and only if $l(\mathfrak{a}\widehat{A}_P + \mathfrak{p}/\mathfrak{p}) < \dim \widehat{A}_P/\mathfrak{p}$ for all prime ideals $P \in As(\mathfrak{a})$ such that $P \cap S \neq \emptyset$ and all $\mathfrak{p} \in \operatorname{Ass} \widehat{A}_P$.*
b) *$R_S(\mathfrak{a})$ is an A-algebra of finite type if and only if there is a $k \in \mathbb{N}$ such that $\mathfrak{b}^n = \mathfrak{b}_S^{(n)}$ for all large $n \in \mathbb{N}$, where $\mathfrak{b} = \mathfrak{a}_S^{(k)}$.*

Proof. Firstly show a). Suppose that $R_S(\mathfrak{a})$ is a finitely generated $R(\mathfrak{a})$-module. Let $P \in \mathrm{As}(\mathfrak{a})$ denote a prime ideal such that $P \cap S \neq \emptyset$. Then $\mathfrak{a}^n A_P : \langle P A_P \rangle \subseteq \mathfrak{a}_S^{(n)} A_P$ for all $n \in \mathbb{Z}$. Therefore $T_{(P A_P, t^{-1})}(R(\mathfrak{a} A_P))$ is – as a submodule of $R_{S A_P}(\mathfrak{a} A_P)$ – a finitely generated $R(\mathfrak{a} A_P)$-module. Therefore 2.14 proves the 'only if' part of the claim.

In order to prove the reverse implication note that by the Artin-Rees Lemma it will be enough to show that there is a $k \in \mathbb{N}$ such that the module

$$E_{k,n} = (\mathfrak{a}_S^{(n+k)} + \mathfrak{a}^n)/\mathfrak{a}^n$$

vanishes for all $n \geq 1$. The set of associated prime ideals of $E_{k,n}$ is contained in the finite set $X = \cup_{n \geq 1}\{P \in \mathrm{Ass}\, A/\mathfrak{a}^n \mid P \cap S \neq \emptyset\}$. Therefore the vanishing of $E_{k,n}$ is a local question for finitely many prime ideals in X. By induction it will be enough to prove the vanishing of $E_{k,n}$ at the localization with respect to a minimal prime ideal in X. By changing the notation let $(A, \mathfrak{m})$ denote the local ring at this localization. Because of the choice of $\mathfrak{m}$ it implies that $\mathfrak{a}_S^{(n)} = \mathfrak{a}^n :_A \langle \mathfrak{m} \rangle$ for all $n \in \mathbb{N}$. By 2.14 it follows that $T_{(\mathfrak{m}, t^{-1})}(R(\mathfrak{a}))$ is a finitely generated $R(\mathfrak{a})$-module. Therefore $\{\mathfrak{m}\} \notin \mathrm{Ass}\, E_{k,n}$ for a certain $k \in \mathbb{N}$ and all $n \geq 1$, i.e. $E_{k,n} = (0)$, as required.

Finally show b). The claim is an easy consequence of 2.11. Recall that $\mathfrak{b}^n = \mathfrak{a}_S^{(nk)}$ if and only if $\mathfrak{b}^n = \mathfrak{b}_S^{(n)}$. □

2.4. The Lichtenbaum-Hartshorne Vanishing Theorem.

The Lichtenbaum-Hartshorne vanishing theorem for local cohomology, see [16], characterizes the vanishing of $H_\mathfrak{a}^d(A)$ for an ideal $\mathfrak{a}$ in a d-dimensional local ring $(A, \mathfrak{m})$. Our proof yields an essential simplification by the use of ideal topologies.

For a finitely generated d-dimensional A-module M let $(\mathrm{Ass}\, M)_d$ denote all the associated prime ideals of M with $\dim A/\mathfrak{p} = d$. For an ideal $\mathfrak{a}$ of A let $\mathfrak{u} = \mathfrak{u}(\mathfrak{a}\widehat{A})$ denote the intersection of those primary components $\mathfrak{q}$ of 0 in $\widehat{A}$ such that $\dim \widehat{A}/(\mathfrak{a}\widehat{A} + \mathfrak{p}) > 0$ for $\mathfrak{p} \in (\mathrm{Ass}\,\widehat{A})_d$, where $\mathfrak{p} = \mathrm{Rad}\,\mathfrak{q}$ and $d = \dim A$.

Theorem 2.20. *Let $\mathfrak{a}$ denote an ideal in a d-dimensional local ring $(A, \mathfrak{m})$. Then $H_\mathfrak{a}^d(A) \simeq \mathrm{Hom}_A(\mathfrak{u}, E)$, where E denotes the injective hull of the residue field $A/\mathfrak{m}$. In particular $H_\mathfrak{a}^d(A)$ is an Artinian A-module and $H_\mathfrak{a}^d(A) = 0$ if and only if $\dim \widehat{A}/(\mathfrak{a}\widehat{A} + \mathfrak{p}) > 0$ for all $\mathfrak{p} \in (\mathrm{Ass}\,\widehat{A})_d$.*

Proof. As above let $T = \mathrm{Hom}_A(\cdot, E)$ denote the Matlis duality functor. Because of the following isomorphisms

$$T(H_\mathfrak{a}^d(A)) \simeq T(H_\mathfrak{a}^d(A) \otimes_A \widehat{A}) \simeq T(H_{\mathfrak{a}\widehat{A}}^d(\widehat{A}))$$

one may assume without loss of generality that A is a complete local ring. So A is the factor ring of a complete local Gorenstein ring $(B, \mathfrak{n})$ with $\dim A = \dim B = d$, say $A = B/\mathfrak{b}$. Replacing $\mathfrak{a}$ by its preimage in B we have to consider $T(H_\mathfrak{a}^d(B/\mathfrak{b}))$. Let $\mathfrak{b}_d$ denote the intersection of all of the primary components $\mathfrak{q}$ of $\mathfrak{b}$ such that $\dim B/\mathfrak{p} = d$ for $\mathfrak{p}$ its associated prime ideal. Because of $\dim \mathfrak{b}_d/\mathfrak{b} < d$ the short exact sequence

$$0 \to \mathfrak{b}_d/\mathfrak{b} \to B/\mathfrak{b} \to B/\mathfrak{b}_d \to 0$$

implies that $H_{\mathfrak{a}}^d(B/\mathfrak{b}) \simeq H_{\mathfrak{a}}^d(B/\mathfrak{b}_d)$. Replacing $\mathfrak{b}_d$ by $\mathfrak{b}$ one may assume that $B/\mathfrak{b}$ is unmixed with respect to the dimension. There is an isomorphism

$$T(H_{\mathfrak{a}}^d(B) \otimes B/\mathfrak{b}) \simeq \operatorname{Hom}_B(B/\mathfrak{b}, T(H_{\mathfrak{a}}^d(B))).$$

Because the Hom-functor transforms direct into inverse limits it turns out that

$$T(H_{\mathfrak{a}}^d(B)) \simeq \varprojlim T(\operatorname{Ext}_B^d(B/\mathfrak{a}^n, B)) \simeq \varprojlim H_{\mathfrak{n}}^0(B/\mathfrak{a}^n),$$

as follows by the local duality, see 1.8. Because of $H_{\mathfrak{n}}^0(B/\mathfrak{a}^n) = \mathfrak{a}^n : \langle \mathfrak{n} \rangle / \mathfrak{a}^n$ and because B is a complete local ring we see that $\varprojlim H_{\mathfrak{n}}^0(B/\mathfrak{a}^n) \simeq \cap_{n \in \mathbb{N}}(\mathfrak{a}^n : \langle \mathfrak{n} \rangle)$. But now the ideal $\cap_{n \in \mathbb{N}}(\mathfrak{a}^n : \langle \mathfrak{n} \rangle)$ is the ideal $\mathfrak{v}$ of B that is the intersection of all primary components $\mathfrak{q}$ of 0 such that $\dim B/(\mathfrak{a} + \mathfrak{p}) > 0$ for $\mathfrak{p}$ the associated prime ideal of $\mathfrak{q}$, see 2.3. Therefore

$$T(H_{\mathfrak{a}}^d(B) \otimes_B B/\mathfrak{b}) \simeq \operatorname{Hom}_B(B/\mathfrak{b}, \mathfrak{v}) \simeq (0 :_B \mathfrak{b}) \cap \mathfrak{v}.$$

Furthermore it follows that $(0 :_B \mathfrak{b}) \cap \mathfrak{b} = 0$ since $\mathfrak{b}$ is an unmixed ideal in a Gorenstein ring B with $\dim B = B/\mathfrak{b}$. Therefore $(0 :_B \mathfrak{b}) \cap \mathfrak{v} \simeq ((0 :_B \mathfrak{b}) \cap \mathfrak{v} + \mathfrak{b})/\mathfrak{b}$. Hence $(0 :_B \mathfrak{b}) \cap \mathfrak{v}$ is isomorphic to an ideal of $B/\mathfrak{b}$. Finally note that $\mathfrak{u} \simeq (0 :_B \mathfrak{b}) \cap \mathfrak{v}$ as follows by considering the set of associated prime ideals. Then the statement is a consequence of the Matlis duality. $\qquad\square$

The vanishing of the ideal $\mathfrak{u}$ is equivalent to the equivalence of certain ideal topologies, see 2.2. So there is another characterization of the vanishing of $H_{\mathfrak{a}}^d(A)$ for certain local rings.

Corollary 2.21. *Suppose that $(A, \mathfrak{m})$ denotes a formally equidimensional local ring. Then $H_{\mathfrak{a}}^d(A) = 0$, $d = \dim A$, if and only if the topology defined by $\{\mathfrak{a}^n : \langle \mathfrak{m} \rangle\}_{n \in \mathbb{N}}$ is equivalent to the $\mathfrak{a}$-adic topology.*

Proof. By virtue of 2.2 $\{\mathfrak{a}^n : \langle \mathfrak{m} \rangle\}_{n \in \mathbb{N}}$ is equivalent to the $\mathfrak{a}$-adic topology if and only if $\dim \widehat{A}/(\mathfrak{a}\widehat{A} + \mathfrak{p}) > 0$ for all $\mathfrak{p} \in \operatorname{Ass} \widehat{A}$. But now $\dim A = \dim \widehat{A}/\mathfrak{p}$ for all $\mathfrak{p} \in \operatorname{Ass} \widehat{A}$ by the assumption on A. So the claim follows by 2.20. $\qquad\square$

2.5. Connectedness Results. Let $\mathfrak{a}, \mathfrak{b}$ denote two ideals of a commutative Noetherian ring A. Then there is a short exact sequence

$$0 \to A/\mathfrak{a} \cap \mathfrak{b} \xrightarrow{i} A/\mathfrak{a} \oplus A/\mathfrak{b} \xrightarrow{p} A/(\mathfrak{a} + \mathfrak{b}) \to 0,$$

where $i(a + \mathfrak{a} \cap \mathfrak{b}) = (a + \mathfrak{a}, -a + \mathfrak{b})$ and $p(a + \mathfrak{a}, b + \mathfrak{b}) = a + b + (\mathfrak{a} + \mathfrak{b})$ for $a, b \in A$. Because of the direct summand in the middle this sequence provides a helpful tool for connecting properties. This short exact sequence is an important ingredient for the next lemma.

In order to prove the connectedness theorem we need some preparations. A basic tool for this section will be the so-called Mayer-Vietoris sequence for local cohomology helpful also for different purposes.

Lemma 2.22. *Let* $\mathfrak{a}, \mathfrak{b}$ *denote two ideals of a commutative Noetherian ring* A. *Then there is a functorial long exact sequence*

$$\dots \to H_{\mathfrak{a}+\mathfrak{b}}^n(M) \to H_{\mathfrak{a}}^n(M) \oplus H_{\mathfrak{b}}^n(M) \to H_{\mathfrak{a}\cap\mathfrak{b}}^n(M) \to H_{\mathfrak{a}+\mathfrak{b}}^{n+1}(M) \to \dots$$

for any A-*module* M.

Proof. Let $\mathfrak{c}$ denote an ideal of the ring A. Then first note that

$$\varinjlim \operatorname{Ext}^n(A/\mathfrak{c}^n, M) \simeq H_{\mathfrak{c}}^n(M)$$

for any A-module M and all $n \in \mathbb{Z}$, see e.g. [14]. Now consider the short exact sequence at the beginning of this subsection for the ideals $\mathfrak{a}^n$ and $\mathfrak{b}^n$. Then take into account that the topologies defined by the families $\{\mathfrak{a}^n + \mathfrak{b}^n\}_{n\in\mathbb{N}}$ and $\{\mathfrak{a}^n \cap \mathfrak{b}^n\}_{n\in\mathbb{N}}$ are equivalent to the $(\mathfrak{a}+\mathfrak{b})$-adic and $\mathfrak{a}\cap\mathfrak{b}$-adic topology resp. Therefore the direct limit of the long exact Ext-sequence proves the claim. $\qquad\square$

Our first connectedness result is the following statement, a slight generalization of Hartshorne's connectedness result, see [15].

Theorem 2.23. *Let* $\mathfrak{c}$ *denote an ideal of a local ring* $(A, \mathfrak{m})$. *Suppose that* grade $\mathfrak{c} >$ 1. *Then the scheme* $\operatorname{Spec} A \setminus V(\mathfrak{c})$ *is connected.*

Proof. Because of grade $\mathfrak{c} > 1$ it follows that $H_{\mathfrak{c}}^i(A) = 0$ for $i = 0, 1$. Assume that $\operatorname{Spec} A \setminus V(\mathfrak{c})$ is not connected. Then there are non-nilpotent ideals $\mathfrak{a}, \mathfrak{b}$ satisfying the following properties:

1) $\mathfrak{a} \cap \mathfrak{b}$ is nilpotent,
2) $\operatorname{Spec} A \setminus V(\mathfrak{a})$ and $\operatorname{Spec} A \setminus V(\mathfrak{b})$ are disjoint and non-empty subsets of $\operatorname{Spec} A$.
3) $\operatorname{Spec} A \setminus V(\mathfrak{c}) = (\operatorname{Spec} A \setminus V(\mathfrak{a})) \cup (\operatorname{Spec} A \setminus V(\mathfrak{b}))$

Note that these conditions imply that $\operatorname{Rad}(\mathfrak{a} + \mathfrak{b}) = \operatorname{Rad}\mathfrak{c}$. Now consider the first part of the Mayer-Vietoris sequence

$$0 \to H_{\mathfrak{a}+\mathfrak{b}}^0(A) \to H_{\mathfrak{a}}^0(A) \oplus H_{\mathfrak{b}}^0(A) \to H_{\mathfrak{a}\cap\mathfrak{b}}^0(A) \to H_{\mathfrak{a}+\mathfrak{b}}^1(A).$$

Because of grade $\mathfrak{c} > 1$ and $\operatorname{Rad}(\mathfrak{a} + \mathfrak{b}) = \operatorname{Rad}\mathfrak{c}$ it turns out that $H_{\mathfrak{a}+\mathfrak{b}}^i(A) = 0$ for $i = 0, 1$. Moreover $\mathfrak{a} \cap \mathfrak{b}$ is nilpotent. Whence it yields that $H_{\mathfrak{a}\cap\mathfrak{b}}^0(A) = A$. So the Mayer-Vietoris sequence implies an isomorphism $H_{\mathfrak{a}}^0(A) \oplus H_{\mathfrak{b}}^0(A) \simeq A$. Since the ring A – as a local ring – is indecomposable it follows either $H_{\mathfrak{a}}^0(A) = A$ and $H_{\mathfrak{b}}^0(A) = 0$ or $H_{\mathfrak{a}}^0(A) = 0$ and $H_{\mathfrak{b}}^0(A) = A$. But this means that $\mathfrak{a}$ resp. $\mathfrak{b}$ is a nilpotent ideal. Therefore we have a contradiction, so $\operatorname{Spec} A \setminus V(\mathfrak{c})$ is connected. $\qquad\square$

The author is grateful to Leif Melkersson for suggesting the above simplification of the original arguments.

Let $\mathfrak{a}$ denote the homogeneous ideal in $A = k[x_0, \dots, x_3]$ describing the union of two disjoint lines in $\mathbb{P}_k^3$. Suppose that $\mathfrak{a}$ is up to the radical equal to an ideal $\mathfrak{c}$ generated by two elements. Then $\operatorname{Spec} A \setminus V(\mathfrak{a}) = \operatorname{Spec} A \setminus V(\mathfrak{c})$ is disconnected. Therefore grade $\mathfrak{c} \leq 1$, contradicting the fact that $\mathfrak{c}$ is an ideal of height 2 in a Cohen-Macaulay ring A. So $\mathfrak{a}$ is not set-theoretically a complete intersection. For further examples of this type see [15].

The previous result implies as a corollary a result on the length of chains of prime ideals in a catenarian local ring.

Corollary 2.24. *Let* $(A, \mathfrak{m})$ *denote a local Noetherian ring satisfying the condition* S_2*. Suppose that* A *is catenarian. Then it is equidimensional, i.e. all of the minimal prime ideals have the same dimension.*

Proof. Let $\mathfrak{p}, \mathfrak{q}$ denote two minimal prime ideals of Spec A. Then it is easily seen that there is a chain of prime ideals

$$\mathfrak{p} = \mathfrak{p}_1, \ldots, \mathfrak{p}_r = \mathfrak{q}$$

such that $\operatorname{height}(\mathfrak{p}_i, \mathfrak{p}_{i+1}) = 1$ for all $i = 1, \ldots, r - 1$. Hence by the catenarian condition

$$\dim A/\mathfrak{p}_i = 1 + \dim A/(\mathfrak{p}_i, \mathfrak{p}_{i+1}) = \dim A/\mathfrak{p}_{i+1}.$$

By iterating this $(r - 1)$-times it follows that $\dim A/\mathfrak{p} = \dim A/\mathfrak{q}$, as required. $\square$

In order to prove the connectedness theorem inspired by G. Faltings we need a lemma first invented by M. Brodmann and J. Rung, see [5].

Lemma 2.25. *Let* $(A, \mathfrak{m})$ *denote an analytically irreducible domain with* $d = \dim A > 1$*. Suppose there are two ideals* $\mathfrak{b}, \mathfrak{c}$ *of* A *such that* $\dim A/\mathfrak{b} > 0, \dim A/\mathfrak{c} > 0$*, and* $H^{d-1}_{\mathfrak{b} \cap \mathfrak{c}}(A) = 0$*. Then* $\dim A/\mathfrak{b} + \mathfrak{c} > 0$*.*

Proof. Suppose the contrary is true. Then $\mathfrak{b} + \mathfrak{c}$ is an $\mathfrak{m}$-primary ideal. The Mayer-Vietoris sequence provides an exact sequence

$$H^{d-1}_{\mathfrak{b} \cap \mathfrak{c}}(A) \to H^d_{\mathfrak{m}}(A) \to H^d_{\mathfrak{b}}(A) \oplus H^d_{\mathfrak{c}}(A) \to H^d_{\mathfrak{b} \cap \mathfrak{c}}(A).$$

Because A is analytically irreducible and because of the vanishing of $H^{d-1}_{\mathfrak{b} \cap \mathfrak{c}}(A)$ the vanishing result 2.20 yields an isomorphism $H^d_{\mathfrak{m}}(A) \simeq H^d_{\mathfrak{b}}(A) \oplus H^d_{\mathfrak{c}}(A)$. Because of the non-vanishing of $H^d_{\mathfrak{m}}(A)$ this provides the non-vanishing of one of the direct summands, say $H^d_{\mathfrak{b}}(A)$. By 2.20 this means that $\mathfrak{b}$ is an $\mathfrak{m}$-primary ideal, contradicting the assumption. $\square$

This lemma is the main technical tool for the connectedness result given in the sequel.

Theorem 2.26. *Suppose* $\mathfrak{a}$ *denotes an ideal of an analytically irreducible domain* $(A, \mathfrak{m})$ *with* $d = \dim A > 1$*. Suppose that* $H^n_{\mathfrak{a}}(A) = 0$ *for* $n = d - 1, d$*. Then* $(\operatorname{Spec} A/\mathfrak{a}) \setminus V(\mathfrak{m}/\mathfrak{a})$ *is connected.*

Proof. Suppose the contrary. Then there exist ideals $\mathfrak{b}, \mathfrak{c}$ of A such that $\operatorname{Rad}(\mathfrak{b} \cap \mathfrak{c}) = \operatorname{Rad} \mathfrak{a}$ and $\mathfrak{b} + \mathfrak{c}$ is $\mathfrak{m}$-primary, but neither $\mathfrak{b}$ nor $\mathfrak{c}$ is an $\mathfrak{m}$-primary ideal. Because of the vanishing of $H^{d-1}_{\mathfrak{a}}(A)$ Lemma 2.25 provides a contradiction. $\square$

The preceding result shows for instance the non-vanishing of $H^3_{\mathfrak{a}}(A)$ for the ideal $\mathfrak{a}$ of the union of two disjoint lines in $\mathbb{P}^3_k$ and $A = k[x_0, \ldots, x_3]$. This yields another proof that $\mathfrak{a}$ is not set-theoretically a complete intersection.

In the following let us generalize this connectedness result to the case of local rings that are not necessarily analytically unmixed. This was obtained by C. Huneke and M. Hochster, see [20], by a different argument.

Theorem 2.27. *Let $(A, \mathfrak{m})$ denote a d-dimensional local ring which is the quotient of a Gorenstein ring. Assume that A satisfies the condition S_2. Suppose that $\mathfrak{a}$ denotes an ideal of A such that $H_{\mathfrak{a}}^n(A) = 0$ for $n = d - 1, d$. Then the scheme $(\operatorname{Spec} A/\mathfrak{a}) \setminus V(\mathfrak{m}/\mathfrak{a})$ is connected.*

Proof. As in the proof of 2.26 suppose the contrary. That is, there exist ideals $\mathfrak{b}, \mathfrak{c}$ of A such that $\operatorname{Rad}(\mathfrak{b} \cap \mathfrak{c}) = \operatorname{Rad} \mathfrak{a}$ and $\mathfrak{b} + \mathfrak{c}$ is $\mathfrak{m}$-primary, but neither $\mathfrak{b}$ nor $\mathfrak{c}$ is an $\mathfrak{m}$-primary ideal. By changing the notation let us assume that A is a complete local ring. Then the Mayer-Vietoris sequence provides an isomorphism

$$H_{\mathfrak{m}}^d(A) \simeq H_{\mathfrak{b}}^d(A) \oplus H_{\mathfrak{c}}^d(A).$$

This implies that $K_A \simeq \mathfrak{u} \oplus \mathfrak{v}$, where $\mathfrak{u}$ resp. $\mathfrak{v}$ denotes the intersection of those primary components $\mathfrak{q}$ of the zero-ideal of A such that $\dim A/\mathfrak{b} + \mathfrak{p} > 0$ resp. $\dim A/\mathfrak{c} + \mathfrak{p} > 0$ for $\mathfrak{p}$, the associated prime ideal of $\mathfrak{q}$, see 2.20. By the definitions, see Section 1.2, it follows now that $K_{K_A} \simeq K_{\mathfrak{u}} \oplus K_{\mathfrak{v}}$. Moreover since A satisfies S_2 it is equidimensional by 2.22. Therefore $A \simeq K_{K_A}$, as turns out by 1.14.

By the Nakayama lemma one of the direct summands, say $K_{\mathfrak{v}}$, is zero, while for the second summand $A \simeq K_{\mathfrak{u}}$. By 1.9 it follows that $\operatorname{Ass} K_{\mathfrak{u}} = \operatorname{Ass} \mathfrak{u}$. Because of

$$\operatorname{Ass}_A \mathfrak{u} = \{\mathfrak{p} \in \operatorname{Ass} A \mid \dim A/\mathfrak{b} + \mathfrak{p} = 0\}$$

the equality $\operatorname{Ass} A = \operatorname{Ass} K_{\mathfrak{u}}$ implies that $\mathfrak{m} \subseteq \operatorname{Rad}(\mathfrak{b} + \mathfrak{p})$ for any associated prime ideal $\mathfrak{p}$ of A. As is easily seen it follows that $\mathfrak{m} \subseteq \operatorname{Rad}(\mathfrak{b})$, a contradiction. Therefore $(\operatorname{Spec} A/\mathfrak{a}) \setminus V(\mathfrak{m}/\mathfrak{a})$ is connected, as required. $\square$

3. Local Cohomology and Syzygies

3.1. Local Cohomology and Tor's. As above let $(A, \mathfrak{m})$ denote a local ring. Let $T(\cdot) = \operatorname{Hom}_A(\cdot, E)$ denote the Matlis duality functor, where $E = E_A(A/\mathfrak{m})$ is the injective hull of the residue field. In the following consider a length estimate for the length of $\operatorname{Tor}_n^A(M, N)$ resp. $\operatorname{Ext}_A^n(M, N)$ under the additional assumption that $M \otimes_A N$ is an A-module of finite length.

Lemma 3.1. *Let M, N denote finitely generated A-modules such that $M \otimes_A N$ is an A-module of finite length. Then*

$$\operatorname{Ext}_A^i(M, H_{\mathfrak{m}}^j(N)) \quad \text{and} \quad \operatorname{Tor}_i^A(M, H_{\mathfrak{m}}^j(N))$$

are A-modules of finite length for all $i, j \in \mathbb{Z}$.

Proof. Without loss of generality we may assume that A is a complete local ring. Then A is a quotient of a local Gorenstein ring B with $\dim B = r$. By the local duality theorem, see 1.8, it turns out that $H_{\mathfrak{m}}^j(N) \simeq T(K_N^j)$ for all $j \in \mathbb{Z}$, where

$K_N^j \simeq \operatorname{Ext}_B^{r-j}(N, B)$ with the natural A-module structure, see 1.8. Moreover there are natural isomorphisms

$$\operatorname{Ext}_A^i(M, H_{\mathfrak{m}}^j(N)) \simeq T(\operatorname{Tor}_i^A(M, K_N^j)) \text{ and } \operatorname{Tor}_i^A(M, H_{\mathfrak{m}}^j(N)) \simeq T(\operatorname{Ext}_A^i(M, K_N^j))$$

for all $i, j \in \mathbb{Z}$. But now $\operatorname{Supp} K_N^j \subseteq V(\operatorname{Ann}_A N)$ for all $j \in \mathbb{Z}$. Therefore

$$\operatorname{Supp} \operatorname{Tor}_i^A(M, K_N^j) \subseteq V(\operatorname{Ann}_A M, \operatorname{Ann}_A N).$$

By the assumption $M \otimes_A N$ is an A-module of finite length. That is,

$$V(\operatorname{Ann}_A M, \operatorname{Ann}_A N) \subseteq V(\mathfrak{m}),$$

which proves that $\operatorname{Tor}_i^A(M, K_N^j)$ is an A-module of finite length for all $i, j \in \mathbb{Z}$ too. By the Matlis duality the first part of the claim is shown. The second part follows by the same argument. $\qquad\square$

The previous result 3.1 provides the desired bounds for the length of $\operatorname{Tor}_n^A(M, N)$ and $\operatorname{Ext}_A^n(M, N)$.

Theorem 3.2. *Let M, N be two finitely generated A-modules such that $M \otimes_A N$ is an A-module of finite length. Then*

  a) $L_A(\operatorname{Ext}_A^n(M, N)) \leq \sum_{i \geq 0} L_A(\operatorname{Ext}_A^i(M, H_{\mathfrak{m}}^{n-i}(N)))$ *and*
  b) $L_A(\operatorname{Tor}_n^A(M, N)) \leq \sum_{i \geq 0} L_A(\operatorname{Tor}_{n+i}^A(M, H_{\mathfrak{m}}^i(N)))$

for all $n \in \mathbb{Z}$.

Proof. First choose $\underline{x} = x_1, \ldots, x_d$, $d = \dim A$, a system of parameters of A. Therefore $\operatorname{Rad} \underline{x} = \mathfrak{m}$. The corresponding Čech complex $K^{\cdot} = K_{\underline{x}}^{\cdot}$ has the property that $H^n(K^{\cdot} \otimes_A N) \simeq H_{\mathfrak{m}}^n(N)$ for all $n \in \mathbb{Z}$, see 1.3. Furthermore choose $F^{\cdot}$ a minimal free resolution of M. In order to show the first claim consider the complex $K^{\cdot} \otimes_A \operatorname{Hom}_A(F^{\cdot}, N)$. Because of the structure of K^i as the direct sum of localizations it turns out that the natural homomorphism

$$\operatorname{Hom}_A(F^{\cdot}, N) \to K^{\cdot} \otimes_A \operatorname{Hom}_A(F^{\cdot}, N)$$

induces an isomorphism in cohomology. Moreover

$$K^{\cdot} \otimes_A \operatorname{Hom}_A(F^{\cdot}, N) \xrightarrow{\sim} \operatorname{Hom}_A(F^{\cdot}, K^{\cdot} \otimes_A N)$$

as it is easily seen. So there is a spectral sequence

$$E_2^{i,j} = \operatorname{Ext}_A^i(M, H_{\mathfrak{m}}^j(N)) \Rightarrow E^n = \operatorname{Ext}_A^n(M, N).$$

Therefore $\operatorname{Ext}_A^n(M, N)$ possesses a finite filtration whose quotients $E_\infty^{i,n-i}$ are modules of finite length such that $L_A(E_\infty^{i,n-i}) \leq L_A(E_2^{i,n-i}) < \infty$, which proves the first bound.

  In order to prove the second bound proceed by a similar argument. Consider the complex $K^{\cdot} \otimes_A (F^{\cdot} \otimes_A N)$. As above the natural map $F^{\cdot} \otimes_A N \to K^{\cdot} \otimes_A (F^{\cdot} \otimes_A N)$ induces an isomorphism in cohomology. In order to continue consider the spectral sequence

$$E_2^{i,j} = \operatorname{Tor}_{-i}^A(M, H_{\mathfrak{m}}^j(N)) \Rightarrow E^n = \operatorname{Tor}_{-n}^A(M, N)$$

for computing the cohomology of $K^{\cdot} \otimes_A (F^{\cdot} \otimes_A N)$. It provides – in a similar way as above – the second claim. $\square$

In the particular case of N a Cohen-Macaulay module with $M \otimes_A N$ of finite length the spectral sequences in the proof of 3.2 degenerate to isomorphisms.

Corollary 3.3. *Let N be a Cohen-Macaulay module. Then there are the following isomorphisms*

$$\operatorname{Ext}_A^n(M, N) \simeq \operatorname{Ext}_A^{n-d}(M, H_{\mathfrak{m}}^d(N)) \ \text{and} \ \operatorname{Tor}_n^A(M, N) \simeq \operatorname{Tor}_{n+d}^A(M, H_{\mathfrak{m}}^d(N))$$

for all $n \in \mathbb{Z}$, where $d = \dim N$.

Under the additional assumption that M is an A-module of finite projective dimension it is of some interest to determine the largest integer n such that $\operatorname{Tor}_n^A(M, N) \neq 0$. This yields an equality of the Auslander-Buchsbaum type, shown by M. Auslander, see [1, Theorem 1.2].

Theorem 3.4. *Let M, N be two non-zero finitely generated A-modules. Suppose that $\operatorname{pd}_A M$ is finite. Then*

$$\sup\{n \in \mathbb{Z} \mid \operatorname{Tor}_n^A(M, N) \neq 0\} + \operatorname{depth}_A N = \operatorname{pd}_A M$$

provided $\operatorname{depth} \operatorname{Tor}_s^A(M, N) = 0$, where $s = \sup\{n \in \mathbb{Z} \mid \operatorname{Tor}_n^A(M, N) \neq 0\}$. In particular the equality holds whenever $M \otimes_A N$ is an A-module of finite length.

Proof. Set $p = \operatorname{pd}_A M$ and $t = \operatorname{depth}_A N$. As in the proof of 3.2 consider the complex $C^{\cdot} := K^{\cdot} \otimes_A F^{\cdot} \otimes_A N$, where $K^{\cdot}$ resp. $F^{\cdot}$ denotes the Čech complex resp. the (finite) minimal free resolution of M. Then there is the following spectral sequence

$$E_2^{-i,j} = \operatorname{Tor}_i^A(M, H_{\mathfrak{m}}^j(N)) \Rightarrow E^{-i+j} = H^{-i+j}(C^{\cdot}).$$

Consider the stages $-i + j =: n \leq -p + t$. In the case $n < -p + t$ it follows that $E_2^{-i,j} = 0$. Note that whenever $j < t$, then $H_{\mathfrak{m}}^j(N) = 0$, and whenever $j \geq t$, then $i > p = \operatorname{pd}_A M$. In the case $n = -p + t$ it follows by a similar consideration that $E_2^{-i,j} = 0$ for $i \neq p$. So there is a partial degeneration to the isomorphism $H^{-p+t}(C^{\cdot}) \simeq \operatorname{Tor}_p^A(M, H_{\mathfrak{m}}^t(N))$ and the vanishing $H^n(C^{\cdot}) = 0$ for all $n < -p + t$.

Next show that $H^{-p+t}(C^{\cdot}) \neq 0$. By 1.5 $H_{\mathfrak{m}}^t(N)$ is an Artinian A-module. Therefore it possesses a submodule which is isomorphic to $k = A/\mathfrak{m}$. The corresponding short exact sequence

$$0 \to k \to H_{\mathfrak{m}}^t(N) \to C \to 0$$

induces an injection $0 \to \operatorname{Tor}_p^A(M, k) \to \operatorname{Tor}_p^A(M, H_{\mathfrak{m}}^t(N))$. Because of $\operatorname{Tor}_p^A(M, k) \neq 0$ this shows the claim.

In order to continue with the proof consider the spectral sequence

$$E_2^{i,-j} = H_{\mathfrak{m}}^i(\operatorname{Tor}_j^A(M, N)) \Rightarrow E^{i-j} = H^{i-j}(C^{\cdot}).$$

Put $i - j =: n$. In the case of $n < -s$ it follows that $E_2^{i,-j} = 0$ by similar arguments as above in the first spectral sequence. Note that $s = \sup\{n \in \mathbb{Z} \mid \operatorname{Tor}_n^A(M, N) \neq$

$0\}$. Therefore $H^n(C^\cdot) = 0$ for all $n < -s$ and $H^{-s}(C^\cdot) \simeq H^0_{\mathfrak{m}}(\mathrm{Tor}^A_s(M, N)) \neq 0$. Recall that $\mathrm{depth}\,\mathrm{Tor}^A_s(M, N) = 0$. This finally proves $-s = -p + t$, as required.
□

Another case of describing $\sup\{n \in \mathbb{Z} \mid \mathrm{Tor}^A_n(M, N) \neq 0\}$ was investigated by M. Auslander, see [1, Theorem 1.2]. It follows in the same way as above by considering both of the spectral sequences.

As an immediate consequence of 3.4 it turns out that $\mathrm{depth}_A N \leq \mathrm{pd}_A M$ provided $M \otimes_A N$ is an A-module of finite length. Under these assumptions a much stronger inequality holds, namely $\dim_A N \leq \mathrm{pd}_A M$. This is the Intersection Theorem proved by C. Peskine and L. Szpiro, see [39], and M. Hochster, see [19], in the equicharacteristic case, and finally by P. Roberts, see [38], in the remaining case. For a summary of these and related results about Cohen-Macaulay rings see also the monograph [7].

In relation to that the following Cohen-Macaulay criterion could be of some interest.

Corollary 3.5. *Let M, N be two finitely generated A-modules such that $M \otimes_A N$ is of finite length. Suppose that $\mathrm{pd}_A M$ is finite. Then N is a Cohen-Macaulay module with $\mathrm{depth}_A N = \mathrm{pd}_A M$ if and only if $\mathrm{Tor}^A_n(M, N) = 0$ for all $n \geq 1$.*

Proof. First assume that N is a Cohen-Macaulay module. Then

$$\mathrm{Tor}^A_n(M, N) \simeq \mathrm{Tor}^A_{n+d}(M, H^d_{\mathfrak{m}}(N)), \ d = \dim N,$$

as follows by 3.3. But now $d = \mathrm{pd}_A M$. Therefore the last module vanishes for all positive n.

For the proof of the reverse implication note that

$$0 = \mathrm{pd}_A M - \mathrm{depth}_A N \geq \mathrm{pd}_A M - \dim_A N$$

as follows by 3.4. But now $\mathrm{pd}_A M - \dim_A N \geq 0$ by view of the Intersection Theorem. This finishes the proof.
□

By view of the formula of M. Auslander and D. Buchsbaum one may interpret the inequality $\mathrm{depth}_A N \leq \mathrm{pd}_A M$ in the following way

$$\mathrm{depth}_A N + \mathrm{depth}_A M \leq \mathrm{depth}\, A,$$

provided $M \otimes_A N$ is of finite length and $\mathrm{pd}_A M$ is finite. One might think of it as a generalization of Serre's inequality $\dim_A N + \dim_A M \leq \dim A$ in the case of A a regular local ring.

In connection to 3.5 the rigidity of Tor could be of some interest. Let $n \in \mathbb{N}$. Then the conjecture says that $\mathrm{Tor}^A_{n+1}(M, N) = 0$ provided $\mathrm{Tor}^A_n(M, N) = 0$. This is true for a regular local ring $(A, \mathfrak{m})$ as shown by J. P. Serre, see [45], in the case of unramified regular local rings, and finally for any regular local ring by S. Lichtenbaum, see [26]. There are also related results in [1]. The rigidity conjecture for a general local ring and $\mathrm{pd}_A M < \infty$ was disproved by R. Heitmann's example, see [18]. For some recent developments on the rigidity in connection to non-regular local rings, see the work of C. Huneke and R. Wiegand in [23].

Under the assumptions of 3.5 J. P. Serre, see [45], considered the Euler characteristic $\chi(M, N) = \sum_{i\geq 0}(-1)^i L_A(\mathrm{Tor}_i^A(M, N))$ as an intersection number. More
generally for $n \in \mathbb{N}$ he defined the partial Euler characteristics

$$\chi_n(M, N) = \sum_{i\geq 0}(-1)^i L_A(\mathrm{Tor}_{n+i}^A(M, N)).$$

Note that $\chi(M, N) = \chi_0(M, N)$. In the case of an unramified regular local ring
$(A, \mathfrak{m})$ J. P. Serre, see [45], proved the non-negativity of $\chi(M, N)$. Moreover he
conjectured that this is true for any regular local ring. Recently O. Gabber, see
[11], proved the non-negativity of $\chi(M, N)$ for any regular local ring. As follows
by view of R. Heitmann's example $\chi_1(M, N) \geq 0$ does not hold in the case of an
arbitrary local ring and $\mathrm{pd}_A M < \infty$.

The Cohen-Macaulay property of N in 3.5 provides that $L_A(M \otimes_A N) =
\chi_0(M, N)$. This equality is equivalent to the vanishing of $\chi_1(M, N)$. Consider the
particular case of a finitely generated A-module N and $M = A/\underline{x}A$, where $\underline{x} =
x_1, \ldots, x_r$ denotes an A-regular sequence. Then $\mathrm{pd}_A M = r$. Suppose that $N/\underline{x}N$ is
an A-module of finite length, i.e. $\dim N \leq r$. Then $\chi_0(M, N) = e_0(\underline{x}; N)$ as follows
since $\mathrm{Tor}_i^A(A/\underline{x}A, N) \simeq H_i(\underline{x}; N), i \in \mathbb{N}$, and $\sum_{i\geq 0}(-1)^i L_A(H_i(\underline{x}; N)) = e_0(\underline{x}; N)$,
where $e_0(\underline{x}; N)$ denotes the multiplicity of N with respect to $\underline{x}$, see [2]. So the
equality $L(N/\underline{x}N) = e_0(\underline{x}; N)$ says that N is a Cohen-Macaulay module with
$\dim N = r$.

Conjecture 3.6. Let M, N be two finitely generated A-modules such that $M \otimes_A N$
is an A-module of finite length and $\mathrm{pd}_A M < \infty$.

a) (Cohen-Macaulay Conjecture) Suppose that $L_A(M \otimes_A N) = \chi_0(M, N)$. Does
   it follows that N is a Cohen-Macaulay module with $\mathrm{pd}_A M = \mathrm{depth}_A N$?
b) (Weak Rigidity Conjecture) Suppose that $\chi_n(M, N) = 0$ for a certain $n \in \mathbb{N}$.
   Does it follows that $\chi_{n+1}(M, N) = 0$?

Suppose that the weak rigidity conjecture is true. Then $\chi_n(M, N) = 0$ implies
inductively that $\mathrm{Tor}_k^A(M, N) = 0$ for all $k \geq n$. To this end recall that $\mathrm{pd}_A M$ is
finite. Let us return to this observation in the following result.

Corollary 3.7. *Let $\mathfrak{a}, \mathfrak{b}$ be two ideals of a local ring $(A, \mathfrak{m})$. Assume that $\mathrm{pd}_A A/\mathfrak{a}$
is finite and $\mathfrak{a} + \mathfrak{b}$ is an $\mathfrak{m}$-primary ideal.*

a) *Suppose that $A/\mathfrak{a}$ is a Cohen-Macaulay ring with $\mathrm{depth}\, A/\mathfrak{a} + \mathrm{depth}\, A/\mathfrak{b} =
   \mathrm{depth}\, A$. Then $\mathfrak{a} \cap \mathfrak{b} = \mathfrak{ab}$ (resp. $\chi(M, N) = L_A(A/(\mathfrak{a} + \mathfrak{b}))$).*
b) *Suppose that $A/\mathfrak{a}$ is rigid (resp. weakly rigid). Then the converse is true.*

Proof. The statement in a) is a consequence of 3.5. Recall that $\mathrm{Tor}_1^A(A/\mathfrak{a}, A/\mathfrak{b}) =
\mathfrak{a} \cap \mathfrak{b}/\mathfrak{ab}$. So its vanishing yields the equality of the intersection with the product.
The statement in b) is clear by the above discussion. ☐

In the case of A a regular local ring this says that $\mathfrak{a} \cap \mathfrak{b} = \mathfrak{ab}$ if and only if
$\dim A/\mathfrak{a} + \dim A/\mathfrak{b} = \dim A$ and both $A/\mathfrak{a}$ and $A/\mathfrak{b}$ are Cohen-Macaulay rings.

This was shown by J. P. Serre, see [45]. So one might think of 3.7 as a generalization to the non-regular case.

3.2. Estimates of Betti Numbers. In the case $M = k$ the second formula shown in 3.2 provides estimates of the Betti numbers of a module in terms of Betti numbers of its local cohomology modules. This point of view is pursued in this subsection.

To this end let the local ring $(A, \mathfrak{m})$ be the quotient of a regular local ring $(B, \mathfrak{n})$ with $r = \dim B$. We are interested in the minimal free resolution of M as a module over B. Because of the local duality, see 1.8, the local cohomology modules of M are the Matlis duals of $K_M^n \simeq \operatorname{Ext}_B^{r-n}(M, B), n \in \mathbb{Z}$, the modules of deficiency of M. Note that $K_M = K_M^d, d = \dim M$, is called the canonical module of M. In the following let

$$\beta_n(M) = \dim_k \operatorname{Tor}_n^B(k, M), \ n \in \mathbb{Z},$$

denote the n-th Betti number of M. Here k denotes the residue field of B.

Theorem 3.8. *Let M denote a finitely generated B-module. Then*

$$\beta_n(M) \leq \begin{cases} \sum_{i=0}^{r-n} \beta_{r-n-i}(K_M^i) & \text{for } n > c, \quad \text{and} \\ \sum_{i=0}^{d} \beta_{r-n-i}(K_M^i) & \text{for } n \leq c, \end{cases}$$

where $c = r - d, d = \dim M$, denotes the codimension of M.

Proof. In order to prove the bounds note that for a B-module X and all $n \in \mathbb{Z}$ there is an isomorphism

$$\operatorname{Tor}_n^B(k, X) \simeq H_n(\underline{x}; X),$$

where $H_n(\underline{x}; X)$ denotes the Koszul homology of X with respect to $\underline{x} = x_1, \ldots, x_r$, a minimal generating set of $\mathfrak{n}$, the maximal ideal of the regular local ring B. This follows because $H_n(\underline{x}; B)$ provides a minimal free resolution of $B/\mathfrak{n}$ over B. Because of the Matlis duality it yields that

$$H_{n+i}(\underline{x}; T(K_M^i)) \simeq T(H^{n+i}(\underline{x}; K_M^i)).$$

By the self-duality of the Koszul complex it turns out that

$$H^{n+i}(\underline{x}; K_M^i) \simeq H_{r-n-i}(\underline{x}; K_M^i).$$

By counting the k-vector space dimension this implies $\beta_{n+i}(H_{\mathfrak{m}}^i(M)) = \beta_{r-n-i}(K_M^i)$ for all $i, n \in \mathbb{Z}$. Therefore the claim follows by virtue of 3.2. $\square$

In the particular case of M a Cohen-Macaulay B-module the underlying spectral sequence degenerates, see 3.3. This proves that $\beta_n(M) = \beta_{c-n}(K_M), 0 \leq n \leq c$. This is well known since $\operatorname{Hom}_B(\cdot, B)$ preserves exactness of $F^{\cdot}$ in this case. Here $F^{\cdot}$ denotes the minimal free resolution of M.

Corollary 3.9. *Let M be a finitely generated B-module with $\operatorname{pd}_A M = p$ and $\operatorname{depth}_B M = t$. Then $\beta_p(M) = \beta_0(K_M^t)$. That is the rank of the last module in a minimal free resolution of M is given by the minimal numbers of generators of the first non-vanishing K_M^i.*

Proof. There is a partial degeneration of the spectral sequence to the isomorphism

$$\operatorname{Tor}^B_p(k, M) \simeq \operatorname{Tor}^B_r(k, H^t_{\mathfrak{m}}(M)).$$

As above $\operatorname{Tor}^B_r(k, H^t_{\mathfrak{m}}(M)) \simeq T(H^r(\underline{x}; K^t_M)) \simeq T(H_0(\underline{x}; K^t_M))$, which proves the claim. Here r denotes the dimension of B or – what is the same – the minimal number of generators of $\mathfrak{n}$, the maximal ideal of the regular local ring B. $\qquad\square$

A case of a particular interest is the situation when $H^i_{\mathfrak{m}}(M), i < d := \dim_A M$, are finite dimensional $A/\mathfrak{m}$-vector spaces. Call a finitely generated A-module M with this property a quasi-Buchsbaum module.

Corollary 3.10. *Let M denote a quasi-Buchsbaum module over the local ring $(A, \mathfrak{m})$. Then*

$$\beta_n(M) \leq \sum_{i=0}^{r-n} \binom{r}{n+i} \dim_k H^i_{\mathfrak{m}}(M)$$

for all $n > c$, where $k = A/\mathfrak{m}$ denotes the residue field.

Proof. It is an immediate consequence of 3.8. Note that $H^i_{\mathfrak{m}}(M), i < d$, are finite-dimensional k-vector spaces. Moreover $\beta_{r-n-i}(k) = \binom{r}{n+i}$ since B is a regular local ring of embedding dimension r. $\qquad\square$

3.3. Castelnuovo-Mumford Regularity.

In order to obtain more precise information about the syzygies it is helpful to have additional structure, e.g. the structure of a graded k-algebra. So let $A = \oplus_{n\geq 0} A_n$ denote a Noetherian graded algebra with $A_0 = k$ a field and $A = A_0[A_1]$. Then A is the epimorphic image of the polynomial ring $B = k[X_1, \ldots, X_r]$, where $r = \dim_k A_1$. In the following let M denote a finitely generated graded A-module. Then one might consider it as a module over B. The finite dimensional k-vector spaces $\operatorname{Tor}^B_n(k, M)$ are graded. They reflect information about the degrees of the minimal generators of the n-th module of syzygies of M.

The Čech complex $K_{\underline{x}}$ of B with respect to $\underline{x} = X_1, \ldots, X_r$ is a complex of graded B-modules. In fact it is a flat resolution of the system of inverse polynomials. So the local cohomology modules of a graded B-module are also graded and therefore $H^n_{\mathfrak{m}}(M) \simeq H^n(K_{\underline{x}} \otimes_A M)$ is a homomorphism of degree zero. Here $\mathfrak{m}$ denotes the homogeneous ideal generated by all variables.

For a graded B-module N let $e(N) = \sup\{n \in \mathbb{Z} \mid N_n \neq 0\}$, where N_n denotes the n-th graded piece of N. In the case of N an Artinian module it follows that $e(N) < \infty$. Recall that $e(N) = -\infty$ in the case $N = 0$.

Then define $\operatorname{reg} M$ the Castelnuovo-Mumford regularity of M a finitely generated graded B-module by

$$\operatorname{reg} M = \max\{e(H^n_{\mathfrak{m}}(M)) + n \mid n \in \mathbb{Z}\}.$$

Note that it is well-defined by 1.5.

The basics for this construction were initiated by D. Mumford, see [31], who attributed it to Castelnuovo. The importance of the regularity lies in the following fact, a relation to the graded Betti numbers of M. There is the equality

$$\operatorname{reg} M = \max\{e(\operatorname{Tor}_n^B(k, M)) - n \mid n \in \mathbb{Z}\},$$

shown by D. Eisenbud and S. Goto, see [10]. In the case of M a Cohen-Macaulay module it turns out that $\operatorname{reg} M = e(\operatorname{Tor}_c^B(k, M)) - c$, $c = \operatorname{codim} M$.

The following provides an improvement by showing that – just as in the Cohen-Macaulay case – the regularity is determined by the tail of the minimal free resolution of M.

Theorem 3.11. *Let M denote a finitely generated graded B-module. Let $s \in \mathbb{N}$ be an integer. Then the following two integers coincide*
a) $\max\{e(H_{\mathfrak{m}}^i(M)) + i \mid 0 \le i \le s\}$ *and*
b) $\max\{e(\operatorname{Tor}_j^B(k, M)) - j \mid r - s \le j \le r\}.$
In particular for $s = \dim_B M$ it follows that

$$\operatorname{reg} M = \max\{e(\operatorname{Tor}_j^B(k, M)) - j \mid c \le j \le r\},$$

where $c = r - \dim_B M$ denotes the codimension of M.

Proof. The proof is based on the following spectral sequence

$$E_2^{i,j} = H_{-i}(\underline{x}; H_{\mathfrak{m}}^j(M)) \Rightarrow E^{i+j} = H_{-i-j}(\underline{x}; M)$$

as it was considered in the proof of 3.2. Here $\underline{x} = X_1, \ldots, X_r$ denotes the set of variables in B. Note that $\operatorname{Tor}_n^B(k, N) \simeq H_n(\underline{x}; N)$ for all $n \in \mathbb{Z}$ and any B-module N. Moreover the spectral sequence is a spectral sequence of graded modules and all the homomorphisms are homogeneous of degree zero.

First show the following claim:

Suppose that $H_s(\underline{x}; M)_{s+t} \ne 0$ for a certain $t \in \mathbb{Z}$ and $r - i \le s \le r$. Then there exists a $j \in \mathbb{Z}$ such that $0 \le j \le i$ and $H_{\mathfrak{m}}^j(M)_{t-j} \ne 0$.

Assume the contrary, i.e., $H_{\mathfrak{m}}^j(M)_{t-j} = 0$ for all $0 \le j \le i$. Then consider the spectral sequence

$$[E_2^{-s-j,j}]_{t+s} = H_{s+j}(\underline{x}; H_{\mathfrak{m}}^j(M))_{t+s} \Rightarrow [E^{-s}]_{t+s} = H_s(\underline{x}; M)_{t+s}.$$

Recall that all the homomorphisms are homogeneous of degree zero. Now the corresponding E_2-term is a subquotient of

$$[\oplus H_{\mathfrak{m}}^j(M)^{\binom{r}{s+j}}(-s-j)]_{t+s}.$$

Let $j \le i$. Then this vectorspace is zero by the assumption about the local cohomology. Let $j > i$. Then $s + j > s + i \ge r$ and $\binom{r}{s+j} = 0$. Therefore the corresponding E_2-term $[E_2^{-s-j,j}]_{t+s}$ is zero for all $j \in \mathbb{Z}$. But then also all the subsequent stages are zero, i.e., $[E_\infty^{-s-j,j}]_{t+s} = 0$ for all $j \in \mathbb{Z}$. Whence $[E^{-s}]_{t+s} = H_s(\underline{x}; M)_{t+s} = 0$, contradicting the assumption.

The second partial result shows that a certain non-vanishing of $H_{\mathfrak{m}}^n(M)$ implies the existence of a certain minimal generator of a higher syzygy module. More precisely we show the following claim:

Let $r = \dim B$ denote the number of variables of B. Suppose that there are integers s, b such that the following conditions are satisfied:
  a) $H_{\mathfrak{m}}^i(M)_{b+1-i} = 0$ *for all* $i < s$ *and*
  b) $H_r(\underline{x}; H_{\mathfrak{m}}^s(M))_{b+r-s} \neq 0$
Then it follows that $H_{r-s}(\underline{x}; M)_{b+r-s} \neq 0$.

Note that the condition in b) means that $H_{\mathfrak{m}}^s(M)$ possesses a socle generator in degree $b - s$. Recall that r denotes the number of generators of $\mathfrak{m}$.

As above we consider the spectral sequence

$$E_2^{-r,s} = H_r(\underline{x}; H_{\mathfrak{m}}^s(M)) \Rightarrow E^{-r+s} = H_{r-s}(\underline{x}; M)$$

in degree $b + r - s$. The subsequent stages of $[E_2^{-r,s}]_{b+r-s}$ are derived by the cohomology of the following sequence

$$[E_n^{-r-n,s+n-1}]_{b+r-s} \to [E_n^{-r,s}]_{b+r-s} \to [E_n^{-r+n,s-n+1}]_{b+r-s}$$

for $n \geq 2$. But now $[E_n^{-r-n,s+n-1}]_{b+r-s}$ resp. $[E_n^{-r+n,s-n+1}]_{b+r-s}$ are subquotients of

$$H_{r+n}(\underline{x}; H_{\mathfrak{m}}^{s+n-1}(M))_{b+r-s} = 0 \text{ resp. } H_{r-n}(\underline{x}; H_{\mathfrak{m}}^{s-n+1}(M))_{b+r-s} = 0.$$

For the second module recall that it is a subquotient of

$$[\oplus H_{\mathfrak{m}}^{s-n+1}(M)^{\binom{r}{r-n}}(-r+n)]_{b+r-s} = 0, \quad n \geq 2.$$

Therefore $[E_2^{-r,s}]_{b+r-s} = [E_\infty^{-r,s}]_{b+r-s} \neq 0$ and

$$[E^{-r+s}]_{b+r-s} \simeq H_{r-s}(\underline{x}; M)_{b+r-s} \neq 0$$

as follows by the filtration with the corresponding E_∞-terms.

Now let us prove the statement of the theorem. First of all introduce two abbreviations. Put

$$a := \max\{e(\mathrm{Tor}_j^B(k, M)) - j \mid r - s \leq j \leq r\}.$$

Then by the first claim it follows that $a \leq b$, where

$$b := \max\{e(H_{\mathfrak{m}}^i(M)) + i \mid 0 \leq i \leq s\}.$$

On the other hand choose j an integer $0 \leq j \leq s$ such that $b = e(H_{\mathfrak{m}}^j(M)) + j$. Then $H_{\mathfrak{m}}^j(M)_{b-j} \neq 0$, $H_{\mathfrak{m}}^j(M)_{c-j} = 0$ for all $c > b$, and $H_{\mathfrak{m}}^i(M)_{b+1-i} = 0$ for all $i < j$. Recall that this means that $H_{\mathfrak{m}}^j(M)$ has a socle generator in degree $b - j$. Therefore the second claim applies and $\mathrm{Tor}_{r-j}^B(K, M)_{b+r-j} \neq 0$. In other words, $b \leq a$, as required. $\square$

An easy byproduct of our investigations is the above mentioned fact that

$$\mathrm{reg}\, M = e(\mathrm{Tor}_c^B(K, M)) - c, \ c = r - \dim M,$$

provided M is a Cohen-Macaulay module.

It is noteworthy to say that P. Jørgensen, see [24], investigated a non-commutative Castelnuovo-Mumford regularity. In fact he generalized 3.11 to the non-commutative situation by an interesting argument.

Theorem 3.12. *Let M be a finitely generated graded B-module with $d = \dim_B M$. Suppose there is an integer $j \in \mathbb{Z}$ such that for all $q \in \mathbb{Z}$ either*
  a) $H_{\mathrm{m}}^q(M)_{j-q} = 0$ *or*
  b) $H_{\mathrm{m}}^p(M)_{j+1-p} = 0$ *for all $p < q$ and $H_{\mathrm{m}}^p(M)_{j-1-p} = 0$ for all $p > q$.*
Then for $s \in \mathbb{Z}$ it follows that
  (1) $\mathrm{Tor}_s^B(k, M)_{s+j} \simeq \oplus_{i=0}^{r-s}(\mathrm{Tor}_{r-s-i}^B(k, K_M^i)_{s+j})^\vee$ *provided $s > c$, and*
  (2) $\mathrm{Tor}_s^B(k, M)_{s+j} \simeq \oplus_{i=0}^{d-1}(\mathrm{Tor}_{s+i}^B(k, K_M^i)_{r-s-j})^\vee \oplus (\mathrm{Tor}_{c-s}^B(k, K_M)_{r-s-j})^\vee$,
  provided $s \leq c$,
where $K_M^i = \mathrm{Ext}_B^{r-i}(M, B(-r))$, $0 \leq i < d$, denote the module of deficiencies and K_M is the canonical module of M.

Proof. As above consider the spectral sequence

$$E_2^{-s-i,i} = H_{s+i}(\underline{x}; H_{\mathrm{m}}^i(M)) \Rightarrow E^{-s} = H_s(\underline{x}; M)$$

in degree $s + j$. Firstly we claim that $[E_2^{-s-i,i}]_{s+j} \simeq [E_\infty^{-s-i,i}]_{s+j}$ for all $s \in \mathbb{Z}$. Because $[E_2^{-s-i,i}]_{s+j}$ is a subquotient of

$$[\oplus H_{\mathrm{m}}^i(M)^{\binom{r}{s+i}}(-s-i)]_{s+j}$$

the claim is true provided $H_{\mathrm{m}}^i(M)_{j-i} = 0$. Suppose that $H_{\mathrm{m}}^i(M)_{j-i} \neq 0$. In order to prove the claim in this case too note that $[E_{n+1}^{-s-i,i}]_{s+j}$ is the cohomology at

$$[E_n^{-s-i-n,i+n-1}]_{s+j} \rightarrow [E_n^{-s-i,i}]_{s+j} \rightarrow [E_n^{-s-i+n,i-n+1}]_{s+j}.$$

Then the module at the left resp. the right is a subquotient of

$$H_{s+i+n}(\underline{x}; H_{\mathrm{m}}^{i+n-1}(M))_{s+j} \text{ resp. } H_{s+i-n}(\underline{x}; H_{\mathrm{m}}^{i-n+1}(M))_{s+j}.$$

Therefore both of them vanish. But this means that the E_2-term coincides with the corresponding E_∞-term. So the target of the spectral sequence $H_s(\underline{x}; M)_{s+j}$ admits a finite filtration whose quotients are $H_{s+i}(\underline{x}; H_{\mathrm{m}}^i(M))_{s+j}$. Because all of these modules are finite dimensional vectorspaces it follows that

$$H_s(\underline{x}; M)_{s+j} \simeq \oplus_{i=0}^{r-s} H_{s+i}(\underline{x}; H_{\mathrm{m}}^i(M))_{s+j}$$

for all $s \in \mathbb{Z}$.

By the Local Duality theorem there are the following isomorphisms $H_{\mathrm{m}}^i(M) \simeq T(K_M^i)$, $0 \leq i \leq d$, where T denotes the Matlis duality functor $\mathrm{Hom}_k(\cdot, k)$ in the case of the graded situation. Therefore we obtain the isomorphisms

$$H_{s+i}(\underline{x}; T(K_M^i))_{s+j} \simeq (T(H^{s+i}(\underline{x}; K_M^i)))_{s+j} \simeq (H_{r-s-i}(\underline{x}; K_M^i)_{r-s-j})^\vee.$$

But the last vector space is isomorphic to $(\mathrm{Tor}_{r-s-i}^B(k, K_M^i)_{r-s-j})^\vee$.

In the case of $s > c$ it is known that $r - s < d$. Hence the first part of the claim is shown to be true. In the remaining case $s \leq c$ the summation is taken from $i = 0, \ldots, d$, which proves the second part of the claim. $\qquad\square$

As an application of 3.12 we derive M. Green's duality theorem [12, Section 2].

Corollary 3.13. *Suppose there exists an integer $j \in \mathbb{Z}$ such that*

$$H_{\mathfrak{m}}^q(M)_{j-q} = H_{\mathfrak{m}}^q(M)_{j+1-q} = 0$$

for all $q < \dim_B M$. Then

$$\mathrm{Tor}_s^B(k, M)_{s+j} \simeq (\mathrm{Tor}_{c-s}^B(k, K_M)_{r-s-j})^\vee,$$

for all $s \in \mathbb{Z}$, where $c = \mathrm{codim}\, M$.

Proof. It follows that the assumptions of Theorem 3.12 are satisfied for j because of $H_{\mathfrak{m}}^p(M)_{j-1-p} = 0$ for all $p > \dim M$. Therefore the isomorphism is a consequence of (1) and (2) in 3.12. To this end recall that

$$\mathrm{Tor}_{s+i}^B(k, H_{\mathfrak{m}}^i(M))_{s+j} \simeq H_{s+i}(\underline{x}; H_{\mathfrak{m}}^i(M))_{s+j} = 0,$$

as follows by the vanishing of $H_{\mathfrak{m}}^i(M)_{j-i}$ for all $i < \dim M$. $\square$

M. Green's duality theorem in 3.13 relates the Betti numbers of M to those of K_M. Because of the strong vanishing assumptions in 3.13 very often it does not give strong information about Betti numbers. Often it says just the vanishing which follows also by different arguments, e.g., the regularity of M.

Theorem 3.12 is more subtle. We shall illustrate its usefulness by the following example.

Example 3.14. Let $C \subset \mathbb{P}_K^n$ denote a reduced integral non-degenerate curve over an algebraically closed field K. Suppose that C is non-singular and of genus $g(C) = 0$. Let $A = B/I$ denote its coordinate ring, i.e., $B = K[x_0, \ldots, x_n]$ and I its homogeneous defining ideal. Then

$$\mathrm{Tor}_s^B(k, B/I)_{s+j} \simeq \mathrm{Tor}_{s+1}^B(k, H_{\mathfrak{m}}^1(B/I))_{s+j}$$

for all $s \geq 1$ and all $j \geq 3$. To this end recall that A is a two-dimensional domain. Moreover it is well-known that $H_{\mathfrak{m}}^q(B/I) = 0$ for all $q \leq 0$ and $q > 2$. Furthermore it is easy to see that $H_{\mathfrak{m}}^1(B/I)_{j-1} = 0$ for all $j \leq 1$. Moreover $H_{\mathfrak{m}}^2(B/I)_{j-1-2} = 0$ for all $j \geq 3$ as follows because of $g(C) = 0$. That is, for $j \geq 3$ one might apply 3.12. In order to conclude we have to show that $\mathrm{Tor}_{c-s}^B(k, K_{B/I})_{r-s-j} = 0$ for $j \geq 3$. To this end note that

$$(H_{c-s}(\underline{x}; K_{B/I})_{r-s-j})^\vee \simeq H_{s+2}(\underline{x}; H_{\mathfrak{m}}^2(B/I))_{s+j}$$

as is shown in the proof of 3.12. But this vanishes for $j \geq 2$ as is easily seen.

3.4. The Local Green Modules. As before let $E = E_A(A/\mathfrak{m})$ denote the injective hull of the residue field of a local ring $(A, \mathfrak{m})$. Let $\underline{x} = x_1, \ldots, x_r$ denote a system of elements of A. Then for all $n \in \mathbb{Z}$ there are canonical isomorphisms

$$H_n(\underline{x}; T(M)) \simeq T(H^n(\underline{x}; M)) \quad \text{and} \quad H^n(\underline{x}; T(M)) \simeq T(H_n(\underline{x}; M)).$$

Here T denotes the duality functor $\mathrm{Hom}_A(\cdot, E)$. In the case $(A, \mathfrak{m})$ is the factor ring of a local Gorenstein ring B, then use the modules of deficiency K_M^n as defined in Section 1.2. In order to continue with our investigations we need a sharpening of the definition of a filter regular sequence. To this end let M denote a finitely generated A-module.

An M-filter regular sequence $\underline{x} = x_1, \ldots, x_r$ is called a strongly M-filter regular sequence provided it is filter regular with respect to $K^n_{M \otimes \widehat{A}}$ for $n = 0, 1, \ldots, \dim M$. Here $\widehat{A}$ denotes the completion of A.

The necessity to pass to the completion is related to the existence of K^n_M. In the case A is the quotient of a Gorenstein ring it is enough to check the filter regularity with respect to K^n_M. This follows because by Cohen's Structure theorem $\widehat{A}$ is the quotient of a Gorenstein ring and $\underline{x}$ is M-filter regular if and only if it is $M \otimes_A \widehat{A}$-filter regular. Because $K^n_{M \otimes \widehat{A}}$ are finitely generated $\widehat{A}$-modules the existence of strongly M-filter regular sequences is a consequence of prime avoidance arguments.

Lemma 3.15. *Suppose that $\underline{x} = x_1, \ldots, x_r$ denotes a strongly M-filter regular sequence. Let $j \in \mathbb{Z}$ denote an integer. Then $H_i(\underline{x}; H^j_{\mathfrak{m}}(M))$ resp. $H^i(\underline{x}; H^j_{\mathfrak{m}}(M))$ are A-modules of finite length in the following two cases:*

a) *for all $i < r$ resp. $i > 0$, and*
b) *for all $i \in \mathbb{Z}$, provided $r \geq j$.*

Proof. Without loss of generality we may assume that A is a complete local ring. Because of the isomorphisms

$$H_i(\underline{x}; H^j_{\mathfrak{m}}(M)) \simeq T(H^i(\underline{x}; K^j_M)), \ i, j \in \mathbb{Z},$$

it will be enough to show that $H^i(\underline{x}; K^j_M))$ is an A-module of finite length. By view of 1.17 is follows that this is of finite length in the case $i < r$. In the case $r > i$ we know that $\dim K^j_M \leq j$, see 1.9. Therefore the Koszul cohomology is also of finite length in the remaining case $i = r$. The rest of the statement is clear by the self-duality of the Koszul complex. $\square$

In a certain sense the modules considered in 3.15 are local analogues to the modules studied by M. Green in [12]. For some results in the graded case see also [44]. In relation to possible further applications it would be of some interest to find interpretations of the modules $H_i(\underline{x}; H^j_{\mathfrak{m}}(M))$. One is given in the following.

Theorem 3.16. *Let $\underline{x} = x_1, \ldots, x_r, r \geq \dim M$, denote a strongly M-filter regular sequence. Let $n \in \mathbb{N}$ be an integer. Then there are the following bounds:*

a) $L_A(H^n(\underline{x}; M)) \leq \sum_{i \geq 0} L_A(H^{n-i}(\underline{x}; H^i_{\mathfrak{m}}(M)))$.
b) $L_A(H_n(\underline{x}; M)) \leq \sum_{i \geq 0} L_A(H_{n+i}(\underline{x}; H^i_{\mathfrak{m}}(M)))$.

Proof. It is enough to prove one of the statements as follows by self-duality of Koszul complexes. Let us prove the claim in b). To this end consider the complex

$$C^\cdot := K^\cdot \otimes_A K.(\underline{x}; M),$$

where $K^\cdot$ denotes the Čech complex with respect to a generating set of the maximal ideal. Now consider the spectral sequences for computing the cohomology of $C^\cdot$.

The first of them is given by

$$H_{\mathfrak{m}}^i(H_j(\underline{x}; M)) \Rightarrow H^{i-j}(C^{\cdot}).$$

Because $H_j(\underline{x}; M)$, $j \in \mathbb{Z}$, is an A-module of finite length we get the vanishing of $H_{\mathfrak{m}}^i(H_j(\underline{x}, M)) = 0$ for all j and $i \neq 0$ and

$$H_{\mathfrak{m}}^0(H_j(\underline{x}; M)) \simeq H_j(\underline{x}; M) \text{ for all } j \in \mathbb{Z}.$$

Therefore there is a partial degeneration of the spectral sequence to the following isomorphisms

$$H^{-n}(C^{\cdot}) \simeq H_n(\underline{x}; M) \text{ for all } n \in \mathbb{Z}.$$

On the other hand there is the spectral sequence

$$E_2^{i,-j} = H_j(\underline{x}; H_{\mathfrak{m}}^i(M)) \Rightarrow E^{i-j} = H^{i-j}(C^{\cdot}).$$

By the assumption all the initial terms $E_2^{i,-j}$ are A-modules of finite length for all $i, j \in \mathbb{Z}$, see 3.15. Therefore also the limit terms $E_\infty^{i,-j}$ are of finite length and $L_A(E_2^{i,-j}) \geq L_A(E_\infty^{i,-j})$ for all $i, j \in \mathbb{Z}$. Whence $E^{-n} = H^{-n}(C^{\cdot})$ admits a filtration with quotients $E_\infty^{i,-n-i}$ for $i \in \mathbb{Z}$. Therefore there is the bound

$$L_A(E^{-n}) \leq \sum_{i \geq 0} L_A(E_2^{i,-n-i}),$$

which proves the result by view of the above estimate. □

The spectral sequence considered in the proof provides also another partial degeneration. This could be helpful for different purposes.

Corollary 3.17. *Let $\underline{x}$ and M be as in 3.16. Then there are the following canonical isomorphisms:*

a) $H^t(\underline{x}; M) \simeq H^0(\underline{x}; H_{\mathfrak{m}}^t(M))$, $t = \operatorname{depth} M$.
b) $H^n(\underline{x}; M) \simeq H^{n-d}(\underline{x}; H_{\mathfrak{m}}^d(M))$ *for all $n \in \mathbb{Z}$, provided M is a d-dimensional Cohen-Macaulay module.*

In the first case of Corollary 3.17 it is possible to compute the Koszul cohomology explicitly.

It turns out that $H^0(\underline{x}; H_{\mathfrak{m}}^t(M)) \simeq (x_1, \ldots, x_t)M :_M \underline{x}/(x_1, \ldots, x_t)M$. It would be of some interest to give further interpretations of some of the modules $H_i(\underline{x}; H_{\mathfrak{m}}^j(M))$.

There is one result in this direction concerning multiplicities. To this end recall the notion of a reducing system of parameters in the sense of M. Auslander and D. A. Buchsbaum, see [2]. Recall that for an arbitrary system of parameters $\underline{x} = x_1, \ldots, x_d$ of M it is known that there is a strongly M-filter regular sequence $\underline{y} = y_1, \ldots, y_d$ such that $(x_1, \ldots, x_i)M = (y_1, \ldots, y_i)M$, $i = 1, \ldots, d = \dim M$. Note that $\underline{y}$ is a reducing system of parameters of M.

In the following denote by $L_A(M/\underline{x}M)$ resp. $e_0(\underline{x}; M)$ the length resp. the multiplicity of M with respect to $\underline{x}$, see [2] for the details.

Theorem 3.18. *Let $\underline{x} = x_1, \ldots, x_d$, $d = \dim M > 1$, denote an arbitrary system of parameters. Choose $\underline{y} = y_1, \ldots, y_d$ as above. Then*

$$L_A(M/\underline{x}M) - e_0(\underline{x}; M) \le \sum_{i=0}^{d-1} L_A(H_i(\underline{y}'; H_{\mathfrak{m}}^i(M))),$$

where $\underline{y}' = y_1, \ldots, y_{d-1}$.

Proof. Because of the previous remark one may replace $\underline{x}$ by $\underline{y}$ without loss of generality. Because $\underline{y}$ is a reducing system of parameters of M it turns out that

$$L_A(M/\underline{y}M) - e_0(\underline{y}; M) = L_A(\underline{y}'M :_M y_d/\underline{y}'M),$$

see [2]. Moreover it follows that $\underline{y}'M :_M y_d/\underline{y}'M \subseteq \underline{y}'M :_M \langle \mathfrak{m} \rangle/\underline{y}'M$. Recall that y_d is a parameter for the one-dimensional quotient module $M/\underline{y}'M$. But now

$$\underline{y}'M :_M \langle \mathfrak{m} \rangle/\underline{y}'M \simeq H_{\mathfrak{m}}^0(M/\underline{y}'M).$$

In order to continue with the proof let $K^{\cdot}$ denote the Čech complex with respect to a system of parameters of $(A, \mathfrak{m})$. Then consider the complex $C^{\cdot} := K^{\cdot} \otimes_A K_{\cdot}(\underline{y}'; M)$, where $K_{\cdot}(\underline{y}'; M)$ denotes the Koszul complex of M with respect to $\underline{y}'$. Then use the spectral sequence

$$E_2^{i,-j} = H_{\mathfrak{m}}^i(H_j(\underline{y}'; M)) \Rightarrow E^{i-j} = H^{i-j}(C^{\cdot}).$$

Because $\underline{y}'$ is an M-filter regular sequence $H_j(\underline{y}'; M), j \ne 0$, is an A-module of finite length. That is, for $j \ne 0$ it follows that $E_2^{i,-j} = 0$ for all $i \ne 0$. So there is a partial degeneration to the isomorphism $H^0(C^{\cdot}) \simeq H_{\mathfrak{m}}^0(M/\underline{y}'M)$. On the other side there is a spectral sequence

$$E_2^{-i,j} = H_j(\underline{y}'; H_{\mathfrak{m}}^i(M)) \Rightarrow E^{i-j} = H^{i-j}(C^{\cdot}).$$

Taking into account that $E_2^{-i,i}$ is of finite length and $E_2^{-i,i} = 0$ for $i < 0$ and $i \ge d$ this provides the estimate of the statement. $\qquad\qquad\square$

In the case that $H_{\mathfrak{m}}^n(M), n = 0, \ldots, d-1$, is an A-module of finite length the result in 3.18 specializes to the following bound

$$L_A(M/\underline{x}M) - e_0(\underline{x}; M) \le \sum_{n=0}^{d-1} \binom{d-1}{i} L_A(H_{\mathfrak{m}}^n(M)).$$

Therefore 3.18 is a generalization of the 'classical' results about Buchsbaum and generalized Cohen-Macaulay modules to an arbitrary situation.

In this context it is noteworthy to say that there is another bound for the length $L_A(M/\underline{x}M)$ of the following type

$$L_A(M/\underline{x}M) \le \sum_{n=0}^{d} L_A(H_n(\underline{x}; H_{\mathfrak{m}}^n(M))).$$

This follows immediately by 3.16 because of $M/\underline{x}M \simeq H^0_{\mathfrak{m}}(M/\underline{x}M)$. In the particular case that $H^n_{\mathfrak{m}}(M)$, $n = 0, \ldots, d-1$, are of finite length it implies that

$$L_A(M/\underline{x}M) \leq \sum_{i=0}^{d-1} \binom{d}{i} L_A(H^n_{\mathfrak{m}}(M)) + L_A(K_M/\underline{x}K_M).$$

To this end note that $H_d(\underline{x}; H^d_{\mathfrak{m}}(M)) \simeq T(H^d(\underline{x}; K_M))$.

Moreover in the case of a Cohen-Macaulay module it yields that $M/\underline{x}M \simeq T(K_M/\underline{x}K_M)$. This implies also the equality of the multiplicities $e_0(\underline{x}; M) = e_0(\underline{x}; K_M)$.

Let us conclude with another application of 3.18.

Corollary 3.19. *Let M denote a finitely generated A-module with $\dim_A M - \operatorname{depth}_A M \leq 1$. Let $\underline{x}, y$, and $\underline{y}'$ be as above. Suppose that A is the quotient of a Gorenstein ring B. Then*

$$L_A(M/\underline{x}M) - e_0(\underline{x}; M) \leq L_A(K_M^{d-1}/\underline{y}'K_M^{d-1}),$$

where $d = \dim_A M$ and $K_M^{d-1} = \operatorname{Ext}_B^{c+1}(M, B)$.

Proof. The proof follows by 3.18 because of $H_{d-1}(\underline{y}'; H^{d-1}_{\mathfrak{m}}(M)) \simeq H_0(\underline{y}'; K_M^{d-1})$. Recall that $T(K_M^{d-1}) \simeq H^{d-1}_{\mathfrak{m}}(M)$ by the Local Duality Theorem. $\square$

References

[1]   M. AUSLANDER: *Modules over unramified regular local rings*, Illinois J. Math. **5** (1961), 631–645.

[2]   M. AUSLANDER, D. A. BUCHSBAUM: *Codimension and multiplicity*, Ann. of Math. **68** (1958), 625–657.

[3]   L. L. AVRAMOV: *Infinite free resolutions*, this volume.

[4]   M. BRODMANN: *Asymptotic stability of* $\operatorname{Ass}(M/I^n M)$, Proc. Amer. Math. Soc. **74** (1979), 16–18.

[5]   M. BRODMANN, J. RUNG: *Local cohomology and the connectedness dimension in algebraic varieties*, Comment. Math. Helv. **61** (1986), 481–490.

[6]   M. BRODMANN, R. Y. SHARP: 'Local Cohomology – An algebraic introduction with geometric applications', Cambr. Univ. Press, to appear.

[7]   W. BRUNS, J. HERZOG: 'Cohen-Macaulay rings', Cambr. Univ. Press, 1993.

[8]   L. BURCH: *Codimension and analytic spread*, Proc. Cambridge Phil. Soc. **72** (1972), 369–373.

[9]   D. EISENBUD: 'Commutative Algebra (with a view towards algebraic geometry)', Springer-Verlag, 1995.

[10]   D. EISENBUD, S. GÔTO: *Linear free resolutions and minimal multiplicity*, J. Algebra **88** (1984), 89–133.

[11]   O. GABBER: *Non negativity of Serre's intersection multiplicities*, Preprint, I.H.E.S., 1995.

[12] M. GREEN: *Koszul homology and the geometry of projective varieties*, J. Diff. Geometry **19** (1984), 125–171.

[13] A. GROTHENDIECK: *Elements de geometrie algebrique, III₁*, Publ. Math., I.H.E.S. **11** (1961).

[14] A. GROTHENDIECK: 'Local cohomology', notes by R. Hartshorne, Lect. Notes in Math., **41**, Springer, 1967.

[15] R. HARTSHORNE: *Complete intersections and connectedness*, Amer. J. Math. **84** (1962), 497–508.

[16] R. HARTSHORNE: *Cohomological dimension of algebraic varieties*, Ann. of Math. **88** (1968), 403–450.

[17] R. HARTSHORNE, A. OGUS: *On the factoriality of local rings of small embedding codimension*, Comm. in Algebra **1** (1974), 415–437.

[18] R. HEITMANN: *A counterexample to the rigidity conjecture for rings*, Bull. Amer. Math. Soc. **29** (1993), 94–97.

[19] M. HOCHSTER: 'Topics in the homological theory of modules over commutative rings', CBMS Regional Conference Series **24**, Amer. Math. Soc., 1975.

[20] M. HOCHSTER, C. HUNEKE: *Indecomposable canonical modules and connectedness*, Proc. Conf. Commutative Algebra (Eds.: W. Heinzer, C. Huneke, J. Sally), Contemporary Math. **159** (1994), 197–208.

[21] C. HUNEKE: *On the associated graded ring of an ideal*, Illinois J. Math. **26** (1982), 121–137.

[22] C. HUNEKE, G. LYUBEZNIK: *On the vanishing of local cohomology modules*, Invent. math. **102** (1990), 73–93.

[23] C. HUNEKE, R. WIEGAND: *Tensor products of modules and the rigidity of* Tor, Math. Ann. **299** (1994), 449–476.

[24] P. JØRGENSEN: *Non-commutative Castelnuovo-Mumford regularity*, Preprint no. 8, Kopenhagen Univ., 1996.

[25] T. KAWASAKI: *On the Macaulayfication of quasi-projective schemes*, Preprint, Tokyo Metrop. Univ., 1996.

[26] S. LICHTENBAUM: *On the vanishing of* Tor *in regular local rings*, Ill. J. Math. **10** (1966), 220–226.

[27] J. LIPMAN: *Equimultiplicity, reduction, and blowing up*, Commutative Algebra: Analytical Methods, Lect. Notes Pure Appl. Math. **68** (1982), 111–147.

[28] H. MATSUMURA: 'Commutative ring theory', Cambridge University Press, 1986.

[29] S. MCADAM: *Asymptotic prime divisors and analytic spreads*, Proc. Amer. Math. Soc. **80** (1980), 555–559.

[30] S. MCADAM, P. EAKIN: *The asymptotic* Ass, J. Algebra **61** (1979), 71–81.

[31] D. MUMFORD: 'Lectures on Curves on an Algebraic Surface', Ann. of Math. Studies No. **59**, Princeton University Press, 1966.

[32] M. P. MURTHY: *A note on factorial rings*, Archiv der Math. **15** (1964), 418–420.

[33] M. NAGATA: *On the chain problem of prime ideals*, Nagoya Math. J. **10** (1956), 51–64.

[34] M. NAGATA: *A treatise on the 14th problem of Hilbert*, Mem. Coll. Sci. Kyoto Univ. **30** (1956), 57–82.

[35] M. NAGATA: 'Local rings', Interscience, 1962.

[36] D. G. NORTHCOTT, D. REES: *Reductions of ideals in local rings*, Proc. Cambr. Phil. Soc. **50** (1954), 145–158.

[37] D. REES: *On a problem of Zariski*, Illinois J. Math. **2** (1958), 145–149.

[38] P. ROBERTS: *Le théorème d'intersection*, C. R. Acad. Sc. Paris, Sér. I, **304** (1987), 177–180.

[39] C. PESKINE, L. SZPIRO: *Dimension projective finie et cohomologie locale*, Publ. Math. I.H.E.S. **42** (1973), 77–119.

[40] L. J. RATLIFF, JR.: *On prime divisors of I^n, n large*, Michigan Math. J. **23** (1976), 337–352.

[41] P. SCHENZEL: 'Dualisierende Komplexe in der lokalen Algebra und Buchsbaum-Ringe', Lect. Notes in Math., **907**, Springer, 1982.

[42] P. SCHENZEL: *Finiteness of relative Rees rings and asymptotic prime divisors*, Math. Nachr. **129** (1986), 123–148.

[43] P. SCHENZEL: *Filtrations and Noetherian symbolic blow-up rings*, Proc. Amer. Math. Soc. **102** (1988), 817–822.

[44] P. SCHENZEL: *Applications of Koszul homology to numbers of generators and syzygies*, J. Pure Appl. Algebra **114** (1997), 287–303.

[45] J. P. SERRE: 'Algèbre locale. Multiplicités', Lect. Notes in Math., **11**, Springer, 1965.

[46] R. Y. SHARP: *Some results on the vanishing of local cohomology modules*, Proc. London Math. Soc. (3) **30** (1975), 177–195.

[47] I. SWANSON: *Linear equivalence of ideal topologies*, Preprint, New Mexico State University, 1997.

[48] C. WEIBEL: 'An Introduction to Homological Algebra', Cambr. Univ. Press, 1994.

Martin-Luther-Universität Halle-Wittenberg
Fachbereich Mathematik und Informatik
D-06 099 Halle (Saale), Germany
E-mail address: schenzel@mathematik.uni-halle.de

Problems and Results on Hilbert Functions of Graded Algebras

Giuseppe Valla

Introduction

Since a projective variety $V = \mathcal{Z}(I) \subseteq \mathbf{P}^n$ is an intersection of hypersurfaces, one of the most basic problems we can pose in relation to V is to describe the hypersurfaces containing it. In particular, one would like to know the maximal number of linearly independent hypersurfaces of each degree containing V, that is to know the dimension of I_d, the vector space of homogeneous polynomials of degree d vanishing on V for various d. Since one knows the dimension $\binom{n+d}{d}$ of the space of all forms of degree d, knowing the dimension of I_d is equivalent to knowing the **Hilbert function** of the homogeneous coordinate ring $A = k[X_0, \ldots, X_n]/I$ of V, which is the vector space dimension of the degree d part of A.

In his famous paper "Über die Theorie der algebraischen Formen" (see [36]) published a century ago, Hilbert proved that a graded module M over a polynomial ring has a finite graded free resolution, and concluded from this fact that its Hilbert function is of polynomial type. The **Hilbert polynomial** of a graded module is thus the polynomial which agrees with the Hilbert function $H_M(s)$ for all large s. Hilbert's insight was that all the information encoded in the infinitely many values of the Hilbert function can be read off from just finitely many of its values.

Hilbert's original motivation for studying these numbers came from invariant theory. Given the action of a group on the linear forms of a polynomial ring, he wanted to understand how the dimension of the space of invariant forms of degree d can vary with d.

The Hilbert function of the homogeneous coordinate ring of a projective variety V, which classically was called the postulation of V, is a rich source of discrete invariants of V and its embedding. The dimension, the degree and the arithmetic genus of V can be immediately computed from the generating function of the Hilbert function of V or from its Hilbert polynomial.

As for the Hilbert polynomial, there are two important geometric contexts in which it appears. The first is the Riemann-Roch theorem which plays an enormously important role in Algebraic Geometry. This celebrated formula arises from the computation of a suitable Hilbert polynomial. Secondly, the information contained in the coefficients of the Hilbert polynomial is usually presented in Algebraic Geometry by giving the Chern classes of the corresponding sheaf, a set of different integers, which can be deduced from the coefficients, and from which the coefficients can also be deduced (see [16], pg. 44).

The influence of Hilbert's paper on Commutative Algebra has been tremendous. Both free resolutions and Hilbert functions have fascinated mathematicians for a long time. In spite of that, many central problems still remain open. The last few decades have witnessed more intense interest in these objects. I believe this is due to two facts. One is the arrival of computers and their breathtaking development. The new power in computation has prompted a renewed interest in the problem of effective construction in Algebra. The ability to compute efficiently with polynomial equations has made it possible to investigate complicated examples that would be impossible to do by hand and has given the right feeling to tackle more difficult questions.

The second is the work of Stanley in connection with Algebraic Combinatorics. In 1978 R. Stanley published the fundamental paper "Hilbert functions of graded algebras" (see [64]) where he related the study of the Hilbert function of standard graded algebras to several basic problems in Combinatorics. An introduction to this aspect of Commutative Algebra is given in Stanley's monograph [66] which is well known as the "green book".

From the point of view of the theory of Hilbert functions, one of the greatest merits of Stanley's work was to restate and explain, in the right setting, a fundamental theorem of Macaulay which characterizes the Hilbert functions of homogeneous k-algebras. This theorem is a great source of inspiration for many researchers in the field.

All the above arguments suffice to justify the choice of "Hilbert functions" as the right topic for a postgraduate course in Commutative Algebra. I would however add some further motivation. One of the delightful things about this subject is that one can begin studying it in an elementary way and, all of a sudden, one can front extremely challenging and interesting problems. An example will help to clarify this. The possible Hilbert functions of a homogeneous Cohen-Macaulay algebra are easily characterized by using Macaulay's theorem and a standard graded prime avoidance theorem. The corresponding problem for a Cohen-Macaulay domain is completely open and we do not even have a guess as to the possible structure of the corresponding numerical functions.

Further, following what is written at the beginning of Eisenbud's excellent book [16], "It has seemed to me for a long time that Commutative Algebra is best practiced with knowledge of the geometric ideas that played a great role in its formation: in short, with a view toward Algebraic Geometry". And what better topic than Hilbert functions to give a concrete example of this way of thinking?

Since I had to choose among the many different scenes which compose the picture, I was very much influenced in my choice by what I know better and what I have worked on recently. This means that the chapters I present here are by no means the most important in the theory of Hilbert functions. However I was also guided by the possibility of inserting open problems and conjectures more than results.

Besides the basic definitions, the first two sections are devoted to the problem of characterizing the possible numerical functions which are the Hilbert functions of some graded algebras with special properties e.g. reduced, Cohen-Macaulay or Gorenstein. Since the main problems arise when the algebras are integral domains, we will show that, by the classical Bertini theorem, the h-vector of a Cohen-Macaulay graded domain of dimension bigger than or equal to two, is the same as that of the homogeneous coordinate ring of an arithmetically Cohen-Macaulay projective curve in a suitable projective space. But, by a more recent result of J. Harris on the generic hyperplane section of a projective curve, this h-vector is also the h-vector of the homogeneous coordinate ring of a set of points in Uniform Position. This gives a very interesting shift from a purely algebraic approach to a more geometric context which has been very useful.

In the third section we introduce a list of conjectures recently made by Eisenbud, Green and Harris and which are closely related to the topics introduced in the first two sections. These conjectures mainly deal with the problem of finding precise bounds on the multiplicity of special Artinian graded algebras, but they can also be read in a more geometric contest. A stronger form of some of these conjectures is a guess which extends, in a very natural way, the main theorem of Macaulay we have introduced above. Despite the fact that we have almost no answer to these questions, they fit very well into the picture, because they are so easy to formulate and so difficult to solve.

A short fourth section is devoted to a longstanding question on the possible Hilbert function of generic graded algebras. A solution in the case of a polynomial ring in two variables is given which uses a nice argument related to the Gröbner basis theory of ideals in the polynomial ring. A very natural problem in this theory closes the section.

In the fifth section we discuss some problems related to the Hilbert function of a scheme of fat points in projective space. We will try to explain how the knowledge of the postulation of these zero-dimensional schemes can be used to study the Waring problem for forms in a polynomial ring and the symplectic packing problem for the four-dimensional sphere.

As for the first problem, following the approach used by T. Iarrobino in his recent work, we will show how a deep theorem of Alexander and Hirschowitz on the Hilbert function of the scheme of generic double fat points gives a complete solution to the Waring problem for forms, which is the old problem of determining the least integer $G(j)$ such that the generic form of degree j in $k[X_0, \ldots, X_n]$ is the sum of $G(j)$ powers of linear forms.

As for the second problem, we just present the relationship between an old conjecture of Nagata on the postulation of a set of fat points in $\mathbf{P}^2$ and the problem on the existence of a full symplectic packing of the four dimensional sphere, as presented in the work of McDuff and Polterovich.

In the last section we come to the Hilbert function of a local Cohen-Macaulay ring and present several results and conjectures on this difficult topic. Since the associated graded ring of a local Cohen-Macaulay ring can be very bad (there do exist local complete intersection domains whose associated graded ring has depth zero), very little is known on the possible Hilbert functions of this kind of ring, even if this has strong connections with the well developed theory of singularities.

Starting from the classical results of S. Abhyankar, D. Northcott and J.Sally, we discuss how Cohen-Macaulay local rings which are extremal with respect to natural numerical constraints on some of their Hilbert coefficients, have good associated graded rings and special Hilbert functions. A recent result of Rossi and Valla, which gives a solution to a longstanding conjecture made by Sally, is also discussed at the end of the section.

I am personally grateful to the organizers of the school for giving me the possibility to teach on my favourite topic. A number of people helped me a great deal in the development of this manuscript. In particular, the section on Waring's problem is very much influenced by a series of talks Tony Geramita gave in Genova last year (see [25]). The last section on the Hilbert function in the local case grew out of a long time cooperation with M. E. Rossi.

Finally I apologize to those whose work I may have failed to cite properly. My feelings are best described by the following sentence which I found in [32]:

"Certainly, the absence of a reference for any particular discussion should be taken simply as an indication of my ignorance in this regard, rather than as a claim of originality."

1. Macaulay's Theorem

Our standard assumption will be that k is a field of characteristic zero, R is the polynomial ring $k[X_1, \ldots, X_n]$ and M a finitely generated graded R-module such that $M_i = 0$ if $i < 0$. If M is such a graded R-module, the homogeneous components M_n of M are k-vector spaces of finite dimension.

Definition 1.1. *Let M be a finitely generated graded R-module. The numerical function*

$$H_M : \mathbf{N} \to \mathbf{N}$$

defined as

$$H_M(t) = dim_k(M_t)$$

for all $t \in \mathbf{N}$ is the Hilbert function of M.

The power series

$$P_M(z) = \sum_{t \in \mathbf{N}} H_M(t) z^t$$

is called the Hilbert Series of M.

For example, for every $t \geq 0$ we have

$$H_R(t) = \binom{n+t-1}{t} \qquad \text{and} \qquad P_R(z) = \frac{1}{(1-z)^n}.$$

The most relevant property of the Hilbert series is the fact that it is additive on short exact sequences of finitely generated R-modules. A classical result of Hilbert says that the series $P_M(z)$ is rational and even more,

Theorem 1.2. (Hilbert-Serre) *Let M be a finitely generated graded R-module. Then there exists a polynomial $f(z) \in \mathbf{Z}[z]$ such that*

$$P_M(z) = \frac{f(z)}{(1-z)^n}.$$

It is easy to see that if $M \neq 0$ then the multiplicity of 1 as a root of $f(z)$ is less than or equal to n so that we can find a unique polynomial

$$h(z) = h_0 + h_1 z + \cdots + h_s z^s \in \mathbf{Z}[z]$$

such that $h(1) \neq 0$ and for some integer d, $0 \leq d \leq n$

$$P_M(z) = \frac{h(z)}{(1-z)^d}.$$

The polynomial $h(z)$ is called *the h-polynomial* of M and the vector $(h_0, h_1, \ldots, h_s)$ *the h-vector* of M.

The integer d is the *Krull dimension* of M.

Now for every $i \geq 0$, let

$$e_i := \frac{h^{(i)}(1)}{i!}$$

and

$$\binom{X+i}{i} := \frac{(X+i)\cdots(X+1)}{i!}.$$

Then it is easy to see that the polynomial

$$p_M(X) := \sum_{i=0}^{d-1} (-1)^i e_i \binom{X+d-i-1}{d-i-1}$$

has rational coefficients and degree $d-1$; further for every $t \gg 0$

$$p_M(t) = H_M(t).$$

The polynomial $p_M(X)$ is called the *Hilbert polynomial* of M and its leading coefficient is

$$\frac{h(1)}{(d-1)!}.$$

This implies that $e_0(M) := h(1)$ is a positive integer which is usually denoted simply by $e(M)$ and called *the multiplicity* of M.

If $d = 0$ we define $e(M) = dim_k(M)$.

Another relevant property of Hilbert series is the so called *sensitivity to regular sequences*. We recall that a sequence $F_1 \ldots, F_r$ of elements of the polynomial ring R is a *regular sequence* on a finitely generated graded R-module M if $F_1, \ldots, F_r$ have positive degrees and F_i is not a zero-divisor on M modulo $(F_1, \ldots, F_{i-1})M$ for $i = 1, \ldots, r$.

If J is the ideal generated by the homogeneous polynomials $F_1, \ldots, F_r$, of degrees $d_1, \ldots, d_r$, one can prove that

$$P_M(z) \leq \frac{P_{M/JM}(z)}{\prod_{i=1}^r (1 - z^{d_i})}$$

with equality holding if and only if the elements $F_1, \ldots, F_r$ form a regular sequence on M. This means that if $L \in R_1$ is regular on M, we have

$$P_{M/LM}(z) = P_M(z)(1 - z).$$

So that the Hilbert function of the module M/LM is given by the so called *first difference function* ΔH_M of H_M which is defined by the formula

$$\Delta H_M(t) = \begin{cases} 1 & \text{if } t = 0 \\ H_M(t) - H_M(t-1) & \text{if } t \geq 1. \end{cases}$$

Now we present a fundamental theorem, due to Macaulay (see [44]), describing exactly those numerical functions which occur as the Hilbert function $H_A(t)$ of a standard homogeneous k-algebra A. Macaulay's theorem says that for each t there is an upper bound for $H_A(t+1)$ in terms of $H_A(t)$, and this bound is sharp in the sense that any numerical function satisfying it can be realized as the Hilbert function of a suitable homogeneous k-algebra.

Let d be a positive integer. One can easily see that any integer a can be written uniquely in the form

$$a = \binom{k(d)}{d} + \binom{k(d-1)}{d-1} + \cdots + \binom{k(j)}{j}$$

where

$$k(d) > k(d-1) > \cdots > k(j) \geq j \geq 1.$$

For example, if $a = 49$, $d = 4$, we get

$$49 = \binom{7}{4} + \binom{5}{3} + \binom{3}{2} + \binom{1}{1}.$$

Given the integers a and d, we let

$$a^{<d>} := \binom{k(d)+1}{d+1} + \binom{k(d-1)+1}{d} + \cdots + \binom{k(j)+1}{j+1}.$$

Hence, for example,

$$49^{<4>} = \binom{8}{5} + \binom{6}{4} + \binom{4}{3} + \binom{2}{2}.$$

Theorem 1.3. (Macaulay) *Let $H : \mathbf{N} \to \mathbf{N}$ be a numerical function. There exists a standard homogeneous k-algebra A with Hilbert function $H_A = H$ if and only if $H(0) = 1$ and $H(t+1) \leq H(t)^{<t>}$ for every $t \geq 1$.*

A numerical function verifying the conditions of the above theorem is called an *admissible* numerical function.

The following example demonstrates the effectiveness of Macaulay's theorem. Let us check that $1+3z+4z^2+5z^3+7z^4$ is not the Hilbert series of a homogeneous k-algebra. We have

$$5 = \binom{4}{3} + \binom{2}{2}$$

$$5^{<3>} = \binom{5}{4} + \binom{3}{3} = 6.$$

In the same paper Macaulay produced an algorithm to construct, given an admissible numerical function H, an homogeneous k-algebra having it as Hilbert function.

Let $n = H(1)$; we fix in the set of monomials of $R = k[X_1, \cdots, X_n]$ a total order compatible with the semigroup structure of this set. Let us fix for example the *degree lexicographic order*. This is the order given by

$$X_1^{a_1} X_2^{a_2} \cdots X_n^{a_n} > X_1^{b_1} X_2^{b_2} \cdots X_n^{b_n}$$

if and only if either $\sum a_i > \sum b_i$ or $\sum a_i = \sum b_i$ and for some integer $j < n$ we have $a_1 = b_1, \cdots, a_j = b_j, a_{j+1} > b_{j+1}$.

Macaulay proved that if for every $t \geq 0$ we delete the smallest $H(t)$ monomials of degree t, the remaining monomials generate an ideal I such that $H_{R/I}(t) = H(t)$ for every $t \geq 0$.

The difficult part of the proof is to show that, due to the upper bound $H(t+1) \leq H(t)^{<t>}$, if a monomial M is in I_t, which means that it has not been erased at level t, then MX_n is not erased at level $t+1$ so that $MX_1, \ldots, MX_n \in I$ as we need.

For example the ideal I in $R = k[X_1, X_2, X_3]$ such that

$$P_{R/I}(z) = \frac{1 + z - 2z^4 + z^5}{(1-z)^2}$$

and constructed by this algorithm, is the ideal

$$I = (X_1^2, X_1 X_2^3, X_1 X_2^2 X_3).$$

The ideal constructed following this method is called a *lex-segment* ideal, in the sense that a k-basis of its homogeneous part of degree t is an initial segment of monomials in the given order. Since it is clearly uniquely determined by the given admissible function, it is called *the lex-segment ideal* associated to the admissible numerical function.

This ideal has some very interesting extremal properties. For example it has the biggest Betti numbers among the perfect ideals with the same multiplicity and codimension (see [70]).

Macaulay's theorem is valid for any homogeneous k-algebra. It is not surprising that additional properties yield further constraints on the Hilbert function.

We will discuss this feature for reduced, Cohen-Macaulay, Gorenstein and domain properties. We start with the reduced case.

Theorem 1.4. *Let* $H : \mathbf{N} \to \mathbf{N}$ *be a numerical function. There exists a reduced homogeneous k-algebra A with Hilbert function $H_A = H$ if and only if either $H = \{1, 0, 0 \dots\}$ or ΔH is admissible.*

Proof. If $A = R/I$ is reduced it is clear that either $depth(A) > 0$, or $I = (X_1, \dots, X_n)$. In the second case all is clear; in the first one we can find a linear form L which is a regular element on A. Hence the difference of the Hilbert function of A is the Hilbert function of the graded algebra A/LA. Conversely, if we are given a function whose difference is admissible, we can construct the lex-segment ideal J in the polynomial ring S such that $H_{S/J} = \Delta H$. This is a monomial ideal which can be deformed to a radical ideal by a general construction due to Hartshorne (see [33] and [26]). This can be achieved in the following way. If $m = X_1^{a_1} X_2^{a_2} \cdots X_n^{a_n}$ is a monomial, we introduce a new variable X_0 and set

$$l(m) := \prod_{j=1}^{n} \prod_{p=0}^{a_j - 1} (X_j - p X_0).$$

If $J = (m_1, \dots, m_s)$ is an ideal generated by monomials, the ideal

$$l(J) := (l(m_1), \dots, l(m_s))$$

is a radical ideal in the polynomial ring $R = k[X_0, \dots, X_n]$, such that X_0 is a regular element on $R/l(J)$ and

$$S/J \simeq (R/l(J))/X_0(R/l(J)).$$

Thus if $I \subseteq R$ is such a radical deformation of J, we have

$$P_{S/J}(z) = P_{(R/I)/X_0(R/I)}(z) = P_{R/I}(z)(1 - z)$$

so that $H_{S/J} = \Delta H_{R/I}$. Since $H_{S/J} = \Delta H = \Delta H_{R/I}$ we get $H = H_{R/I}$. $\qquad\square$

For example the numerical function $H = \{1, 2, 1, 1, 1, \dots\}$ is admissible so it is the Hilbert function of a suitable graded k-algebra, but it is not the Hilbert function of a reduced k-algebra since its difference function $\{1, 1, -1, 0, 0, \dots\}$ is not admissible.

We pass now to the Cohen-Macaulay case and obtain the following characterization of the Hilbert Series of Cohen-Macaulay homogeneous algebras.

Theorem 1.5. *Let $h_0, \dots, h_s$ be a finite sequence of positive integers. There exists an integer d and a Cohen-Macaulay homogeneous k-algebra A of dimension d such that*

$$P_A(z) = \frac{h_0 + h_1 z + \cdots + h_s z^s}{(1 - z)^d}$$

if and only if $(h_0, \dots, h_s)$ is admissible.

Proof. If R/I is Cohen-Macaulay, by prime avoidance in the graded case, we can find a maximal regular sequence of linear forms $L_1, \ldots, L_d$ such that R/I and its Artinian reduction $(R/I)/(L_1, \ldots, L_d)(R/I)$ share the same h-polynomial. □

In particular the h-vector of a Cohen-Macaulay graded algebra has positive coordinates.

2. The Perfect Codimension Two and Gorenstein Codimension Three Case

The problem we have studied in the first section becomes much more difficult if one tries to deal with Cohen-Macaulay graded *domains*. In this case very little is known. Nevertheless, in the codimension two case, thanks to the structure theorem of Hilbert and Burch, we have a complete solution of the problem.

Let us assume that $A = R/I$ is a *non degenerate* (i.e. $I_1 = (0)$) codimension two homogeneous k-algebra which is Cohen-Macaulay. Its Hilbert Series is given by

$$P_A(z) = \frac{1 + 2z + h_2 z^2 + \cdots + h_s z^s}{(1-z)^{n-2}}.$$

The admissibility of the h-polynomial simply means, in this particular case, that if we let a be the *initial degree* of A, which is the least integer j such that $h_j < j+1$, then

$$h_{t+1} - h_t \leq 0$$

for every $t \geq a - 1$. This is a trivial consequence of the formula

$$n^{<k>} = n \text{ for every } k \leq n.$$

Now the Hilbert-Burch theorem says that given a minimal graded free resolution of R/I

$$0 \to \oplus_{i=1}^{s} R(-b_i) \xrightarrow{f} \oplus_{i=1}^{s+1} R(-a_i) \to R \to R/I \to 0$$

the ideal I can be generated by the maximal minors of the $(s+1) \times s$ matrix M associated to the map of free modules f. We can assume that

$$a_1 \leq a_2 \leq \cdots \leq a_{s+1}$$

and

$$b_1 \leq b_2 \leq \cdots \leq b_s.$$

If we let

$$u_{ij} := b_j - a_i \text{ for every } i \text{ and } j$$

the matrix (u_{ij}) is called *the degree matrix* associated to A. The reason for this is that if $M := (G_{ij})$, then $deg(G_{ij}) = u_{ij}$ so that $G_{ij} = 0$ if $u_{ij} \leq 0$.

The matrix (u_{ij}) has the following properties.
• For every i and j we have

$$u_{ij} \leq u_{i\ j+1}$$
$$u_{ij} \geq u_{i+1\ j}.$$

This means that going up or going right in the degree matrix the integers do not decrease. Hence the smallest integer is in the lower left hand corner. • For every i, j, h, k we have

$$u_{ij} + u_{hk} = u_{ik} + u_{hj}.$$

• For every $i = 1, \ldots, s$

$$u_{i+1\ i} = b_i - a_{i+1} > 0$$

The first two properties are immediate; the third one depends on the fact that if for some i we have $u_{i+1\ i} \leq 0$, then in the Hilbert-Burch matrix $G_{hk} = 0$ for $h \geq i$ and $k \leq j$. But this easily implies that the minor which is obtained by deleting the first row has to be zero, contrary to the minimality of the resolution.

• If we further assume I to be prime, then for every $i = 1, \ldots, s - 1$ we have

$$u_{i+2\ i} = b_i - a_{i+2} > 0.$$

This is clear and easily understood by looking at this picture in which we are assuming $s = 5$ and $u_{42} \leq 0$.

$$\begin{pmatrix} \bullet & \bullet & \bullet & \bullet & \bullet \\ \bullet & \bullet & \bullet & \bullet & \bullet \\ \bullet & \bullet & \bullet & \bullet & \bullet \\ 0 & 0 & \bullet & \bullet & \bullet \\ 0 & 0 & \bullet & \bullet & \bullet \\ 0 & 0 & \bullet & \bullet & \bullet \end{pmatrix}$$

Here the minor obtained by deleting the first row, which among the other has minimal degree, splits in two factors, a contradiction to the primality of I.

We clearly have

$$P_A(z) = \frac{1 + \sum_{i=1}^{s} z^{b_i} - \sum_{i=1}^{s+1} z^{a_i}}{(1 - z)^n},$$

hence the h-polynomial of A is

$$h(z) = \frac{1 + \sum_{i=1}^{s} z^{b_i} - \sum_{i=1}^{s+1} z^{a_i}}{(1 - z)^2}.$$

Since the initial degree of A is a_1, the condition

$$h_{t+1} - h_t \leq 0$$

for every $t \geq a_1 - 1$ can better be written through the difference function

$$\Delta h(z) = h(z)(1 - z) = \frac{1 + \sum_{i=1}^{s} z^{b_i} - \sum_{i=1}^{s+1} z^{a_i}}{(1 - z)}.$$

This is a polynomial of degree $b_s - 1$ which we denote by $\sum_{j=0}^{b_s-1} p_j z^j$. With this notation, the admissibility condition of $h(z)$ becomes

$$p_j \leq 0 \text{ for every } j \geq a_1.$$

Now the crucial remark is that we have

$$p_j = 1 + \#\{m | b_m \leq j\} - \#\{m | a_m \leq j\}$$

where $\sharp(X)$ denotes the cardinality of the set X. We claim that if $u_{i+2\ i} > 0$ for every $i = 1, \ldots, s - 1$ then

$$p_j < 0 \Longrightarrow p_{j+1} < 0 \tag{1}$$

unless $j = b_s - 1$.

This can be proved in the following way: let $p_j < 0$ with $j + 1 < b_s$; if $j + 1 < b_1$, then clearly $p_{j+1} \le p_j < 0$. If instead $b_1 \le j + 1$, we can find an integer t such that $1 \le t \le s - 1$ and

$$a_{t+2} < b_t \le j + 1 < b_{t+1}.$$

This means that we have at least $t + 2$ $a_i's$ and exactly t $b_i's$ which are less than or equal to $j + 1$. Hence we get

$$p_{j+1} = 1 + \sharp\{m|b_m \le j+1\} - \sharp\{m|a_m \le j+1\} \le 1 + t - (t+2) = -1,$$

as claimed.

We remark that the condition of the claim simply means that the h-polynomial is of *decreasing type* or *strictly decreasing* i.e. after the first step down there is no flat.

We have proved the following result (see [28]).

Theorem 2.1. *Let $A = R/I$ be a codimension two non-degenerate Cohen-Macaulay homogeneous domain. Then the h-polynomial of A is of decreasing type.*

For example there does not exist a graded Cohen-Macaulay domain A of dimension d and Hilbert series

$$P_A(z) = \frac{1 + 2z + z^2 + z^3}{(1 - z)^d}.$$

But the graded algebra

$$A = k[X_1, X_2]/(X_1^2, X_1 X_2, X_2^3)$$

has Hilbert series

$$P_A(z) = 1 + 2z + z^2 + z^3.$$

If we start with a polynomial

$$h(z) = 1 + 2z + h_2 z^2 + \cdots + h_s z^s$$

which is of decreasing type, one can ask whether we can construct a Cohen-Macaulay graded domain of dimension d such that

$$P_A(z) = \frac{h(z)}{(1 - z)^d}.$$

Peskine and Gruson proved in [29] that the answer is positive and that the projective coordinate ring of a smooth arithmetically Cohen-Macaulay curve in $\mathbf{P}^3$ does the job. This result can be obtained also by using a sort of deformation which we are going to explain in the following particular case (see [34]).

Let

$$h(z) = 1 + 2z + 3z^2 + z^3.$$

This is of decreasing type but if we consider the lex-segment ideal with this Hilbert series, we get the algebra

$$A = k[X_1, X_2]/(X_1^3, X_1^2 X_2, X_1 X_2^2, X_2^4).$$

This is an Artinian codimension two algebra whose degree matrix is

$$\begin{pmatrix} 1 & 1 & 2 \\ 1 & 1 & 2 \\ 1 & 1 & 2 \\ 0 & 0 & 1 \end{pmatrix}$$

Since $u_{42} = 0$ we have no hope to deform the corresponding Hilbert Burch matrix in order to get a domain. Better, we recall that

$$h(z) = \frac{1 + \sum_{i=1}^{s} z^{b_i} - \sum_{i=1}^{s+1} z^{a_i}}{(1-z)^2},$$

hence we get

$$(1 + 2z + 3z^2 + z^3)(1-z)^2 = 1 - 3z^3 + z^4 + z^5 = 1 + \sum_{i=1}^{s} z^{b_i} - \sum_{i=1}^{s+1} z^{a_i}$$

and we can let

$$a_1 = a_2 = a_3 = 3, \; b_1 = 4, \; b_2 = 5.$$

The corresponding degree matrix is

$$\begin{pmatrix} 1 & 2 \\ 1 & 2 \\ 1 & 2 \end{pmatrix}.$$

We then consider the ideal J of $k[X_1, X_2]$ which is generated by the maximal minors of the matrix

$$\begin{pmatrix} X_1 & 0 \\ X_2 & X_1^2 \\ 0 & X_2^2 \end{pmatrix}$$

It is clear that the algebra $B = k[X_1, X_2]/J$ is an Artinian graded algebra such that

$$P_B(z) = \frac{1 - 3z^3 + z^4 + z^5}{(1-z)^2} = 1 + 2z + 3z^2 + z^3.$$

If we are able to deform this matrix in order to get a domain, we are done. Let us consider the polynomial ring

$$R = k[X_1, X_2, X_3, X_4]$$

and the ideal I generated by the maximal minors of the matrix obtained by deforming the above matrix in the following way. We replace the zeroes by forms in the new variables and of suitable degrees, accordingly to the degree matrix. For example we can consider the matrix

$$\begin{pmatrix} X_1 & X_3X_4 \\ X_2 & X_1^2 \\ X_4 & X_2^2 \end{pmatrix}.$$

Then it is easy to see that I is the Kernel of the map

$$\phi : R \to k[u, v]$$

given by $\phi(X_1) = u^3v^4$, $\phi(X_2) = u^2v^5$, $\phi(X_3) = u^7$ and $\phi(X_4) = v^7$. Hence $A = R/I$ is a codimension two Cohen-Macaulay domain such that

$$P_A(z) = \frac{1 + 2z + 3z^2 + z^3}{(1 - z)^2}.$$

We remark that in order to deform the matrix to get a prime ideal, we can need many new variables so that the embedding dimension and the dimension can grow arbitrarily. But the classical Bertini's theorem tells us that we may find an example with $dim(A) = 2$.

Theorem 2.2. (Bertini) *Let A be a Cohen-Macaulay graded domain of dimension greater or equal than two. Then the h-vector of A is the h-vector of a Cohen-Macaulay graded domain of dimension two.*

For a proof of this result see Corollary 3.3 in [65].
We collect the above remarks in the following theorem.

Theorem 2.3. *Let $h(z) = 1 + 2z + h_2z^2 + \cdots + h_sz^s$ be a polynomial with integer coefficients. There exists an integer d and a Cohen-Macaulay graded domain A of dimension d such that*

$$P_A(z) = \frac{1 + 2z + h_2 + \cdots + h_sz^s}{(1 - z)^d}$$

if and only if $h(z)$ is admissible and of decreasing type.

If one deletes the codimension two hypothesis very little is known on the possible Hilbert function of graded Cohen-Macaulay domain. Several conjectures have been made but very few results are known. For example Hibi made the following conjecture in [35].

Recall that a finite sequence $(h_o, h_1, \ldots, h_s) \in \mathbf{Z}^{s+1}$ is called *flawless* if
i) $h_i \leq h_{s-i}$ for every $0 \leq i \leq [s/2]$
ii) $h_0 \leq h_1 \leq \cdots \leq h_{[s/2]}$.
It is called *unimodal* if

$$h_0 \leq h_1 \leq \cdots \leq h_j \geq h_{j+1} \geq \cdots \geq h_s$$

for some j, $0 \leq j \leq s$.
Hibi conjectured that the h-vector of a Cohen-Macaulay domain is flawless.

This conjecture has been recently disproved by Niesi and Robbiano. They constructed a Cohen-Macaulay graded domain A with

$$P_A(z) = \frac{1 + 3z + 5z^2 + 4z^3 + 4z^4 + z^5}{(1-z)^2}.$$

This domain is the homogeneous coordinate ring of a reduced irreducible curve in $\mathbf{P}^4$ (see [50]).

The following conjecture has been stated by Stanley.

Conjecture 2.4. (Unimodal Conjecture) *The h-vector of a graded algebra which is a Cohen-Macaulay integral domain is unimodal.*

The best information we know on the Hilbert function of a graded Cohen-Macaulay domain comes from a very crucial device introduced by J.Harris in his fundamental approach to Castelnuovo theory (see [31]). This result could be considered as a further step after Bertini's theorem. It tells us that the h-vector of a Cohen-Macaulay graded domain of dimension at least two is also the h-vector of the homogeneous coordinate ring of a set of points with a good uniformity property.

More precisely, let C be a reduced, irreducible and non-degenerate curve in $\mathbf{P}^n$; the general hyperplane section of C is a set Γ of $e = deg(C)$ distinct points in $\mathbf{P}^{n-1}$. The key idea of Harris is the uniform position principle: *The points of Γ are indistinguishable from one another.*

Theorem 2.5. *Let Γ be a set of points which are the general hyperplane section of a reduced, irreducible projective curve. For every subset Γ' of d' points of Γ and for every $n \geq 0$*

$$H_{\Gamma'}(n) = \min\{d', H_\Gamma(n)\}.$$

A set of points which verifies the above condition is said to be *in uniform position.* We will write UP for short.

If we start with a non-degenerate graded Cohen-Macaulay domain A of positive dimension $d \geq 2$ and codimension g, then, as we have seen before, the h-vector of A is the same as that of a graded Cohen-Macaulay domain of dimension two and codimension g. This can be seen as the homogeneous coordinate ring of a reduced, irreducible projective curve. Hence the given h-vector is also the h-vector of a set of points in uniform position in $\mathbf{P}^g$. As suggested by Harris, the question becomes

Problem 2.6. *What could be the Hilbert function of a set of points in Uniform Position?*

We need good algebraic properties of a set of points in Uniform Position. The first easy result is the following one (see [45]).

Proposition 2.7. *Let I be the defining ideal of a set X of points in $\mathbf{P}^n$ in UP. If a is the initial degree of I, then I_a has no fixed components.*

We remark that if $dim(I_a) = 1$ this means that the generator of this vector space is an irreducible polynomial. We will use later this result to prove that the h-vector of a set of points in UP in $\mathbf{P}^2$ is of decreasing type.

One of the most important result on Harris problem is a byproduct of a very elementary inequality one can prove for points in UP.

In the following, given a set X of points in the projective space $\mathbf{P}^n$, we denote by $H_X(t)$ the Hilbert function of the homogeneous coordinate ring of X.

Theorem 2.8. *Let X be a set of r distinct points in $\mathbf{P}^n$ in UP. Then we have*

$$H_X(m + t) \geq \min\{r, H_X(m) + H_X(t) - 1\}$$

for every $m, t \geq 0$.

We remark that this property is not so deep. For example does not skip the h-vector $(1, 2, 3, 2, 2)$ which is not strictly decreasing. Namely, the corresponding Hilbert function is $(1, 3, 6, 8, 10, 10, \dots)$ which does not contradicts the above inequality.

In [8] further interesting conditions that have to be satisfied by points in UP in $\mathbf{P}^n$, $n \geq 3$, are given.

As a trivial application of this result we get the following theorem (see Theorem 2.1 in [65]).

Theorem 2.9. (Stanley inequalities) *Let A be a graded Cohen-Macaulay domain of dimension $d \geq 2$. Let $(1, h_1, \dots, h_s)$ be the h-vector of A and $m \geq 0$, $n \geq 1$ with $m + n < s$. Then*

$$h_1 + h_2 + \cdots + h_n \leq h_{m+1} + h_{m+2} + \cdots + h_{m+n}.$$

Proof. The h-vector of A is the same as the h-vector of a set X of points in UP. Hence

$$P_X(z) = \frac{1 + h_1 z + \cdots + h_s z^s}{1 - z}$$

so that $H_X(p) = \sum_{j=0}^{p} h_j$ for every $p \geq 0$ and the degree of X is $1 + h_1 + \cdots + h_s$. By the above theorem we get

$$H_X(m + n) \geq \min\{1 + h_1 + \cdots + h_s, H_X(m) + H_X(n) - 1\}$$

hence

$$1 + h_1 + \cdots + h_{m+n} \geq \min\{1 + h_1 + \cdots + h_s, 1 + h_1 + \cdots + h_m + 1 + h_1 + \cdots + h_n - 1\}.$$

Since $m + n < s$, this implies

$$h_1 + \cdots + h_{m+n} \geq h_1 + \cdots + h_m + h_1 + \cdots + h_n$$

so

$$h_{m+1} + \cdots + h_{m+n} \geq h_1 + \cdots + h_n$$

as desired.   □

We now wish to examine what can be said about the Hilbert function of a Gorenstein graded algebra. Since the h-vector does not change by passing to an Artinian reduction, we may assume that A is Artinian and s is *the socle degree* of A, which is the biggest integer t such that $A_t \neq 0$. With these notations we have the following theorem (see [13]).

Theorem 2.10. (Hilbert Function by Liaison) *Let A be an Artinian Gorenstein algebra with socle degree s and let I be an homogeneous ideal in A. If we let $J = 0 : I$, we have for every t such that $0 \leq t \leq s$,*

$$H_A(t) = H_A(s - t) = H_I(t) + H_J(s - t).$$

In particular

$$H_A(t) = H_{A/I}(t) + H_{A/J}(s - t).$$

Proof. Since A is Gorenstein we have

$$0 : A_1 = A_s.$$

Hence if i and j are nonnegative integers with $i + j \leq s$ and $F \in A_i$ is such that $FA_j = 0$, we get $FA_{j-1}A_1 = 0$ so that $FA_{j-1} \subseteq A_s$. This implies $FA_{j-1} = 0$ and going on in this way we get $F = 0$. This means that the k-bilinear map induced by multiplication

$$A_i \times A_j \to A_{i+j}$$

is nonsingular.

Now let I be an homogeneous ideal of A and let $J := 0 : I$. It is easy to see that we have

$$(0 : I_t)_{s-t} = J_{s-t}$$

and this implies that J_{s-t} is the kernel of the canonical map

$$A_{s-t} \to Hom(I_t, A_s).$$

This means that

$$H_A(s - t) \leq H_I(t) + H_J(s - t).$$

On the other hand, by the nonsingularity of the pairing

$$A_{s-t} \times A_t \to A_s,$$

we have an embedding

$$J_{s-t} \to Hom(A_t/I_t, A_s)$$

which gives

$$H_J(s - t) \leq H_{A/I}(t).$$

Hence

$$H_A(s - t) \leq H_I(t) + H_J(s - t) \leq H_A(t) = H_A(s - (s - t)) \leq H_A(s - t).$$

Thus we have equality above and the conclusion follows.   □

A trivial consequence of this result is that

the h-vector of a Gorenstein algebra is symmetric.

It is easy to see that the converse does not hold. Let us consider the graded algebra $A = k[X_1, X_2]/(X_1^2, X_1 X_2, X_2^3)$. Then the h vector of A is $(1, 2, 1)$ which is symmetric, but A is not Gorenstein since the socle is not a one-dimensional vector space.

However, somewhat surprisingly, it is sufficient to assume that A is a Cohen-Macaulay integral domain (see [64]).

This is a consequence of a more general result which we want now to describe.

Definition 2.11. *Given a set X of s points in $\mathbf{P}^n$, we say that X is a Cayley-Bacharach scheme (CB for short), if for every subset Y of $s - 1$ points of X we have for every $t \geq 0$,*

$$H_Y(t) = \min\{s - 1, H_X(t)\}.$$

With this definition, one can prove easily, by mean of the main theorem in [27], the following result.

Theorem 2.12. *Let X be a set of s points in $\mathbf{P}^n$. If X is a CB scheme, then the h-vector of X is symmetric if and only if the homogeneous coordinate ring of X is Gorenstein.*

Since it is clear that UP implies CB, we have as a corollary that if the h-vector of a graded Cohen-Macaulay domain A is symmetric, then A is Gorenstein.

We want now to prove that the h-vector of a set of points in UP in $\mathbf{P}^2$ is strictly decreasing (see [45]). This generalize the result we have proved on the h-vector of a codimension two graded Cohen-Macaulay domain.

Theorem 2.13. *The h-vector of the homogeneous coordinate ring of a set of points in UP in $\mathbf{P}^2$ is of decreasing type.*

Proof. Let X be a set of points in UP in $\mathbf{P}^2$ and I the defining ideal of X in $R = k[X_0, X_1, X_2]$. Since we have seen that the initial part of I has no fixed components, we may assume that $I = (F_1, \ldots, F_r)$ where $\deg(F_1) \leq \deg(F_2) \leq \cdots \leq \deg(F_r)$ and F_1, F_2 is a regular sequence in R. Let $J = (F_1, F_2)$, $a = \deg(F_1)$ and $b = \deg(F_2)$. We may choose a linear form L which is a regular element both on R/I and R/J. As usual we may assume $L = X_0$. If we let $S := R/X_0 R$ we have

$$(1 - z)P_{R/J}(z) = P_{(R/J)/X_0(R/J)}(z) = P_{S/JS}(z),$$

hence

$$(1 - z)P_{R/J}(z) = (1 - z)P_R(z)(1 - z^a)(1 - z^b) = P_{S(z)}(1 - z^a)(1 - z^b).$$

This proves that the residue classes of F_1 and F_2 in $R/X_0 R$ form a regular sequence.

We are left to the case where I is an ideal in $S = k[X_1, X_2]$ which contains a regular sequence F, G of forms of degree a and b and we need to prove that the Hilbert function of S/I is decreasing for every $t \geq b - 1$.

This is now a consequence of a forthcoming result on the tail of the Hilbert function of suitable zero-dimensional ideals. □

This proof does not use the structure theorem of Hilbert-Burch, hence, it seems to be the right way towards an extension of the above result for points in UP in $\mathbf{P}^n$ $(n \geq 3)$. Thus one is led to study the following problem which will be discussed also later.

Problem 2.14. *What can be said about the Hilbert function of an ideal in a polynomial ring R if we know the degrees of a maximal regular sequence contained in I?*

The following theorem gives some control on the tail of the Hilbert function of R/I in terms of the degrees of a regular sequence inside the ideal I. However it is not enough even to guess the possible shape of the Hilbert function of a set of points in UP in $\mathbf{P}^3$.

Theorem 2.15. *Let I be a zero-dimensional ideal of R such that I contains a regular sequence $F_1, \ldots, F_n$ of forms of degrees $d_1 \leq d_2 \leq \cdots \leq d_n$. Set $d := \sum_{i=1}^{n} d_i - n$ and $s := socdeg(R/I)$.*
  a) For every $t \geq d - d_n + 1$,
$$H_{R/I}(t) \geq H_{R/I}(t+1)$$
  b) For $d - d_{n-1} + 1 \leq t \leq s$,
$$H_{R/I}(t) > H_{R/I}(t+1).$$
  c) If $(F_1, \ldots, F_{n-2})$ is a prime ideal, then for $d - d_{n-1} + 1 \leq t < s$,
$$H_{R/I}(t) \geq H_{R/I}(t+1) + n - 1.$$

A proof of this result can be found in [68].

Coming back to the Gorenstein case, it is natural to ask for other restrictions on the Hilbert function of a Gorenstein algebra, besides the symmetry of the h-vector. Since a rational function $Q(z)$ is the Hilbert series of some graded Gorenstein algebra of dimension d if and only if $(1-z)^d Q(z)$ is the Hilbert series of some 0-dimensional graded Gorenstein algebra, it suffices to consider the case when A is Artinian.

Problem 2.16. *What sequences $\{1, h_1, h_2, \ldots, h_s\}$ with $h_s \neq 0$ satisfies $h_i = H_A(i)$ for some Artinian graded Gorenstein algebra A?*

We call such a sequence a *Gorenstein sequence*.

The conditions that the sequence be admissible and symmetric (i.e. $h_i = h_{s-i}$) are by no means sufficient. For example the sequence $(1, 3, 6, 7, 9, 7, 6, 3, 1)$ is admissible and symmetric but is not a Gorenstein sequence as we will see in the next theorem.

By using the structure theorem, due to Buchsbaum and Eisenbud, of codimension three Gorenstein ideals, one can prove, as we did in the perfect codimension two case, the following theorem (see [64]).

Theorem 2.17. *Let* $\{1, 3, h_2, \ldots, h_s\}$ *be a sequence of nonnegative integers. Then it is a Gorenstein sequence if and only if it is symmetric and the sequence* $\{1, 2, h_2 - h_1, , \ldots, h_{[s/2]} - h_{[s/2]-1}\}$ *is admissible.*

By using more or less the same ideas as in the perfect codimension two case, one can prove the following result (see [14]).

Theorem 2.18. *Let* $h(z) = 1 + 3z + h_2 z^2 + \cdots + h_s z^s$ *be a polynomial with integer coefficients. There exists an integer d and a Gorenstein graded domain A of dimension d such that*

$$P_A(z) = \frac{1 + 3z + h_2 + \cdots + h_s z^s}{(1 - z)^d}$$

if and only if $h(z)$ is symmetric and $\{1, 2, h_2 - h_1, , \ldots, h_{[s/2]} - h_{[s/2]-1}\}$ *is admissible and of decreasing type.*

For example the vector $(1, 3, 4, 5, 5, 4, 3, 1)$ is symmetric and $s = 7$. The vector $(1, 2, 1, 1)$ is admissible but is not of decreasing type. Hence the given vector $(1, 3, 4, 5, 5, 4, 3, 1)$ is the h-vector of a Gorenstein algebra, but not of a Gorenstein domain.

Instead, the vector $(1, 3, 6, 10, 13, 14, 14, 13, 10, 6, 3, 1)$ is symmetric and $s = 11$. Since the vector $(1, 2, 3, 4, 3, 1)$ is admissible of decreasing type, the given vector is the h-vector of a Gorenstein domain.

Coming back to the original question about Gorenstein sequences, Stanley and Iarrobino independently made the conjecture that $\{1, h_1, h_2, \ldots, h_s\}$ is a Gorenstein sequence if and only if it is symmetric and moreover the sequence $\{1, h_1 - 1, h_2 - h_1, , \ldots, h_{[s/2]} - h_{[s/2]-1}\}$ is admissible. Unfortunately this conjecture, or even the weaker conjecture that the h-vector is unimodal, is false.

Stanley constructed in [64] an Artinian Gorenstein graded algebra whose h-vector is $(1, 13, 12, 13, 1)$. Bernstein and Iarrobino showed examples of Artinian Gorenstein algebras with a non unimodal h-vector and embedding dimension 5 (see [7]).

Recently Boij and Laksov showed that the examples of Stanley and Bernstein-Iarrobino are extreme cases in large classes of counterexamples (see [9]). Namely they have minimal socle degree and embedding dimension, respectively, in this class.

We close with the following question:

Problem 2.19. *If $(1, 4, h_2, \ldots, 4, 1)$ is a Gorenstein sequence, is it unimodal?*

3. The EGH Conjecture

Recently Eisenbud, Green and Harris, in the framework of the so called Higher Castelnuovo Theory, set some very nice conjectures which can be related to a number of questions we have introduced in the first two sections.

Following the idea that "curves of rather large genus for their degree are quite special" they came up with the following conjecture on the Hilbert function of a set of points in the projective space (see [17]).

Conjecture 3.1. *If $X \subseteq \mathbf{P}^n$ is a nondegenerate collection of s points lying on m independent quadrics whose intersection Y is zero-dimensional, then*

$$s \le 2^a + 2^b + n - a - 1$$

where $\binom{a}{2} + b$ is the 2-binomial expansion of $\binom{n+1}{2} - m$.

In the above setting, it is clear that $deg(X) \le deg(Y)$. By passing to the homogeneous coordinate ring of Y modulo a general linear form, i.e. by passing to an Artinian reduction, the above conjecture becomes the following :

Conjecture 3.2. *($\mathcal{C}_n$) Let $A = k[X_1, \ldots, X_n]/I$ where I is a zero-dimensional ideal generated by quadrics. If $\binom{a}{2} + b$ is the 2-binomial expansion of $H_A(2)$, then*

$$e(A) = dim_k(A) \le 2^a + 2^b + n - a - 1.$$

We will prove later that this bound is sharp, if indeed it holds.

One can relate the above conjecture to the classical Cayley-Bacharach theory. Recall that a special case of a modern version of the classical Cayley-Bacharach theorem is the following result which is also an immediate consequence of the result on the tail of the Hilbert Function we have seen before (see Theorem 2.15).

Theorem 3.3. *Let X be a zero-dimensional scheme in $\mathbf{P}^n$ which is the complete intersection of n hypersurfaces of degree $d_1, \ldots, d_n$. Let $d := \sum_{i=1}^{n} d_i - n - 1$, and Y a subscheme of X with $deg(Y) = deg(X) - 1$. Then every hypersurface of degree d containing Y must contain X.*

If $d_1 = \cdots = d_n = 2$, then $d = n - 1$ and

$$deg(X) - 1 = 2^n - 1 > 2^n - 2 = 2^n - 2^{n-(n-1)}.$$

Hence one can ask more generally the following question.

Conjecture 3.4. *($\mathcal{CB}$) Let X be a zero-dimensional scheme in $\mathbf{P}^n$ which is a complete intersection of quadrics. Let Y be a subscheme of X with $deg(Y) > deg(X) - 2^{n-m} = 2^n - 2^{n-m}$. Then every hypersurface of degree m containing Y must contain X.*

Let us consider in $S := k[X_0, \ldots, X_n]$ the defining ideals $\beta \subseteq \alpha$ of X and Y respectively. The conjecture can be read as follows: if $e(S/\alpha) > 2^n - 2^{n-m}$, then $H_{S/\alpha}(m) = H_{S/\beta}(m)$.

As before, we may assume that X_0 is S/α and S/β regular. If we let $R := S/X_0 S$, $I = \alpha + (X_0)/(X_0)$ and $J = \beta + (X_0)/(X_0)$, we have

$$P_{S/\alpha}(z) = P_{R/I}(z)/(1 - z)$$

and

$$P_{S/\beta}(z) = P_{R/J}(z)/(1 - z).$$

From this it follows that

$$H_{S/\alpha}(m) = \sum_{j=0}^{m} H_{R/I}(j) \qquad \text{and} \qquad H_{S/\beta}(m) = \sum_{j=0}^{m} H_{R/J}(j).$$

Further $e(S/\alpha) = e(R/I)$ and $J \subseteq I$ is a zero-dimensional ideal generated by a regular sequence of quadrics. To prove the above conjecture we need to prove that if $e(S/\alpha) = e(R/I) > 2^n - 2^{n-m}$, then $H_{R/I}(j) = H_{R/J}(j)$ for every $j = 0, \ldots, m$. But if $H_{R/I}(j) < H_{R/J}(j)$ for some j, then $I_j \supset J_j$ and we can find an element $F \in I_j$, $F \notin J$. Hence $I \supseteq (J, F) \supset J$ and we have

$$2^n - 2^{n-m} < e(R/I) \le e(R/(J, F)).$$

Thus the above conjecture is proved if one can prove the following zero-dimensional version.

Conjecture 3.5. $(\mathcal{A}_{nj})$ Let $A = R/I$ where $I = J + (F)$ with J a zero-dimensional ideal generated by a regular sequence of quadrics and $F \notin J$ an element of degree j. Then

$$e(A) \le 2^n - 2^{n-j}.$$

This conjecture is related to $\mathcal{C}_n$ in the following way. The first non trivial case in $\mathcal{C}_n$ is when I is an almost complete intersection, i.e. when I is generated by $n + 1$ quadrics. This means that

$$H_A(2) = \binom{n+1}{2} - (n+1) = \binom{n-1}{2} + (n-2).$$

Since

$$2^{n-1} + 2^{n-2} + n - (n-1) - 1 = 3 \cdot 2^{n-2} = 2^n - 2^{n-2},$$

it is clear that the first non trivial case of $\mathcal{C}_n$, namely the case $H_A(2) = \binom{n-1}{2} + (n-2)$, is equivalent to $\mathcal{A}_{n2}$.

This is the conjecture stated several years ago by Rossi and Valla in [55] and proved under the very restrictive assumption $n \le 6$. Despite the fact that much time has gone by and many variations of the theme have been introduced in [17], this is the unique positive result we know on these conjectures. We will prove this theorem later.

A different way to think about the classical Cayley-Bacharach theorem is the following. Here, for a zero-dimensional scheme X in $\mathbf{P}^n$, the number

$$\omega_X(t) := deg(X) - H_X(t)$$

is called the *superabundance* of the linear system of the hypersurfaces of degree t passing through X. It is well known that

$$\omega_X(t) = h^1 \mathcal{I}_X(t)$$

where $\mathcal{I}$ is the ideal sheaf of X. With this notation, the classical Cayley-Bacharach theorem takes the following form.

Theorem 3.6. *Let X be a zero dimensional scheme in $\mathbf{P}^n$ which is the complete intersection of n hypersurfaces of degree $d_1, \ldots, d_n$. Let $d := \sum_{i=1}^{n} d_i - n$, and Y a subscheme of X with $\omega_Y(d-1) > 0$. Then $Y = X$.*

Accordingly, we have the following new version of the above conjecture.

Conjecture 3.7. *($\mathcal{GCB}_m$) Let X be a zero dimensional scheme in $\mathbf{P}^n$ which is a complete intersection of quadrics and let Y be a subscheme of X with $\omega_Y(m) > 0$. Then*

$$deg(Y) \geq 2^{m+1}.$$

Note that this conjecture is independent of the embedding dimension.

By using the same notation as before, the condition $\omega_Y(m) > 0$ clearly implies that the socle degree of R/I is bigger or equal than $m + 1$. Hence we may state the following conjecture which implies $\mathcal{GCB}_m$ for every m.

Conjecture 3.8. *($\mathcal{A}$) Let $A = R/I$ be an Artinian graded algebra where I is an ideal containing a zero-dimensional regular sequence of quadrics. If $s = socdeg(A)$ then*

$$e(A) \geq 2^s.$$

It is not so difficult to see that in order to prove the above conjecture, we may assume that A is Gorenstein. Hence we state yet another conjecture.

Conjecture 3.9. *($\mathcal{G}$) Let $A = R/I$ be an Artinian graded Gorenstein algebra where I is an ideal containing a zero-dimensional regular sequence of quadrics. If $s = socdeg(A)$ then*

$$e(A) \geq 2^s.$$

It turns out that $\mathcal{A}$, $\mathcal{G}$ and $\mathcal{A}_{nj}$ are equivalent and we have the promised theorem.

Theorem 3.10. *Conjecture $\mathcal{G}$ holds if the socle degree is less than or equal to 4.*

Proof. Let $R = k[X_1, \ldots, X_n]$ and $H_A(1) = r \leq n$. Then $dim(I_1) = n - r$ and we can find in I_1 linear forms $L_1, \ldots, L_{n-r}$ which are linearly independent. Since I contains a zero-dimensional regular sequence of quadrics, it is easy to see that we can find in I_2 elements $F_{n-r+1}, \ldots, F_n$ such that $\{L_1, \ldots, L_{n-r}, F_{n-r+1}, \ldots, F_n\}$ is a regular sequence in R. If we let $K := (L_1, \ldots, L_{n-r}, F_{n-r+1}, \ldots, F_n)$, then $K \subseteq I$ and R/K is a Gorenstein Artinian algebra with $e(R/K) = 2^r$ and $socdeg(R/K) = r \geq s$.

Since $H_A(1) = r$, the condition $r \geq s$ gives the conclusion for $s = 1, 2, 3$. Let $s = 4$ so that $r = H_A(1) = H_A(3)$. If $r = 4$ then by theorem 2.10 we get

$$H_{R/K}(4) = H_{R/I}(4) + H_{R/K:I}(0)$$

so that $K : I = R$ and $K = I$. This gives $e(R/I) = e(R/K) = 2^4$ as wanted. Let $r \geq 5$; since $3^{<2>} = 4$ and by Macaulay theorem $r = H_A(3) \leq H_A(2)^{<2>}$, we must have $H_A(2) \geq 4$ so that

$$e(A) \geq 1 + 5 + 4 + 5 + 1 = 16 = 2^4$$

and the conclusion follows. □

Let us see how the above theorem implies conjecture $\mathcal{A}_{n2}$ for $n \leq 6$. We have an exact sequence of Artinian rings

$$0 \to R/(J : F) \to R/J \to R/I \to 0$$

from which we deduce

$$e(R/I) = e(R/J) - e(R/(J : F)) = 2^n - e(R/(J : F)).$$

The conclusion follows if we can prove that

$$e(R/(J : F)) \geq 2^{n-2}.$$

Hence we must prove that

$$socdeg(R/(J : F)) = n - 2.$$

Now $socdeg(R/J) = n$, hence $R_{n+1} \subseteq J$ and $F \in J : R_{n-1}$ so that $R_{n-1} \subseteq J : F$. On the other hand, if $R_{n-2} \subseteq J : F$, then $FR_{n-3} \subseteq J : R_1$ and $F \in R_2$, a contradiction.

Finally we state a last conjecture which is stronger than $\mathcal{C}_n$. This conjecture is very interesting since it can be seen as a natural extension of Macaulay's theorem.

Recall that if d is a positive integer, any integer a can be written uniquely in the form

$$a = \binom{k(d)}{d} + \binom{k(d-1)}{d-1} + \cdots + \binom{k(j)}{j}$$

where

$$k(d) > k(d-1) > \cdots > k(j) \geq j \geq 1.$$

In the following we let

$$a_{(d)} := \binom{k(d)}{d+1} + \binom{k(d-1)}{d} + \cdots + \binom{k(j)}{j+1}.$$

Conjecture 3.11. ($\mathcal{M}$) *Let* $A = R/I$ *where* I *is an homogeneous ideal of* R *which contains a zero-dimensional regular sequence of quadrics. Then for every* $t \geq 1$ *we have*

$$H_A(t+1) \leq H_A(t)_{(t)}.$$

Let us prove that this conjecture implies conjecture $\mathcal{C}_n$.

Let A as in conjecture $\mathcal{C}_n$. Then I contains a zero-dimensional regular sequence of quadrics, hence if $H_A(2) = \binom{a}{2} + b$, $a > b \geq 0$, then $H_A(0) = 1$, $H_A(1) = n$, $H_A(2) = \binom{a}{2} + b$, $H_A(3) \leq \binom{a}{3} + \binom{b}{2}$, and so on. Summing up we get

$$e(A) = dim_k(A) \leq 1 + n + \left[\binom{a}{2} + \cdots + \binom{a}{a}\right] + \left[\binom{b}{1} + \binom{b}{2} + \cdots + \binom{b}{b}\right] = n$$

$$= 1 + n + 2^a - a - 1 + 2^b - 1 = 2^a + 2^b + n - a - 1.$$

We finish this section by proving that the bound in the above statement is sharp, if true.

Let h be an integer $0 < h \leq \binom{n}{2}$ and $t := \binom{n}{2} - h$. We claim that the ideal

$$I = (X_1^2, \ldots X_n^2, X_i X_j)$$

where $X_i X_j$ runs among the biggest t square free monomials in the lexicographic order, has Hilbert function

$$(1, n, h, h_{(2)}, h_{(2)_{(3)}}, \ldots).$$

In order to prove the claim, we write $h = \binom{a}{2} + b$ with $a > b \geq 0$. Then

$$t = \binom{n}{2} - \binom{a}{2} - b.$$

Now

$$X_1 X_2, \ldots, X_1 X_n, X_2 X_3, \ldots, X_2 X_n, \ldots, X_{n-a} X_{n-a+1}, \ldots, X_{n-a} X_n$$

are the biggest square free monomials of degree 2 and their number is $\binom{n}{2} - \binom{a}{2}$. We have to delete the b smallest monomials in the above set and these are the monomials

$$X_{n-a} X_{n-b+1}, \ldots, X_{n-a} X_n$$

which, accordingly, are not in I. Hence the square free monomials of degree t in the variables $X_{n-a-1}, \ldots, X_n$ are not in I_t and their number is $\binom{a}{t}$. Also the monomials of degree t of the kind $X_{n-a} M$ where M runs among the monomials of degree $t-1$ in the variables $X_{n-b-1}, \ldots, X_n$ are not in I and their number is $\binom{b}{t-1}$. But these are the only monomials of degree t which are not in I, and we have proved

$$H_{R/I}(t) = \binom{a}{t} + \binom{b}{t-1} = H_{R/I}(t-1)_{(t)}.$$

It is clear that one can state similar conjectures for ideals generated by forms of the same degree not necessarily equal to two. For example in [55] we stated the following conjecture.

Conjecture 3.12. *Let $A = R/I$ where I is an homogeneous codimension h ideal generated by forms of degree t. Then*

$$e(A) \leq t^{h-2}(t^2 - t + 1).$$

For example if $h = 2$ we can prove the conjecture quite easily. Namely it is easy to see that we may assume $R = k[X, Y]$ and $I = (F_1, F_2, F)$ is an almost complete intersection ideal generated by forms of degree t such that F_1, F_2 form a regular sequence in R. Let $J = (F_1, F_2)$; then R/J is an Artinian Gorenstein ring to which we may apply theorem 2.10 and get

$$e(R/J : F) \geq t - 1.$$

From this we get

$$e(R/I) = t^2 - e(R/J : F) \leq t^2 - t + 1.$$

4. Hilbert Function of Generic Algebras

In this section we introduce and discuss a longstanding conjecture on the Hilbert function of generic algebras. As before we let $R = k[X_1, \ldots, X_n]$ and $A = R/I$ an homogeneous graded algebra. If $F \in R_j$ is a generic form, it is very natural to guess that for every $t \geq 0$, the multiplication map

$$A_t \xrightarrow{F} A_{t+j}$$

is of maximal rank, which means that it is injective if $dim_k(A_t) \leq dim_k(A_{t+j})$, it is surjective if $dim_k(A_t) \geq dim_k(A_{t+j})$. Since for every $F \in R_j$ we have an exact sequence

$$0 \to (0 : F)(-j) \to A(-j) \xrightarrow{F} A \to A/FA \to 0,$$

we get

$$H_{A/FA}(t) = H_A(t) - H_A(t - j) + H_{0:F}(t - j).$$

Hence we guess the following equality in the case F is generic:

$$H_{A/FA}(t) = \max\{0, H_A(t) - H_A(t - j)\}.$$

By using the Hilbert series, this can be rewritten as

$$P_{A/FA}(z) = \left|(1 - z^j)P_A(z)\right|,$$

where for a power series $\sum a_i z^i$ with integers coefficients we let $\left|\sum a_i z^i\right| = \sum b_i z^i$ with $b_i = a_i$ if $a_0, \ldots, a_i > 0$, and $b_i = 0$ if $a_j \leq 0$ for some $j \leq i$.

More generally we can fix positive integers r and $d_1, \ldots, d_r$ and state the following conjecture.

Conjecture 4.1. *Let* $F_1, \ldots, F_r$ *be generic forms in* R *of degrees* $d_1, \ldots, d_r$. *If* $I = (F_1, \ldots, F_r)$, *then*

$$P_{R/I}(z) = \left|\frac{\prod(1 - z^{d_i})}{(1 - z)^n}\right|.$$

We let

$$P_e(z) := \left|\frac{\prod(1 - z^{d_i})}{(1 - z)^n}\right|$$

and call it the *expected Hilbert series* of the generic algebra of type $(r; d_1, \ldots, d_r)$.

By the above remark the conjecture holds if and only if for every $i = 1, \ldots, r$ the multiplication map

$$[A/(F_1, \ldots, F_{i-1})]_t \xrightarrow{F_i} [A/(F_1, \ldots, F_{i-1})]_{t+d_i}$$

is of maximal rank. Further it is easy to prove that if we let $P_g(z)$ be the Hilbert series of a generic graded algebra of type $(r; d_1, \ldots, d_r)$, given an ideal I generated by elements $F_1, \ldots, F_r$ of degrees $d_1, \ldots, d_r$, one has

$$P_{R/I}(z) \geq P_g(z) \quad \text{and} \quad P_{R/I}(z) \overset{lex}{\geq} P_e(z)$$

where the first inequality is coefficientwise, the second is lexicographic. Both the conjecture and the inequalities above can be found in [24].

This has an important consequence: if we can find an example of a graded algebra R/I such that $P_{R/I}(z) = P_e(z)$, then we have

$$P_e(z) = P_{R/I}(z) \geq P_g(z) \overset{lex}{\geq} P_e(z).$$

This clearly implies that $P_g(z) = P_e(z)$ and the conjecture holds. This means that in order to prove the conjecture, it is enough to produce one example with the expected Hilbert series.

The conjecture is true if $r \leq n$ since in this case a set of generic forms is a regular sequence. It was proved to be true for $r = n + 1$ by Stanley by using the so called "Hard Lefschetz theorem", for $n = 2$ by Fröberg, for $n = 3$ by Anick (see [5]). Finally one can prove the conjecture for the first terms in the Hilbert series. The first non trivial statement comes for degree $d + 1$ where $d = \min\{d_i\}$ and was proved by Hochster and Laksov (see [38]).

We give here a proof of the conjecture in the case $n = 2$, by using a device related to the Gröbner basis Theory of ideals in the polynomial ring.

Since the conjecture is true if $r \leq n$, we may argue by induction on r and assume that a graded algebra $B := R/(F_1, \ldots, F_{r-1})$ is given with the expected Hilbert series. We must show that we can find a form F of degree $d := d_r$ in R such that

$$B_t \overset{F}{\to} B_{t+d}$$

is of maximal rank. We fix in the set of monomials of R the degree reverse lexicographic order. This is the order given by

$$X_1^{a_1} X_2^{a_2} \cdots X_n^{a_n} > X_1^{b_1} X_2^{b_2} \cdots X_n^{b_n}$$

if and only if either $\sum a_i > \sum b_i$ or $\sum a_i = \sum b_i$ and for some integer $j \leq n$ we have $a_n = b_n, a_{n-1} = b_{n-1}, \cdots, a_j = b_j, a_{j-1} < b_{j-1}$. Now for a given polynomial F in R we let $M(F)$ be the biggest among the monomials of F. For a given ideal I we let $M(I)$ be the monomial ideal generated by $M(F)$, F in I. A *Gröbner basis* of the ideal I with respect to the given order, is a set of polynomials $F_1, \ldots, F_r \in I$ such that

$$M(I) = (M(F_1), \ldots, M(F_r)).$$

A classical result of Macaulay says that the monomials which are not in $M(I)$ form a k-vector basis for the k-vector space R/I.

Now the crucial information comes from an old result by A. Galligo (see [6]). Here $Gl(n, k)$ is the general linear group acting on R and its Borel subgroup is the subgroup

$$\mathcal{B} := \{g \in Gl(n, k) \,|\, g_{ij} = 0 \,\forall j < i\}.$$

Theorem 4.2. (Galligo) *Let I be an homogeneous ideal of R; there exists a Zariski open set $\mathcal{U} \subseteq Gl(n, k)$, such that the monomial ideal $M(gI)$ is invariant under the action of $\mathcal{B}$.*

Now it is easy to see that if J is a monomial ideal in R then J is *Borel-fixed* if and only if the condition $X_1^{p_1} \cdots X_n^{p_n} \in J$ implies $X_1^{p_1} \cdots X_j^{p_j+q} \cdots X_i^{p_i-q} \cdots X_n^{p_n} \in J$ for every j, i and q such that $1 \le j < i \le n$ and $0 \le q \le p_i$.

If we restrict ourselves to the case when $n = 2$, it is clear that the ideal J is Borel-fixed if and only if J as a k-vector space is generated in each degree by an initial segment in the given order.

We come now to the proof of the conjecture in $k[X_1, X_2]$.

Theorem 4.3. *Let $F_1, \ldots, F_r$ be generic forms in two variables of degrees $d_1, \ldots, d_r$. If $I = (F_1, \ldots, F_r)$, then*

$$P_{R/I}(z) = \left| \frac{\prod(1 - z^{d_i})}{(1 - z)^n} \right|.$$

Proof. Let $J = (F_1, \ldots, F_{r-1})$ and $B = R/J$. Since the generators are generic we have that $M(J)$ is Borel-fixed. Hence, for every degree t, if $H_{R/J}(t) = r$, we have

$$\overbrace{X_1^t, X_1^{t-1}X_2, \ldots, X_1^r X_2^{t-r}}^{M(I)_t} \overbrace{X_1^{r-1}X_2^{t-r+1}, \ldots, X_2^t}^{k-base\ of\ B_t}$$

We let $d := d_r$ $F = F_r = X_2^d$. We have two possibilities. Either $r := dim(B_t) < dim(B_{t+d})$ or $r := dim(B_t) \ge dim(B_{t+d})$. In the first case, by Macaulay's theorem, we must have $r = t + 1$, so that $I_t = 0$ and a base of B_t is

$$\{X_1^t, X_{t-1}X_2, \ldots, X_1 X_2^{t-1}, X_2^t\}.$$

By multiplying these monomials with X_2^d, we get the smallest $t + 1$ monomials of degree $t + d$. Since $dim(B_{t+d}) > t + 1$, these monomials are linearly independent being part of a k-vector base of B_{t+d}.

In the second case, a k-vector base of B_t is

$$\{X_1^{r-1}X_2^{t-r+1}, \ldots, X_2^t\}.$$

By multiplying these monomials with X_2^d, we get the smallest r monomials of degree $t + d$. Since $dim(B_{t+d}) \le r$, the conclusion follows. $\square$

Unfortunately, if $n \ge 3$, a Borel fixed monomial ideal is no longer an initial segment. Here the main problem is:

Problem 4.4. *If we fix the reverse lexicographic order in the set of monomials of a polynomial ring R, what is the shape of the Gröbner basis of an ideal generated by generic forms?*

5. Fat Points: Waring's Problem and Symplectic Packing

A famous theorem of Lagrange states that every natural number is the sum of at most four squares. For example

$$7 = 2^2 + 1^2 + 1^2 + 1^2$$

and one cannot do better. In 1770 Waring stated, without a proof, the following:

- Every natural number is the sum of (at most) 9 positive cubes;
- Every natural number is the sum of (at most) 19 biquadratics.

Much later Hilbert proved that for every $j \geq 2$, there exists an integer $g(j)$ such that every natural number is the sum of (at most) $g(j)$ j^{th}-powers of positive integers.

So Waring was asserting that $g(3) = 9$, $g(4) = 19$, while Lagrange's theorem says that $g(2) = 4$.

The problem of determining $g(j)$ seems close to a final resolution.

However the problem above is only one of the Waring problems, the so called *Little Waring Problem*.

The *Big Waring Problem* comes from the following remark. We know that $g(3) = 9$ but only the integers 23 and 239 actually require 9 cubes and only 15 integers (≤ 8042) actually require 8 cubes. So one is naturally lead to the following

Definition 5.1. *Let $G(j)$ be the least integer such that all sufficiently large integers are the sum of (at most) $G(j)$ j^{th} powers of positive integers.*

It is clear that $G(j) \leq g(j)$, and since we know that every number congruent to 7 mod(8) is a sum of 4 squares and not 3, we have $G(2) = g(2) = 4$. We also know that

$G(3) \leq 7$, $G(4) = 16$, $G(6) \leq 27$, $G(7) \leq 36$.

The Big Waring Problem is the problem of determining $G(j)$ for every j.

We can ask the analogous questions in the context of forms in the polynomial ring $\mathbf{C}[X_0, \ldots, X_n]$, $\mathbf{C}$ the complex numbers. In this way we enter into the classical theory of canonical forms of polynomials.

The problem of canonical forms, stated rather loosely, is the following:

Problem 5.2. *How much can one form be "simplified" by linear changes of variables (that is under the action of the general linear group in n variables)?*

The history of this problem is interesting and it might be useful to begin with a specific example. Let us consider a generic quadratic form in $\mathbf{C}[X_1, X_2, X_3]$. We have to deal with 6 coefficients; hence, if we are allowed two linear forms each containing 3 coefficients, we have all together 6 coefficients. Thus, if we rely merely on a count of constants, we are led to the erroneous conclusion that a generic quadratic form in three variables can be written as the sum of two squares of linear forms. But this assertion is false, since the sum of two squares is a reducible polynomial.

More delicate is the case of a generic quartic in three variables. By a similar counting constants method one can guess that it could be written as the sum of five fourth powers of linear forms. Clebsch was the first to prove in 1860 that this is false. We will see later an elementary explanation for this.

We can state now the corresponding Waring problems in the polynomial case.

Problem 5.3. (LWP) *For every $j \geq 2$, determine the least integer $g(j)$ such that every form of degree j in $\mathbf{C}[X_0, \ldots, X_n]$ can be written as the sum of (at most) $g(j)$ powers of linear forms.*

Problem 5.4. (BWP) *For every $j \geq 2$, determine the least integer $G(j)$ such that the generic form of degree j in $R := \mathbf{C}[X_0, \ldots, X_n]$ can be written as the sum of $G(j)$ powers of linear forms.*

Let us start with the classical case where $j = 2$. To each quadratic form in $n+1$ variables we may associate its symmetric matrix M. Since this matrix can be diagonalized, the quadratic form is the sum of $rank(M)$ squares of linear forms. This proves that

$$G(2) \leq n + 1.$$

But the matrices of rank $< n + 1$ form a closed subspace, so that

$$G(2) = n + 1.$$

Beside this, very little was known on the BWP and in [15] Ehrenborg and Rota write:

"It is our purpose to give in this paper a complete, self-contained, updated introduction to the theory of canonical forms of polynomials, as it has been known to this day, to the best of our knowledge ... We believe to yield the complete solution of the analog of Waring's problem for forms and we hope to present the complete solution elsewhere. In the present work, we have limited our exposition to all cases thus far considered in the literature, supplemented by a few new cases that caught our fancy."

But recently, as observed in [42], R. Lazarsfeld noted that the solution of the BWP is a consequence of a result of Alexander and Hirschowitz on the Hilbert function of fat points in the projective space (see [2], [3], [4], [37]). Here we present a more algebraic approach to the problem as described in the paper by Iarrobino. We will prove that all the examples worked out in [15] can be obtained by using a very elementary result of Catalisano, Trung and Valla (see [11]), thus avoiding the difficult papers of Alexander and Hirschowitz (more than hundred pages of journal articles!).

We start with a very classical result by A. Terracini (see [67]) which states that in the case the basic field is $\mathbf{C}$,

• The integer $G(j)$ is the smallest integer s for which there are linear forms $L_1, \ldots, L_s$ in R_1 such that

$$R_j = (L_1^{j-1}, \ldots, L_s^{j-1})_j.$$

It is clear that, above, we can restrict ourselves to consider linear forms which are generic.

We now denote by S the polynomial ring $S = k[Y_0, \ldots, Y_n]$ upon which the elements of the polynomial ring R act as higher order partial differential operators. This action is sometimes called the "apolarity" action of R on S. We can describe this action by saying

$$X_i \circ Y_j = \frac{\partial}{\partial Y_i}(Y_j) = \begin{cases} 0 & \text{if } i \neq j \\ 1 & \text{if } i = j. \end{cases}$$

The pairing

$$R_j \times S_j \to k$$

is a perfect pairing for each $j \geq 0$. This is due to the fact that the matrix of the corresponding bilinear form is non singular.

Given an ideal I in R we can consider for each $t \geq 0$ the vector space $(I_t)^\perp$ in S which is the orthogonal of I_t. We know that

$$dim_k((I_t)^\perp) = dim_k(S_t) - dim_k(I_t) = dim_k(R_t) - dim_K(I_t) = H_{R/I}(t).$$

For example if $I = (X_1, \ldots, X_n)^2$, then it is clear that

$$(I_t)^\perp = \{\text{monomials of } S_t \text{ not in } I_t\}.$$

Of course when we write "not in I_t" we mean "not in the degree t part of the ideal of S obtained by changing X_i with Y_i in the generators of I."

Now suppose we are given a set of s distinct points $P_1, \ldots, P_s$ in $\mathbf{P}^n$ whose corresponding ideals are $\wp_1, \ldots, \wp_s$, and a sequence $m_1, \ldots, m_s$ of positive integers. The subscheme of $\mathbf{P}^n$ defined by the ideal

$$I = \wp^{m_1} \cap \wp^{m_2} \cap \cdots \cap \wp^{m_s}$$

is called the subscheme of *fat points*

$$\mathcal{Z} = m_1 P_1 + \cdots + m_s P_s.$$

The scheme $\mathcal{Z}$ is a zero-dimensional scheme which has support on the $P_i's$ and multiplicity $\binom{n+m_i-1}{n}$ at P_i. What makes these schemes interesting is the fact that each vector space I_t gives the linear systems on $\mathbf{P}^n$ consisting of all hypersurfaces of degree t having at least multiplicity $\binom{n+m_i-1}{n}$ at each P_i. Since R/I is a one dimensional graded Cohen-Macaulay ring, its Hilbert Function is strictly increasing and becomes constant when it reaches the degree of $\mathcal{Z}$ which is

$$deg(\mathcal{Z}) = \sum_{i=1}^{s} \binom{n+m_i-1}{n}.$$

One can easily compute the Hilbert function of a scheme of generic fat points in the case when $m_1 = \cdots = m_s = 1$. With this assumption we have

$$H_{R/I}(t) = \min\left\{s, \binom{n+t}{n}\right\}.$$

If this is the case, we say that R/I has *maximal Hilbert function*.

In the case of higher multiplicities, the problem is much more complicated.

We use the following notation. If $P_1, \ldots, P_s$ in $\mathbf{P}^n$ is a set of distinct points, such that $P_i := (a_{i0}, \ldots, a_{in})$, we let

$$L_{P_i} := a_{i0} Y_0 + \cdots + a_{in} Y_n$$

be the corresponding linear form in S_1.

The following result has been proved in [22].

Theorem 5.5. *Let I be the defining ideal of the scheme of double fat points*

$$\mathcal{Z} = 2P_1 + \cdots + 2P_s.$$

For every $j > 0$ we have

$$(I_j)^{\perp} = (L_{P_1}^{j-1}, \ldots, L_{P_s}^{j-1})_j.$$

By using this theorem we get:
- $G(j)$ is the least integer s such that there exist s generic points in $\mathbf{P}^n$ with

$$H_{R/I}(j) = H_R(j) = \binom{n+j}{j}$$

where I is the defining ideal of the scheme of double fat points $\mathcal{Z} = 2P_1 + \cdots + 2P_s$.

We can immediately use this theorem to get most of the results of [15]. We remark that since $m_1 = \cdots = m_s = 2$, we have

$$\deg(\mathcal{Z}) = e(R/I) = s(n+1).$$

Let us start with the classical case: $G(2) = n+1$. Let I be the defining ideal of the scheme of double fat points

$$\mathcal{Z} = 2P_1 + \cdots + 2P_{n+1};$$

since $n+1$ generic points are not on an hyperplane, we have

$$H_{R/I}(2) = \binom{n+2}{2}.$$

Since a set of j points with $j \leq n$ is always on an hyperplane, we have proved the claim.

Let us pass to the easy case when we are dealing with forms in two variables. In this case $n = 1$ and we prove that

$$G(j) = \text{ the least integer bigger or equal than } \tfrac{j+1}{2}.$$

We must prove that

$$H_{R/I}(j) = j+1 \iff t \geq \frac{j+1}{2}$$

where $I = \wp_1^2 \cap \cdots \cap \wp_t^2$. But $R = k[X_0, X_1]$, so that the Artinian reduction of R/I is an Artinian ring S/J where $S = k[X]$. This implies that the Hilbert function of R/I is

$$H_{R/I}(j) = \min\{j+1, e(R/I)\}.$$

Since $e(R/I) = 2t$, we get the conclusion.

We have proved that
- A generic form of degree $2j - 1$ in two variables can be written as a sum of j powers of linear forms.

This is the classical Sylvester's theorem.

We present now some easy cases where the expected value for $G(j)$ is not attained.

- The generic form of degree 4 in 3 variables is not the sum of 5 fourth powers.

    We need to prove that

$$H_{R/I}(4) < 15 = H_R(4)$$

where I is the defining ideal of the scheme of double fat points with support 5 points of $\mathbf{P}^2$. Since 5 points in $\mathbf{P}^2$ are on a conic, it is clear that the conclusion follows.

    This is the result of Clebsch we have seen before.

- *The generic form of degree 4 in 4 variables is not the sum of 9 fourth powers.*

    We need to prove that

$$H_{R/I}(4) < 35 = H_R(4)$$

where I is the defining ideal of the scheme of double fat points with support 9 points of $\mathbf{P}^3$. Since 9 points in $\mathbf{P}^3$ are on a quadric, it is clear that the conclusion follows.

- *The generic form of degree 4 in 5 variables is not the sum of 14 cubes.*

    We need to prove that

$$H_{R/I}(4) < 70 = H_R(4)$$

where I is the defining ideal of the scheme of double fat points with support 14 points of $\mathbf{P}^4$. Since 14 points in $\mathbf{P}^4$ are on a quadric hypersurface, it is clear that the conclusion follows.

    To study further examples, we need a very elementary result proved by Catalisano, Trung and Valla in [11].

    Here, for a one-dimensional graded Cohen-Macaulay ring A, the *regularity index* $r(A)$ of A is the least integer t such that $H_A(t) = e(A)$. Also we say that the points are in *general position* if $n + 1$ of them are not on an hyperplane of $\mathbf{P}^n$.

    It is clear that generic points are in general position.

Theorem 5.6. *Let* $\mathcal{Z} = 2P_1 + \cdots + 2P_s$ *where* $P_1, \ldots, P_s$ *are points in general position in* $\mathbf{P}^n$. *If* A *denotes the homogeneous coordinate ring of* $\mathcal{Z}$, *then*

$$r(A) \leq \max\left\{3, \left[\frac{n - 2 + 2s}{n}\right]\right\}.$$

Further, if the support of $\mathcal{Z}$ *lies on a rational normal curve, then equality holds above.*

    This very elementary result gives a fourth case where the expected value for $G(j)$ is not attained.

- *The generic form of degree 3 in 5 variables is not the sum of seven cubes.*

    We must show that

$$H_{R/I}(3) < 35$$

where I is the defining ideal of the scheme of double fat points with support 7 points of $\mathbf{P}^4$. By a theorem of Bertini, if the points are in general position, they are on a rational normal curve, so, by the above result, we have $r(R/I) = \max\{3, 4\} = 4$. Hence

$$H_{R/I}(3) < e(R/I) = 35 = H_R(3).$$

The following example end up the list worked out in [15].

- *The generic form of degree 3 in 4 variables is the sum of 5 cubes.*

We need to prove that

$$H_{R/I}(3) = 20 = e(R/I)$$

where I is the defining ideal of the scheme of double fat points with support 5 generic points of $\mathbf{P}^3$. Since we have

$$\max\left\{3, \left[\frac{3 - 2 + 10}{3}\right]\right\} = 3,$$

the conclusion follows.

Despite the fact that all the above results easily follow by elementary arguments, the solution of the BWP is a consequence of a very difficult theorem proved by Alexander and Hirschowitz.

Theorem 5.7. *Let* $\mathcal{Z} = 2P_1 + \cdots + 2P_s$ *where* $P_1, \ldots, P_s$ *are generic points in* $\mathbf{P}^n$. *If A denotes the homogeneous coordinate ring of A, then, except for the four pathological cases considered above, we have*

$$H_A(t) = \min\left\{s(n+1), \binom{n+t}{n}\right\}.$$

This theorem proves that, with four exceptions, the Hilbert function of a set of double fat points with generic support is maximal, i.e. behaves as if the scheme were reduced.

As a consequence we have the following solution of the BWP.

Theorem 5.8. *For every* $j \geq 3$ *we have*

$$G(j) = \min\left\{t \mid (n+1)t \geq \binom{n+j}{j}\right\}$$

except for the following four cases:
$j = 4$ $n = 2$ where $G = 6$, instead of $G = 5$,
$j = 3$ $n = 4$ where $G = 8$, instead of $G = 7$,
$j = 4$ $n = 3$ where $G = 10$, instead of $G = 9$,
$j = 4$ $n = 4$ where $G = 15$, instead of $G = 14$,

We remark that the number of forms of degree j in $n + 1$ variables is $\binom{n+j}{j}$, while the numbers of coefficients in a sum of G powers of linear forms

$$L_1^j + \cdots + L_G^j$$

is $(n+1)G$ so that the following guess is confirmed with the well known four exceptions:

- As soon as the number of coefficients is bigger than or equal to the number of equations, we do have a solution.

Now we want to describe a more geometric approach to the BWP involving some interesting questions related to the dimension of the secant variety of the Veronese embeddings.

Given the integers n and j we let

$$N := \binom{n+j}{j} - 1.$$

Then it is clear that we have a trivial identification

$$\mathbf{P}(R_j) \simeq \mathbf{P}^N$$

where $\mathbf{P}(R_j)$ is the projectivization of the vector space R_j of the forms of degree j in R. After this identification, let us consider the Veronese immersion

$$\nu_j : \mathbf{P}^n \to \mathbf{P}^N$$

where if $P = (\alpha_0, \ldots, \alpha_n) \in \mathbf{P}^n$ we let

$$\nu_j(P) := (\alpha_0^{i_0} \cdots \alpha_n^{i_n})_{i_0 + \cdots + i_n = j}.$$

The image $\nu_j(\mathbf{P}^n)$ is called the j-Veronese immersion of $\mathbf{P}^n$ into $\mathbf{P}^N$.

The following remark will be crucial.

Remark 5.9. $\nu_j(\mathbf{P}^n)$ *may be seen in* $\mathbf{P}(R_j)$ *as the projectivization* $\mathbf{P}(W)$ *of the subspace* W *of* R_j *of all forms consisting of powers of linear forms.*

For example let $n = 2$ and $j = 2$ so that $N = 5$. Given the quadratic form

$$Q = \sum_{0 \le i \le j \le 2} a_{ij} X_i X_j$$

we consider the symmetric matrix

$$\begin{pmatrix} a_{00} & a_{01} & a_{02} \\ a_{01} & a_{11} & a_{12} \\ a_{02} & a_{12} & a_{22} \end{pmatrix}.$$

This matrix has rank ≤ 1 if and only if Q is the square of a linear form. Since the Veronese surface $\nu_2(\mathbf{P}^2)$ is the locus of the points in $\mathbf{P}^5$ such that the matrix

$$\begin{pmatrix} X_{00} & X_{01} & X_{02} \\ X_{01} & X_{11} & X_{12} \\ X_{02} & X_{12} & X_{22} \end{pmatrix}$$

has rank one, Q is the square of a linear form if and only if

$$(a_{00}, a_{01}, a_{02}, a_{11}, a_{12}, a_{22}) \in \nu_2(\mathbf{P}^2).$$

The above result does not hold if the characteristic of the base field is positive. For example if $ch(k) = 2$ and we consider the case $n = 1$, $j = 2$ so that $N = 2$, then $\nu_2(\mathbf{P}^1) \subseteq \mathbf{P}^2$ is the conic of equation

$$X_{00} X_{11} - X_{01}^2.$$

Instead, the set of quadratic forms in two variables which are squares of linear forms correspond to the line $X_{01} = 0$.

Now it is clear that going through this idea, one can see how the forms in R_j which are sum of two powers correspond to the lines which meet two distinct points of the Veronese variety $\nu_j(\mathbf{P}^n)$. There is a classical construction in Algebraic Geometry which gives us the right setting.

Definition 5.10. *Let $X \subseteq \mathbf{P}^d$ be a projective variety and let $k \geq 2$. $Sec_{k-1}(X)$ is defined as the closure of the union of the $\mathbf{P}^{k-1}$ in $\mathbf{P}^d$ which cut X in k distinct points. This variety is called the $(k-1)$ Secant Variety to X.*

It is a classical result that $Sec_{k-1}(X)$ is a locally closed subscheme of $\mathbf{P}^d$. The main remark here is the following.

Remark 5.11. *A generic point of $Sec_{k-1}(\nu_j(\mathbf{P}^n))$ corresponds to a form of degree j which can be written as the sum of k powers of linear forms.*

Hence we may rewrite the big Waring problem as follows.

Problem 5.12. (BWP) *Given the integer j determine the least integer $G(j)$ such that*

$$Sec_{G(j)-1}(\nu_j(\mathbf{P}^n)) = \mathbf{P}^N.$$

This way of rereading the BWP is classical. One can find more details in the book by J. Harris (see [31]).

By using the arguments of Emsalem and Iarrobino, we get a way for computing the dimension of these secant varieties by means of the Hilbert Function of a set of generic fat points in the projective space $\mathbf{P}^n$.

More precisely we have the following theorem.

Theorem 5.13. *For every n, j and k, let I be the defining ideal of the scheme of double fat points $\mathcal{Z} = 2P_1 + \cdots + 2P_k$ where $P_1, \ldots, P_k$ are generic points in $\mathbf{P}^n$. Then*

$$dim(Sec_{k-1}(\nu_j(\mathbf{P}^n))) = H_{R/I}(j) - 1.$$

By using this result, we want to give now a contribution to the LWP by proving that in the case we are dealing with forms in two variables we have $g(j) = j$. This means that every form of degree j in 2 variables is the sum of at most j powers of linear forms. Now, given any point in $\mathbf{P}^j$, we can find an hyperplane through this point cutting the rational normal curve in j distinct points. This proves that $g(j) \leq j$. On the other hand one can easily see that $g(j) = j$. Let us consider the case $j = 3$. Let P be a point in $\mathbf{P}^3$ which lies on a tangent to the rational normal curve $\mathcal{C}$ with parametric equations $X_0 = t^3$, $X_1 = t^2 u$, $X_2 = tu^2$, $X_3 = u^3$. Take for example the point $P := (0, 1, 0, 0)$ which lies on the line $X_2 = X_3 = 0$ which is tangent to $\mathcal{C}$ in the point $(1, 0, 0, 0)$. This point cannot be on a secant to $\mathcal{C}$, otherwise the plane through the tangent and the secant line would meet $\mathcal{C}$ in four points. This shows that

$$Sec_1(\nu_3(\mathbf{P}^1)) = \mathbf{P}^3.$$

However the union of the secant lines is not the whole $Sec_1(\nu_3(\mathbf{P}^1))$. We have

$$Sec_1(\nu_3(\mathbf{P}^1)) = \{\cup(\text{secant lines})\} \cup \{\cup(\text{tangent lines})\}.$$

We also proved that the form $X_0^2 X_1$, which correspond to the point $(0,1,0,0)$, is not the sum of two cubes of linear forms.

The problem of determining the Hilbert function of a scheme of fat points has other unexpected and interesting applications. First of all we remark that it can be seen as the *Problem of infinitesimal generic interpolation* by which we mean the following.

Problem 5.14. *At how many points is it possible to prescribe values to a polynomial of given degree, together with derivatives up to a given order?*

Another recent and interesting application of the arithmetical properties of schemes of fat points is due to McDuff and Polterovich (see [47]). Let

$$B^4 := \{z \in \mathbf{C}^2 \mid |z| \le 1\}$$

be the standard four dimensional ball; consider all possible symplectic embeddings of k disjoint standard 4-dimensional balls of equal radii into B^4. Denote by $\hat{\nu}(B^4, k)$ the supremum of volumes which can be filled by such embeddings and define

$$\nu(B^4, k) := \frac{\hat{\nu}(B^4, k)}{Vol(B^4)}.$$

A basic aspect of the symplectic packing problem is to distinguish between the following cases:

- $\nu(B^4, k) = 1$, that is there exists a *full filling*;
- $\nu(B^4, k) < 1$, that is there is a *packing obstruction*.

The history of this problem goes back to Fefferman and Phong who raised in [23] a somewhat similar question in connection with the uncertainty principle in quantum mechanics.

The result of McDuff and Polterovich relates this question with a conjecture made by Nagata in connection with his construction of a counterexample to the 14-th problem of Hilbert (see [48]).

Conjecture 5.15. (Nagata) *Let $P_1, \ldots, P_k$ be generic points in $\mathbf{P}^2$ and $m_1, \ldots, m_k$ be fixed non negative integers. For every $k \ge 9$, if d is the initial degree of the defining ideal of the scheme of fat points*

$$Z = m_1 P_1 + \cdots + m_k P_k,$$

we have

$$d \ge \frac{\sum_{i=1}^{k} m_i}{\sqrt{k}}.$$

For $k = 2, 3, 5, 6, 7, 8$ the assertion of the conjecture is wrong. Nagata showed that the conjecture is true for the case $k = p^2$ where p is an integer. Except for this special case, Nagata's conjecture is still quite open.

The connection with the Hilbert function of fat points is given in the following theorem.

Theorem 5.16. (McDuff-Polterovich) *Assume Nagata's conjecture is true for some k. Then there exists a full symplectic packing of B^4 by k equal standard balls, that is,*

$$\nu(B^4, k) = 1.$$

By these different reasons the problem of determining properties of the Hilbert Function of fat points becomes very interesting. Even if we assume that the support consists of generic points in $\mathbf{P}^2$, we have no clear idea on the possible exceptions to the guess that the Hilbert function should be the same as for reduced zero-dimensional schemes of the same degree.

Problem 5.17. *What is the Hilbert function of a scheme of fat points in $\mathbf{P}^n$?*

A very special case is when we have a support which is contained on a rational normal curve. In this case we have seen that the regularity index is as large as possible. This should imply that the Hilbert function is as small as possible and leads one to make the following conjecture.

Conjecture 5.18. *Given a set of s distinct points on a rational normal curve in $\mathbf{P}^n$, and given a set of natural numbers $m_1, \ldots, m_s$ let*

$$\mathcal{Z} = m_1 P_1 + \cdots + m_s P_s$$

be the corresponding scheme of fat points. Then
- *The Hilbert function of $\mathcal{Z}$ does not depend on the points.*
- *For every scheme $\mathcal{Y}$ of fat points in general position and with the same multiplicities, we have*

$$H_{\mathcal{Z}}(t) \leq H_{\mathcal{Y}}(t)$$

for every $t \geq 0$,

The conjecture holds if $m_1 = \cdots = m_s = 2$ (see [12]); a proof of the conjecture if $n = 2$ and of the first part if $n = 3$ is in [10].

6. The HF of a CM Local Ring

In this section $(A, \mathcal{M})$ will denote a local ring with maximal ideal $\mathcal{M}$. The Hilbert function of A is, by definition, the Hilbert function of the *associated graded ring* of A which is the homogeneous k-algebra

$$gr_{\mathcal{M}}(A) = \oplus_{t \geq 0} \mathcal{M}^t / \mathcal{M}^{t+1}.$$

Hence

$$H_A(t) = dim_k(\mathcal{M}^t / \mathcal{M}^{t+1})$$

and

$$P_A(z) = P_{gr_{\mathcal{M}}(A)}(z).$$

This graded algebra $gr_{\mathcal{M}}(A)$ corresponds to a relevant geometric construction. If A is the localization at the origin 0 of the coordinate ring of an affine variety V passing through 0, then $gr_{\mathcal{M}}(A)$ is the coordinate ring of the *tangent cone* of V at

0, which is the cone composed of all lines that are the limiting positions of secant lines to V in 0.

The *Proj* of this algebra can also be seen as the *exceptional set* of the *blowing up* of V in 0.

Despite the fact that the Hilbert function of a graded k-algebra A is well understood in the case A is Cohen-Macaulay, very little is known in the local case. This because the associated graded ring of a local Cohen-Macaulay ring can be very bad. Consider, for example, the power series ring

$$A = k[[t^4, t^5, t^{11}]].$$

This is a one-dimensional Cohen-Macaulay local ring and its associated graded ring is

$$gr_{\mathcal{M}}(A) = k[X, Y, Z]/(XZ, YZ, Z^2, Y^4)$$

which is not Cohen-Macaulay. This follows from the fact that

$$(X, Y, Z) = 0 \; : \; \bar{Z} \in Ass(gr_{\mathcal{M}}(A))$$

and also from the Hilbert series of A which is

$$P_A(z) = \frac{1 + 2z + z^3}{1 - z}.$$

In the above example, one can compute the defining equations of the tangent cone in the following way. It is easy to see that

$$A = k[[X, Y, Z]]/(X^4 - YZ, Y^3 - XZ, Z^2 - X^3Y^2).$$

Hence $gr_{\mathcal{M}}(A) = k[X, Y, Z]/I^*$ where $I = (X^4 - YZ, Y^3 - XZ, Z^2 - X^3Y^2)$ and I^* is the ideal generated by the initial forms of the elements of I. Here, if $a \in A$ is a non zero element and n is the greatest integer such that $a \in \mathcal{M}^n$, we let

$$a^* := \bar{a} \in \mathcal{M}^n/\mathcal{M}^{n+1}$$

and call it *the initial form of* a *in* $gr_{\mathcal{M}}(A)$.

Due to the pioneering work made by Northcott in the 50's (see for example [51]), several efforts have been made to better understand the Hilbert function of a Cohen-Macaulay local ring, also in relation with its Hilbert coefficients and the numerical characters of the corresponding tangent cone.

We start with the observation that, unlike in the graded case, the Hilbert function is not sensitive to regular sequence, but rather to strong regular sequences. The second part of the following theorem is due to Valabrega and Valla (see [69]), while the first assertion is a consequence of a theorem by Singh (see [63]).

Theorem 6.1. *Let* $(A, \mathcal{M})$ *be a local ring and* $I = (x_1, \ldots, x_r)$ *an ideal in* A. *Let* $\bar{x}_1, \ldots, \bar{x}_r$ *be the residue class of* $x_1, \ldots, x_r$ *in* $\mathcal{M}/\mathcal{M}^2$ *and* $(B, \mathcal{N}) := (A/I, \mathcal{M}/I)$. *Then we have :*

- $P_B(z)/(1 - z)^r \geq P_A(z)$.
- *The following conditions are equivalent:*

a) $P_B(z) = P_A(z)(1 - z)^r$.

b) $\bar{x}_1, \ldots, \bar{x}_r$ *form a regular sequence in* $gr_{\mathcal{M}}(A)$.

c) $x_1, \ldots, x_r$ *form a regular sequence in A and*

$$gr_{\mathcal{N}}(B) = gr_{\mathcal{M}}(A)/(\bar{x}_1, \ldots, \bar{x}_r).$$

d) $x_1, \ldots, x_r$ *form a regular sequence in A and for every $n \geq 1$*

$$\mathcal{M}^n \cap I = \mathcal{M}^{n-1}I.$$

A crucial notion in local algebra is that of superficial element.

Definition 6.2. *Let $(A, \mathcal{M})$ be a local ring; an element x in $\mathcal{M}$ is called superficial for A if there exists an integer $c > 0$ such that*

$$(\mathcal{M}^n : x) \cap \mathcal{M}^c = \mathcal{M}^{n-1}$$

for every $n > c$.

It is easy to see that a superficial element x is not in $\mathcal{M}^2$ and that x is superficial for A if and only if $x^* \in \mathcal{M}/\mathcal{M}^2$ does not belong to the relevant associated prime of $gr_{\mathcal{M}}(A)$. Hence, if the residue field is infinite, superficial elements always exist. Further, if $depth(A) > 0$, then every superficial element is also a regular element in A.

Definition 6.3. *A set of elements $x_1, \ldots, x_r$ in the local ring $(A, \mathcal{M})$ is said to be a superficial sequence if x_1 is superficial for A, $\overline{x_2}$ is superficial for $A/x_1 A$ and so on.*

We collect in the following theorem two main properties of superficial sequences.

Proposition 6.4. *Let $(A, \mathcal{M})$ be a local ring and $x_1, \ldots, x_r$ a superficial sequence. If we let $J := (x_1, \ldots, x_r)$ and $(B, \mathcal{N}) := (A/J, \mathcal{M}/J)$, then*
- $depth(gr_{\mathcal{M}}(A)) \geq r \iff x_1^*, \ldots, x_r^*$ *is a regular sequence in $gr_{\mathcal{M}}(A)$.*
- $depth(gr_{\mathcal{M}}(A)) \geq r + 1 \iff depth(gr_{\mathcal{N}}(B)) \geq 1$.

The second property in the above theorem is *Sally's machine* which is a very important trick to reduce dimension in questions relating to depth properties of $gr_{\mathcal{M}}(A)$.

Sally proved this result in the case $r = d - 1$ in [58]; a complete and nice proof of the general case can be found in [39].

We have seen that we can always find a superficial element inside a local ring A with infinite residue field. Now we remark that we may always assume the residue field has this property by passing, if needed, to the local ring $A[X]_{(\mathcal{M},X)}$. Hence if we assume moreover that A has positive depth, every superficial element is also a regular element in A. This has the right properties to control most of the numerical invariants under reduction modulo the ideal it generates.

Here we denote by $\lambda_A(M)$ the length of an A-module M and by $embcod(A)$ the *embedding codimension* of A which is the integer $H_A(1) - d$. It is clear that $H_A(1) - d$ is the coefficient h_1 of z in the h-polynomial of A. Further $embcod(A) = 0$ if and only if A is a regular local ring. Since in this case our problems are clear, we may assume in the following that A has positive embedding codimension.

Proposition 6.5. *Let $(A, \mathcal{M})$ be a d-dimensional local ring and x a superficial element which is also regular on A. If $B := A/xA$, we have :*
- $dim(B) = d - 1$.
- $H_A(1) = H_B(1) + 1$.
- $embcod(A) = embcod(B)$.
- $e_k(B) = e_k(A)$ *for every* $k = 0, \ldots, d - 1$.
- $(-1)^d e_d(A) = (-1)^d e_d(B) - \sum_{j=0}^{n} \lambda_A(\mathcal{M}^{j+1} : x/\mathcal{M}^j)$ *for every* $n \gg 0$.
- $e_d(A) = e_d(B)$ *if and only if x^* is regular in $gr_{\mathcal{M}}(A)$.*

We immediately get two classical and basic results due to Abhyankar and Northcott respectively (see [1] and [51]). In the following e will denote the multiplicity of A and h its embedding codimension. Further we will write e_i instead of $e_i(A)$ and G instead of $gr_{\mathcal{M}}(A)$. Since henceforth A is assumed to be Cohen-Macaulay, for every r, $1 \le r \le d$, every superficial sequence of length r is also a regular sequence in A.

Theorem 6.6. (Abhyankar) *Let $(A, \mathcal{M})$ be a d-dimensional Cohen-Macaulay local ring; then*

$$e \ge h + 1.$$

Proof. Since e and h does not change modulo a superficial element, we may assume $d = 0$. Then

$$P_A(z) = 1 + hz + \cdots + h_s z^s$$

where $s \ge 1$ and all the $h_i's$ are positive integers. Thus

$$e = 1 + h + h_2 + \cdots + h_s \ge h + 1. \qquad \square$$

We remark that the assumption A is Cohen-Macaulay in the above theorem is essential.

Let $A = k[[X, Y]]/(X^2, XY)$; then A is not Cohen-Macaulay and we have

$$P_A(z) = \frac{1 + z - z^2}{1 - z}$$

so that $e = 1 < h + 1 = 2$.

Theorem 6.7. (Northcott) *Let $(A, \mathcal{M})$ be a d-dimensional Cohen-Macaulay local ring; then*

$$e_1 \ge e - 1.$$

Proof. If $d = 0$ we have

$$P_A(z) = 1 + hz + \cdots + h_s z^s$$

and we need to prove

$$e_1 = h + 2h_2 + \cdots + sh_s \ge e - 1 = 1 + h + h_2 + \cdots + h_s - 1.$$

Since all the $h_i's$ are non negative integers, we get the conclusion.

If $d \geq 1$, then $e_1(A) = e_1(B) \geq e_1(C)$ and $e(A) = e(B) = e(C)$ where if $x_1, \ldots, x_d$ is a superficial sequence, we let

$$B = A/(x_1, \ldots, x_{d-1})$$

and

$$C = A/(x_1, \ldots, x_d).$$

The conclusion follows.   □

We have an extension of the above theorems which is easy and useful in applications.

Theorem 6.8. *Let $(A, \mathcal{M})$ be a d-dimensional Cohen-Macaulay local ring; then*

$$e_1 \geq 2e - h - 2.$$

Proof. If $d = 0$ we have

$$P_A(z) = 1 + hz + \cdots + h_s z^s$$

and we need to prove

$$e_1 = h + 2h_2 + \cdots + sh_s \geq 2(1 + h + \cdots + h_s) - h - 2$$

which is clearly true. If $d \geq 1$, then $e_1(A) = e_1(B) \geq e_1(C)$ and $e(A) = e(B) = e(C)$ where B and C are Cohen-Macaulay local rings with the same embedding codimension as A and with dimension 1 and 0 respectively. The conclusion follows by the case $d = 0$.   □

We prove now that equality holds in the above theorem if and only if the h-vector of A is short enough.

Theorem 6.9. *Let $(A, \mathcal{M})$ be a d-dimensional Cohen-Macaulay local ring and let s be the degree of the h-polynomial of A. The following conditions are equivalent:*
  - *$s \leq 2$.*
  - *$e_1 = 2e - h - 2$.*
Further, if either of the above conditions holds, then G is Cohen-Macaulay.

Proof. If

$$P_A(z) = \frac{1 + hz + h_2 z^2}{(1 - z)^d},$$

we have $e = 1 + h + h_2$ and $e_1 = h + 2h_2$. Hence

$$2e - h - 2 = 2(1 + h + h_2) - h - 2 = h + 2h_2 = e_1.$$

Conversely, let $e_1 = 2e - h - 2$. If $d = 0$, we have $P_A(z) = 1 + hz + \cdots + h_s z^s$ where $s \geq 1$. The condition $e_1 = 2e - h - 2$ can be read as

$$h_3 + 2h_4 + \cdots + (s - 2)h_s = 0$$

which implies $s \leq 2$.

If $d \geq 1$, then

$$2e - h - 2 = e_1(A) = e_1(B) \geq e_1(C) \geq 2e - h - 2$$

and $e(A) = e(B) = e(C)$ where B and C are Cohen-Macaulay local ring with the same embedding codimension as A and with dimension 1 and 0 respectively. Hence $e_1(B) = e_1(C) = 2e - h - 2$ so that

$$P_B(z) = \frac{P_C(z)}{1 - z} = \frac{1 + hz + h_2 z^2}{1 - z}$$

and $gr(B)$ has depth 1. By using Sally's machine we get $depth(G) = d$ so that G is Cohen-Macaulay and

$$P_A(z) = \frac{P_C(z)}{(1 - z)^d} = \frac{1 + hz + h_2 z^2}{(1 - z)^d}$$

as wanted.   □

By using the above result, we can see that the extremal cases $s \leq 1$, $e = h + 1$ and $e_1 = e - 1$ are equivalent and force the Hilbert function of A. We remark here that results of this kind are not so expected since the Hilbert coefficients give partial information on the Hilbert polynomial which, in turn, gives asymptotic information on the Hilbert function.

Theorem 6.10. *Let $(A, \mathcal{M})$ be a d-dimensional Cohen-Macaulay local ring and let s be the degree of the h-polynomial of A. The following conditions are equivalent:*

1. $s \leq 1$.
2. $P_A(z) = (1 + hz)/(1 - z)^d$.
3. $e_1 = e - 1$.
4. $e_1 = h$.
5. $e = h + 1$.

Further, if either of the above conditions holds, then G is Cohen-Macaulay.

Proof. It is clear that 1 and 2 are equivalent. Further 2 implies all the other conditions. Since we have $e_1 \geq 2e - h - 2$, we get

$$e_1 - e + 1 \geq e - h - 1 \geq 0, \quad \text{and} \quad e_1 - h \geq 2(e - h - 1) \geq 0$$

so that 3 implies 5 and 4 implies 5. We need only to prove that 5 implies 1. This is clear if $d = 0$. Let $d \geq 1$ and $e = h + 1$. If J denote the ideal generated by a maximal superficial sequence in $\mathcal{M}$, we have the following square

$$\begin{array}{ccc} \mathcal{M} & \supset & \mathcal{M}^2 \\ \cup & & \cup \\ J & \supset & J\mathcal{M} \end{array}$$

Since $\lambda(\mathcal{M}/\mathcal{M}^2) = h + d$, $\lambda(\mathcal{M}/J) = h$ and $\lambda(J/J\mathcal{M}) = d$, we get $\mathcal{M}^2 = J\mathcal{M}$ so that $\mathcal{M}^n = J\mathcal{M}^{n-1}$ for every $n \geq 2$. By Theorem 6.1 this implies that A and its Artinian reduction have the same h-vector and the conclusion follows by the case $d = 0$.   □

The following theorem shows that the next steps, $e_1 = e$, $e = h + 2$, are no longer equivalent. The condition $e_1 = e$ is still *rigid* in the sense that it forces the Hilbert function of A and implies that G is Cohen-Macaulay. On the other hand, despite the fact that $e = h + 2$ is not rigid, a recent result of Rossi and Valla describes all the possible Hilbert functions of Cohen-Macaulay local rings verifying this condition (see [56]).

Theorem 6.11. *Let $(A, \mathcal{M})$ be a d-dimensional Cohen-Macaulay local ring. The following conditions are equivalent:*
  1. $P_A(z) = (1 + hz + z^2)/(1 - z)^d$.
  2. $e_1 = e$.
  3. $e_1 = h + 2$.
  4. $e = h + 2$ *and* G *is Cohen-Macaulay.*

Proof. It is clear that 1 implies 2. If $e_1 = e$, we have by the above theorem $e_1 = e \geq h + 2$. On the other hand $e_1 \geq 2e - h - 2 = 2e_1 - h - 2$, so that $e_1 = h + 2$ and 2 implies 3.

Let $e_1 = h + 2$; then by the above theorem

$$h + 2 = e_1 \geq e \geq h + 2, \qquad \text{hence} \qquad e_1 = e = h + 2 = 2e - h - 2$$

and G is Cohen-Macaulay by theorem 6.9. This proves that 3 implies 4.

If $e = h + 2$ and G is Cohen-Macaulay, we get

$$P_A(z) = \frac{P_C(z)}{(1 - z)^d}$$

where C is an Artinian local ring with the same multiplicity and embedding codimension as A. Since C is Artinian and $e = h + 2$ we clearly have

$$P_A(z) = \frac{P_C(z)}{(1 - z)^d} = \frac{1 + hz + z^2}{(1 - z)^d}. \qquad \square$$

Now we come to the very recent result of Rossi and Valla which gives a positive answer to a longstanding conjecture made by J. Sally (see [60]).

For a local Cohen-Macaulay ring $(A, \mathcal{M})$ we denote by $\tau(A)$ the *Cohen-Macaulay type* of A which the dimension of the socle of any Artinian reduction of A.

Theorem 6.12. *Let $(A, \mathcal{M})$ be a d-dimensional Cohen-Macaulay local ring. The following are equivalent:*
- $e = h + 2$.
- $P_A(z) = (1 + hz + z^s)/(1 - z)^d$ *for some integer s, $2 \leq s \leq h + 1$.*
   Further, if either of the above conditions holds, then we have:
- $\tau(A) \leq h$.
- $depth(G) \geq d - 1$.
- G *is Cohen-Macaulay* $\iff s = 2 \iff e_1 = e \iff e_1 = h + 2 \iff \tau(A) < h$.
Finally if A is Gorenstein, such is G.

For every s in the range $2 \leq s \leq h+1$, we can exhibit, as in [60], the following examples of local Cohen-Macaulay ring A_s with $e = h + 2$ and

$$P_{A_s}(z) = \frac{1 + hz + z^s}{1 - z}.$$

Let

$$A_2 = k[[t^e, t^{e+1}, t^{e+3}, \dots, t^{2e-1}]],$$

$$A_s = k[[t^e, t^{e+1}, t^{e+s+1}, t^{e+s+2}, \dots, t^{2e-1}, t^{2e+3}, t^{2e+4}, \dots, t^{2e+s}]]$$

for every s, $3 \leq s \leq e - 2$ and

$$A_{e-1} = k[[t^e, t^{e+1}, t^{2e+3}, t^{2e+4}, \dots, t^{3e-1}]].$$

By using the above theorem we can settle the case $e_1 = e + 1$. It turns out that this condition does not force the Hilbert function of A.

Theorem 6.13. *Let $(A, \mathcal{M})$ be a d-dimensional Cohen-Macaulay local ring. The following conditions are equivalent:*
- *$e_1 = e + 1$.*
- *Either $P_A(z) = (1 + hz + 2z^2)/(1 - z)^d$ or $P_A(z) = (1 + hz + z^3)/(1 - z)^d$.*
  In the first case G is Cohen-Macaulay, in the second case G has depth $d - 1$.

Proof. It is clear that both Hilbert series imply $e_1 = e + 1$. If $e_1 = e + 1$, then

$$e_1 = e + 1 \geq 2e - h - 2$$

implies $h + 1 \leq e \leq h + 3$. If $e = h + 1$, then $e_1 = e - 1$, hence we have two possibilities: either $e = h + 2$, and $e_1 = h + 3$, or $e = h + 3$ and $e_1 = h + 4$. In the first case, by the above theorem, we get

$$P_A(z) = \frac{1 + hz + z^3}{(1 - z)^d}$$

and $depth(G) = d - 1$.
     In the second case

$$2e - h - 2 = 2(h + 3) - (h - 2) = h + 4 = e_1$$

so that by Theorem 6.9

$$P_A(z) = \frac{1 + hz + 2z^2}{(1 - z)^d}$$

and G is Cohen-Macaulay. $\square$

The next case is still quite open.

Problem 6.14. *Find a structure theory for local Cohen-Macaulay rings verifying*

$$e \geq h + 3.$$

The unique result we know was given by Sally. She proved in [59] that if A is Gorenstein and $e = h + 3$, then G is Cohen-Macaulay. In the same paper Sally gave an example of a Gorenstein local ring with $e = h + 4$ such that G is not Cohen-Macaulay. This is the ring

$$A = k[[t^6, t^7, t^{15}]]$$

which is even a complete intersection. We have

$$A = k[X, Y, Z]/(Y^3 - XZ, X^5 - Z^2)$$
$$G = k[X, Y, Z]/(XZ, Y^6, Y^3 Z, Z^2)$$

and

$$P_A(z) = \frac{1 + 2z + z^2 + z^3 + z^5}{1 - z}.$$

On the other side we have an equivalent problem.

Problem 6.15. *Find the possible Hilbert functions of a local Cohen-Macaulay ring with*

$$e_1 \geq e + 2.$$

After proving the rigidity of e and e_1 near to the lower bound, we now want to consider the same problem near the upper bound. As for e_1, a result of Elias (see [18]) says that for every local Cohen-Macaulay ring of dimension greater than or equal to 1, we have

$$e - h - 2 \leq e_1 \leq \binom{e}{2} - \binom{h}{2}.$$

Further, the bound is sharp in the sense that for every integer t such that $e - h - 2 \leq t \leq \binom{e}{2} - \binom{h}{2}$ there exists a Cohen-Macaulay local ring A of dimension 1 such that the Hilbert polynomial of A is

$$p_A(X) = eX - t.$$

We remark that this gives all the possible Hilbert polynomials of such rings but the problem of determining all the possible Hilbert functions is far from a solution.

We proved recently in [21] the following result.

Theorem 6.16. *Let $(A, \mathcal{M})$ be a 1-dimensional Cohen-Macaulay local ring. The following conditions are equivalent:*
- $e_1 = \binom{e}{2} - \binom{h}{2}$.
- $P_A(z) = (1 + hz + \sum_{i=h+1}^{e-1} z^i)/(1 - z)$.

The proof is quite involved and we are not going to present it here. However we could not extend this result to the higher dimensional case. Hence we formulate the following problem.

Problem 6.17. *Let $(A, \mathcal{M})$ be a d-dimensional Cohen-Macaulay local ring. Are the following conditions equivalent?*
- $e_1 = \binom{e}{2} - \binom{h}{2}$.
- $P_A(z) = (1 + hz + \sum_{i=h+1}^{e-1} z^i)/(1 - z)^d$.

To get a lower bound for e_2, we introduce a device due to J.Sally (see [61]) which was the main tool in the proof of Sally's conjecture.

We start by remarking that if A is zero-dimensional then

$$e_1 = h + 2h_2 + \cdots + sh_s \quad \text{and} \quad e_2 = h_2 + 3h_3 + \cdots + \binom{s}{2} h_s$$

so that we can express e_1 and e_2 in terms of the integers $h_0, \ldots, h_s$ which are non negative. The same conclusion holds if we are in the one-dimensional case. For this we need the following classical result on the Hilbert function of a one-dimensional Cohen-Macaulay local ring.

Theorem 6.18. *Let $(A, \mathcal{M})$ be a 1-dimensional Cohen-Macaulay local ring. If x is a superficial element in $\mathcal{M}$, then for every $j \geq 0$*

$$H_A(j) = e - \rho_j$$

where $\rho_j := \lambda(\mathcal{M}^{j+1}/x\mathcal{M}^j)$.

In particular, if s is the degree of the h-polynomial of A,

$$\rho_0 = e - 1, \quad \rho_1 = e - h - 1, \quad \rho_j = 0 \quad \forall j \geq s,$$

$$e_1 = \sum_{j=0}^{s-1} \rho_j \quad \text{and} \quad e_2 = \sum_{j=1}^{s-1} j\rho_j$$

Proof. We have

$$
\begin{array}{ccccc}
A & \supset & \mathcal{M}^j & \supset & \mathcal{M}^{j+1} \\
\cup & & \cup & & \cup \\
xA & \supset & x\mathcal{M}^j & = & x\mathcal{M}^j
\end{array}
$$

From this we get

$$H_A(j) = \lambda(\mathcal{M}^j/x\mathcal{M}^j) - \lambda(\mathcal{M}^{j+1}/x\mathcal{M}^j).$$

But $\lambda(A/xA) = e$ and $A/\mathcal{M}^j \simeq xA/x\mathcal{M}^j$ so that

$$H_A(j) = e - \lambda(\mathcal{M}^{j+1}/x\mathcal{M}^j)$$

for every $j \geq 0$. Further $\rho_0 = e - H_A(0) = e - 1$ and $\rho_1 = e - H_A(1) = e - h - 1$.
Finally, since we have

$$P_A(z) = \frac{1 + (e - \rho_1 - 1)z + (\rho_1 - \rho_2)z^2 + \cdots + (\rho_{s-2} - \rho_{s-1})z^{s-1} + \rho_{s-1}z^s}{1 - z}$$

we get the desired formulas for e_1 and e_2. □

Similar formulas can be found in the two-dimensional case. For example in [40] it has been proved that for a two-dimensional Cohen-Macaulay local ring $(A, \mathcal{M})$ one has

$$e_1 = \sum_{j \geq 0} v_j \quad \text{and} \quad e_2 = \sum_{j \geq 1} j v_j$$

where, if we let J be the ideal generated by a maximal superficial sequence in $\mathcal{M}$, we put

$$v_j := \lambda(\mathcal{M}^{j+1}/J\mathcal{M}^j) - \lambda(\mathcal{M}^j : J\mathcal{M}^{j-1}).$$

Unfortunately the integers v_i can be negative; however, the following construction due to Ratliff and Rush (see [54]), gives a way to overcome the problem.

Let $(A, \mathcal{M})$ be a Cohen-Macaulay local ring. For every n we consider the chain of ideals

$$\mathcal{M}^n \subseteq \mathcal{M}^{n+1} : \mathcal{M} \subseteq \mathcal{M}^{n+2} : \mathcal{M}^2 \subseteq \cdots \subseteq \mathcal{M}^{n+k} : \mathcal{M}^k \subseteq \cdots$$

This chain stabilizes at an ideal which was denoted by Ratliff and Rush as

$$\widetilde{\mathcal{M}^n} := \bigcup_{k \geq 1} (\mathcal{M}^{n+k} : \mathcal{M}^k).$$

We have

$$\widetilde{\mathcal{M}} = \mathcal{M},$$

and for every i, j and n

$$\mathcal{M}^n \subseteq \widetilde{\mathcal{M}^n} \quad \text{and} \quad \widetilde{\mathcal{M}^i}\widetilde{\mathcal{M}^j} \subseteq \widetilde{\mathcal{M}^{i+j}}.$$

Further if x is superficial for A,

$$\widetilde{\mathcal{M}^{n+1}} : x = \widetilde{\mathcal{M}^n}$$

for every $n \geq 0$.

In the following we let J be the ideal generated by a maximal superficial sequence of elements in A and we define for every $n \geq 0$

$$\sigma_n := \lambda(\widetilde{\mathcal{M}^{n+1}}/J\widetilde{\mathcal{M}^n}).$$

For example we get $\sigma_0 = e - 1$.

With these notation one can easily prove (see also [39]) that in the case A has dimension two, we have

$$e_1 = \sum_{j \geq 0} \sigma_j \quad \text{and} \quad e_2 = \sum_{j \geq 1} j\sigma_j.$$

As a consequence we get the following lower bound for e_2 (see [62]).

Proposition 6.19. *If* $(A, \mathcal{M})$ *is a Cohen-Macaulay local ring, we have*

$$e_2 \geq e_1 - e + 1 \geq 0.$$

Proof. If A has dimension greater than or equal to two, $e_2(A) = e_2(B)$ and $e_1(A) = e_1(B)$ where B is a local Cohen-Macaulay ring of dimension two. Hence we have

$$e_2 = \sum_{j \geq 1} j\sigma_j = \sum_{j \geq 1} \sigma_j + \sum_{j \geq 2}(j-1)\sigma_j \geq e_1 - \sigma_0 = e_1 - e + 1.$$

If the dimension is one we can use the same chain of inequalities with ρ_j instead of σ_j. Finally if $d = 0$, then

$$e_2 - e_1 + e - 1 = h_3 + 3h_4 + \cdots + \binom{s-1}{2}h_s$$

and the conclusion follows. $\qquad\qquad\square$

Several partial results and computations suggest the following conjecture.

Conjecture 6.20. *Let $(A, \mathcal{M})$ be a d-dimensional Cohen-Macaulay local ring and let s be the degree of its h-polynomial. The following conditions are equivalent:*
- $e_2 = e_1 - e + 1.$
- $s \leq 2.$

Further if this is the case, then G is Cohen-Macaulay.

It is clear that if $s \leq 2$, then $e = 1 + h + h_2$, $e_1 = h + 2h_2$ and $e_2 = h_2$. Hence

$$e_1 - e + 1 = h + 2h_2 - (1 + h + h_2) + 1 = h_2 = e_2.$$

Also the conjecture holds if $e_2 = 0, 1, 2$. Namely, in the case $e_2 = 0$, we have $e_1 = e - 1$ and we use 6.10, in the case $e_2 = 1$ we have $e_1 = e$ and we use 6.11, finally in the case $e_2 = 2$ we have $e_1 = e + 1$ and we use 6.13.

Here we prove that the conjecture holds if $d = 0, 1$.

Proposition 6.21. *Let $(A, \mathcal{M})$ be a Cohen-Macaulay local ring of dimension less than or equal to 1. If $e_2 = e_1 - e + 1$ then $s \leq 2$.*
Proof. If $d = 0$ and $e_2 = e_1 - e + 1$, by the proof of Proposition 6.19 we have

$$h_3 + 3h_4 + \cdots + \binom{s-1}{2} h_s = 0.$$

This clearly forces s to be less than or equal to two.

If $d = 1$, our condition implies

$$\rho_1 + 2\rho_2 + \cdots + (s-1)\rho_{s-1} = \rho_1 + \rho_2 + \cdots + \rho_{s-1}$$

so that

$$\rho_2 + 2\rho_3 + \cdots + (s-2)\rho_{s-1} = 0.$$

If $s \geq 3$, this would imply $\rho_{s-1} = 0$, a contradiction.   □

We remark that if $e_2 = 3$ we would get a proof of the conjecture if we could prove that there does not exist a two-dimensional Cohen-Macaulay local ring A such that

$$P_A(z) = \frac{1 + hz + 3z^3 - z^4}{(1 - z)^2}.$$

In [21] we found an upper bound for e_2 and proved that, if equality holds, the Hilbert function is forced and G is Cohen-Macaulay. Since in this case the Hilbert function is quite complicated, we are not going to describe it. However we note that a consequence of that result gives the inequality

$$e_2 \leq \binom{e_1}{2} - \binom{h}{2}$$

which is nice if it is compared with the inequality found by Elias for e_1.

We finish with the following very general question.

Problem 6.22. *What can be said about the Hilbert function of a Cohen-Macaulay local ring A?*

Very little is know even if the ring has dimension one. However, in this case, Elias proved that the Hilbert function is not decreasing if the embedding dimension is three (see [19]). Orecchia gave in [53] examples of decreasing Hilbert function with Cohen-Macaulay local rings of embedding dimension five, and finally in [30] one can find similar examples with embedding dimension four.

It is clear that all the problems we have studied until now for the Hilbert function of the maximal ideal of a local Cohen-Macaulay ring, can be extended to the case of an $\mathcal{M}$-primary ideal I. Several results have been given in this setting showing that the behaviour is quite the same. But for example Sally gave in [62] an example of a primary ideal I in a local Cohen-Macaulay ring $(A, \mathcal{M})$, such that the associated graded ring is not Cohen-Macaulay even if $e_2 = e_1 - e + \lambda(A/I)$. This shows that the last conjecture does not hold in the $\mathcal{M}$-primary case. However we do not know of any counterexample in the case $I = \mathcal{M}$ or even with the weaker assumption $I = \tilde{I}$.

References

[1] S. Abhyankar, *Local rings of high embedding dimension*, Amer. J. Math., **89** (1967), 1073–1077.

[2] J. Alexander, *Singularités imposable en position general á une hypersurface projective*, Compositio Math, **68** (1988), 305–354.

[3] J. Alexander and A. Hirschowitz, *La méthode d'Horace éclaté: Application á l'interpolation en degré quatre*, Inv. Math, **107** (1992), 585–602.

[4] J. Alexander and A. Hirschowitz, *Un Lemme d'Horace différentiel: Application aux singularités hyperquartiques de $\mathbf{P}^5$*, J. Alge. Geo. 1 (1992), 411–426.

[5] D. Anick, *Thin algebras of embedding dimension three*, J. Algebra, **100** (1986), 235–259.

[6] D. Bayer and M. Stillman, *A criterion for detecting m-regularity*, Inv. Math, **87** (1987), 1–11.

[7] D. Bernstein and A. Iarrobino, *A nonunimodal graded Gorenstein Artin algebra in codimension five*, Comm. Algebra, **20** (1992), 2323–2336.

[8] A. M. Bigatti, A. V. Geramita and J. Migliore, *Geometric consequences of extremal behaviour in a Theorem of Macaulay*, Trans. A.M.S., **346 (1)** (1994), 203–235.

[9] M. Boij and D. Laksov, *Nonunimodality of graded Gorenstein Artin algebras*, Proc. A. M. S., **120** (1994), 1083–1092.

[10] M. V. Catalisano and A. Gimigliano, *On the Hilbert function of fat points on a rational normal cubic*, Preprint,(1996)

[11] M. V. Catalisano, N. V. Trung and G. Valla *A sharp bound for the regularity index of fat points in general position*, Proc. A. M. S., **118** (1993), 717–724.

[12] K. Chandler, *Hilbert functions of dots in linear general position*, Zero-dimensional Schemes, Proceedings of the International Conference held in Ravello, June 8–13, 1992, Edited by F. Orecchia and L. Chiantini, Walter De Gruyter (Berlin-New York) **1994**, 65–79.

[13] E. D. Davis, A. V. Geramita and F. Orecchia, *Gorenstein Algebras and the Cayley-Bacharach theorem*, Proc. A. M. S., **93** (1985), 593–598.

[14] E. De Negri and G. Valla, *The h-vector of a Gorenstein codimension three domain*, Nagoya Math. J., **138** (1995), 113–140.

[15] R. Ehrenborg and G. C. Rota, *Apolarity and canonical forms for homogeneous polynomials*, Europ. J. Comb., **14** (1993), 157–182.

[16] D. Eisenbud, *Commutative Algebra with a view toward Algebraic Geometry*, Graduate Texts in Mathematics, **150** 1995, Springer-Verlag, New York.

[17] D. Eisenbud, M. Green and J. Harris, *Higher Castelnuovo Theory*, Asterisque, **218** (1993), 187–202.

[18] J. Elias, *Characterization of the Hilbert-Samuel polynomial of curve singularities*, Comp. Math, **74** (1990), 135–155.

[19] J. Elias, *The conjecture of Sally on the Hilbert function for curve singularities*, J. of Algebra, **160** (1993), 42–49.

[20] J. Elias and G. Valla, *Rigid Hilbert Functions*, J. Pure and Appl. Algebra, **71** (1991), 19–41.

[21] Elias, J., Rossi, M.E., Valla, G., *On the coefficients of the Hilbert polynomial*, J. Pure Appl. Algebra., **108** (1996), 35–60.

[22] J. Emsalem and A. Iarrobino, *Inverse System of a symbolic power, I*, J. of Algebra, **174** (1995), 1080–1090.

[23] C. Fefferman and D. Phong, *The uncertainty principle and sharp Garding inequalities*, Comm. Pure Appl.Math, **34** (1981), 285–331.

[24] R. Fröberg, *An inequality for Hilbert series of graded algebras*, Math. Scand., **56** (1985), 117–144.

[25] A. V. Geramita, *Inverse Systems of fat points: Waring's Problem, Secant Variety of Veronese Varieties and Parameter Spaces for Gorenstein Ideals*, Queen's Papers Pure App. Math., Vol. X, The Curves Seminar at Queen's, **102** (1996), 1–104.

[26] A. V. Geramita, D. Gregory and L. Roberts, *Monomial ideals and points in projective space*, J. Pure and Appl. Alg., **40** (1986),33–62

[27] A. V. Geramita, M. Kreuzer and L. Robbiano, *Cayley-Bacharach schemes and their canonical modules*, Trans. A. M. S. , **339** (1993),163–189.

[28] A. V. Geramita and J. C. Migliore, *Hyperplane sections of a smooth curve in* $\mathbf{P}^3$, Comm. in Alg., **17** (1989), 3129–3164.

[29] L. Gruson and C. Peskine, *Genre des courbes de l'espace projectif*, Algebraic Geometry, Lect. Notes Math., **687** (1978) Springer.

[30] S. K. Gupta and L.G. Roberts, *Cartesian squares and ordinary singularities of curves*, Comm. in Alg., **11(1)** (1983), 127–182.

[31] J. Harris, *Curves in projective space*, Montreal: Les Presses de l'Universite' de Montreal, **1982**.

[32] J. Harris, *Algebraic Geometry. A first course*, Graduate Texts in Mathematics, **133** (1992), Springer-Verlag, New York.

[33] R. Hartshorne , *Connectedness of the Hilbert scheme*, Math. Inst. des Hautes Etudes Sci., **29** (1966), 261–304.

[34] J. Herzog, N. V. Trung and G. Valla, *On hyperplane sections of reduced irreducible varieties of low codimension,* J. Math. Kyoto Univ, **34** (1994), 47–72.

[35] T. Hibi, *Flawless O-sequences and Hilbert functions of Cohen-Macaulay integral domains,* J. Pure and Appl. Algebra, **60** (1989), 245–251.

[36] D. Hilbert, *Über die Theorie der algebraischen Formen,* Math. Ann., **36** (1890), 473–534.

[37] A. Hirschowitz, *La méthode d'Horace pour l'interpolation á plusieurs variables,* Manus. Math, **50** (1985), 337–388.

[38] M. Hochster and D. Laksov, *The linear syzygies of generic forms,* Comm. Algebra, **15** (1987), 227–239.

[39] S. Huckaba and T. Marley, *Hilbert coefficients and the depths of associated graded rings,* Preprint, (1996).

[40] C. Huneke, *Hilbert functions and symbolic powers,* Michigan Math. J., **34** (1987), 293–318.

[41] K. Kubota, *On the Hilbert-Samuel function,* Tokyo J. Math., **8** (1985), 439–448.

[42] A. Iarrobino, *Inverse System of a symbolic power, II. The Waring Problem for forms,* J. of Algebra, **174** (1995), 1091–1110.

[43] F. S. Macaulay, *Algebraic theory of modular systems,* Cambridge Tracts **16,** Cambridge University Press, Cambridge, U. K., 1916.

[44] F. S. Macaulay, *Some properties of enumeration in the theory of modular systems,* Proc. London Math. Soc., **26** (1927), 531–555.

[45] R. Maggioni and A. Ragusa, *The Hilbert Function of generic plane sections of curves in* $\mathbf{P}^3$, Inv. Math., **91** (1988), 253–258.

[46] T. Marley, *The coefficients of the Hilbert polynomial and the reduction number of an ideal,* J. London Math. Soc., **40** (1989), 1–8.

[47] D. McDuff and L. Polterovich, *Symplectic packings and Algebraic Geometry,* Inv. Math., **115** (1994), 405–429.

[48] M. Nagata, *On the 14-th problem of Hilbert,* Am. J. Math., **81** (1959), 766–772.

[49] M. Narita, *A note on the coefficients of Hilbert characteristic functions in semi-regular local rings,* Proc. Cambridge Philos. Soc., **59** (1963), 269–275.

[50] G. Niesi and L. Robbiano, *Disproving Hibi's conjecture with CoCoA or projective curves with bad Hilbert functions,* in: F. Eyssette and A. Galligo, Eds., Proceedings of Mega 92 (Birkhäuser, Boston) (1993), 195–201.

[51] D. G. Northcott, *A note on the coefficients of the abstract Hilbert Function,* J. London Math. Soc., **35** (1960), 209–214.

[52] A. Ooishi, *Δ-genera and sectional genera of commutative rings,* Hiroshima Math. J., **17** (1987), 361–372.

[53] F. Orecchia, *One-dimensional local rings with reduced associated graded ring and their Hilbert functions,* Manus. Math., **32** (1980), 391–405.

[54] L. J. Ratliff and D. Rush, *Two notes on reductions of ideals,* Indiana Univ. Math. J, **27** (1978), 929–934.

[55] M. E. Rossi and G. Valla, *Multiplicity and t-isomultiple ideals,* Nagoya Math. J., **110** (1988), 81–111.

[56] M. E. Rossi and G. Valla, *A conjecture of J. Sally*, Comm. in Algebra, **24(13)** (1996), 4249–4261.

[57] J. Sally, *On the associated graded ring of a local Cohen-Macaulay ring*, J. Math Kyoto Univ, **17** (1977), 19–21.

[58] J. Sally, *Super regular sequences*, Pac. J. of Math , **84** (1979), 475–481.

[59] J. Sally, *Good embedding dimension for Gorenstein singularities*, Math. Ann., **249** (1980), 95–106.

[60] J. Sally, *Cohen-Macaulay local ring of embedding dimension $e + d - 2$*, J. Algebra, **83** (1983), 325–333.

[61] J. Sally, *Hilbert coefficients and reduction number 2*, J. Algebraic Geometry, **1** (1992), 325–333.

[62] J. Sally, *Ideals whose Hilbert function and Hilbert polynomial agree at $n = 1$*, J. Algebra, **157** (1993), 534–547.

[63] B. Singh *Effect of a permissible blowing-up on the local Hilbert function*, Inv. Math., **26** (1974), 201–212.

[64] R. Stanley, *Hilbert functions of graded algebras*, Adv. in Math, **28** (1978), 57–83.

[65] R. Stanley, *On the Hilbert function of a graded Cohen-Macaulay domain*, J. Pure Appl. Algebra, **73** (1991), 307–314.

[66] R. Stanley, *Combinatorics and Commutative Algebra*, Birkhäuser, Boston, **1983**.

[67] A. Terracini, *Sulla rappresenatzione di coppie di forme ternarie mediante somme di potenze di forme lineari*, Ann. Mat. Pura Appl. Ser. (3), **24** (1915), 1–10.

[68] N. V. Trung and G. Valla, *On zero-dimensional subschemes of a complete intersection*, Math. Z, **219** (1995), 187–201.

[69] P. Valabrega and G. Valla, *Form rings and regular sequences*, Nagoya Math. J., **72** (1978), 93–101.

[70] G. Valla, *On the Betti numbers of perfect ideals*, Comp. Math., **91** (1994), 305–319.

Department of Mathematics, University of Genoa
Via Dodecaneso 35, 16146 Genoa, Italy
E-mail address: valla@dima.unige.it

Cohomological Degrees of Graded Modules

Wolmer V. Vasconcelos

Introduction

By a *degree* of a module M we mean a numerical measure of information carried by M. It must serve the purposes of allowing comparisons between modules and to exhibit flexible calculus rules that track the degree under some basic constructions in module theory.

The premier example of a degree (vector space dimension excluded) is that of the *multiplicity* of a module. Let $(S, \mathfrak{m})$ be a local ring and let M be a finitely generated S–module. The Hilbert function of M is

$$H_M : n \mapsto \ell(M/\mathfrak{m}^n M),$$

which for $n \gg 0$ is given by a polynomial:

$$H_M(n) = \frac{\deg(M)}{d!} n^d + \text{lower order terms.}$$

The integer $\deg(M)$ is the *multiplicity* of the module M, while the integer $d = \dim M$ is its *dimension*.

An important property of this degree is its computability. When S is a standard graded algebra, say $S = k[x_1, \ldots, x_n]/I$, and $<$ is a term ordering for the ring of polynomials, then

$$\deg(S) = \deg(k[x_1, \ldots, x_n]/in(I)),$$

where $in(I)$ is the corresponding initial ideal of I.

On the other hand $\deg(\cdot)$ may fail to capture features which are significant for M. For instance, if M is filtered by a chain of submodules

$$M = M_1 \supset M_2 \supset \cdots \supset M_r \supset 0, \quad M_i/M_{i+1} \simeq S/\mathfrak{p}_i, \tag{1}$$

where $\mathfrak{p}_i$ is a prime ideal, then

$$\deg(M) = \sum_{\dim S/\mathfrak{p}_i = \dim M} \deg(S/\mathfrak{p}_i).$$

Alternatively, if $\mathfrak{p}$ is an associated prime of M of dimension $\dim S/\mathfrak{p} = \dim M$, the localization $M_\mathfrak{p}$ is an Artinian module whose length $\ell(M_\mathfrak{p})$ is the number of times $\mathfrak{p}$ occurs in (1).

---

[0]This research was partially supported by the NSF.

This formulation still ignores the contributions of the lower dimensional components. Partly to address this, another degree was defined that collects information about the associated primes of M in all codimensions. First one attaches to any associated prime $\mathfrak{p}$ of M a number $\text{mult}_M(\mathfrak{p})$ similar to $\ell_\mathfrak{p}(M)$ (see section 2 for all details). The *arithmetic degree* of M is the integer

$$\text{adeg}(M) = \sum_{\mathfrak{p} \in \text{Ass}(M)} \text{mult}_M(\mathfrak{p}) \cdot \deg S/\mathfrak{p}.$$

A related degree is $\text{gdeg}(M)$, the *geometric degree* of M, where in this sum one adds only the terms corresponding to minimal associated primes of M.

One feature of these degrees is that they can be expressed in a way which does not require the actual knowledge of the associated primes of M. Both $\text{adeg}(M)$ and $\text{gdeg}(M)$ are put together from a sum of multiplicities of certain modules derived from M.

These numbers can be used for many purposes, for instance for the estimation of the exponent in the Nullstellensatz. More precisely, given an ideal $I = (f_1, \ldots, f_m)$ of a polynomial ring $S = k[x_1, \ldots, x_n]$, consider the integers s such that

$$(\sqrt{I})^s \subset I.$$

The *index of nilpotency* or *degree of nilpotency* $nil(I)$ of an ideal I is the smallest such integer s. A related index of nilpotency is $nil_0(I)$, the smallest integer t such that $x^t \in I$, $\forall x \in \sqrt{I}$. Although we shall not pursue it here (see [25], [34], [41]) it can be seen that $nil(I) \leq \deg(S/I)$. In fact, the degrees of the associated primes of I do not play a role and the notion of multiplicity has to be replaced by another one, not larger, that carries structure better.

Our main use of these degrees is to make predictions about the outcome of carrying out Noether normalization on a graded algebra and as numerical predictors of the properties of an ideal as seen from its primary decomposition.

Let A be a finitely generated, positively graded algebra over a field k,

$$A = k + A_1 + A_2 + \cdots = k + A_+,$$

where A_i denotes the space of homogeneous elements of degree i. We further assume that A is generated by its 1–forms, $A = k[A_1]$, in which case $A_i = A_1{}^i$. If k is sufficiently large and $\dim A = d$, there are forms $x_1, \ldots, x_d \in A_1$, such that

$$R = k[x_1, \ldots, x_d] = k[\mathbf{z}] \hookrightarrow A = S/I, \quad S = k[x_1, \ldots, x_d, x_{d+1}, \ldots, x_n]$$

is a Noether normalization, that is, the x_i are algebraically independent over k and A is a finite R–module. Let $b_1, b_2, \ldots, b_s$, be a minimal set of homogeneous generators of A as an R–module

$$A = \sum_{1 \leq i \leq s} Rb_i, \quad \deg(b_i) = r_i.$$

The distribution of the r_i's, particularly of the largest $r_R(A)$ of these degrees, may depend on the choice of R. For the 'best' of all Noether normalizations, it will be called the *reduction number* of A: $r(A)$. It turns out that $\mathrm{adeg}(A)$ (in large characteristics) give estimates for $r(A)$.

The Castelnuovo–Mumford's *regularity* of a graded module M is a commonly used yardstick for it. It can be defined in terms of the vanishing of the local cohomology modules of M. If $\mathfrak{m}$ is the irrelevant maximal ideal of the graded algebra S, then

$$\mathrm{reg}(M) = \sup_i\{\alpha_+(H^i_{\mathfrak{m}}(M)) + i\},$$

where $\alpha_+(L)$ of a graded module L is the supremum of n for which L_n is non–trivial.

If A is a standard graded algebra, $\mathrm{reg}(A) \geq r(A)$, but its relationship to $\mathrm{adeg}(A)$ is irregular. Nevertheless $\mathrm{reg}(A)$ is a wonderful measure since it can be read from projective resolutions.

Most of our effort will lie in the development of *degrees* with the following aspects. S will be either a local ring or a standard graded algebra. Denote by $\mathcal{M}(S)$ the appropriate category of finitely generated S–modules. We shall define a class of numerical functions

$$\mathrm{Deg}(\cdot) : \mathcal{M}(S) \mapsto \mathbb{R},$$

that satisfy the following conditions:

(i) If $L = \Gamma_{\mathfrak{m}}(M)$ is the submodule of elements of finite support of M and $\overline{M} = M/L$, then

$$\mathrm{Deg}(M) = \mathrm{Deg}(\overline{M}) + \ell(L), \tag{2}$$

where $\ell(\cdot)$ is the ordinary length function.

(ii) If $h \in S$ is a regular, generic hyperplane section on M, then

$$\mathrm{Deg}(M) \geq \mathrm{Deg}(M/hM). \tag{3}$$

(iii) (The calibration rule) If M is a Cohen–Macaulay module, then

$$\mathrm{Deg}(M) = \mathrm{deg}(M), \tag{4}$$

where $\mathrm{deg}(M)$ is the ordinary multiplicity of the module M.

Any such function will satisfy $\mathrm{Deg}(M) \geq \mathrm{deg}(M)$, with equality holding if and only if M is Cohen–Macaulay.

A first issue that arises is whether such functions exist at all in all dimensions. This is settled in Section 3 when we introduce the notion of the *homological degree* $\mathrm{hdeg}(M)$ ([40]). Suppose that S is either a graded algebra over an Artinian ring or a local ring, and let M be a finitely generated S–module. The simplest case where to look at $\mathrm{hdeg}(M)$ is when $\dim M = \dim S = d > 0$:

$$\mathrm{hdeg}(M) = \mathrm{deg}(M) + \sum_{i=1}^{d} \binom{d-1}{i-1} \cdot \mathrm{hdeg}(\mathrm{Ext}^i_S(M, S)).$$

Notice that this is basically an iterative definition and therefore can be implemented as a procedure whenever the computation of Exts and multiplicities is possible.

This definition points at possible uses for this degree. Let I be an ideal with an irredundant primary decomposition

$$I = Q_1 \cap \cdots \cap Q_m.$$

When applied to S/I, hdeg($\cdot$) will capture two main aspects of this decomposition. First, the local contribution of each associated prime ideal, in other words all of adeg(S/I). Second, in the iterated Exts it will look for interactions amongst the components.

The most significant aspect of this particular degree lies in the fact that it satisfies a number of rules of calculation with regard to certain exact sequences. They are looser than those followed by reg($\cdot$). One of the comparisons that will be made shows that for any standard graded algebra A (Theorem 4.6):

$$\mathrm{Deg}(A) > \mathrm{reg}(A),$$

for any Deg($\cdot$) function, in particular for hdeg($\cdot$). The fact that it comes coded by an explicit formula will permit making many *a priori* estimates for the number of generators in terms of multiplicities.

Another set of applications concerns the relationship between the number of generators of an ideal I and the degrees of R/I. Suppose that R is a Cohen–Macaulay ring which is either a graded algebra over a field or a homomorphic image of a Gorenstein ring. Then for any ideal I (homogeneous when R is graded), one has (Theorem 5.3):

$$\nu(I) \le \deg(R) + (g-1)\deg(R/I) + (d-r-1)(\mathrm{Deg}(R/I) - \deg(R/I)),$$

where $d = \dim R$, $g = \mathrm{height}\ (I)$ and $r = \mathrm{depth}\ (R/I)$.

Another set of results concerns Hilbert functions of local rings. For a Cohen–Macaulay local ring $(R, \mathfrak{m})$, it was established in [27] that for any integer n,

$$\nu(\mathfrak{m}^n) \le \deg(R)n^{d-1} + d - 1,$$

where $d = \dim R \ge 1$. In [40], for a local ring, not necessarily Cohen–Macaulay, which is a homomorphic image of a Gorenstein ring, it is proved that this estimate will hold replacing $\deg(R)$ by hdeg(R). Following [9], one can show that in the Cohen–Macaulay case one has the considerably sharper estimate (Theorem 6.5):

$$\nu(\mathfrak{m}^n) \le \deg(R)\binom{n+d-2}{d-1} + \binom{n+d-2}{d-2}.$$

In particular this establishes the rational function

$$H_R(\mathbf{t}) = \frac{1 + (\deg(R) - 1)\mathbf{t}}{(1 - \mathbf{t})^d},$$

as the maximal Hilbert function for Cohen–Macaulay local rings of dimension d and multiplicity $\deg(R)$.

There are similar bounds when R is an arbitrary local ring equipped with a $\mathrm{Deg}(\cdot)$ function with $\deg(R)$ simply replaced by $\mathrm{Deg}(R)$. With its extensions to $\mathfrak{m}$–primary ideals, these estimates provide for sharp predictions on the outcome of effecting Noether normalizations of certain algebras.

These are text notes for lectures at the *Summer School on Commutative Algebra*, held at the Centre de Recerca Matematica, in July 1996. They are intended to be understood without much technical background with the exception of familiarity with basic local cohomology. The material itself comes from several sources in the literature, but particularly from [40], [9] and [41, Chapter 9]. The latter contains a more elementary discussion of these degrees and their constructive aspects.

We are all very grateful to the organizers of SSCA, Professors J. Elias, J. M. Giral, R. M. Miró-Roig and S. Zarzuela, and the direction and staff of CRM for creating a pleasant and invigorating framework for the meeting. I am personally thankful to several people but particularly Alberto Corso, Luisa R. Doering and Tor Gunston.

1. Arithmetic Degree of a Module

In the following, and mostly for convenience, M will be a finitely generated module over a standard graded algebra, which may be taken to be a polynomial ring $S = k[x_1, \ldots, x_n]$, or alternatively a module over a local ring.

We first introduce all the degrees except for one that will come onto stage later. We shall draw repeatedly from general facts on Hilbert functions and local cohomology; they all can be found in [6].

Multiplicity
We begin by recalling the definition of multiplicity in Local Algebra. Let $(R, \mathfrak{m})$ be a local ring and let M be a finitely generated R–module. The Hilbert function of M is

$$H_M : n \mapsto \ell(M/\mathfrak{m}^n M),$$

which for $n \gg 0$ is given by a polynomial:

$$H_M(n) = \frac{\deg(M)}{d!} n^d + \text{lower order terms}.$$

For convenience we record:

Definition 1.1. The integer $\deg(M)$ is the *multiplicity* of the module M, while the integer $d = \dim M$ is its *dimension*.

The behavior of $\deg(M)$ and $\dim M$ with regard to submodules and primary decomposition follows from:

Proposition 1.2. *Let R be a local ring and let*

$$0 \to L \longrightarrow M \longrightarrow N \to 0$$

be an exact sequence of R–modules. Then

$$\deg(M) = \begin{cases} \deg(L) + \deg(N) & \text{if } \dim L = \dim N = \dim M, \\ \deg(L) & \text{if } \dim N < \dim M, \\ \deg(N) & \text{if } \dim L < \dim M. \end{cases}$$

The proof is easy and left to the reader. Used in tandem with the following definition it gives some flexible rules of computation.

Definition 1.3. Let R be a Noetherian ring and M a finitely R–module. For a prime ideal $\mathfrak{p} \subset R$, the integer

$$\text{mult}_M(\mathfrak{p}) = \ell(\Gamma_{\mathfrak{p}}(M_{\mathfrak{p}}))$$

is the *length multiplicity* of $\mathfrak{p}$ with respect to M.

This number $\text{mult}_M(\mathfrak{p})$, which vanishes if $\mathfrak{p}$ is not an associated prime of M, is a measure of the contribution of $\mathfrak{p}$ to the primary decomposition of the null submodule of M. It is not usually accessible through direct computation except in exceptional cases.

Castelnuovo–Mumford regularity

Let k be a field, $S = k[x_1, \dots, x_m]$ a polynomial ring over k, $R = S/I$ a homogeneous k-algebra, and M a finitely generated graded R-module. Then M, as an S-module, admits a finite graded free resolution:

$$0 \to \oplus_j S(-j)^{b_{pj}} \longrightarrow \cdots \longrightarrow \oplus_j S(-j)^{b_{0j}} \longrightarrow M \to 0.$$

Definition 1.4. The *Castelnuovo-Mumford* regularity, or simply the *Castelnuovo regularity*, of M is the integer

$$\text{reg } M = \max\{j - i \colon b_{ij} \neq 0\}.$$

In other words, $\text{reg } M = \max\{\alpha_+(\text{Tor}_i^S(M, k)) - i \colon i \in \mathbb{Z}\}$ where for a graded module N with $N_j = 0$ for large j, we set $\alpha_+(N) = \max\{j \colon N_j \neq 0\}$. Let q be an integer. The module M is called q-*regular* if $q \geq \text{reg }(M)$, equivalently, if $\text{Tor}_i^S(M, k)_{j+i} = 0$ for all i and all $j > q$.

Eisenbud and Goto [12] gave an interesting interpretation of regularity. Denoting by $M_{\geq q}$ the truncated graded R-module $\oplus_{j \geq q} M_j$, one has (see also [41, Appendix B]):

Theorem 1.5. *The following conditions are equivalent:*
(a) *M is q-regular;*
(b) *$H_{\mathfrak{m}}^i(M)_{j-i} = 0$ for all i and all $j > q$;*
(c) *$M_{\geq q}$ admits a linear S-resolution, i.e., a graded resolution of the form*

$$0 \to S(-q - l)^{c_l} \longrightarrow \cdots \longrightarrow S(-q - 1)^{c_1} \longrightarrow S(-q)^{c_0} \longrightarrow M_{\geq q} \to 0.$$

Proof. (a) $\Leftrightarrow$ (c): By definition, the module $M_{\geq q}$ has a linear resolution if and only if for all i

$$\mathrm{Tor}_i^S(M_{\geq q}, k)_r = H_i(\mathbf{x}; M_{\geq q})_r = 0$$

for $r \neq i + q$. Here $H(\mathbf{x}; M)$ denotes the Koszul homology of M with respect to the sequence $\mathbf{x} = x_1, \ldots, x_m$.

Since $(M_{\geq q})_j = 0$ for $j < q$, we always have $H_i(\mathbf{x}; M_{\geq q})_r = 0$ for $r < i + q$, while for $r > i + q$

$$H_i(\mathbf{x}; M_{\geq q})_r = H_i(\mathbf{x}; M)_r = \mathrm{Tor}_i^S(M, k)_r.$$

Thus the desired result follows.

(b) $\Rightarrow$ (c): We may assume $q = 0$, and $M = M_{\geq 0}$. Then it is immediate that $H_\mathfrak{m}^0(M)$ is concentrated in degree 0. This implies that $M = H_\mathfrak{m}^0(M) \oplus M/H_\mathfrak{m}^0(M)$. The first summand is a direct sum of copies of k. Hence M is 0-regular if and only if $M/H_\mathfrak{m}^0(M)$ is 0-regular. In other words we may assume that depth $M > 0$. Without any problems we may further assume that k is infinite. Then there exists an element $y \in S$ of degree 1 which is M-regular. From the cohomology exact sequence associated with

$$0 \to M(-1) \xrightarrow{y} M \longrightarrow M/yM \to 0$$

we see that M/yM is 0-regular. By induction on the dimension on M, we may suppose that M/yM has linear S/yS-resolution. But if F is a minimal graded free S-resolution, the F/yF is a minimal graded S/yS-resolution of M/yM. This implies that F is a linear S-resolution of M.

(c) $\Rightarrow$ (b): Again we may assume $q = 0$, and $M = M_{\geq 0}$. Then M has a linear resolution

$$\cdots \longrightarrow S(-2)^{c_2} \longrightarrow S(-1)^{c_1} \longrightarrow S^{c_0} \longrightarrow M \to 0.$$

Computing $\mathrm{Ext}_S^i(M, S)$ with this resolution we see at once that $\mathrm{Ext}_S^i(M, S)_j = 0$ for $j < -i$. By duality (see [6, Section 3.6]) there exists an isomorphism of graded R-modules

$$H_\mathfrak{m}^i(M) \simeq \mathrm{Hom}_k(\mathrm{Ext}_S^{m-i}(M, S(-m)), k).$$

Therefore, $H_\mathfrak{m}^i(M)_{j-i} = 0$ for all $j > 0$, as desired. $\qquad\square$

Arithmetic degree of a module

We now define a key notion of this section. Throughout the ring A will be as before, either a standard graded algebra or a local ring.

Definition 1.6. Let M be a finitely generated A–module. The *arithmetic degree* of M is the integer

$$\mathrm{adeg}(M) = \sum_{\mathfrak{p} \in \mathrm{Ass}(M)} \mathrm{mult}_M(\mathfrak{p}) \cdot \deg A/\mathfrak{p}. \tag{5}$$

When applied to a module $M = S/I$, it is clear that this number gives a numerical signature for the primary decomposition of I.

The definition applies to general modules, not just graded modules, although its main use is for graded modules over a standard algebra A. It only requires the definition of *degree* of integral domains.

If all the associated primes of M have the same dimension, then adeg(M) is just the multiplicity deg(M) of M, which is obtained from its Hilbert polynomial.

In general, adeg(M) can be put together as follows. Collecting the associated primes of M by their dimensions:

$$\dim M = d_1 > d_2 > \cdots > d_n,$$

$$\mathrm{adeg}(M) = a_{d_1}(M) + a_{d_2}(M) + \cdots + a_{d_n}(M),$$

where $a_{d_i}(M)$ is contribution of all primes in Ass(M) of dimension d_i.

The integer $a_{d_1}(M)$, in the case of a graded module M, is its multiplicity deg(M). Bayer and Mumford ([2]) have refined deg(M) into the following integer:

Definition 1.7. The *geometric degree* of M is the integer

$$\mathrm{gdeg}(M) = \sum_{\mathfrak{p}\ \mathrm{minimal}\ \in \mathrm{Ass}(M)} \mathrm{mult}_M(\mathfrak{p}) \cdot \deg A/\mathfrak{p}. \tag{6}$$

If $A = S/I$, we want to view reg(A) and adeg(A) as two basic measures of complexity of A. [2] has a far-flung survey of reg(A) and adeg(A) in terms of the degree data of a presentation $A = k[x_1, \ldots, x_n]/I$.

Stanley–Reisner rings

An important class of rings arises from monomial ideals generated by squarefree elements. They are elegantly coded in the following notion.

Definition 1.8. A *simplicial complex* $\triangle$ on vertex set V is a family of subsets of V satisfying

1. If $x \in V$ then $\{x\} \in \triangle$;
2. If $F \in \triangle$ and $G \subset F$ then $G \in \triangle$.

The elements of $\triangle$ are called *faces* or *simplices*. If $|F| = p+1$ then F has dimension p. We define the dimension of the complex as $\dim(\triangle) = \max_{F \in \triangle}(\dim F)$.

Definition 1.9. Given any field k and any simplicial complex $\triangle$ on the finite vertex set $V = \{x_1, \ldots, x_n\}$ define the *face ring* or *Stanley-Reisner ring* $k[\triangle]$ by

$$k[\triangle] = k[x_1, \ldots, x_n]/I_\triangle,$$

where

$$I_\triangle = (x_{i_1} x_{i_2} \cdots x_{i_r} \mid i_1 < i_2 < \ldots < i_r \text{ and } \{x_{i_1}, \ldots, x_{i_r}\} \notin \triangle).$$

We need only consider the minimal non-faces of $\triangle$ to arrive at a set of generators for $I_\triangle$.

Let f_p be the number of p-simplices. Since $\emptyset \in \triangle$ and $\dim \emptyset = -1$, $f_{-1} = 1$. Also f_0 is the number of vertices, thus $f_0 = n$. We call the d-tuple $\mathbf{f}(\triangle) = (f_0, \ldots, f_{d-1})$, consisting of the number of faces in each dimension, the $\mathbf{f}$–*vector* of $\triangle$. The Hilbert series of $k[\triangle]$ is determined entirely by $\mathbf{f}(\triangle)$,

$$H_{k[\triangle]}(\mathbf{t}) = \frac{h_0 + h_1\mathbf{t} + \cdots + h_d\mathbf{t}^d}{(1 - \mathbf{t})^d}$$

where the h_i are certain linear combinations of the f_j, in particular

$$h_d = (-1)^{d-1}(\chi(\triangle) - 1), \quad \chi(\triangle) = \sum_{i=0}^{d-1}(-1)^i f_i. \tag{7}$$

We quote two elementary facts about these rings (see [6]).

Proposition 1.10. *Let $\triangle$ be a simplicial complex on the vertex set $V = \{x_1, \ldots, x_n\}$. Then:*

(a) $\dim k[\triangle] = 1 + \dim \triangle = d$.
(b) $I_\triangle = \bigcap_F (x_{i_1}, \ldots, x_{i_r}, x_{i_j} \notin F)$.

As a consequence we have that

$$\gdeg(k[\triangle]) = \adeg(k[\triangle]) \quad = \quad \text{number of maximal faces of } \triangle;$$
$$\deg(k[\triangle]) \quad = \quad \text{number of faces of } \triangle \text{ of maximal dimension.}$$

Computation of the arithmetic degree of a module
We show how a program with the capabilities of *CoCoA* ([7]) or *Macaulay* ([3]) can be used to compute the arithmetic degree of a graded module M without availing itself of any primary decomposition. Let $S = k[x_1, \ldots, x_n]$ and suppose $\dim M = d \le n$. It suffices to construct graded modules M_i, $i = 1 \ldots n$, such that

$$a_i(M) = \deg(M_i).$$

For each integer $i \ge 0$, denote $L_i = \operatorname{Ext}_S^i(M, S)$. By local duality ([6, Section 3.5]), a prime ideal $\mathfrak{p} \subset S$ of height i is associated to M if and only if $(L_i)_\mathfrak{p} \ne 0$; furthermore $\ell((L_i)_\mathfrak{p}) = \operatorname{mult}_M(\mathfrak{p})$.

We are set to find a path to $\adeg(M)$: Compute for each L_i its degree $e_1(L_i)$ and codimension c_i. Then choose M_i according to

$$M_i = \begin{cases} 0 & \text{if } c_i > i \\ L_i & \text{otherwise.} \end{cases}$$

Proposition 1.11. *For a graded S–module M and for each integer i denote by c_i the codimension of $\operatorname{Ext}_S^i(M, S)$. Then*

$$\adeg(M) = \sum_{i=0}^{n} \lfloor \frac{i}{c_i} \rfloor \, \deg(\operatorname{Ext}_S^i(M, S)). \tag{8}$$

This can also be expressed as

$$\mathrm{adeg}(M) = \sum_{i=0}^{n} \deg(\mathrm{Ext}_S^i(\mathrm{Ext}_S^i(M,S),S)). \tag{9}$$

Note that this formula (one sets $\lfloor \frac{0}{0} \rfloor = 1$) gives a sum of sums of terms some of which are not always available.

Degrees and hyperplane sections

Definition 1.12. Let k be a field and let A be a finitely generated $\mathbb{N}$-graded k-algebra,

$$A = k + A_1 + \cdots + A_i + \cdots .$$

A is a *standard algebra* if it is generated by its elements of degree 1, that is $A_i = (A_1)^i$. A *hyperplane section* h of A is a form in A_1 that is not contained in any minimal prime of A (sometimes stricter conditions are imposed).

If k is an infinite field, most elements of A_1 are hyperplane sections. A fruitful method to probe A is to compare the properties of A with those of $A/(h)$ for some hyperplane section h.

Theorem 1.13. *Let $(R, \mathfrak{m})$ be a Noetherian local ring of dimension $d > 0$ and let M be a finitely generated R–module. Let $x \in \mathfrak{m}$ and consider the short exact sequence induced by multiplication by x,*

$$0 \to L \longrightarrow M \xrightarrow{x} M \longrightarrow G \to 0.$$

If L is a module of finite length, then $\ell(H_{\mathfrak{m}}^0(G)) \geq \ell(L)$. Moreover, if $d = 1$ then

$$\ell(H_{\mathfrak{m}}^0(G)) = \ell(L) + \ell(\overline{M}/x\overline{M}),$$

where $\overline{M} = M/\mathrm{torsion}$.

Note that the last formula follows immediately if M decomposes into a direct sum of a torsionfree plus torsion summands. In case R is a standard graded algebra over a field of characteristic zero, the first assertion is contained in the refined statement of [19, Proposition 3.5].

Arithmetic degree and hyperplane sections

To examine the critical behavior of $\mathrm{adeg}(M)$ under hyperplane sections we focus on the following (see [2], [25], [40]):

Theorem 1.14. *Let A be a standard graded algebra, let M be a graded algebra and let $h \in A_1$ be a regular element on M. Then*

$$\mathrm{adeg}(M/hM) \geq \mathrm{adeg}(M). \tag{10}$$

Proof. We first show that if M is a finitely generated R–module and $\mathfrak{p}$ is an associated prime of M, then for any $h \in R$ that is regular on M, any minimal prime P of $(\mathfrak{p}, h)$ is an associated prime of M/hM.

We may assume that P is the unique maximal ideal of R. Let $L = \Gamma_{\mathfrak{p}}(M)$; note that we cannot have $L \subset hM$ as otherwise since h is regular on M we would have $L = hL$. This means that there is a mapping

$$R/\mathfrak{p} \longrightarrow M,$$

which on reduction modulo h does not have a trivial image. Since $R/(\mathfrak{p}, h)$ is annihilated by a power of P, its image will also be of finite length and non-trivial, and therefore $P \in \mathrm{Ass}(M/hM)$.

We take stock of the relationship between the associated primes of a module M and the associated primes of M/hM, where h is a regular element on M. According to the above, for each associated prime $\mathfrak{p}$ of M, for which $(\mathfrak{p}, h)$ is not the unity ideal (which will be the case when M is a graded module and h is a homogeneous form of positive degree) there is at least one associated prime of M/hM containing $\mathfrak{p}$: any minimal prime of $(\mathfrak{p}, h)$ will do.

There may be associated primes of M/hM, such as $\mathfrak{n}$, which do not arise in this manner. We indicate this in the diagram below:

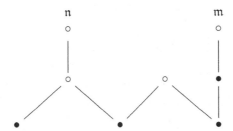

In the expression for $\mathrm{adeg}(M/hM)$ we are going to keep apart primes such as $\mathfrak{m}$ (which we call associated primes of the first kind) and primes as $\mathfrak{n}$, which are not minimal over $(\mathfrak{p}, h)$, for any $\mathfrak{p} \in \mathrm{Ass}(M)$.

Finally we consider the proof proper. For $\mathfrak{m} \in \mathrm{Ass}(M/hM)$ of the first kind, let

$$A(\mathfrak{m}) = \{\mathfrak{p}_1, \dots, \mathfrak{p}_r\}$$

be the set of associated primes of M such that $\mathfrak{m}$ is minimal over each ideal $(\mathfrak{p}_i, h)$. Note that these prime ideals have the same dimension, $\dim(A/\mathfrak{m}) + 1$. Denote by I the product of the $\mathfrak{p}_i$ and let $L(\mathfrak{m}) = L = \Gamma_I(M)$. We have an embedding

$$L/hL \hookrightarrow M/hM$$

since $L \cap hM = hL$. As a consequence, $\mathrm{mult}_{L/hL}(\mathfrak{m}) \leq \mathrm{mult}_{M/hM}(\mathfrak{m})$. On the other hand,

$$\mathrm{mult}_L(\mathfrak{p}_i) \quad = \quad \mathrm{mult}_M(\mathfrak{p}_i) \tag{11}$$

for any prime in the set $A(\mathfrak{m})$.

Let now

$$L = L_0 \supset L_1 \supset L_2 \supset \cdots \tag{12}$$

be a filtration of L whose factors are of the form $A/\mathfrak{q}$ for some homogeneous prime ideal $\mathfrak{q}$. The number of times in which the prime $\mathfrak{p} = \mathfrak{p}_i$ occurs in the filtration is $\mathrm{mult}_L(\mathfrak{p})$. Consider the effect of multiplication by h on the terms of this filtration: By the snake lemma, we have exact sequences

$$0 \to {}_hL_{i+1} \longrightarrow {}_hL_i \longrightarrow {}_hA/\mathfrak{q} \longrightarrow L_{i+1}/hL_{i+1} \longrightarrow L_i/hL_i \longrightarrow A/(\mathfrak{q}, h) \to 0.$$

Localizing at $\mathfrak{m}$ we get sequences of modules of finite length over $A_\mathfrak{m}$. Adding these lengths, and taking into account the collapsing that may occur, we get

$$\ell(L_\mathfrak{m}/hL_\mathfrak{m}) = \sum \ell((A/(\mathfrak{q}, h))_\mathfrak{m}) - \ell(({}_hA/\mathfrak{q})_\mathfrak{m}), \tag{13}$$

where ${}_hA/\mathfrak{q}$ is $A/\mathfrak{q}$ if $h \in \mathfrak{q}$ and zero otherwise. Note that some of the $\mathfrak{q}$ may occur repeatedly according to the previous observation. After the localization at $\mathfrak{m}$ only the $\mathfrak{q}$ corresponding to the $\mathfrak{p}_i$'s survive:

$$\mathrm{mult}_{L/hL}(\mathfrak{m}) = \ell(L_\mathfrak{m}/hL_\mathfrak{m}) = \sum_{\mathfrak{p}_i \subset \mathfrak{m}} \mathrm{mult}_M(\mathfrak{p}_i)\ell((A/(\mathfrak{p}_i, h))_\mathfrak{m}). \tag{14}$$

We thus have

$$
\begin{aligned}
\mathrm{adeg}(M/hM) &= \sum_\mathfrak{m} \mathrm{mult}_{M/hM}(\mathfrak{m}) \deg(A/\mathfrak{m}) \\
&+ \sum_\mathfrak{n} \mathrm{mult}_{M/hM}(\mathfrak{n}) \deg(A/\mathfrak{n}) \\
&\geq \sum_\mathfrak{m} \mathrm{mult}_{M/hM}(\mathfrak{m}) \deg(A/\mathfrak{m}) \\
&\geq \sum_\mathfrak{m} \mathrm{mult}_{L(\mathfrak{m})/hL(\mathfrak{m})}(\mathfrak{m}) \deg(A/\mathfrak{m}) \\
&= \sum_\mathfrak{m} \Big(\sum_{\mathfrak{p}_i \in A(\mathfrak{m})} \mathrm{mult}_M(\mathfrak{p}_i)\ell((A/(\mathfrak{p}_i, h))_\mathfrak{m}) \Big) \deg(A/\mathfrak{m}) \tag{15} \\
&= \sum_{\mathfrak{p}_i} \mathrm{mult}_M(\mathfrak{p}_i) \Big(\sum_{\mathfrak{p}_i \in A(\mathfrak{m})} \ell((A/(\mathfrak{p}_i, h))_\mathfrak{m}) \deg(A/\mathfrak{m}) \Big) \\
&= \sum_{\mathfrak{p}_i} \mathrm{mult}_M(\mathfrak{p}_i) \deg(A/\mathfrak{p}_i) \\
&= \mathrm{adeg}(M),
\end{aligned}
$$

where in equation (15) we first used the equality (11), and then derived from each $L(\mathfrak{m})$ the equality provided by the computation of multiplicities in (13) and (14), while taking into account that

$$\deg(A/\mathfrak{p}_i) = \deg(A/(\mathfrak{p}_i, h)) = \sum_{\mathfrak{p}_i \in A(\mathfrak{m})} \ell((A/(\mathfrak{p}_i, h))_\mathfrak{m}) \deg(A/\mathfrak{m}),$$

the last equality by [6, Corollary 4.6.8]. □

2. Reduction Number of an Algebra

The primary goal here is to develop a number of techniques designed to help in making predictions on the outcome of carrying out Noether normalization on a graded algebra. It is a mix of homological and combinatorial tools assembled from several sources. Unfortunately some of them are sensitive to the characteristic of the field.

Let A be a finitely generated, positively graded algebra over a field k,

$$A = k + A_1 + A_2 + \cdots = k + A_+,$$

where A_i denotes the space of homogeneous elements of degree i. Let

$$R = k[x_1, \ldots, x_d] = k[\mathbf{z}] \hookrightarrow A = S/I, \ S = k[x_1, \ldots, x_d, x_{d+1}, \ldots, x_n]$$

be a Noether normalization, that is, the x_i are algebraically independent over k and A is a finite R–module. Let $b_1, b_2, \ldots, b_s$ be a minimal set of homogeneous generators of A as an R–module.

$$A = \sum_{1 \le i \le s} Rb_i, \ \deg(b_i) = r_i.$$

These integers are very hard to predict for which reason we focus on:

Definition 2.1. The *reduction number* $r_R(A)$ of A with respect to R is the supremum of all $\deg(b_i)$. The (absolute) reduction number $\mathrm{r}(A)$ is the infimum of $r_R(A)$ over all possible Noether normalizations of A.

One of our aims is to make predictions about these integers, but without availing ourselves of any Noether normalization. We emphasize this by saying that the Noether normalizations are invisible to us, and the information we may have about A comes from the presentation $A = S/I$.

The integer $\mathrm{r}(A)$ is a measure of complexity of the algebra A. It has been compared to another index of complexity of A, the Castelnuovo–Mumford regularity of A.

Castelnuovo–Mumford regularity and reduction number

Theorem 2.2. *Let A be a standard graded algebra over an infinite field k and let $R \hookrightarrow A$ be a standard Noether normalization. Then $r_R(A) \le \mathrm{reg}(A)$.*

Proof. A more general statement is given in [35] and we content ourselves here for algebras such as A.

Let $R \hookrightarrow A$ be a (graded) Noether normalization of A. From Theorem 1.5(b), $\mathrm{reg}(A)$ can be determined by the degrees where the local cohomology modules $H^i_{\mathfrak{m}}(A)$ vanish: If we set

$$a_i(A) = a_+(H^i_{\mathfrak{m}}(A)) = \sup\{n \mid H^i_{\mathfrak{m}}(A)_n \ne 0\},$$

then

$$\mathrm{reg}(A) = \sup\{a_i(A) + i\}.$$

Note that these modules are also given by $H_{\mathfrak{p}}^i(A)$, where $\mathfrak{p}$ is the irrelevant maximal ideal of R. If we now apply the definition of $\mathrm{reg}(A)$, to a minimal resolution of A as an R-module, we gather that $r_R(A) \leq \mathrm{reg}(A)$.   □

This result indicates why $\mathrm{reg}(A)$ tends to overshoot $r_R(A)$ if A is not Cohen–Macaulay.

Hilbert function and the reduction number of an algebra

Let A be an standard graded algebra over a field k and let

$$HP_A(\mathbf{t}) = \frac{f(\mathbf{t})}{(1-\mathbf{t})^d}, \ \ d = \dim A, \ f(\mathbf{t}) = 1 + a_1 \mathbf{t} + \cdots + a_b \mathbf{t}^b, \ a_b \neq 0$$

be its Hilbert–Poincaré series.

If A is Cohen–Macaulay, its reduction number can be read from $HP_A(\mathbf{t})$,

$$\mathrm{r}(A) = b.$$

One can still roughly estimate $\mathrm{r}(A)$ given $HP_A(\mathbf{t})$.

The number $a(A) = b - d$ is the *index of regularity* or *a–invariant* of A: For $n > a(A)$, the Hilbert function of A, $H_A(n)$, and its Hilbert polynomial, $P_A(n)$, agree (see [6, Chapter 4] for fuller details). Since $P_A(n)$ can be obtained from $f(\mathbf{t})$ and d,

$$P_A(n) = \sum_{i=0}^{d-1} (-1)^i e_i \binom{n+d-i-1}{d-i-1}, \ \ e_i = \frac{f^{(i)}(1)}{i!},$$

we can derive a very crude bound for $\mathrm{r}(A)$ from an abstract yet sharp result about reductions given by the following theorem of Eakin and Sathaye ([10]), which we state in our setting as follows:

Theorem 2.3. (Eakin–Sathaye) *Let A be a standard graded algebra over an infinite field k. For positive integers n and r, suppose $\dim_k A_n < \binom{n+r}{r}$. Then there exist $z_1, \ldots, z_r \in A_1$ such that*

$$A_n = (z_1, \ldots, z_r) A_{n-1}.$$

Moreover, if $x_1, \ldots, x_p \in A_1$ are such that $(x_1, \ldots, x_p)^n = A_n$, then r generic linear combinations of $x_1, \ldots, x_p$ will define such sets.

Proposition 2.4. *Let A be a standard graded algebra. It is always possible to bound $\mathrm{r}(A)$ given the Hilbert series of A.*

Example 2.5. Let us consider two examples. If $\dim A = 1$, $P_A(n) = e_0$, so that taking $\binom{n+1}{1} > e_0$, we have

$$\mathrm{r}(A) \leq \sup\{a(A), e_0 - 1\}.$$

If $\dim A = 2$, its Hilbert polynomial is

$$P_A(n) = e_0(n+1) - e_1.$$

According to the observation above, if we pick n such that

$$\binom{n+2}{2} > e_0(n+1) - e_1,$$

that is,

$$n^2 + (3 - 2e_0)n + (2 + 2e_0 - 2e_1) > 0,$$

then

$$r(A) \le \sup\{a(A), n - 1\}.$$

Remark 2.6. It would be interesting to find similar estimations for $\mathrm{reg}(A)$ and $\mathrm{adeg}(A)$. In other words, from the Hilbert function to predict estimates for $\mathrm{reg}(A)$ and/or $\mathrm{adeg}(A)$. We are not raising similar possibilities for the Hilbert polynomial.

The relation type of an algebra

Definition 2.7. Let $A = k[x_1, \ldots, x_n]/I$ be a standard algebra over a field k. Suppose that $I \subset (x_1, \ldots, x_n)^2$. The *relation type* of A is the least integer s such that $I = (I_1, \ldots, I_s)$, where I_i is the ith graded component of I. We will denote it by $\mathrm{rt}(A)$.

This integer $\mathrm{rt}(A)$ is independent of the presentation. The notion can be extended to the cases of graded algebras over a commutative ring k.

Proposition 2.8. *Let k be an infinite field and let A be a standard Cohen–Macaulay k-algebra. Then $\mathrm{rt}(A) \le r(A) + 1$.*

Proof. Let

$$R = k[z_1, \ldots, z_d] \hookrightarrow A = k[x_1, \ldots, x_n]/I$$

be a standard Noether normalization ($\deg z_i = 1$). Since A is Cohen–Macaulay, $A = \oplus_q R b_q$ and $r(A) = \max_q\{\deg(b_q)\}$.

Computing the Castelnuovo regularity $\mathrm{reg}(A)$ with respect to the ring R gives $\mathrm{reg}(A) = r(A)$, while computing it with respect to the ring $k[x_1, \ldots, x_n]$ gives $\mathrm{reg}(A) \ge \mathrm{rt}(A) - 1$. $\qquad\square$

Remark 2.9. It is usually the case that $\mathrm{rt}(A)$ is much smaller than $r(A)$. For example, let $\mathbf{T}$ be a $m \times n$ matrix of distinct indeterminates, and let $A = k[T_{ij}\text{'s}]/I$, where I is the ideal generated by all minors of order 2 of $\mathbf{T}$. It can be shown that $r(A) = \min\{m, n\}$, while $\mathrm{rt}(A) = 2$ by definition.

More general relationships between $\mathrm{rt}(A)$ and $r(A)$ are not known, one difficulty being that they behave differently with respect to operations such as taking hyperplane section.

Cayley–Hamilton theorem

From now on A is a standard graded ring and $R = k[\mathbf{z}] \hookrightarrow A$ is a fixed Noether normalization. To determine $r(A)$, we look for equations of integral dependence of the elements of A with respect to R.

A simple approach is to find graded R–modules on which A acts as endomorphisms (e.g. A itself). The most naive path to the equation is through the Cayley–Hamilton theorem developed by G. Almkvist ([1]), working as follows. Let E be a finitely generated R–module and let

$$f : E \mapsto E$$

be an endomorphism. Map a free graded module over E and lift f:

$$
\begin{array}{ccc}
F & \xrightarrow{\ \pi\ } & E \\
\varphi \downarrow & & \downarrow f \\
F & \xrightarrow{\ \pi\ } & E
\end{array}
$$

Let

$$P_\varphi(t) = \det(tI + \varphi) = t^n + \cdots + a_n$$

be the characteristic polynomial of φ, $n = \mathrm{rank}(F)$. By the usual Cayley–Hamilton theorem, we have that $P_\varphi(f) = 0$. The drawback is that n, which is at least the minimal number of generators of E, may be too large. One should do much better using a trick of [1]. Lift f to a mapping from a projective resolution of E into itself,

$$
\begin{array}{ccccccccccc}
0 & \longrightarrow & F_s & \longrightarrow & \cdots & \longrightarrow & F_1 & \longrightarrow & F_0 & \longrightarrow & E & \longrightarrow & 0 \\
& & \downarrow \varphi_s & & & & \downarrow \varphi_1 & & \downarrow \varphi_0 & & \downarrow f \\
0 & \longrightarrow & F_s & \longrightarrow & \cdots & \longrightarrow & F_1 & \longrightarrow & F_0 & \longrightarrow & E & \longrightarrow & 0
\end{array}
$$

and define $\quad P_f(t) = \prod_{i=0}^{s} (P_{\varphi_i}(t))^{(-1)^i}.$

This rational function is actually a polynomial in $R[t]$ ([1]). If E is a graded module and f is homogeneous, then $P_f(t)$ is a homogeneous polynomial, $\deg E = \deg P_f(t)$.

Theorem 2.10. (Cayley-Hamilton theorem) *Let E be a graded module over a ring of polynomials and let f be an endomorphism of E. If the rank of E over R is e, $P_f(t)$ is a monic polynomial of degree e. Moreover, if E is torsion–free over R then $P_f(f) \cdot E = 0$. Furthermore, if E is a faithful A–module and $f \in A_1$ then $f^e \in (\mathbf{z})A_+$.*

Proof. Most of these properties are proved in [1]. Passing over to the field of fractions of R, the characteristic polynomial of the vector space mapping

$$f \otimes K : E \otimes K \longrightarrow E \otimes K$$

is precisely $P_f(t)$.

In case the module E is A itself, we do not need the device of Theorem 2.10, as we can argue directly as follows. For $u \in A_1$, $R[u] \simeq R[t]/I$, where $I = f \cdot J$, height$(J) \geq 2$. But if A is torsion-free over R, $R[u]$ will have the same property and necessarily $J = (1)$. This means that the rank of $R[u]$, which is the degree of f, is at most $\deg(A)$.

A question of independent interest is to find R–modules of small multiplicity that afford embeddings

$$A \hookrightarrow \mathrm{Hom}_R(E, E).$$

For example, the relationship between these multiplicities may be as large as $\deg(E) = n$ and $\deg(A) = \lfloor \frac{n^2}{2} \rfloor + 1$ ([29]). There are however certain restrictions to be overcome: If the Cohen–Macaulay type of the localization of A at its minimal primes is at most 3, then $\deg(E) \geq \deg(A)$ (see [14]).

The arithmetic degree of an algebra versus its reduction number

The Cayley–Hamilton theorem proved above is restricted by the requirement on the module that all associated primes have the same dimension. To be able to overcome this, we proceed as follows. Given

$$\varphi : E \mapsto E,$$

we show that there is a filtration by characteristic submodules

$$E = E_1 \supset E_2 \supset \cdots \supset E_n = 0$$

such that

$$\mathrm{adeg}(E) = \sum \deg(E_i/E_{i+1}),$$

where the factors are torsion free over appropriate subrings of a Noether normalization of $A/ann(E)$. We then combine the various characteristic polynomials.

Theorem 2.11. *Let A be an affine algebra over an infinite field k, let $k[\mathbf{z}]$ be a Noether normalization of A, and let M be a finitely generated graded, faithful A–module. Then every element of A satisfies a monic equation over $k[\mathbf{z}]$ of degree at most $\mathrm{adeg}(M)$.*

Proof. Let

$$(0) = L_1 \cap L_2 \cap \cdots \cap L_n$$

be an equidimensional decomposition of the trivial submodule of M, derived from an indecomposable primary decomposition by collecting the components of the same dimension. If

$$I_i = \text{annihilator } (M/L_i),$$

then each ring A/I_i is unmixed, equidimensional and

$$\dim A/I_i > \dim A/I_{i+1}.$$

Since k is infinite, there exists a Noether normalization $k[z_1, \ldots, z_d]$ of A such that for each ideal I_i, a subset of the $\{z_1, \ldots, z_d\}$ generates a Noether normalization for A/I_i.

First, we are going to check that $\mathrm{adeg}(M)$ can be determined by adding the arithmetic degrees of the factors of the filtration

$$M \supset L_1 \supset L_1 \cap L_2 \supset \cdots \supset L_1 \cap L_2 \cap \cdots \cap L_n = (0),$$

at the same time that we use the Cayley–Hamilton theorem.

We write the arithmetic degree of M as

$$\mathrm{adeg}(M) = a_1(M) + a_2(M) + \cdots + a_n(M),$$

where $a_i(M)$ is the contribution of the prime ideals minimal over I_i. (Warning: This does not mean that $a_i(M) = \mathrm{adeg}(M/L_i)$.) We first claim that (set $L_0 = M$)

$$\mathrm{adeg}(M) = \sum_{i=1}^{n} \mathrm{adeg}(L_1 \cap \cdots \cap L_{i-1}/L_1 \cap \cdots \cap L_i).$$

Indeed, there is an embedding

$$F_i = L_1 \cap \cdots \cap L_{i-1}/L_1 \cap \cdots \cap L_i \hookrightarrow M_i = M/L_i,$$

showing that F_i is equidimensional of the same dimension as M_i. If $\mathfrak{p}$ is an associated prime of I_i, localizing we get $(L_1 \cap \cdots \cap L_i)_\mathfrak{p} = (0)$ which shows that

$$\Gamma_\mathfrak{p}(M_\mathfrak{p}) \subset \Gamma_\mathfrak{p}((L_1 \cap \cdots \cap L_{i-1})_\mathfrak{p}),$$

while the converse is clear. This shows that the geometric degree of the module F_i is exactly the contribution of $e_i = a_i(M)$ to $\mathrm{adeg}(M)$.

We are now ready to use Proposition 2.10 on the modules F_i. Let $f \in A$ act on each F_i. For each integer i, we have a polynomial

$$P_i(t) = t^{e_i} + c_1 t^{e_i - 1} + \cdots + c_{e_i},$$

with $c_j \in (\mathbf{z})^j$, and such that $P_i(f) \cdot F_i = (0)$. Consider the polynomial

$$P_f(t) = \prod_{i=1}^{n} P_i(t),$$

and evaluate it on f from left to right. As

$$P_i(f) \cdot F_i = 0,$$

meaning that $P_i(f)$ maps $(L_1 \cap \cdots \cap L_{i-1})$ into $(L_1 \cap \cdots \cap L_i)$, a simple inspection shows that $P_f(f) = 0$, since M is a faithful module. $\qquad\square$

Observe that if A is a standard algebra and f is an element of A_1, then $P_f(t)$ gives an equation of integrality of f relative to the ideal generated by $\mathbf{z}$.

Corollary 2.12. *Let A be a standard graded algebra and denote*

$$\mathrm{edeg}(A) = \inf\{\ \mathrm{adeg}(M) \mid M \text{ faithful graded module }\}.$$

For any standard Noether normalization R of A, every element of A satisfies an equation of degree $\mathrm{edeg}(A)$ *over R.*

Reduction equations from integrality equations
If A is an standard algebra, for a given element $u \in A_1$, a typical equation of reduction looks like

$$u^e \in (\mathbf{z})A_1{}^{e-1},$$

which is less restrictive than an equation of integrality. One should therefore expect these equations to have lower degrees. Unfortunately we do not yet see how to approach it.

The following argument shows how to pass from integrality equation to some reduction equations, but unfortunately injects the issue of characteristic into the fray.

Proposition 2.13. *Let $A = k[A_1]$ be a standard algebra over a field k of characteristic zero. Let $R = k[\mathbf{z}] \hookrightarrow A$ be a Noether normalization, and suppose that every element of A_1 satisfies a monic equation of degree e over $k[\mathbf{z}]$. Then $\mathrm{r}(A) \leq e - 1$.*

Proof. Let $u_1, \ldots, u_n$ be a set of generators of A_1 over k, and consider the integrality equation of

$$u = x_1 u_1 + \cdots + x_n u_n,$$

where the x_i are elements of k. By assumption, we have

$$u^e = (x_1 u_1 + \cdots + x_n u_n)^e = a_1 u^{e-1} + \cdots + a_e,$$

where $a_i \in (\mathbf{z})^i$. Expanding u^e we obtain

$$\sum_\alpha a_\alpha m_\alpha u^\alpha \in (\mathbf{z})A_1^{e-1},$$

where $\alpha = (\alpha_1, \ldots, \alpha_n)$ is an exponent of total degree e, m_α is the multinomial coefficient $\begin{pmatrix} e \\ \alpha \end{pmatrix}$, and a_α is the corresponding 'monomial' in the x_i. We must show

$$u^\alpha \in (\mathbf{z})A_1^{e-1}$$

for each α.

To prove the assertion, it suffices to show that the span of the vectors $(a_\alpha m_\alpha)$, indexed by the set of all monomials of degree e in n variables, has the dimension of the space of all such monomials. Indeed, if these vectors lie on a hyperplane

$$\sum_\alpha c_\alpha T_\alpha = 0,$$

we would have a homogeneous polynomial

$$f(X_1, \ldots, X_n) = \sum_\alpha c_\alpha m_\alpha X^\alpha$$

which vanishes on k^n. This means that all the coefficients $c_\alpha m_\alpha$ are zero, and therefore each c_α is zero since the m_α do not vanish in characteristic zero.   □

3. Cohomological Degree of a Module

In this section we introduce a family of refinements of the arithmetic degree of a module. The key property we seek is that it behaves well under generic hyperplane section.

Big degs

Let S be either a standard graded algebra over a field or an Artinian ring, or a local ring. For reasons of convenience, to be later lifted, S will be assumed to be a Gorenstein ring. Denote by $\mathcal{M}(S)$ the category of finitely generated S–modules (graded in case S is a standard graded algebra).

Definition 3.1. A *cohomological degree* on $\mathcal{M}(S)$ is a numerical function

$$\mathrm{Deg}(\cdot) : \mathcal{M}(S) \mapsto \mathbb{R},$$

satisfying the following conditions:

(i) If $L = \Gamma_\mathfrak{m}(M)$ is the submodule of elements of finite support of M and $\overline{M} = M/L$, then

$$\mathrm{Deg}(M) = \mathrm{Deg}(\overline{M}) + \ell(L), \tag{16}$$

where $\ell(\cdot)$ is the ordinary length function.

(ii) (Bertini's rule) If $h \in S$ is a regular, generic hyperplane section on M, then

$$\mathrm{Deg}(M) \geq \mathrm{Deg}(M/hM). \tag{17}$$

(iii) (The calibration rule) If M is a Cohen–Macaulay module, then

$$\mathrm{Deg}(M) = \deg(M), \tag{18}$$

where $\deg(M)$ is the ordinary multiplicity of the module M.

There is a great deal of independence among these conditions. We just make some comments by defining a related function. Suppose that S is a standard graded algebra over an infinite field. Given a finitely generated graded module M, of dimension d, there exists a subalgebra $R \hookrightarrow S$, generated by d forms of degree 1, such that M is a finitely generated R–module. Define

$$\mathrm{Bdeg}_R(M) = \sum_{i \geq 0} \beta_i(M),$$

where the $\beta_i(M)$'s are the Betti numbers of M as an R–module. The infimum, among all such Noether 'normalizations' of M provide a function similar to $\mathrm{Deg}(\cdot)$ but which differs from it already in dimension 1.

In the local case, $(R, \mathfrak{m})$, $\dim M = \dim R$, one takes a minimal reduction $x_1, \ldots, x_d$ of $\mathfrak{m}$ (the residue field assumed infinite) and define

$$\mathrm{Kdeg}(M) = \sum_{i \geq 0} \ell(H_i(\mathbb{K} \otimes M)),$$

where $\mathbb{K}$ is the Koszul complex on the x_i. This is the usual formula for multiplicity but with the signs ignored!

There are two main issues regarding $\mathrm{Deg}(\cdot)$ functions:

- Do they exist in all dimensions?
- What are their properties?

In this section we construct one such function, according to [40]. In the next 3 sections we develop some of their properties and consider several applications, following [9].

Dimension one

There is just one such degree in dimension at most 1.

Proposition 3.2. *Let* $\mathrm{Deg}(\cdot)$ *be a cohomological degree on* $\mathcal{M}(S)$*. Then*

  (i) *If* $\dim M = 0$, $\mathrm{Deg}(M) = \ell(M)$.
  (ii) *If* $\dim M = 1$, $\mathrm{Deg}(M) = \mathrm{adeg}(M) = \deg(M) + \ell(\Gamma_\mathfrak{m}(M))$.

Proof. Follows directly from (16) and (18).   □

This means that if $(S, \mathfrak{p})$ is a discrete valuation ring and M is a finitely generated S–module,

$$M = S^r \bigoplus_{1 \leq i \leq s} S/\mathfrak{p}^{e_i},$$

then

$$\mathrm{Deg}(M) = r + \sum_{1 \leq i \leq s} e_i$$

is taken as the full degree of M. It obviously mixes oranges and pineapples.

Homological degree of a module

To give a proper generality to the next degree to be introduced, let $(R, \mathfrak{m})$ be an Artinian local ring and let A be a finitely generated, graded R algebra, $A = A_0[A_1]$, where A_0 is a finite R–algebra. Such algebras are homomorphic images of polynomial rings $S = R'[x_1, \ldots, x_n]$ where R' is a Gorenstein local ring which is finite over R.

Definition 3.3. Let M be a finitely generated graded module over the graded algebra A and let S be a Gorenstein graded algebra mapping onto A. Assume that $\dim S = r$, $\dim M = d$. The *homological degree* of M is the integer

$$\mathrm{hdeg}(M) = \deg(M) + \sum_{i=r-d+1}^{r} \binom{d-1}{i-r+d-1} \cdot \mathrm{hdeg}(\mathrm{Ext}_S^i(M, S)). \quad (19)$$

It becomes more compact when $\dim M = \dim S = d > 0$:

$$\mathrm{hdeg}(M) = \deg(M) + \sum_{i=1}^{d} \binom{d-1}{i-1} \cdot \mathrm{hdeg}(\mathrm{Ext}_S^i(M,S)). \tag{20}$$

Note that $\mathrm{hdeg}(\cdot)$ has been defined recursively on the dimension of the support of the module. The explanation for the binomial coefficients will appear later when we explore this notion under the effect of hyperplane sections.

Remark 3.4. The notation $\deg(\cdot)$ occurs also in [20] to denote an entirely distinct notion of degree.

Dimension two

Proposition 3.5. *Let M be a finitely generated S–module. If $\dim M = 2$,*

$$\mathrm{hdeg}(M) = \mathrm{adeg}(M) + \ell(\mathrm{Ext}_S^2(\mathrm{Ext}_S^1(M,S),S)).$$

Proof. By definition $(d - 1 = 1)$,

$$\mathrm{hdeg}(M) = \deg(M) + \mathrm{hdeg}(\mathrm{Ext}_S^1(M,S)) + \mathrm{hdeg}(\mathrm{Ext}_S^2(M,S)).$$

By local duality,

$$\ell(\Gamma_{\mathfrak{m}}(M)) = \ell(\mathrm{Ext}_S^2(M,S)),$$

so we only have to recognize $\mathrm{hdeg}(\mathrm{Ext}_S^1(M,S))$. But $\mathrm{Ext}_S^1(M,S)$ has dimension at most 1. If the dimension is 1, $\mathrm{hdeg}(\mathrm{Ext}_S^1(M,S))$ consists of two summands, one that represents the contributions to $\mathrm{adeg}(M)$ of the prime ideals to codimension 1 and of $\ell(\mathrm{Ext}_S^2(\mathrm{Ext}_S^1(M,S),S))$. When the dimension is 0 the situation is even simpler. □

Example 3.6. To see how the various degrees compare, let $S = k[x,y,z,w]$, and J an unmixed, non-Cohen–Macaulay ideal of codimension 2. Define $I = (x,y,z,w)J$ and $A = S/I$. From the sequence

$$0 \to J/I \simeq k^{\nu(J)} \longrightarrow S/I \longrightarrow S/J \to 0,$$

the long sequence of Ext gives

$$\begin{aligned} \mathrm{adeg}(A) &= \mathrm{gdeg}(A) + \nu(J), \\ \mathrm{hdeg}(A) &= \mathrm{adeg}(A) + \ell(\mathrm{Ext}_S^3(S/J,S)). \end{aligned}$$

Thus $\mathrm{hdeg}(A)$ and $\mathrm{adeg}(A)$ differ by the length of the Hartshorne-Rao module $\mathrm{Ext}_S^3(S/J,S)$.

For instance, if $J = (x,y) \cap (z,w)$, $\nu(J) = 4$ and $\mathrm{Ext}_S^3(S/J,S) \simeq k$, which gives $\deg(A) = \mathrm{gdeg}(A) = 2$, $\mathrm{adeg}(A) = 6$ and $\mathrm{hdeg}(A) = 7$.

Example 3.7. Let $A = S/I$, with S a Gorenstein local ring. Suppose that $\dim A = 2$ and unmixed. It is not difficult to see that

$$\mathrm{hdeg}(A) = \deg(A) + \ell(\widetilde{A}/A),$$

where $\widetilde{A}$ is the S_2–ification of A. We note that the module $\widetilde{A}/A$ is a 'concrete' realization of the Hartshorne–Rao module of A.

Hyperplane section

Let R be a Gorenstein local ring, let M be an R–module and h a regular element on it. The long exact sequence of cohomology associated to the exact sequence

$$0 \to M \xrightarrow{h} M \longrightarrow \overline{M} \to 0$$

shows which aspects must be attended to.

   We saw in the previous section that the modules $\text{Ext}_R^i(M, R)$ carried all the information required to define $\text{adeg}(M)$. Looking at

$$\text{Ext}_R^i(M, R) \xrightarrow{h} \text{Ext}_R^i(M, R) \longrightarrow \text{Ext}_R^{i+1}(\overline{M}, R) \longrightarrow \text{Ext}_R^{i+1}(M, R) \xrightarrow{h} \text{Ext}_R^{i+1}(M, R)$$

shows that the action of multiplication by h on the modules $M_i = \text{Ext}_R^i(M, R)$ is responsible for how $\text{adeg}(M)$ and $\text{adeg}(M/hM)$ fail to agree. We must therefore examine not just the associated primes of the M_i's but also those of their cohomology modules. We begin by taking a look at basic features of these M_i's.

Proposition 3.8. *Let R be a (locally) Gorenstein ring and let M be a finitely generated R–module. Suppose the annihilator of M has codimension r. Then the module $\text{Ext}_R^r(M, R)$ is nonzero, and all of its associated primes are of codimension r.*

Proof. This is a consequence of the following: If $\mathbf{x}$ is a R–regular sequence of length r, contained in the annihilator of M, then

$$\text{Ext}_R^r(M, R) \simeq \text{Hom}_{R/(\mathbf{x})}(M, R/(\mathbf{x})),$$

which is the dual of a module over the Cohen–Macaulay ring $R/(\mathbf{x})$. In particular the associated prime ideals of $\text{Ext}_R^r(M, R)$ are contained in the set of associated primes of $R/(\mathbf{x})$. (Even more strongly, $\text{Ext}_R^r(M, R)$ satisfies the condition S_2 of Serre as a module over $R/(\mathbf{x})$.)   □

   Let us begin listing elementary properties of this $\text{hdeg}(\cdot)$.

Proposition 3.9. *Let M be a graded module over the graded algebra A and let S be a Gorenstein graded algebra such that $S/I = A$.*

   (a) $\text{hdeg}(M)$ *is independent of S.*
   (b) $\deg(M) \leq \text{gdeg}(M) \leq \text{adeg}(M) \leq \text{hdeg}(M)$, *and equality holds throughout if and only if M is Cohen–Macaulay.*

Proof. (a) Given two Gorenstein graded algebras over R, S_1 and S_2, mapping onto A, we can find another Gorenstein graded algebra S mapping onto S_1 and onto S_2 which means that to prove (a) we may consider the case of algebras S and S' such that $S = S'/I$.

   We are going to show that up to shifts,

$$\text{Ext}_S^i(M, S) = \text{Ext}_{S'}^{i+r}(M, S'),$$

where $r = \dim S' - \dim S$.

   There exists a regular sequence $\mathbf{x}$ consisting of r homogeneous elements of S' contained in the ideal I, and therefore annihilating the module M.

Before we go on with the proof, we recall

Proposition 3.10. (Rees lemma) *Let R be a commutative ring, E and F R–modules and $x \in R$. If x is a regular element on R and F and $xE = 0$, then*

$$\mathrm{Ext}_R^n(E, F) \simeq \mathrm{Ext}_{R/(x)}^{n-1}(E, F/xF), \forall n \geq 1. \tag{21}$$

In our case, this means that

$$\mathrm{Ext}_{S'/(\mathbf{x})}^i(M, S'/(\mathbf{x})) = \mathrm{Ext}_{S'}^{i+r}(M, S'),$$

so that we may assume that S and S' have the same dimension.

We may further assume that $\dim S = \dim S' = \dim M$. Let us show that in this case,

$$\mathrm{Ext}_S^i(M, S) = \mathrm{Ext}_{S'}^i(M, S').$$

Let us begin with the right hand side of this equation. Let

$$\mathbb{E}_\bullet : \qquad 0 \to S' \longrightarrow E_0 \longrightarrow \cdots \longrightarrow E_d \to 0,$$

be an injective resolution of S'. Applying $\mathrm{Hom}_{S'}(M, \cdot)$, we get a complex

$$\mathrm{Hom}_{S'}(M, \mathbb{E}_\bullet) = \mathrm{Hom}_S(M, \mathrm{Hom}_{S'}(S, \mathbb{E}_\bullet)) \tag{22}$$

of S–modules, in which the cohomology of $\mathrm{Hom}_{S'}(S, \mathbb{E}_\bullet)$ is $\mathrm{Ext}_{S'}^i(S, S')$, that vanishes in all dimensions but 0, since S is a maximal Cohen–Macaulay module for S'. It follows that $\mathrm{Hom}_{S'}(S, \mathbb{E}_\bullet)$ is an S–injective resolution of $\mathrm{Hom}_{S'}(S, S') = S$, since S is a Gorenstein ring. The cohomology of both sides of (22) gives the desired assertion.

(b) The inequalities are clear and the Cohen–Macaulayness of M is, by local duality, expressed by the vanishing of all $\mathrm{Ext}_S^i(M, S)$ for $i > 0$. $\square$

Remark 3.11. By the theorem of local duality, $\mathrm{hdeg}(M)$ can be in the following setting. Let $R \mapsto A$ be a finite homomorphism, where R is a Gorenstein algebra. Then

$$\mathrm{Ext}_R^i(M, R) \simeq \mathrm{Ext}_S^i(M, S),$$

where S is as above.

Given the recursive character of the definition of $\mathrm{hdeg}(\cdot)$ it is not difficult to extend the notion to arbitrary local rings. One passes to the completion of the ring and uses any presentation afforded by Cohen's structure theorem. We leave the details to the reader.

The next properties, some of which have attached restrictions, begin to explore the means to obtain *a priori* estimates for these degrees.

Proposition 3.12. *Let M be a graded module over the graded algebra A and let S be a Gorenstein graded algebra such that $S/I = A$.*

(a) *Let $L = H^0_{\mathfrak{m}}(M)$ be the submodule of M of elements with finite support. Then*

$$\mathrm{hdeg}(M) = \mathrm{hdeg}(M/L) + \ell(L).$$

In particular, if $\dim M = 1$, then $\mathrm{adeg}(M) = \mathrm{hdeg}(M)$.
(b) *If $\dim M = 2$ and h is a regular hyperplane section on M, then*

$$\mathrm{hdeg}(M) \geq \mathrm{hdeg}(M/hM).$$

Proof. We may again assume that $\dim S = \dim M = d$. Part (a) is clear since

$$\begin{aligned}
\mathrm{Ext}^i_S(M, S) &= \mathrm{Ext}^i_S(M/L, S), \ 1 \leq i < d \\
\mathrm{Ext}^d_S(M/L, S) &= 0 \\
\ell(\mathrm{Ext}^d_S(M, S)) &= \ell(L).
\end{aligned}$$

To prove (b), starting from the exact sequence

$$0 \to M \xrightarrow{h} M \longrightarrow \overline{M} \to 0,$$

we obtain the long exact sequence of Ext,

$$\begin{aligned}
0 \to \mathrm{Hom}_S(M, S) &\longrightarrow \mathrm{Hom}_S(M, S) \longrightarrow \mathrm{Ext}^1_S(\overline{M}, S) \longrightarrow \\
\mathrm{Ext}^1_S(M, S) &\longrightarrow \mathrm{Ext}^1_S(M, S) \longrightarrow \mathrm{Ext}^2_S(\overline{M}, S) \to 0,
\end{aligned}$$

since $\mathrm{Hom}_S(\overline{M}, S) = \mathrm{Ext}^2_S(M, S) = 0$. As for the other modules of this sequence, both $\mathrm{Hom}_S(M, S)$ and $\mathrm{Ext}^1_S(\overline{M}, S)$ have no embedded primes, and $F = \mathrm{Ext}^2_S(\overline{M}, S)$ is a module of finite length which is the cokernel of the endomorphism induced by multiplication by h on the module $E = \mathrm{Ext}^1_S(M, S)$, of Krull dimension at most 1.

In arbitrary dimensions, we have by standard properties,

$$\begin{aligned}
\deg(\mathrm{Ext}^1_S(\overline{M}, S)) &= \deg(\overline{M}) \\
&= \deg(M) \\
&= \mathrm{adeg}(\mathrm{Hom}_S(M, S)).
\end{aligned}$$

If $\dim E = 0$, since

$$\ell(F) \leq \ell(E),$$

by adding we get the assertion.

Suppose $\dim E = 1$, and let L denote its submodule of finite support and put $G = E/L$. One has $\mathrm{adeg}(E) = \mathrm{adeg}(G) + \ell(L)$. Consider the exact sequence, spawned by the snake lemma, obtained by multiplication on E by h; note that it induces endomorphisms on L and on G:

$$0 \to {}_hL \longrightarrow {}_hE \longrightarrow {}_hG \longrightarrow L/hL \longrightarrow E/hE \longrightarrow G/hG \to 0.$$

From above, we have that $E/hE = \mathrm{Ext}^2_S(\overline{M}, S)$, which is a module of finite length, and thus G/hG has finite length as well. But G is a module of dimension 1 without embedded primes, which means that h is a system of parameters for it and thus must be a regular element on it; this means that $_hG = 0$, and therefore $\ell(_hE) < \infty$, which we take into the exact sequence

$$0 \to \mathrm{Hom}_S(M, S)/h \cdot \mathrm{Hom}_S(M, S) \longrightarrow \mathrm{Ext}^1_S(\overline{M}, S) \longrightarrow {_hE} \to 0,$$

and get

$$\deg(\mathrm{Hom}_S(M, S)) = \deg(\mathrm{Ext}^1_S(\overline{M}, S)),$$

as in the previous case. Since we also have

$$
\begin{aligned}
\mathrm{adeg}(E) &= \ell(L) + \deg(G) \\
&= \ell(L) + \deg(G/hG) \\
&\geq \ell(L/hL) + \ell(G/hG) \\
&\geq \ell(E/hE) \\
&= \ell(F),
\end{aligned}
$$

we add as earlier to prove the assertion.   □

Generalized Cohen–Macaulay modules

For some special modules, Buchsbaum modules being noteworthy, it is possible to have a more explicit expression of its homological degree. Let R be as above and assume $\dim M = d$. If M is a Cohen–Macaulay module on the punctured spectrum, we have

$$
\begin{aligned}
\mathrm{hdeg}(M) &= \deg(M) + \sum_{i=1}^{d} \binom{d-1}{i-1} \cdot \mathrm{hdeg}(\mathrm{Ext}^i_R(M, R)) \\
&= \deg(M) + \sum_{i=1}^{d} \binom{d-1}{i-1} \cdot \ell(\mathrm{Ext}^i_R(M, R)) \\
&= \deg(M) + \sum_{i=1}^{d} \binom{d-1}{i-1} \cdot \ell(H^{d-i}_{\mathfrak{m}}(M)) \\
&= \deg(M) + \sum_{i=0}^{d-1} \binom{d-1}{i} \cdot \ell(H^i_{\mathfrak{m}}(M)).
\end{aligned}
$$

Note that the binomial term is the invariant of Stückrad–Vogel in the theory of Buchsbaum modules ([33, Chap. 1, Proposition 2.6]).

These modules also behave nicely under regular hyperplane sections. Indeed suppose

$$0 \to M \xrightarrow{h} M \longrightarrow \overline{M} \to 0$$

is exact and consider the long exact sequence of cohomology

$$0 \to M_0 \xrightarrow{h} M_0 \longrightarrow \overline{M}_0 \longrightarrow M_1 \xrightarrow{h} M_1 \longrightarrow \overline{M}_1 \longrightarrow \cdots$$

$$M_{d-2} \xrightarrow{h} M_{d-2} \longrightarrow \overline{M}_{d-2} \longrightarrow M_{d-1} \xrightarrow{h} M_{d-1} \longrightarrow \overline{M}_{d-1} \to 0,$$

where we may assume $\dim M = \dim R = d$ and set $M_i = \operatorname{Ext}^i_R(M, R)$ and $\overline{M}_i = \operatorname{Ext}^{i+1}_R(\overline{M}, R)$.

We break up this sequence into two families of shorter sequences

$$0 \to L_i \longrightarrow M_i \xrightarrow{\ h\ } M_i \longrightarrow N_i \to 0$$

$$0 \to N_i \longrightarrow \overline{M}_i \longrightarrow L_{i+1} \to 0$$

for $0 \le i \le d - 1$.

If $i = 0$, $L_0 = 0$ and $\deg(M) = \deg(\overline{M})$. For $i \ge 1$, the modules of both sequences have finite lengths and thus

$$\begin{aligned} \ell(N_i) &= \ell(L_i) \\ \ell(\overline{M}_i) &= \ell(L_i) + \ell(L_{i+1}). \end{aligned}$$

As $\ell(M_i) \ge \ell(L_i)$, we get that

$$\ell(\overline{M}_i) \le \ell(M_i) + \ell(M_{i+1}),$$

so multiplying by $\dbinom{d-2}{i-1}$ and adding we get

$$\operatorname{hdeg}(M) \ge \operatorname{hdeg}(M/hM).$$

Strict equality will result if $h \cdot M_i = 0$, as in the case of a Buchsbaum module.

Homologically associated primes of a module

One way to look at the associated primes of a module M is as a set of visible obstructions to carrying out on M constructions inspired from linear algebra (e.g. building duals). Some other obstructions may only become visible when we look at objects derived from a projective resolution of M.

Motivated by the notion of the homological degree of a graded module, we introduce the following definition.

Definition 3.13. Let R be a Noetherian local ring and let M be a finitely generated R–module. Suppose S is a Gorenstein local ring with a surjective morphism $S \mapsto R$. The *homologically associated primes* of M are the prime ideals of R

$$\operatorname{h-Ass}(M) \quad = \quad \bigcup_{i \ge 0} \operatorname{Ass}(\operatorname{Ext}^i_S(M, S)). \tag{23}$$

There remains to prove that this definition is independent of the Gorenstein ring S. In the main case of interest, when R is a graded algebra, the independence is assured if S is also taken graded as earlier in this section.

Definition 3.14. The prime ideals in

$$\operatorname{h-Ass}(M) \setminus \operatorname{Ass}(M),$$

are the *hidden associated primes* of M.

This definition can be iterated: one can consider the associated primes of all modules

$$\mathrm{Ext}_S^{i_1}\,(\mathrm{Ext}_S^{i_2}\,(\cdots(\mathrm{Ext}_S^{i_r}\,(M,S),S),\cdots,S)).$$

In fact we are going to use this particular form in the main result of this section. Note that it naturally leads to the notion of *well–hidden* associated prime of a module!

Problem 3.15. Let R be a Noetherian local ring, not necessarily Gorenstein, and let M be a finitely generated R–module. Is the set

$$\bigcup_{i\geq 0}\mathrm{Ass}(\mathrm{Ext}_R^i(M,R))$$

always finite?

Homological degree and hyperplane sections

Let S be a Gorenstein standard graded ring with infinite residue field and M be a finitely generated graded module over S. We recall that a *superficial* element of order r for M is an element $z \in S_r$ such that $0:_M z$ is a submodule of M of finite length.

Definition 3.16. A *generic hyperplane section* of M is an element $h \in S_1$ that is superficial for all the iterated Exts

$$M_{i_1,i_2,\ldots,i_p} = \mathrm{Ext}_S^{i_1}\,(\mathrm{Ext}_S^{i_2}\,(\cdots(\mathrm{Ext}_S^{i_p-1}\,(\mathrm{Ext}_S^{i_p}\,(M,S),S),\cdots,S))),$$

and all sequences of integers $i_1 \geq i_2 \geq \cdots \geq i_p \geq 0$.

By local duality it follows that, up to shifts in grading, there are only finitely many such modules. Actually, it is enough to consider those sequences in which $i_1 \leq \dim S$ and $p \leq 2\dim S$, which ensures the existence of such 1–forms as h.

The following result proves a case of the conjecture which will suffice for all our applications.

Theorem 3.17. *Let S be a standard Gorenstein graded algebra and let M be a finitely generated graded module. If $h \in S$ is a regular, generic hyperplane section then*

$$\mathrm{hdeg}(M) \geq \mathrm{hdeg}(M/hM).$$

Proof. It will require several technical reductions. We assume that h is a regular, generic hyperplane section for the module M which is regular on S. We also assume that $\dim M = \dim S = d$, and derive several exact sequences from

$$0 \to M \xrightarrow{h} M \longrightarrow N \to 0. \tag{24}$$

For simplicity, we write $M_i = \mathrm{Ext}_S^i(M,S)$, and in the case of N, $N_i = \mathrm{Ext}_S^{i+1}(N,S)$. (The latter because N is a module of dimension $\dim S - 1$ and $N_i = \mathrm{Ext}_{S/(h)}^i(N,S/(h))$.)

Using this notation, in view of the binomial coefficients in the definition of hdeg($\cdot$), it will be enough to show that

$$\text{hdeg}(N_i) \leq \text{hdeg}(M_i) + \text{hdeg}(M_{i+1}), \text{ for } i \geq 1.$$

The sequence (24) gives rises to the long sequence of cohomology

$$0 \to M_0 \longrightarrow M_0 \longrightarrow N_0 \longrightarrow M_1 \longrightarrow M_1 \longrightarrow N_1 \longrightarrow M_2 \longrightarrow \cdots$$
$$\cdots \longrightarrow M_{d-2} \longrightarrow M_{d-2} \longrightarrow N_{d-2} \longrightarrow M_{d-1} \longrightarrow M_{d-1} \longrightarrow N_{d-1} \to 0,$$

which are broken up into shorter exact sequences as follows:

$$0 \to L_i \longrightarrow M_i \longrightarrow \widetilde{M}_i \to 0 \tag{25}$$
$$0 \to \widetilde{M}_i \longrightarrow M_i \longrightarrow G_i \to 0 \tag{26}$$
$$0 \to G_i \longrightarrow N_i \longrightarrow L_{i+1} \to 0. \tag{27}$$

We note that all L_i have finite length from the condition on h. For $i = 0$, we have the usual relation, $\deg(M) = \deg(N)$. In case $\widetilde{M}_i$ has finite length, then M_i, G_i and N_i have finite length, and

$$\begin{aligned}
\text{hdeg}(N_i) = \ell(N_i) &= \ell(G_i) + \ell(L_{i+1}) \\
&\leq \text{hdeg}(M_i) + \text{hdeg}(M_{i+1}).
\end{aligned}$$

It is a similar relation that we want to establish for all other cases.

Proposition 3.18. *Let S be a Gorenstein graded algebra and let*

$$0 \to A \longrightarrow B \longrightarrow C \to 0$$

be an exact sequence of graded modules. Then
  (a) *If A is a module of finite length, then*

$$\text{hdeg}(B) = \text{hdeg}(A) + \text{hdeg}(C).$$

  (b) *If C is a module of finite length, then*

$$\text{hdeg}(B) \leq \text{hdeg}(A) + \text{hdeg}(C).$$

Proof. They are both clear if B is a module of finite length so we assume $\dim B = d \geq 1$.
  (a): This is immediate since $\deg(C) = \deg(B)$ and the cohomology sequence gives

$$\begin{aligned}
\text{Ext}_S^i(B, S) &= \text{Ext}_S^i(C, S), \ 1 \leq i \leq d-1, \text{ and} \\
\ell(\text{Ext}_S^d(B, S)) &= \ell(\text{Ext}_S^d(A, S)) + \ell(\text{Ext}_S^d(C, S)).
\end{aligned}$$

  (b): Similarly we have

$$\text{Ext}_S^i(B, S) = \text{Ext}_S^i(A, S), \ 1 \leq i < d-1,$$

and the exact sequence

$$0 \to \text{Ext}_S^{d-1}(B, S) \to \text{Ext}_S^{d-1}(A, S) \to \text{Ext}_S^d(C, S) \to \text{Ext}_S^d(B, S) \to \text{Ext}_S^d(A, S) \to 0.$$

If $\text{Ext}_S^{d-1}(A, S)$ has finite length,

$$\text{hdeg}(\text{Ext}_S^{d-1}(B, S)) \leq \text{hdeg}(\text{Ext}_S^{d-1}(A, S)) \text{ and}$$
$$\text{hdeg}(\text{Ext}_S^d(B, S)) \leq \text{hdeg}(\text{Ext}_S^d(A, S)) + \text{hdeg}(\text{Ext}_S^d(, S)).$$

Otherwise, $\dim \text{Ext}_S^{d-1}(A, S) = 1$, and

$$\text{hdeg}(\text{Ext}_S^{d-1}(A, S)) = \deg(\text{Ext}_S^{d-1}(A, S)) + \ell(\Gamma_m(\text{Ext}_S^{d-1}(A, S))).$$

Since we also have

$$\deg(\text{Ext}_S^{d-1}(B, S)) = \deg(\text{Ext}_S^{d-1}(A, S)), \text{ and}$$
$$\ell(\Gamma_m(\text{Ext}_S^{d-1}(B, S))) \leq \ell(\Gamma_m(\text{Ext}_S^{d-1}(A, S))),$$

we again obtain the stated bound.   □

Suppose $\dim \widetilde{M}_i \geq 1$. From Proposition 3.18(b) we have

$$\text{hdeg}(N_i) \leq \text{hdeg}(G_i) + \ell(L_{i+1}). \tag{28}$$

We must now relate $\text{hdeg}(G_i)$ to $\deg(M_i)$. Apply the functor $\Gamma_m(\cdot)$ to the sequence (26) and consider the commutative diagram

$$
\begin{array}{ccccccccc}
0 & \to & \widetilde{M}_i & \longrightarrow & M_i & \longrightarrow & G_i & \to & 0 \\
& & \uparrow & & \uparrow & & \uparrow & & \\
0 & \to & \Gamma_m(\widetilde{M}_i) & \longrightarrow & \Gamma_m(M_i) & \longrightarrow & \Gamma_m(G_i) & &
\end{array}
$$

in which we denote by H_i the image of

$$\Gamma_m(M_i) \longrightarrow \Gamma_m(G_i).$$

Through the snake lemma, we obtain the exact sequence

$$0 \to \widetilde{M}_i/\Gamma_m(\widetilde{M}_i) \xrightarrow{\alpha} M_i/\Gamma_m(M_i) \longrightarrow G_i/H_i \to 0. \tag{29}$$

Furthermore, from (25) there is a natural isomorphism,

$$\beta : M_i/\Gamma_m(M_i) \simeq \widetilde{M}_i/\Gamma_m(\widetilde{M}_i)$$

while from (26) there is a natural injection

$$\widetilde{M}_i/\Gamma_m(\widetilde{M}_i) \hookrightarrow M_i/\Gamma_m(M_i),$$

whose composite with β is induced by multiplication by h on $M_i/\Gamma_m(M_i)$. We may thus replace $\widetilde{M}_i/\Gamma_m(\widetilde{M}_i)$ by $M_i/\Gamma_m(M_i)$ in (29) and take α as multiplication by h:

$$0 \to M_i/\Gamma_m(M_i) \xrightarrow{h} M_i/\Gamma_m(M_i) \longrightarrow G_i/H_i \to 0.$$

Observe that since

$$\text{Ext}_S^j(M_i/\Gamma_m(M_i), S) = \text{Ext}_S^j(M_i, S), \quad j < \dim S,$$

h is still a regular, generic hyperplane section for $M_i/\Gamma_m(M_i)$. By induction on the dimension of the module, we have

$$\text{hdeg}(M_i/\Gamma_m(M_i)) \geq \text{hdeg}(G_i/H_i).$$

Now from Proposition 3.18(a), we have

$$\mathrm{hdeg}(G_i) = \mathrm{hdeg}(G_i/H_i) + \ell(H_i).$$

Since these summands are bounded, respectively, by $\mathrm{hdeg}(M_i/\Gamma_{\mathfrak{m}}(M_i))$ and $\ell(\Gamma_{\mathfrak{m}}(M_i))$ (in fact, $\ell(H_i) = \ell(L_i)$), we have

$$\mathrm{hdeg}(G_i) \le \mathrm{hdeg}(M_i/\Gamma_{\mathfrak{m}}(M_i)) + \ell(\Gamma_{\mathfrak{m}}(M_i)) = \mathrm{hdeg}(M_i),$$

the last equality by Proposition 3.18(a) again. Finally, taking this estimate into (28) we get

$$\begin{aligned}
\mathrm{hdeg}(N_i) \; &\le \; \mathrm{hdeg}(G_i) + \ell(L_{i+1}) & (30)\\
&\le \; \mathrm{hdeg}(M_i) + \mathrm{hdeg}(M_{i+1}),
\end{aligned}$$

to establish the claim.   □

Remark 3.19. That equality does not always hold is shown by the following example. Let $R = k[x, y]$ and $M = (x, y)^2$. Then $\mathrm{hdeg}(M) = 4$ but for any hyperplane section h, $\mathrm{hdeg}(M/hM) = 3$. To get an example of a ring one takes the idealization of M.

We shall need the following consequence of the technical details in the proof:

Corollary 3.20. *Let M and h be as above and let r be a positive integer. Then*

$$r \cdot \mathrm{hdeg}(M) \ge \mathrm{hdeg}(M/h^r M).$$

Proof. We will again argue by induction on the dimension of M, keeping the notation above on the exact sequence

$$0 \to M \xrightarrow{\; h^r \;} M \longrightarrow N \to 0.$$

To begin with, we always have $\deg(M/h^r M) = r \cdot \deg(M)$. In all of the previous proof there are just two places where h, instead of h^r, has significance. First, from (30) we have

$$\mathrm{hdeg}(N_i) \le \mathrm{hdeg}(G_i) + \ell(\widetilde{L}_{i+1}), \; i \ge 1.$$

Note that $\widetilde{L}_i$ denotes the kernel of the multiplication by h^r on M_i; by induction on r and the snake lemma, $\ell(\widetilde{L}_{i+1}) \le r \cdot \ell(L_{i+1})$. One also has

$$\mathrm{hdeg}(G_i) = \mathrm{hdeg}(G_i/H_i) + \ell(H_i),$$

while from (29) and induction, we have that

$$\begin{aligned}
\mathrm{hdeg}(G_i/H_i) \; &\le \; r \cdot \mathrm{hdeg}(M_i/\Gamma_{\mathfrak{m}}(M_i))\\
&= \; r \cdot (\mathrm{hdeg}(M_i) - \ell(\Gamma_{\mathfrak{m}}(M_i))).
\end{aligned}$$

Finally, noting that H_i is a homomorphic image of $\Gamma_{\mathfrak{m}}(M_i)$ (as remarked above, tracing thru it holds that $\ell(H_i) = \ell(\widetilde{L}_i)$, which is in any event bounded by $\ell(\Gamma_{\mathfrak{m}}(M_i))$ since h, and therefore h^r, is also a superficial element).

Adding all pieces together,

$$\begin{aligned} \text{hdeg}(N_i) \;\leq\; & r \cdot \text{hdeg}(M_i) + r \cdot \ell(L_{i+1}) \\ & -r \cdot \ell(\Gamma_{\mathfrak{m}}(M_i)) + \ell(H_i), \end{aligned}$$

which on dropping the second line and replacing $\ell(L_{i+1})$ by $\text{hdeg}(M_{i+1})$ gives

$$\text{hdeg}(N_i) \leq r \cdot (\text{hdeg}(M_i) + \text{hdeg}(M_{i+1})),$$

and we finish as in the theorem. □

It is to be expected that a similar result holds for a generic form of degree r but not necessarily of type h^r.

Remark 3.21. The arguments given show that in trying to plug the leaks that occur in the formula for the arithmetic degree under hyperplane section

$$\text{adeg}(M) \leq \text{adeg}(M/hM),$$

we may have gone too far in establishing

$$\text{hdeg}(M) \geq \text{hdeg}(M/hM).$$

Perhaps, somewhere in-between, there exists a degree function $\text{Deg}(M)$ that gives equality. Although $\text{hdeg}(\cdot)$ fulfills the key requisite of giving *a priori* estimates, the proofs show several places when degree counts may have been overstated.

Homological multiplicity of a local ring

We begin by observing that it still makes sense to define the arithmetic degree, $\text{adeg}(R)$, of the local ring R: In the formula (5), the multiplicity $\text{deg}(R/\mathfrak{p})$ of the local ring $R/\mathfrak{p}$ replaces $\text{deg}(A/\mathfrak{p})$. The geometric degree, $\text{gdeg}(R)$, is defined similarly, but $\text{reg}(A)$ has no obvious extension.

In order to define $\text{hdeg}(R)$ we must assume that R is the homomorphic image of a Gorenstein ring if we want a more or less explicit formula.

In the general case we define $\text{hdeg}(R)$ as $\text{hdeg}(\widehat{R})$ in view of the following which re-states a point made earlier.

Theorem 3.22. *Let R be a local ring that is a homomorphic image of a Gorenstein ring. Then*

$$\text{hdeg}(R) = \text{hdeg}(\widehat{R}).$$

Definition 3.23. Let R be a local ring that is the homomorphic image of a Gorenstein ring. The *homological multiplicity* is the integer $e_h(R) = \text{hdeg}(R)$. More generally, for any $\text{Deg}(\cdot)$ function, $e_D(R) = \text{Deg}(R)$.

4. Regularity versus Cohomological Degrees

We assume throughout that R is a standard graded algebra over a field k, or a local ring. In either case $\mathfrak{m}$ will denote the irrelevant maximal ideal. We shall assume also that at least one extended degree function $\mathrm{Deg}(\cdot)$ has been defined. The set of all extended degree functions will be denoted by $\mathcal{D}(R)$. Terminology, except for that introduced here, will be found in [6].

Our main purpose in this section is to compare $\mathrm{Deg}(R)$ to the Castelnuovo regularity of R. We begin by collecting general properties of $\mathrm{Deg}(\cdot)$ functions.

Proposition 4.1. *Let* $\mathrm{Deg}(\cdot)$ *be a degree function satisfying* (16), (17) *and* (18). *Let* M *be a finitely generated module.*

(i) *If* $\deg(\cdot)$ *denotes the ordinary multiplicity as given by the Hilbert function of* M,

$$\mathrm{Deg}(M) \geq \deg(M),$$

with equality if and only if M *is Cohen–Macaulay.*

(ii) *If* $\nu_R(\cdot)$ *denotes the minimal number of generators function,*

$$\nu_R(M) \leq \mathrm{Deg}(M).$$

(iii) *If* L *is any submodule of finite length of* M *then*

$$\mathrm{Deg}(M) = \mathrm{Deg}(M/L) + \ell(L).$$

(iv) $\mathcal{D}(R)$ *is a convex set.*

Proof. Most assertions following directly from the definitions or elementary properties of the multiplicity (see [6, Section 4.6]) we prove only (ii).

On induct on $d = \dim(M)$. If $d = 0$, $\mathrm{Deg}(M) = \ell(M) \geq \nu_R(M)$. In case $d > 0$, if $L = \Gamma_{\mathfrak{m}}(M) \neq 0$, the condition (2) implies that it will be enough to prove the assertion for $N = M/L$, a module of depth > 0. Let h be a generic hyperplane section of M; from (3),

$$\mathrm{Deg}(N) \geq \mathrm{Deg}(N/hN) \geq \nu_R(N/hN) = \nu_R(N),$$

the second inequality follows by the induction hypothesis, and the last equality by Nakayama lemma. $\qquad\square$

Definition 4.2. The difference $\mathrm{Deg}(M) - \deg(M)$ will be called (for $\mathrm{Deg}(\cdot)$) the Cohen–Macaulay deviation of M.

Remark 4.3. We saw that if if $\dim R = 1$, there is a single $\mathrm{Deg}(\cdot)$ function on the category of finitely generated R–modules,

$$\mathrm{Deg}(M) = \deg(M) + \ell(\Gamma_{\mathfrak{m}}(M)).$$

It follows easily that if $\dim R = 1$, $\mathrm{Deg}(\cdot)$ is semi–additive: For any exact sequence of modules in $\mathcal{M}(R)$,

$$0 \to M \longrightarrow N \longrightarrow P \to 0,$$
$$\mathrm{Deg}(N) \leq \mathrm{Deg}(M) + \mathrm{Deg}(P).$$

In addition it holds that for any ideal I of R, $\nu(I) \leq \mathrm{Deg}(R)$. This follows by considering the exact sequence

$$0 \to I \cap L \longrightarrow I \longrightarrow (I+L)/L \to 0,$$

where $L = H^0_{\mathfrak{m}}(R)$. The module $(I+L)/L$ is an ideal of the one–dimensional Cohen–Macaulay local ring R/L, whereby $\deg(R/L)$ bounds the minimal number of generators of any ideal.

For $\dim R \geq 2$, there are several such functions. This is easy to see when $R = k[x,y]$.

Proposition 4.4. *Let M be a finitely generated R–module and let x be a generic hyperplane section. Then*

$$\mathrm{Deg}(M) \geq \mathrm{Deg}(M/xM) + \mathrm{Deg}(x \cdot \Gamma_x(M)).$$

Proof. By the choice of x, $\Gamma_x(M) = \Gamma_{\mathfrak{m}}(M)$, which we denote by L. Consider the exact sequence

$$0 \to L \longrightarrow M \longrightarrow M_0 \to 0.$$

Multiplication by x induces by the snake lemma the exact sequence

$$0 \to {}_xL \longrightarrow {}_xM \longrightarrow {}_xM_0 \longrightarrow L/xL \longrightarrow M/xM \longrightarrow M_0/xM_0 \to 0,$$

where ${}_xM = \{m \in M \mid xm = 0\}$; note that ${}_xM_0 = 0$. We have the inequalities of degrees

$$\begin{aligned}
\mathrm{Deg}(M) &= \mathrm{Deg}(M_0) + \mathrm{Deg}(L) \\
&\geq \mathrm{Deg}(M_0/xM_0) + \mathrm{Deg}(L),
\end{aligned}$$

while on the other hand

$$\begin{aligned}
\mathrm{Deg}(M/xM) &= \mathrm{Deg}(M_0/xM_0) + \mathrm{Deg}(L/xL) \\
&= \mathrm{Deg}(M_0/xM_0) + \mathrm{Deg}(L) - \mathrm{Deg}(xL),
\end{aligned}$$

from which the assertion follows. $\square$

Castelnuovo regularity

The general rules for the processing of Castelnuovo regularity are codified by the following:

Proposition 4.5. *Let $(A, \mathfrak{m})$ be a graded algebra and let*

$$0 \to M \longrightarrow N \longrightarrow P \to 0$$

be an exact sequence of graded modules and homomorphisms. Then

$$\begin{aligned}
\mathrm{reg}(N) &\leq \max\{\mathrm{reg}(M), \mathrm{reg}(P)\} \\
\mathrm{reg}(M) &\leq \max\{\mathrm{reg}(N), \mathrm{reg}(P) + 1\} \\
\mathrm{reg}(P) &\leq \max\{\mathrm{reg}(M) - 1, \mathrm{reg}(N)\} \\
\mathrm{reg}(N) &= \max\{\mathrm{reg}(M), \mathrm{reg}(P)\}, \quad \textit{if } M \textit{ has finite length.}
\end{aligned}$$

In particular, if N is a graded module and h is a regular hyperplane section on N, then

$$\mathrm{reg}(N) = \mathrm{reg}(N/hN).$$

Proof. It follows from the exact sequence of local cohomology

$$H_{\mathfrak{m}}^{i-1}(P) \longrightarrow H_{\mathfrak{m}}^{i}(M) \longrightarrow H_{\mathfrak{m}}^{i}(N) \longrightarrow H_{\mathfrak{m}}^{i}(P) \longrightarrow H_{\mathfrak{m}}^{i+1}(M),$$

and inspecting the degrees where the terms do not vanish. □

The following gives a basic comparison between any $\mathrm{Deg}(\cdot)$ function and the Castelnuovo–Mumford index of regularity of a standard graded algebra.

Theorem 4.6. *Let A be a standard graded algebra over an infinite field k. For any function $\mathrm{Deg}(\cdot)$, it holds*

$$\mathrm{reg}(A) < \mathrm{Deg}(A).$$

Proof. Set $d = \dim A$. We argue by induction on d. The case $d = 0$ is clear since A is a standard algebra.

Let $L = H_{\mathfrak{m}}^{0}(A)$; consider the exact sequence

$$0 \to L \longrightarrow A \longrightarrow \overline{A} \to 0.$$

Since $\mathrm{reg}(A) = \max\{\,\mathrm{reg}(L), \mathrm{reg}(\overline{A})\,\}$ (see [11, Corollary 20.19(d)]), it will suffice to show that $\mathrm{Deg}(A)-1$ bounds $\mathrm{reg}(L)$ and $\mathrm{reg}(\overline{A})$. We first consider the case where $L = 0$. This means $A = \overline{A}$, when we can choose a generic hyperplane section h for A. Since $\mathrm{Deg}(A) \geq \mathrm{Deg}(A/hA)$ by (3) and $\mathrm{reg}(A) = \mathrm{reg}(A/hA)$, we are done by the induction hypothesis.

Suppose now that $d \geq 1$ and $L \neq 0$. We must show that L has no component in degrees $\mathrm{Deg}(A)$ or higher. Let h be a generic hyperplane section and consider the exact sequence

$$0 \to L_0 \longrightarrow A \overset{h}{\longrightarrow} A \longrightarrow \overline{A} \to 0.$$

Taking local cohomology, we have the induced exact sequence

$$0 \to L_0 \longrightarrow L \overset{h}{\longrightarrow} L \longrightarrow H_{\mathfrak{m}}^{0}(\overline{A}).$$

By induction, $H_{\mathfrak{m}}^{0}(\overline{A})$ has no components in degrees higher than $r = \mathrm{Deg}(\overline{A})-1$. In particular for $n > r$ we must have $L_n = hL_{n-1}$.

Finally, from Proposition 4.4, we have that $s = \ell(hL) \leq \mathrm{Deg}(A) - \mathrm{Deg}(\overline{A})$. This implies that

$$L_{r+s+1} = h^{s+1} L_r = 0,$$

since the chain of submodules

$$hAL_r \supset h^2 AL_r \supset \cdots \supset h^{s+1} AL_r \supset 0$$

has length at most s. Thus $\mathrm{reg}(L) \leq r + s < \mathrm{Deg}(A)$, as claimed. □

What remains to be clarified is the extent by which $\mathrm{Deg}(A)$, particularly $\mathrm{hdeg}(A)$, exceeds $\mathrm{reg}(A)$.

5. Cohomological Degrees and Numbers of Generators

In this section we show how an extended degree function leads to estimates for the number of generators of an ideal I of a Cohen–Macaulay ring R in terms of the degrees of R/I.

Let $(R, \mathfrak{m})$ be a local ring and let M be a finitely generated R-module. There are no particularly general relationship between $\nu(M)$, the minimal number of generators of M, and its ordinary multiplicity $\deg(M)$, except when M is Cohen–Macaulay. One of these situations is ([6, Corollary 4.6.11]):

Proposition 5.1. *Suppose that M is a module of generic rank* $\mathrm{rank}(M) = r > 0$ *and let I be a minimal reduction of* $\mathfrak{m}$. *Then*

$$\ell(M/IM) \geq r \cdot \deg(R),$$

and equality holds if and only if M is Cohen–Macaulay.

This is useful since $\ell(M/IM)$ is often an approximation for $\ell(M/\mathfrak{m}M) = \nu(M)$. Another instance, very far away from this context, is that of a module that is 'almost free'. We have here in mind the case of an $\mathfrak{m}$–primary ideal.

The following is a slight extension of a result of [37] and [40]:

Proposition 5.2. *Let $(R, \mathfrak{m})$ be a Cohen–Macaulay local ring of dimension $d > 0$ and let M be a submodule of a free module*

$$0 \to M \longrightarrow R^r \longrightarrow C \to 0,$$

such that C has finite length. Then

$$\nu(M) \leq r \cdot \deg(R) + (d-1)\ell(C).$$

Proof. We use $\mathrm{hdeg}(M)$ to estimate $\nu(M)$. Without loss of generality, say by passing to its completion, we may assume that R has a canonical module ω.

To assemble $\mathrm{hdeg}(M)$, note that

$$\mathrm{Ext}_R^i(M, \omega) = \mathrm{Ext}_R^{i+1}(C, \omega),$$

which vanishes if $i < d - 1$ and is a module of length $\ell(C)$ when $i = d - 1$. This gives

$$\begin{aligned}
\mathrm{hdeg}(M) &= \deg(M) + \binom{d-1}{d-(d-1)}\ell(C) \\
&= r \cdot \deg(R) + (d-1)\ell(C),
\end{aligned}$$

since $\deg(M) = \deg(R^r)$. Finally we use that $\mathrm{Deg}(M) \geq \nu(M)$, for any $\mathrm{Deg}(\cdot)$ function. $\qquad\square$

Theorem 5.3. *Let $(R, \mathfrak{m})$ be a Cohen–Macaulay ring of dimension d, and let I be an ideal of codimension $g > 0$. If* depth $R/I = r$, *then*

$$\begin{aligned}
\nu(I) &\leq \deg(R) + (g-1)\mathrm{Deg}(R/I) + (d-g-r)(\mathrm{Deg}(R/I) - \deg(R/I)) \\
&= \deg(R) + (g-1)\deg(R/I) + (d-r-1)(\mathrm{Deg}(R/I) - \deg(R/I)). \quad (31)
\end{aligned}$$

Proof. Without loss of generality we may assume that R has infinite residue field. Consider the exact sequence

$$0 \to I \longrightarrow R \longrightarrow R/I \to 0.$$

If $g = d$, we are in the setting of Proposition 5.2 and we have

$$\nu(I) \leq \deg(R) + (d-1)\ell(R/I).$$

Thus suppose $g < d$ and let x be a generic hyperplane section for both R/I and R. First assume that $r > 0$. Reducing the sequence modulo (x) gives another sequence

$$0 \to I/xI \longrightarrow R/(x) \longrightarrow R/(I,x) \to 0, \tag{32}$$

where I/xI can be identified to an ideal I' of $R' = R/(x)$, with the same number of generators as I. Furthermore, by (3) $\deg(R'/I') = \deg(R/I)$ and $\mathrm{Deg}(R'/I') \leq \mathrm{Deg}(R/I)$. When taken in (31), any changes would only reinforce the inequality.

We may continue in this manner until we exhaust the depth of R/I. This means that instead of (32) we have the exact sequence

$$0 \to {}_x(R/I) \longrightarrow I/xI \longrightarrow R/(x) \longrightarrow R/(I,x) \to 0. \tag{33}$$

Note that

$$_x(R/I) \subset L = H^0_{\mathfrak{m}}(R/I),$$

and therefore $\ell(_x(R/I)) \leq \mathrm{Deg}(R/I) - \deg(R/I)$. This leads to the inequality

$$\nu(I) \leq \nu(I') + (\mathrm{Deg}(R/I) - \deg(R/I)),$$

where I' is the image of I/xI in the ring $R' = R/(x)$.

On the other hand, according to Proposition 4.4, $\mathrm{Deg}(R'/I') \leq \mathrm{Deg}(R/I) - \mathrm{Deg}(xL)$. In particular we also have that $\deg(R'/I') \leq \mathrm{Deg}(R/I)$. This means that the reduction from the case $d > g$ to the case $d = g$ can be accomplished with the addition of $(d-g)(\mathrm{Deg}(R/I) - \deg(R/I))$. Making use of the first reduction on r we obtain the desired estimate. $\qquad\square$

Remark 5.4. An elementary application is the well-known result that if both R and R/I are regular local rings then I is generated by a regular sequence.

6. Hilbert Functions of Local Rings

Let $(R, \mathfrak{m})$ be a local ring of dimension $d > 0$ and let I be an $\mathfrak{m}$–primary ideal. There are at least three graded algebras that have been used to make infinitesimal studies of I:

$$
\begin{aligned}
R[It] &= R + It + I^2 t^2 + \cdots, &\text{Rees algebra of } I\\
\mathrm{gr}_I(R) &= R/I \oplus I/I^2 \oplus I^2/I^3 \oplus \cdots, &\text{associated graded ring of } I\\
T(I) = F_I(\mathfrak{m}) &= R/\mathfrak{m} \oplus I/\mathfrak{m}I \oplus I^2/\mathfrak{m}I^2 \oplus \cdots, &\text{the special fiber of } R[It]
\end{aligned}
$$

The last two algebras come equipped with Hilbert functions whose properties impact directly on the arithmetic of $R[It]$. Our aim here is to derive general bounds for these functions, specially in the case of $\mathrm{gr}_{\mathfrak{m}}(R)$.

In the Cohen–Macaulay case, bounds for $H_R(n)$, valid for all n, found in the literature ([4], [5], [18], [27], [28], [30]) have tended to be too pessimistic. Here we derive a bound that is much sharper and then extend it to arbitrary local rings.

Bounding rules

One general approach to finding estimates for the number of generators of an ideal I of a local ring R is made up of simple rules. They seek to position I in a diagram

$$I/J \hookrightarrow R/J,$$

in which J is a Cohen–Macaulay ideal and the multiplicity of R/J is small.

Proposition 6.1. *Let $(R, \mathfrak{m})$ be a one–dimensional local ring and let I be an ideal. Then*

$$\nu(I) \leq \deg(R) + \ell(H^0_{\mathfrak{m}}(R)). \tag{34}$$

Proof. Denote $L = H^0_{\mathfrak{m}}(R)$. For any ideal I, the exact sequence

$$0 \to I \cap L \longrightarrow I \longrightarrow I(R/L) \to 0$$

implies that it suffices to show that $\nu(I(R/L)) \leq \deg(R/L)$.

Noting that $\overline{R} = R/L$ is a Cohen–Macaulay ring of the same multiplicity as R, we have that $\deg(I\overline{R}) \leq \deg(R)$. Since $I\overline{R}$ is a Cohen–Macaulay module, the multiplicity bounds the number of generators. $\square$

Another simple 'rule' is:

Proposition 6.2. *Let I be an ideal and let $J \subset I$ be a subideal such that one of the following conditions hold:*

(i) *R/J is a ring of dimension one.*
(ii) *R/J and I/J are Cohen–Macaulay modules.*

Then

$$\nu(I) \leq \nu(J) + \mathrm{Deg}(R/J).$$

Note that in the two cases all the $\mathrm{Deg}(\cdot)$ functions coincide.

When $\dim R = 2$ there is the following general result to bound $\nu(I)$ ([5]):

Theorem 6.3. *Let $(R, \mathfrak{m})$ be a Cohen–Macaulay local ring of dimension 2 and multiplicity $e(R)$. If the ideal I has an irreducible representation $I = \cap I_i$ in which r of the I_i's are $\mathfrak{m}$–primary, then*

$$\nu(I) \leq (1 + r)e(R).$$

Proof. Let x, y be a system of parameters so that $\deg(R) = \ell(R/(x,y))$. If $\mathbb{K}$ is the Koszul complex on these elements, the multiplicity of any module M of dimension 2 is given by the Euler characteristic of the complex $\mathbb{K} \otimes M$. In the case of I we have

$$
\begin{aligned}
\deg(I) &= \ell(H_0(\mathbb{K} \otimes I)) - \ell(H_1(\mathbb{K} \otimes I)) \\
&= \ell(I/(x,y)I) - \ell(\mathrm{Tor}_1(R/(x,y), I)) \\
&= \ell(I/(x,y)I) - \ell(\mathrm{Tor}_2(R/(x,y), R/I)) \\
&= \ell(I/(x,y)I) - \ell(H_2(\mathbb{K} \otimes R/I)) \\
&= \ell(I/(x,y)I) - \ell(\mathrm{Hom}_R(R/(x,y), R/I)).
\end{aligned}
$$

Since $\ell(\mathrm{Hom}_R(R/\mathfrak{m}, R/I)) = r$ by hypothesis and as $R/(x,y)$ has length $e(R)$ it follows by the half-exactness of $\mathrm{Hom}_R(\cdot, R/I)$ that

$$
\ell(\mathrm{Hom}_R(R/(x,y), R/I)) \leq r \cdot e(R).
$$

The assertion of the theorem follows because $\ell(I/(x,y)I) \geq \nu(I)$ and $\deg(I) \leq \deg(R)$. $\qquad \square$

Maximal Hilbert functions

We are going to use the bounding rules for other choices of the ideal J. In the case above $\nu(J)$ was very small but $\deg(R/J)$ extremely large. Our choices will seek J's in such a manner that the increased number of generators is more than offset by a decrease in the multiplicity of R/J. We are going to show one case where this gain can be realized.

Let us recall the notion of reduction of an ideal. A reduction of an ideal I is a subideal $J \subset I$ such that $I^{r+1} = JI^r$ for some integer r. The minimum exponent $r_J(I)$ is called the reduction number of I with respect to J. The reductions of I are ordered by inclusion with the smallest ones referred to as minimal reductions. If $(R, \mathfrak{m})$ is a local ring with infinite residue field then every ideal I has minimal reductions. They correspond to lifts of Noether normalizations of the algebra

$$
T(I) = \bigoplus_{n \geq 0} I^n / \mathfrak{m} I^n,
$$

by forms of degree 1. The smallest reduction number attained among all minimal reductions is called the reduction number of I; it will be noted by $\mathrm{r}(I)$.

We first illustrate how this setup can be used by deriving some results of [27].

Theorem 6.4. *Let $(R, \mathfrak{m})$ be a Cohen–Macaulay ring of dimension $d > 0$ and let I be an $\mathfrak{m}$–primary ideal such that the index of nilpotency of R/I is s. Then*

$$
\nu(I) \leq \deg(R)s^{d-1} + d - 1.
$$

Proof. Let $K = (a_1, \dots, a_d)$ be a minimal reduction of the maximal ideal $\mathfrak{m}$. Set

$$
J = (a_1^s, \dots, a_{d-1}^s).
$$

We have that R/J is a Cohen–Macaulay ring of dimension 1 and multiplicity $\deg(R)s^{d-1}$. Applying Proposition 6.2(ii), we get the assertion since $\nu(J) = d-1$. $\qquad \square$

Before examining arbitrary local ring we deal with the case of Cohen–Macaulay rings.

Theorem 6.5. *Let* $(R, \mathfrak{m})$ *be a Cohen–Macaulay local ring of dimension* $d > 1$. *Let* I *be an* $\mathfrak{m}$*–primary ideal and* s *the index of nilpotency of* R/I. *Then*

$$\nu(I) \le e(R)\binom{s+d-2}{d-1} + \binom{s+d-2}{d-2}. \tag{35}$$

Proof. We may assume that the residue field of R is infinite. Let $J = (a_1, \dots, a_d)$ be a minimal reduction of $\mathfrak{m}$. By assumption, $J^s \subset I$.

Set $J_0 = (a_1, \dots, a_{d-1})^s$. This is a Cohen–Macaulay ideal of height $d-1$, and the multiplicity of R/J_0 is (easy exercise)

$$e(R/J_0) = e(R)\binom{s+d-2}{d-1}.$$

Consider the exact sequence

$$0 \to I/J_0 \longrightarrow R/J_0 \longrightarrow R/I \to 0.$$

We have that I/J_0 is a Cohen–Macaulay ideal of the one–dimensional Cohen–Macaulay ring R/J_0. This implies that

$$\nu(I/J_0) \le e(R/J_0).$$

One the other hand, we have

$$\nu(I) \le \nu(J_0) + \nu(I/J_0) \le \binom{s+d-2}{d-2} + e(R/J_0),$$

to establish the claim. $\square$

Corollary 6.6. *Let* $(R, \mathfrak{m})$ *be a Cohen–Macaulay local ring of dimension* $d > 0$. *Then for all* n,

$$H_R(n) \le e(R)\binom{n+d-2}{d-1} + \binom{n+d-2}{d-2}. \tag{36}$$

Putting the information together into the Hilbert series of $T(\mathfrak{m})$ we have:

Theorem 6.7. *Let* $(R, \mathfrak{m})$ *of a Cohen–Macaulay local ring of dimension* $d > 0$ *and multiplicity* $e(R)$. *The Hilbert series of* R *is bounded by the rational function*

$$\frac{1 + (e(R) - 1)\mathbf{t}}{(1 - \mathbf{t})^d}.$$

The bounding it refers to is coefficient by coefficient. There is an interpretation of the maximality of this function (that arose in conversation with Luisa Doering) in terms of Sally modules. We sketch it leaving the verification of details (and one conjecture!) to the reader.

Let $(R, \mathfrak{m})$ be a Cohen–Macaulay local ring of dimension $d > 0$ and let I be an $\mathfrak{m}$–primary ideal. For a given minimal reduction of I, $J \subset I$, the Sally module of I with respect to J is defined by the exact sequence ([38, Chapter 5])

$$0 \to I \cdot R[Jt] \longrightarrow I \cdot R[It] \longrightarrow S_J(I) \to 0.$$

The module $S_J(I)$ (including the degrees of its minimal generators) may depend on J. However it is easy to verify that the Hilbert function $\ell(S_J(I)_n)$ is independent of J.

The theorem above is equivalent to the following:

Theorem 6.8. *Let J be a minimal reduction of the maximal ideal. The Hilbert function of $S_J(\mathfrak{m})$ is monotonic.*

Conjecture 6.9. This assertion holds for the Sally module of any $\mathfrak{m}$–primary ideal.

Gorenstein ideals
If R/I is a Gorenstein ring one has similar bounds.

Theorem 6.10. *Let $(R, \mathfrak{m})$ be a Cohen–Macaulay local ring of dimension $d > 1$. Let I be an $\mathfrak{m}$–primary irreducible ideal and let s be the index of nilpotency of R/I. Then*

$$\nu(I) \leq 2e(R)\binom{s + d - 3}{d - 2} + \binom{s + d - 3}{d - 3}. \tag{37}$$

Proof. The case of $\dim R \leq 2$ follows from Theorem 6.3. For higher dimension, we use the same reduction as above with one distinction: set $J_0 = (a_1, \ldots, a_{d-2})^s$. The ideal

$$I/J_0 \hookrightarrow R/J_0$$

is still irreducible, and R/J_0 is a Cohen–Macaulay ring of dimension 2 and multiplicity

$$e(R/J_0) = e(R)\binom{s + d - 3}{d - 2}.$$

Applying Theorem 6.3 again to I/J_0 and adding generators as above gives the stated estimate. $\qquad \square$

General local rings
We now extend Theorem 6.5 to arbitrary local rings.

Theorem 6.11. *Let $(R, \mathfrak{m})$ be a local ring of dimension $d > 0$ and infinite residue field. Let I be an $\mathfrak{m}$–primary ideal and let s be the index of nilpotency of R/I. Then for any degree function $\mathrm{Deg}(\cdot)$ defined on $\mathcal{M}(R)$, one has*

$$\nu(I) \leq \mathrm{Deg}(R)\binom{s + d - 2}{d - 1} + \binom{s + d - 2}{d - 2}. \tag{38}$$

Proof. We limit ourselves to address the points in the proof of Theorem 6.5 where the Cohen–Macaulay condition is actually used.

First, the construction of the superficial sequence $a_1, \ldots, a_{d-1}$ must ensure that they form a regular sequence at each localization $R_{\mathfrak{p}}$ where $\mathfrak{p}$ is a prime ideal of height $d - 1$. Since R has an infinite residue field this step is clear. Set $L = (a_1, \ldots, a_{d-1})$ and $J_0 = L^s$.

Next, we claim (using the previous notation) that

$$\mathrm{Deg}(R/J_0) \leq \mathrm{Deg}(R) \binom{s+d-2}{d-1}.$$

We prove this by induction on s, where the case $s = 1$ follows from Proposition 4.4. Since these modules have dimension at most 1, by Remark 4.3

$$\mathrm{Deg}(R/L^s) \leq \mathrm{Deg}(R/L^{s-1}) + \mathrm{Deg}(L^{s-1}/L^s).$$

It suffices to bound $\mathrm{Deg}(L^{s-1}/L^s)$ by $r \cdot \mathrm{Deg}(R)$ where $r = \binom{s+d-3}{d-2}$. This however follows from the exact sequence

$$0 \to K \longrightarrow (R/L)^r \longrightarrow L^{s-1}/L^s \to 0$$

since L^{s-1}/L^s is an R/L–module that can be generated by r elements. Furthermore, since L is a complete intersection at the primes of height $d - 1$, the module K must have finite length, and therefore

$$\mathrm{Deg}(L^{s-1}/L^s) + \mathrm{Deg}(K) = \mathrm{Deg}((R/L)^r) = r \cdot \mathrm{Deg}(R/L) \leq r \cdot \mathrm{Deg}(R).$$

The rest of the argument is similar:

$$\nu(I) \leq \nu(L^s) + \nu(I/L^s),$$

and from Remark 4.3 we have that the number of generators of the ideal

$$I/L^s \hookrightarrow R/L^s$$

of a one–dimensional ring is bounded by its degree $\mathrm{Deg}(R/L^s)$.   □

Bounding reduction numbers

Theorem 6.12. *Let $(R, \mathfrak{m})$ be a Cohen–Macaulay local ring of dimension $d > 0$, with infinite residue field. Then*

$$\mathrm{r}(\mathfrak{m}) \leq d \cdot e(R) - 2d + 1.$$

Proof. We apply Theorem 6.5 to the powers of the ideal $\mathfrak{m}$. One has

$$\nu(\mathfrak{m}^n) \leq e(R) \binom{n+d-2}{d-1} + \binom{n+d-2}{d-2}.$$

According to the main theorem of [10], it suffices to find n such that

$$\nu(\mathfrak{m}^n) < \binom{n+d}{d},$$

since it will imply that $\mathrm{r}(\mathfrak{m}) \leq n - 1$.

To this end, choose n so that

$$e(R)\binom{n+d-2}{d-1} + \binom{n+d-2}{d-2} < \binom{n+d}{d},$$

which is equivalent with

$$e(R)\frac{n}{n+d-1} + \frac{d-1}{n+d-1} < \frac{n+d}{d}.$$

This inequality will be satisfied for $n = d \cdot e(R) - 2d + 2$, as desired. □

Remark 6.13. By comparison, from which we may exclude the regular local ring case, the best known estimate for the reduction number of $\mathfrak{m}$ ([27]) is $r(\mathfrak{m}) \le d! \cdot e(R) - 1$.

Corollary 6.14. *For any local ring $(R, \mathfrak{m})$ of dimension $d > 0$ with infinite residue field and equipped with a degree function $\mathrm{Deg}(\cdot)$, one has*

$$r(\mathfrak{m}) \le \max\{d \cdot \mathrm{Deg}(R) - 2d + 1, \ 0\}.$$

Primary ideals

An as yet unsolved problem is to bound the Hilbert function of an $\mathfrak{m}$–primary ideal I. In [8] there is an approach to this issue but the results are preliminary.

Here we consider the Hilbert function of $T(I)$, the special fiber of $\mathrm{gr}_I(R)$. It will suffice to give estimates for the reduction number of I.

Theorem 6.15. *Let $(R, \mathfrak{m})$ be a Cohen–Macaulay local ring of dimension $d \ge 1$ and infinite residue field and let I be an $\mathfrak{m}$–primary ideal. Then*

$$\nu(I^n) \le e(I)\binom{n+d-2}{d-1} + \binom{n+d-2}{d-2}, \tag{39}$$

where $e(I)$ is the multiplicity of the ideal I.

Proof. The argument here is similar to the previous ones except at one place. Let $J = (a_1, \dots, a_{d-1}, a_d)$ be a minimal reduction of I. Consider the embedding

$$I^n/J_0^s \hookrightarrow R/J_0^n,$$

where $J_0 = (a_1, \dots, a_{d-1})$. To bound the number of generators we need a bound on the multiplicity of R/J_0^n. Since

$$\deg(R/J_0^n) = \deg(R/J_0)\binom{n+d-2}{d-1},$$

we only need to get hold of $\deg(R/J_0)$.

Finally, observe that

$$\deg(R/J_0) = \inf\{ \ell(R/(J_0, x)), \quad x \text{ regular on } R/J_0 \},$$

in particular $\deg(R/J_0) \le \ell(R/(J_0, a_d)) = \ell(R/J) = e(I)$. □

Remark 6.16. In a similar manner, one can show that if I is an $\mathfrak{m}$–primary ideal such that the index of nilpotency of R/I is s and whose multiplicity is $e(I)$, then

$$r(I) \le \min\{ d \cdot s^{d-1}e(R) - d - 1, d \cdot e(I) - 2d + 1 \}.$$

We may also, in analogy to $e_h(R)$, define the *homological multiplicity of a primary ideal I*. We first suit up the local ring R so that it has an infinite residue field and is the homomorphic image of a Gorenstein ring. The details are left for the reader.

Remark 6.17. Lech ([21]) gave a rough estimate of the multiplicity $e(I)$ in terms of $\ell(R/I)$,

$$e(I) \leq (\dim R)! \cdot e(R) \cdot \ell(R/I).$$

It would be interesting to have an (indirect) independent proof of this formula along with possible improvements for special ideals.

Depth conditions

The presence of good depth properties in the Rees algebra of I or on its special fiber $F(I)$ cuts down these estimates for the reduction of I considerably. Let us give an instance of this.

Theorem 6.18. *Let $(R, \mathfrak{m})$ be an analytically equidimensional local ring with residue field of characteristic zero. Let I be an $\mathfrak{m}$–primary ideal. Suppose that the Rees algebra $R[It]$ of I satisfies the condition S_2 of Serre (e.g. $R[It]$ is normal). Then $r(I) \leq e(I; R) - 1$. ($\deg(\mathrm{gr}_I(R)) = e(I; R)$ is the multiplicity of the ideal I.)*

Proof. The hypothesis implies that the associated graded ring of I,

$$\mathrm{gr}_I(R) = R[It] \otimes_R (R/I),$$

has the condition S_1 of Serre and is equidimensional.

Since the characteristic of the residue field $R/\mathfrak{m}$ is zero, the Artinian ring R/I will contain a field k that maps onto the residue field $R/\mathfrak{m}$. Let $k[t_1, \ldots, t_d]$ be a standard Noether normalization of the special fiber $F(I) = \mathrm{gr}_I(R) \otimes (R/\mathfrak{m})$. We can lift it to $\mathrm{gr}_I(R)$ and therefore assume that $\mathrm{gr}_I(R)$ is a finitely generated, graded module over $A = k[t_1, \ldots, t_d]$ of rank equal to the multiplicity $e(I; R)$. As all associated primes of $\mathrm{gr}_I(R)$ have the same dimension, we can bound reduction number by arithmetic degree (see [39]),

$$r(I) = r(\mathrm{gr}_I(R)) < \mathrm{adeg}(\mathrm{gr}_I(R)) = \deg(\mathrm{gr}_I(R)) = e(I; R),$$

giving the estimate.   □

The following (see [35]) places a different kind of constraint on $\mathrm{gr}_I(R)$:

Theorem 6.19. *Let R be a Buchsbaum (resp. Cohen–Macaulay) local ring of dimension $d \geq 1$, let I be an $\mathfrak{m}$–primary ideal and suppose that depth $\mathrm{gr}_I(R) \geq d - 1$. Then $r(I) \leq \deg(R)t^{d-1}$ (resp. $r(I) \leq \deg(R)t^{d-1} - 1$), where $t = \mathrm{nil}(R/I)$.*

Question 6.20. Let $(R, \mathfrak{m})$ be a local ring that is the homomorphic image of a Gorenstein ring and let I be an ideal of R. How are $e_h(R[It])$ and $e_h(\mathrm{gr}_I(R))$ related? (At least in case I is $\mathfrak{m}$–primary?)

7. Open Questions

There are at least two sets of open problems about these degrees, those related to comparisons of different degrees and those associated to the unknown properties of cohomological degrees.

Bounds problems
There are several conjectures that purport to connect the various measures of complexity of an algebra A. These questions arise from numerous examples and special cases when one seeks to link the degree data of the algebra with multiplicity data.

One of long-standing is ([12]):

Conjecture 7.1. (Eisenbud-Goto) If A is a standard graded domain over an algebraically closed field k then

$$\operatorname{reg}(A) \leq \deg(A) - \operatorname{codim} A + 2. \qquad (40)$$

A weaker question asks whether $\operatorname{r}(A) \leq \deg(A) - \operatorname{codim} A + 2$.

Another set of questions are about observed comparisons between the 'degrees' of an ideal and those of an initial ideal. Here is a typical one:

Conjecture 7.2. Let I be a homogeneous ideal of $S = k[x_1, \ldots, x_n]$ and $in(I)$ its initial ideal with respect to some term order. Then[1]

$$\operatorname{r}(S/I) \leq \operatorname{r}(S/in(I)). \qquad (41)$$

Let us display these questions and others in a diagram. The notation employed is: $A = S/I$, $A' = S/in(I)$, a solid arrow denotes an established inequality (some only in characteristic zero), while a broken arrow signifies a conjectural one (sometimes corrected by $\pm$).

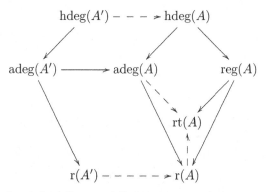

The relationship $\operatorname{hdeg}(A) > \operatorname{reg}(A)$ (Theorem 4.6) is too one–sided. It would be of interest to estimate the gap between these numbers. In low dimensions this is done in [15].

---

[1] *Added in proof:* Bresinsky and Hoa have established this conjecture for generic coordinates.

Cohomological degrees problems

One advantage of the definition of homological degree is its localization property:

Proposition 7.3. *Let M be a finitely generated module over the Noetherian local ring R. For any prime ideal $\mathfrak{p}$*

$$\mathrm{hdeg}(M) \geq \mathrm{hdeg}(M_\mathfrak{p}).$$

Proof. From the recursive character of the definition, this only requires [26, Theorem 40.1]. □

Problem 7.4. Given an arbitrary $\mathrm{Deg}(\cdot)$ on $\mathcal{M}(S)$ and a prime ideal $\mathfrak{p} \subset S$, construct a corresponding function on $\mathcal{M}(S_\mathfrak{p})$.

Since $\mathrm{hdeg}(\cdot)$ is given explicitly, it would be interesting if its Bertini's condition was valid without the appeal to genericity:

Conjecture 7.5. Let M be a graded module and let h be a regular hyperplane section. Then

$$\mathrm{hdeg}(M) \geq \mathrm{hdeg}(M/hM). \tag{42}$$

Remark 7.6. Two general lines of investigation yet to be exploited of all of these degrees are their behavior under linkage and the role played by Koszul homology.

Problem 7.7. Is $\mathcal{D}(A)$, the set of all cohomological degree functions, a convex set of finite dimension?

Problem 7.8. Is there a function $\mathrm{Deg}(\cdot)$ that can be used as an intersection of multiplicity of pairs of modules? To use the analogy, let A be a regular local ring and let M and N be finitely generated A–modules. What are the properties of the function

$$\mathrm{Deg}(M, N) = \sum_{i \geq 0} c_i \mathrm{Deg}(\mathrm{Tor}_i^A(M, N)),$$

where the c_i are weights? How are $\mathrm{Deg}(M, N), \mathrm{Deg}(M)$ and $\mathrm{Deg}(N)$ related? In a Bézout embrace?

References

[1] G. Almkvist, K–theory of endomorphisms, *J. Algebra* **55** (1978), 308–340; *Erratum*, J. Algebra **68** (1981), 520–521.

[2] D. Bayer and D. Mumford, What can be computed in Algebraic Geometry?, in *Computational Algebraic Geometry and Commutative Algebra*, Proceedings, Cortona 1991 (D. Eisenbud and L. Robbiano, Eds.), Cambridge University Press, 1993, 1–48.

[3] D. Bayer and M. Stillman, *Macaulay: A system for computation in algebraic geometry and commutative algebra*, 1992. Available via anonymous ftp from `math.harvard.edu`.

[4] J. Becker, On the boundedness and unboundedness of the number of generators of ideals and multiplicity, *J. Algebra* **48** (1977), 447–453.

[5] M. Boratynski, D. Eisenbud and D. Rees, On the number of generators of ideals in local Cohen–Macaulay rings, *J. Algebra* **57** (1979), 77–81.

[6] W. Bruns and J. Herzog, *Cohen–Macaulay Rings*, Cambridge University Press, Cambridge, 1993.

[7] A. Capani, G. Niesi and L. Robbiano, *CoCoA*: A system for doing computations in commutative algebra, 1995. Available via anonymous ftp from `lancelot.dima.unige.it`.

[8] L. R. Doering, *Multiplicities, Cohomological Degrees and Generalized Hilbert Functions*, Ph.D. Thesis, 1997, Rutgers University.

[9] L. R. Doering, T. Gunston and W. V. Vasconcelos, Cohomological degrees and Hilbert functions of graded modules, Preprint, 1996.

[10] P. Eakin and A. Sathaye, Prestable ideals, *J. Algebra* **41** (1976), 439–454.

[11] D. Eisenbud, *Commutative Algebra with a view toward Algebraic Geometry*, Springer–Verlag, Berlin–Heidelberg–New York, 1995.

[12] D. Eisenbud and S. Goto, Linear free resolutions and minimal multiplicities, *J. Algebra* **88** (1984), 89–133.

[13] A. Grothendieck, *Local Cohomology*, Lecture Notes in Mathematics **41**, Springer–Verlag, Berlin–Heidelberg–New York, 1967.

[14] T. H. Gulliksen, On the length of faithful modules over Artinian local rings, *Math. Scand.* **31** (1972), 78–82.

[15] T. Gunston, Ph.D. Thesis in Progress, Rutgers University.

[16] R. Hartshorne, Connectedness of the Hilbert scheme, Publications Math. I.H.E.S. **29** (1966), 261–304.

[17] J. Herzog and E. Kunz, *Der kanonische Modul eines Cohen–Macaulay Rings*, Lecture Notes in Mathematics **238**, Springer–Verlag, Berlin–Heidelberg–New York, 1971.

[18] L. T. Hoa, Reduction numbers of equimultiple ideals, *J. Pure & Applied Algebra* **109** (1996), 111–126.

[19] C. Huneke and B. Ulrich, General hyperplane sections of algebraic varieties, *J. Algebraic Geometry* **2** (1993), 487–505.

[20] D. Kirby and D. Rees, Multiplicities in graded rings I: The general theory, *Contemporary Math.* **159** (1994), 209–267.

[21] C. Lech, Note on multiplicities of ideals, *Arkiv för Matematik* **4**, (1960), 63–86.

[22] F. S. Macaulay, *The Algebraic Theory of Modular Systems*, Cambridge University Press, Cambridge, 1916.

[23] F. S. Macaulay, Some properties of enumeration in the theory of modular systems, *Proc. London Math. Soc.* **26** (1927), 531–555.

[24] H. Matsumura, *Commutative Algebra*, Benjamin/Cummings, Reading, 1980.

[25] C. Miyazaki, K. Yanagawa and W. Vogel, Associated primes and arithmetic degree, *J. Algebra*, to appear.

[26] M. Nagata, *Local Rings*, Interscience, New York, 1962.

[27] J. D. Sally, Bounds for numbers of generators for Cohen–Macaulay ideals, *Pacific J. Math.* **63** (1976), 517–520.

[28] J. D. Sally, *Number of Generators of Ideals in Local Rings*, Lecture Notes in Pure & Applied Math. **35**, Marcel Dekker, New York, 1978.

[29] I. Schur, Zur Theorie der vertauschbaren Matrizen, *J. reine angew. Math.* **130** (1905), 66–76.

[30] A. Shalev, On the number of generators for ideals in local rings, *Advances in Math.* **59** (1986), 82–94.

[31] R. P. Stanley, Hilbert functions of graded algebras, *Advances in Math.* **28** (1978), 57–83.

[32] R. P. Stanley, On the Hilbert function of a graded Cohen-Macaulay domain, *J. Pure & Applied Algebra* **73** (1991), 307–314.

[33] J. Stückrad and W. Vogel, *Buchsbaum Rings and Applications*, Springer-Verlag, Vienna-New York, 1986.

[34] B. Sturmfels, N. V. Trung and W. Vogel, Bounds on degrees of projective schemes, *Math. Annalen* **302** (1995), 417–432.

[35] N. V. Trung, Reduction exponent and degree bound for the defining equations of graded rings, *Proc. Amer. Math. Soc.* **101** (1987), 229–236.

[36] N. V. Trung, Bounds for the minimum number of generators of generalized Cohen–Macaulay ideals, *J. Algebra* **90** (1984), 1–9.

[37] G. Valla, Generators of ideals and multiplicities, *Comm. in Algebra* **9** (1981), 1541–1549.

[38] W. V. Vasconcelos, *Arithmetic of Blowup Algebras*, London Math. Soc., Lecture Note Series **195**, Cambridge University Press, Cambridge, 1994.

[39] W. V. Vasconcelos, The reduction number of an algebra, *Compositio Math.* **104** (1996), 189–197.

[40] W. V. Vasconcelos, The homological degree of a module, *Trans. Amer. Math. Soc.*, to appear.

[41] W. V. Vasconcelos, *Computational Methods in Commutative Algebra and Algebraic Geometry*, Springer-Verlag, to appear in 1997.

[42] C. Weibel, *An Introduction to Homological Algebra*, Cambridge University Press, Cambridge, 1994.

Department of Mathematics, Rutgers University
New Brunswick, NJ 08903
E-mail address: vasconce@rings.rutgers.edu

Index